HANDBOOK OF

Statistical Data Editing and Imputation

WILEY HANDBOOKS IN

Survey Methodology

The *Wiley Handbooks in Survey Methodology* is a series of books that present both established techniques and cutting-edge developments in the field of survey research. The goal of each handbook is to supply a practical, one-stop reference that treats the statistical theory, formulae, and applications that, together, make up the cornerstones of a particular topic in the field. A self-contained presentation allows each volume to serve as a quick reference on ideas and methods for practitioners, while providing an accessible introduction to key concepts for students. The result is a high-quality, comprehensive collection that is sure to serve as a mainstay for novices and professionals alike.

De Waal, Pannekoek, and Scholtus—*Handbook of Statistical Data Editing and Imputation*

Forthcoming *Wiley Handbooks in Survey Methodology*

Bethlehem, Cobben, and Schouten—*Handbook of Nonresponse in Household Surveys*

Bethlehem and Biffignandi—*Handbook of Web Surveys*

Alwin—*Handbook of Measurement and Reliability in the Social and Behavioral Sciences*

Larsen and Winkler—*Handbook of Record Linkage Methods*

Johnson—*Handbook of Health Survey Methods*

HANDBOOK OF

Statistical Data Editing and Imputation

WILEY HANDBOOKS IN

Survey Methodology

The *Wiley Handbooks in Survey Methodology* is a series of books that present both established techniques and cutting-edge developments in the field of survey research. The goal of each handbook is to supply a practical, one-stop reference that treats the statistical theory, formulae, and applications that, together, make up the cornerstones of a particular topic in the field. A self-contained presentation allows each volume to serve as a quick reference on ideas and methods for practitioners, while providing an accessible introduction to key concepts for students. The result is a high-quality, comprehensive collection that is sure to serve as a mainstay for novices and professionals alike.

De Waal, Pannekoek, and Scholtus—*Handbook of Statistical Data Editing and Imputation*

Forthcoming *Wiley Handbooks in Survey Methodology*

Bethlehem, Cobben, and Schouten—*Handbook of Nonresponse in Household Surveys*

Bethlehem and Biffignandi—*Handbook of Web Surveys*

Alwin—*Handbook of Measurement and Reliability in the Social and Behavioral Sciences*

Larsen and Winkler—*Handbook of Record Linkage Methods*

Johnson—*Handbook of Health Survey Methods*

HANDBOOK OF

Statistical Data Editing and Imputation

TON DE WAAL

JEROEN PANNEKOEK

SANDER SCHOLTUS

Statistics Netherlands

WILEY

A John Wiley & Sons, Inc., Publication

Published by John Wiley & Sons, Inc., Hoboken, New Jersey
Published simultaneously in Canada

Library of Congress Cataloging-in-Publication Data:

Waal, Ton de.
Handbook of statistical data editing and imputation / Ton de Waal, Jeroen Pannekoek, Sander Scholtus.
p. cm.
Includes bibliographical references and index.
ISBN 978-0-470-54280-4 (cloth)
1. Statistics—Standards. 2. Data editing. 3. Data integrity. 4. Quality control. 5. Statistical services—Evaluation. I. Pannekoek, Jeroen, 1951-II. Scholtus, Sander, 1983-III. Title.
HA29.W23 2011
001.4′22—dc22

2010018483

oBook: 978-0-470-90484-8
eBook: 978-0-470-90483-1

MIX
Paper from responsible sources
FSC® C013604

Contents

Preface

Collected survey data generally contain errors and missing values. In particular, the data collection stage is a potential source of errors and of nonresponse. For instance, a respondent may give a wrong answer (intentionally or not), or no answer at all (either because he does not know the answer or because he does not want to answer the question). Errors and missing values can also be introduced later on, for instance when the data are transferred from the original questionnaires to a computer system. The occurrence of nonresponse and, especially, errors in the observed data makes it necessary to carry out an extensive process of checking the collected data, and, when necessary, correcting them. This checking and correction process is referred to as "statistical data editing and imputation."

In this book, we discuss theory and practical applications of statistical data editing and imputation; important topics for all institutes that produce survey data, such as national statistical institutes (NSIs). In fact, it has been estimated that NSIs spend approximately 40% of their resources on editing and imputing data. Any improvement in the efficiency of the editing and imputation process should therefore be highly welcomed by NSIs. Besides NSIs, other producers of statistical data, such as market researchers, also apply edit and imputation techniques.

The importance of statistical data editing and imputation for NSIs and academic researchers is reflected by the sessions on statistical data editing and imputation that are regularly organized at international conferences on statistics. The United Nations consider statistical data editing to be such an important topic they organize a so-called work session on statistical data editing every 18 months. This work session, well-attended by experts from all over the world, addresses modern techniques for statistical data editing and imputation and discusses its practical applications.

As far as we are aware, this is the first book to treat statistical data editing in detail. Existing books treat statistical data editing only a secondary topic, and the discussion of statistical data editing is limited to just a few pages. In this book, statistical data editing is a main topic, and we discuss both general theoretical results and practical applications.

The other main topic of this book is imputation. Since several well-known books on imputation of missing data are available, this raises the question why we deemed a further treatment of imputation in this book useful. We can give two reasons. First, in practice—in any case in practice at NSIs—statistical data editing and imputation are often applied in combination. In some cases it is even hard to point out where the detection of errors ends and where imputation of better values starts. This is, for instance, the case for the so-called Nearest-neighbor Imputation Methodology discussed in Chapter 4. It would therefore be somewhat contrived to reserve attention only to data editing and leave out a discussion of imputation methods altogether.

The second reason why we have included imputation as a topic in this book is that data often have to satisfy certain consistency rules, the so-called edit rules. In practice it is quite common that imputed values have to satisfy at least some consistency rules. However, as far as we are aware, no other book on imputation examines this particular topic. NSIs and other producers of statistical data generally apply ad hoc operations to adapt the imputation methods described in the literature so imputed data will satisfy the edit rules. We hope that our book will be a valuable guide to applying more advanced imputation methods that can take edit rules into account.

In fact a close connection exists between statistical data editing and imputation. Statistical data editing is used to detect erroneous data and imputation is used to correct these data. During these processes often the same edit rules are applied, i.e., the same consistency checks that are used to detect errors in the original data must be satisfied by the imputed data later on. We feel that this close connection between statistical data editing and imputation merits one book describing both topics.

The intended audience of this book consists of researchers at NSIs and other data producers, and students at universities. Since the overall aim of both statistical data editing and imputation is to obtain data of high statistical quality, the book obviously treats many statistical topics, and is therefore primarily of interest to students and experts in the field of (mathematical) statistics.

Some readers might be surprised to find that the book also treats topics from operations research, such as optimization algorithms, as well as some techniques from general mathematics. However, an interest in these topics arises quite naturally from the problems that we discuss. An important example is the problem of finding the erroneous values in a record of data that does not satisfy certain edit rules. This so-called error localization problem can be cast as a mathematical programming problem, which may then be solved using techniques from operations research. Another example concerns the adjustment of data to attain consistency with edit rules or with data from other sources (benchmarking). These problems too can be cast as mathematical programming problems.

Broadly speaking, applications of operations research techniques frequently occur in the chapters on statistical data editing, whereas the chapters on imputation are of a more statistical nature. Some parts of the material on statistical data editing (Chapters 2 to 6 and 10) may therefore be more appealing to students and experts in operations research or general mathematics rather than to students and experts in the field of statistics.

Acknowledgments

There are many people who have at some point contributed to this book, often without being aware of this at the time. We would like to thank all students from several Dutch universities who did an internship on statistical data editing or imputation at Statistics Netherlands, all former and current colleagues at Statistics Netherlands we had the pleasure of collaborating with on these topics, and all colleagues at universities and statistical institutes around the world we have worked with in international research projects or met at one of the work sessions on statistical data editing. It has been a privilege for us to work with, and get to know, all of you.

There are too many people we would like to thank to name all of them here, at least not without expanding this book to twice its current size. We restrict ourselves to naming a few of them. We would like to thank Jacco Daalmans, Jeffrey Hoogland, Abby Israëls, Mark van der Loo, and Caren Tempelman, for collaborating with us over the years on many practical applications of statistical data editing and imputation at Statistics Netherlands. The experiences gathered from all these projects have somehow made their way into this book. We should also mention that Caren Tempelman's Ph.D. thesis on imputation methods for restricted data, which she wrote while working at Statistics Netherlands, was a wonderful source of material for Chapter 9.

We also would like to thank co-authors of several articles that form the basis of some chapters in this book: Ronan Quere, Marco Remmerswaal, and especially Wieger Coutinho and Natalie Shlomo.

We want to thank a few people who have been more directly involved with this book. Firstly, we want to thank Natalie Shlomo again. This time for carefully identifying errors and occurrences of missing data in an early version of the manuscript. Your remarks left us with many imputations to carry out, but they have only improved the contents of this book. Any errors that remain are of course entirely our fault. We also want to thank the staff at John Wiley & Sons, especially Jackie Palmieri and Lisa Van Horn, for giving us the opportunity to write this book in the first place, and for always reminding us of approaching deadlines. Without you this book would never have been finished.

Finally, we want to thank our families for their support and love.

Ton de Waal
Jeroen Pannekoek
Sander Scholtus

The Hague, The Netherlands
August 2010

CHAPTER ONE

Introduction to Statistical Data Editing and Imputation

1.1 Introduction

It is the task of National Statistical Institutes (NSIs) and other official statistical institutes to provide high-quality statistical information on many aspects of society, as up-to-date and as accurately as possible. One of the difficulties in performing this task arises from the fact that the data sources that are used for the production of statistical output, both traditional surveys as well as administrative data, inevitably contain errors that may influence the estimates of publication figures. In order to prevent substantial bias and inconsistencies in publication figures, NSIs therefore carry out an extensive process of checking the collected data and correcting them if necessary. This process of improving the data quality by detecting and correcting errors encompasses a variety of procedures, both manual and automatic, that are referred to as *statistical data editing*. The effects of statistical data editing on the errors have been examined since the mid-1950s [see Nordbotten (1955)].

Besides errors in the data, another complicating factor in order to fulfill the task of NSIs and other statistical institutes successfully is that data are often missing. This can be seen as a simple form of erroneous data, simple in the sense that missing values are easy to identify; estimating good values for these missing values may, however, be hard.

Handbook of Statistical Data Editing and Imputation, by T. de Waal, J. Pannekoek, and S. Scholtus

Errors and missing data can arise during the measurement process. Errors arise during the measurement process when the reported values differ from the true values. A possible reason for a measurement error can be that the true value is unknown to the respondent or difficult to obtain. Another reason could be that questions are misinterpreted or misread by the respondent. An example is the so-called unity measurement error that occurs if the respondent reports in euros when it was required to report in thousands of euros. Another example is a respondent reporting his own income when asked for the household income. For business surveys, errors also occur due to differences in definitions used by the statistical office and the accounting system of the responding unit. There may, for instance, be differences in the reference period used by the business and the requested period (financial year versus calender year is an example). After the data have been collected, they will pass through several other processes, such as keying, coding, editing, and imputation. Errors that arise during this further processing are referred to as processing errors. Note that although the purpose of editing is to correct errors, it is mentioned here also as a process that may occasionally introduce errors. This undesirable situation arises if an item value is adjusted because it appeared to be in error but it is actually correct. Missing data can arise when a respondent does not know the answer to a question or refuses to give the answer to a certain question.

Traditionally, NSIs have always put a lot of effort and resources into statistical data editing, because they considered it a prerequisite for publishing accurate statistics. In traditional survey processing, statistical data editing was mainly an interactive activity intended to correct all data in every detail. Detected errors or inconsistencies were reported and explained on a computer screen and corrected after consulting the questionnaire or contacting respondents, which are time- and labor-intensive procedures. In this book we examine more efficient statistical data editing methods.

It has long been recognized that it is not necessary to correct all data in every detail. Several studies [see, for example, Granquist (1984, 1997) and Granquist and Kovar (1997)] have shown that in general it is not necessary to remove all errors from a data set in order to obtain reliable publication figures. The main products of statistical offices are tables containing aggregate data, which are often based on samples of the population. This implies that small errors in individual records are acceptable. First, because small errors in individual records tend to cancel out when aggregated. Second, because if the data are obtained from a sample of the population, there will always be a sampling error in the published figures, even when all collected data are completely correct. In this case an error in the results caused by incorrect data is acceptable as long as it is small in comparison to the sampling error. In order to obtain data of sufficiently high quality, it is usually enough to remove only the most influential errors. The above-mentioned studies have been confirmed by many years of practical experience at several statistical offices.

In the past, and often even in the present, too much effort was spent on correcting errors that did not have a noticeable impact on the ultimately published figures. This has been referred to as "over-editing." Over-editing not only costs money, but also takes a considerable amount of time, making the period between data collection and publication unnecessarily long. Sometimes over-editing even becomes "creative editing"; the editing process is then continued for such a length of time that unlikely, but correct, data are "corrected." Such unjustified alterations can be detrimental for data quality. For more about the danger of over-editing and creative editing, see, for example, Granquist (1995, 1997) and Granquist and Kovar (1997).

It has been argued that the role of statistical data editing should be broader than only error localization and correction. Granquist (1995) identifies the following main objectives:

1. Identify error sources in order to provide feedback on the entire survey process.
2. Provide information about the quality of the incoming and outgoing data.
3. Identify and treat influential errors and outliers in individual data.
4. When needed, provide complete and consistent individual data.

During the last few years, the first two goals—providing feedback on the other survey phases, such as the data collection phase, and providing information on the quality of the final results—have gained in importance. The feedback on other survey phases can be used to improve those phases and reduce the amount of errors arising in these phases. In the next few years the first two goals of data editing are likely to become even more important. The main focus in this book is, however, on the latter, more traditional, two goals of statistical data editing. Statistical data editing is examined in Chapters 2 to 6.

Missing data is a well-known problem that has to be faced by basically all institutes that collect data on persons or enterprises. In the statistical literature, ample attention is hence paid to missing data. The most common solution to handle missing data in data sets is imputation, where missing values are estimated and filled in. An important problem of imputation is to preserve the statistical distribution of the data set. This is a complicated problem, especially for high-dimensional data. Chapters 7 and 8 examine this aspect of imputation.

At NSIs the imputation problem is further complicated owing to the existence of constraints in the form of edit restrictions, or edits for short, that have to be satisfied by the data. Examples of such edits are that the profit and the costs of an enterprise have to sum up to its turnover and that the turnover of an enterprise should be at least zero. Records that do not satisfy these edits are inconsistent and are hence considered incorrect. Details about imputation and adjustment techniques that ensure that edits are satisfied can be found in Chapters 9 and 10.

The rest of this chapter is organized as follows. In Section 1.2 we examine the statistical process at NSIs and other statistical organizations, and especially the role that statistical data editing and imputation play in this process. In Section 1.3 we examine (kinds of) data, errors, missing data, and edits. Section 1.4 briefly describes the editing methods that will be explored in more detail later in this book. Finally, Section 1.5 concludes this chapter by describing a basic editing strategy.

1.2 Statistical Data Editing and Imputation in the Statistical Process

1.2.1 OVERVIEW OF THE STATISTICAL PROCESS

The processes of detecting and correcting errors and handling missing data form a part of the process of producing statistical information as practiced at NSIs. This process of producing statistical information can be broken down into a number of steps. Willeboordse (1998) distinguishes the following phases in the statistical process for business surveys:

- Setting survey objectives.
- Questionnaire design and sampling design.
- Data collection and data entry.
- Data processing and data analysis.
- Publication and data dissemination.

A similar division can be made for social surveys. Each phase itself can be subdivided into several steps.

Setting Survey Objectives. In the first phase, user groups for the statistical information under consideration are identified, user needs are assessed, available data sources are explored, potential respondents are consulted about their willingness to cooperate, the survey is embedded in the general framework for business surveys, the target population and the target variables of the intended output are specified, and the output table is designed.

Questionnaire Design and Sampling Design. In the second phase the potential usefulness of available administrative registers is determined, the frame population in the so-called Statistical Business Register is compared with the target population, the sampling frame is defined, the sampling design and estimation method are selected, and the questionnaire is designed. There is a decision process on how to collect the data: paper questionnaires, personal interviews, telephone interviews, or electronic data interchange.

Data Collection and Data Entry. In the third phase the sample is drawn, data are collected from the sampled units and entered into the computer system at the statistical office. During this phase the statistical office tries to minimize the response burden for businesses and to minimize nonresponse.

Data Processing and Data Analysis. In the fourth phase the collected data are edited, missing and erroneous data are imputed, raising weights are determined, population figures are estimated, the data are incorporated in the integration framework, and the data are analysed (for example, to adjust for seasonal effects). The process of detecting and correcting errors and handling missing data forms an important part of this phase. The bulk of the work at NSIs and other statistical agencies that collect and process data is spent on this phase, especially on statistical data editing.

Publication and Data Dissemination. The final phase includes setting out a publication and dissemination strategy, protecting the final data (both tabular data and microdata, i.e. the data of individual respondents) against disclosure of sensitive information, and lastly publication of the protected data.

1.2.2 THE EDIT AND IMPUTATION PROCESS

During statistical data editing and the imputation process, erroneous records—and erroneous values within these records—are localized and new values are estimated for the erroneous values and values missing in the data set. To edit an erroneous record, two steps have to be carried out. First, the incorrect values in such a record have to be localized. This is often called *error localization*. Second, after the faulty fields in an erroneous record have been identified, these faulty fields have to be *imputed*; that is, the values of these fields have to be replaced by better, preferably the correct, values.

For erroneous records, error localization and imputation are closely related. Often it is hard to distinguish where the error localization phase ends and where the imputation phase starts. For instance, when humans edit data, they frequently look at possible ways of imputing a record before completing the error localization phase. Another example is that the localization of erroneous values might be based on estimating values first and then determining the deviation between the observed and estimated values. The observed values that differ most from their estimated counterparts are then considered erroneous, or in any case suspicious. In this approach the detection of erroneous values and the estimation of better values are highly intertwined. A third example is that during manual review (see Chapter 6) the detection of erroneous values and the "estimation" of better values are highly intertwined. This "estimation" often simply consists of filling in correct answers obtained by recontacting the respondents. Despite the fact that error localization and imputation can be closely related, we will treat them as two separate processes throughout most of this book. This is a simplification of the edit and imputation problem, but one that has shown to work well for most cases arising in practice.

In principle, it is not necessary to impute missing or erroneous values in order to obtain valid estimates for the target variables. Instead, one can estimate the target variables directly during an estimation phase, without imputing the missing and erroneous values first. However, this approach would in most practical cases become extremely complex and very demanding from a computational point of view. By first imputing the missing and erroneous values, a complete data set is obtained. From this complete data set, estimates can be obtained by standard estimation methods. In other words, imputation is often applied to simplify the estimation process.

1.3 Data, Errors, Missing Data, and Edits

1.3.1 KINDS OF DATA

Edit and imputation techniques can be divided into two main classes, depending on the kind of data to be edited or imputed: techniques for numerical data and techniques for categorical data. Generally, there are major differences between techniques for these kinds of data. At NSIs and other statistical institutes, numerical data occur mainly in surveys on businesses whereas categorical data occur mainly in social surveys—for instance, surveys on persons or households.

At Statistics Netherlands and other NSIs, editing of business surveys is a much bigger problem than editing of most social surveys on households and persons. The main reason is that for business surveys, generally much more edit rules (see below) are defined than for social surveys, and business surveys generally contain much more errors than social surveys. Typically, business surveys are not very large. Large and complicated business surveys may have somewhat over 100 variables and 100 edits. In a small country such as the Netherlands, the number of records in a business survey is usually a few thousand.

Population censuses form an important exception to the general rule that editing is easier for social surveys than for business surveys. Census data do not contain a high percentage of errors, but the number of edits (a few hundred), the number of variables (a few hundred), and especially the number of records can be high (several millions). Due to the sheer volume of the data, editing of data from a population census forms a major problem. The only efficient way to edit such volumes of data is to edit these data in an automatic manner whenever possible.

A recent development at NSIs is the increasing use of administrative (or register-based) data, as opposed to the more traditional data collection by means of sample surveys. The editing and imputation of administrative data for statistical purposes has certain specific features not shared by sample surveys. For instance, if data from several registers are combined, apart from the errors that are present in the individual registers, additional inconsistencies may occur between data from different registers due to matching errors or diverging metadata definitions. Because this is a relatively new topic, suitable methodology for the statistical data editing and imputation of administrative data has not yet been fully developed. We refer to Wallgren and Wallgren (2007) for an overview of current methods for register-based statistics.

1.3.2 KINDS OF ERRORS

One of the important goals of statistical data editing is the detection and correction of errors. Errors can be subdivided in several ways. A first important distinction we shall make is between systematic and random errors. A second important distinction we shall make is between influential errors and noninfluential errors. The final distinction is between outliers and nonoutliers.

Systematic Errors. A systematic error is an error that occurs frequently between responding units. This type of error can occur when a respondent misunderstands or misreads a survey question. A well-known type of systematic error is the so-called unity measure error, which is the error of, for example, reporting financial amounts in euros instead of the requested thousands of euros. Systematic errors can lead to substantial bias in aggregates. Once detected, systematic errors can easily be corrected because the underlying error mechanism is known.

Systematic errors, such as unity measure errors, can often be detected by comparing a respondent's present values with those from previous years, by comparing the responses to questionnaire variables with values of register variables, or by using subject-matter knowledge. Other systematic errors, such as transpositions of returns and costs and redundant minus signs, can be detected and corrected by systematically exploring all possible transpositions and inclusions/omissions of minus signs. Rounding errors—a class of systematic errors where balance edits (see Section 1.3.4) are violated because the values of the involved variables have been rounded—can be detected by testing whether failed balance edits can be satisfied by slightly changing the values of the involved variables. We treat systematic errors in more detail in Chapter 2.

Random Errors. Random errors are not caused by a systematic deficiency, but by accident. An example is an observed value where a respondent by mistake typed in a digit too many. In general statistics, the expectation of a random error is typically zero. In our case, however, the expectation of a random error may also differ from zero. This is, for instance, the case in the above-mentioned example.

Random errors can result in outlying values. In such a case they can be detected by outlier detection techniques or by selective editing techniques (see Chapter 6). Random errors can also be influential (see below), in which case they may again be detected by selective editing techniques. In many cases, random errors do not lead to outlying values or influential errors. In such cases, random errors can often be corrected automatically, assuming that they do lead to violated edit restrictions. Automatic editing of random errors is treated in detail in Chapters 3 to 5.

Influential Errors. Errors that have a substantial influence on publication figures are called influential errors. They may be detected by selective editing techniques (see Chapter 6).

The fact that a value has a substantial influence on publication figures does not necessarily imply that this value is erroneous. It may also be a correct value. In fact, in business surveys, influential observations are quite common, because many variables of businesses, such as turnover and costs, are often highly skewed.

Outliers. A value, or a record, is called an outlier if it is not fitted well by a model that is posited for the observed data. If a single value is an outlier, this is called a univariate outlier. If an entire record, or at least a subset consisting of several values in a record, is an outlier when the values are considered simultaneously—that is, if they do not fit the posited model well when considered simultaneously—this is called a multivariate outlier. Again we have that the mere fact that a value (or a record) is an outlier does not necessarily imply that this value (set of values) contains an error. It may also be a correct value (set of values).

Outliers are related to influential values. An influential value is often also an outlier, and vice versa. However, an outlier may also be a noninfluential value and an influential value may also be a nonoutlying value. In the statistical editing process, outliers are often detected during so-called macro-editing (see Chapter 6). In this book we do not examine general outlier detection techniques, except for a very brief discussion in Chapter 3. We refer to the literature for descriptions of these techniques [see, e.g., Rousseeuw and Leroy (1987), Barnett and Lewis (1994), Rocke and Woodruff (1996), Chambers, Hentges and Zhao (2004), and Todorov, Templ, and Filzmoser (2009)].

1.3.3 KINDS OF MISSING DATA

The occurrence of missing data implies a reduction of the effective sample size and consequently an increase in the standard error of parameter estimates. This loss of precision is often not the main problem with nonresponse. Survey organizations can anticipate the occurrence of nonresponse by oversampling, and moreover the loss of precision can be quantified when standard errors are estimated. A more problematic effect, which cannot be measured easily, is that nonresponse may result in biased estimates.

Missing data can be subdivided in several ways according to the underlying nonresponse mechanism. Whether the problem of biased estimates due to nonresponse actually occurs will depend on the nonresponse mechanism. Informally speaking, if the nonresponse mechanism does not depend on unobserved data (conditionally on the observed data), imputation may lead to unbiased estimates without making further assumptions. If the nonresponse mechanism does depend on unobserved data, then further—unverifiable—assumptions are necessary to reduce bias by means of imputation. We shall now make these statements more precise.

A well-known and very often used classification of nonresponse mechanisms is: "missing completely at random" (MCAR), "missing at random" (MAR), and "not missing at random" (NMAR); see Rubin (1987), Schafer (1997), and Little and Rubin (2002).

MCAR. When missing data are MCAR, the probability that a value is missing does not depend on the value(s) of the target variable(s) to be imputed or on the values of auxiliary variables. This situation can occur when a respondent forgets to answer a question or when a random part of the data is lost while processing it. MCAR is the simplest nonresponse mechanism, because the item nonrespondents (i.e., the units that did not respond to the target variable) are similar to the item respondents (i.e., the units that did respond to the target variable). Under MCAR, the observed data may simply be regarded as a random subset of the complete data. Unfortunately, MCAR rarely occurs in practice.

More formally, a nonresponse mechanism is called MCAR if

$$P(r_j \mid y_j, \mathbf{x}, \xi) = P(r_j \mid \xi). \tag{1.1}$$

In this notation, r_j is the response indicator of target variable y_j, where $r_{ij} = 1$ means that record i contains a response for variable y_j, and $r_{ij} = 0$ that the value of variable y_j is missing for record i, $\mathbf{x}$ is a vector of always observed auxiliary variables, and ξ is a parameter of the nonresponse mechanism.

MAR. When missing data are MAR, the probability that a value is missing does depend on the values of auxiliary variables, but not on the value(s) of the target variable(s) to be imputed. Within appropriately defined groups of population units, the nonresponse mechanism is again MCAR. This situation can occur, for instance, when the nonresponse mechanism of elderly people differs from that of younger people, but within the group of elderly people and the group of younger people the probability that a value is missing does not depend on the value(s) of the target variable(s) or on the values of other auxiliary variables. Similarly, for business surveys, larger businesses may exhibit a different nonresponse mechanism than small businesses, but within each group of larger businesses, respectively small businesses, the nonresponse mechanism may be MCAR.

In more formal terms, a nonresponse mechanism is called MAR if

$$P(r_j \mid y_j, \mathbf{x}, \xi) = P(r_j \mid \mathbf{x}, \xi), \tag{1.2}$$

using the same notation as in (1.1).

MAR is a more complicated situation than MCAR. In the case of MAR, one needs to find appropriate groups of population units to reduce MAR to MCAR for these groups. Once these groups of population units have been found, it is simple to correct for missing data because within these groups all units may be assumed to have the same probability to respond.

In practice, one usually assumes the nonresponse mechanism to be MAR and tries to construct appropriate groups of population units. These groups are then used to correct for missing data.

NMAR. When missing data are NMAR, the probability that a value is missing does depend on the value(s) of the target variable(s) to be imputed, and possibly also on the values of auxiliary variables. This situation can occur, for instance, when reported values of income are more likely to be missing for persons with a high income, when the value of ethnicity is more likely to be missing for certain ethnic groups, or—for business surveys—when the probability that the value of turnover is missing depends on the value of turnover itself.

In more formal terms, a nonresponse mechanism is called NMAR if

$$P(r_j \mid y_j, \mathbf{x}, \xi)$$

cannot be simplified, that is, if both (1.1) and (1.2) do not hold.

NMAR is the most complicated case. In order to correct for NMAR, one cannot use only the observed data. Instead, one also has to make model assumptions in order to model the dependence of the nonresponse mechanism on the value(s) of the target variable(s).

A related classification of nonresponse mechanisms is: "ignorable" and "nonignorable."

Ignorable. A nonresponse mechanism is called ignorable if it is MAR (or MCAR) and the parameters to be estimated are distinct from the parameter ξ. Here, distinctness means that knowledge of ξ (which could be inferred from the response indicator r_j that is available for all units, responding or not) is not helpful in estimating the parameters of interest. However, as noted by Little and Rubin (2002, Chapter 6), the MAR condition is more important than distinctness, because MAR alone is sufficient to make valid inference possible. If the parameters are not distinct, this can merely result in some loss of efficiency.

Nonignorable. A nonresponse mechanism is called nonignorable if the conditions for ignorability do not hold. That is, either the nonresponse mechanism is NMAR or the parameter ξ is not distinct from the parameters of interest, or both.

For more details on MCAR, MAR, NMAR, and (non-)ignorability we refer to Rubin (1987), Schafer (1997), and Little and Rubin (2002).

1.3.4 EDIT RULES

Errors are most often detected by edit rules. Edit rules, or edits for short, define the admissible (or plausible) values and combinations of values of the variables in each record. Errors are detected by verifying whether the values are admissible according to the edits—that is, by checking whether the edits are violated or satisfied. An edit e can be formulated as

$$e : x \in S_x,$$

with S_x the set of admissible values of x. As we shall see below, x can refer to a single variable as well as multiple variables. If e is false, the edit is violated and otherwise the edit is satisfied.

Edits can be divided into *hard* (or *fatal*) edits and *soft* (or *query*) edits. Hard edits are edits that must be satisfied in order for a record to qualify as a valid record. As an example, a hard edit for a business survey specifies that the variable *Total costs* needs to be equal to the sum of the variables *Employee costs*, *Capital costs*, *Transport costs*, and *Other costs*. Records that violate one or more hard edits are considered to be inconsistent and it is deduced that some variable(s) in such a record must be in error. Soft edits are used to identify unlikely or deviating values that are suspected to be in error, although this is not a logical necessity. Examples are (a) an edit specifying that the yearly income of employees must be less than 10 million euros or (b) an edit specifying that the turnover per employee of a firm may not be larger than 10 times the value of the previous year. The violation of soft edits can be a trigger for further investigation of these edit failures, to either confirm or reject the suspected values.

To illustrate the kind of edits that are often applied in practice, examples of a number of typical classes of edits are given below.

Univariate Edits or Range Restrictions. An edit describing the admissible values of a single variable is called a univariate edit or a range restriction. For categorical variables, a range restriction simply verifies whether the observed category codes for the variable belong to the specified set of codes. The set of allowable values S_x is

$$S_x = \{x_1, x_2, \dots, x_C\}$$

and consists of an enumeration of the C allowed codes. For instance, for the variable *Sex* we could have $S_x = \{0, 1\}$ and for a date variable in the conventional *yyyy-mm-dd* notation the set S_x would consist of all allowed integer combinations describing the year, month, and day. Range restrictions for continuous variables are usually specified using inequalities. The simplest, but often encountered, range restrictions of this type are nonnegativity constraints, that is,

$$S_x = \{x \mid x \geq 0\}.$$

Examples are *Age*, *Rent*, and many of the financial variables in business surveys (costs of various types, turnover and revenues of various activities and so on). Range restrictions describing an interval as

$$S_x = \{x \mid l \leq x \leq u\}$$

are also common. Examples are setting lower (l) and upper (u) bounds on the allowable values of age, income, or working hours per week. Range restrictions can be hard edits (for instance, if S_x is an enumeration of allowable codes), but they can also be soft edits if the bounds set on the allowable range are not a logical necessity (for instance, if the maximum number of weekly working hours is set to 100).

Bivariate Edits. In this case the set of admissible values of a variable x depends on the value of another variable, say y, observed on the same unit. The set of admissible values is then the set of admissible pairs of values (x, y). For instance, if x is *Marital status* with values 0 (never married), 1 (married) and 2 (previously married) and y is *Age*, we may have that

$$S_{xy} = \{(x, y) \mid x = 0 \wedge y < 15\} \cup \{(x, y) \mid y \geq 15\},$$

reflecting the rule that persons younger than 15 are not allowed to be married, while for persons of 15 years or more all marital states are allowed. Another example of a bivariate edit is

$$S_{xy} = \{(x, y) \mid x - y > 14\},$$

with x the age, in years, of a mother and y the age of her child. This example reflects the perhaps not "hard" edit that a mother must be at least 14 years older than her child. Finally, an important and often encountered class of bivariate edits is the so-called *ratio edit* which sets bounds on the allowable range of a ratio between two variables and is defined by

$$S_{xy} = \{(x, y) \mid l \leq \frac{x}{y} \leq u\}.$$

A ratio edit could, for example, specify bounds on the ratio of the turnover and the number of employees of firms in a certain branch of industry. Ratio edits are often used to compare data on the same units from different sources, such as values reported in the current survey (x) with values for the same variables reported in last year's survey (y) or values of variables from a tax register with similarly defined variables from a survey.

Balance Edits. Balance edits are multivariate edits that state that the admissible values of a number of variables are related by a linear equality. They occur mainly in business statistics where they are linear equations that should be satisfied according to accounting rules. Two examples are

$$\textit{Profit} = \textit{Turnover} - \textit{Total costs} \tag{1.3}$$

and

$$\textit{Total costs} = \textit{Employee costs} + \textit{Other costs}. \tag{1.4}$$

These rules are related because they have the variable *Total costs* in common. If the first rule is satisfied but the second is not, it may seem more likely that *Employee costs* or *Other costs* are in error than *Total costs*, because in the last case the first rule should probably also be violated. Balance edits are of great importance for editing economic surveys where there are often a large number of such edits. For instance, in the yearly structural business statistics there are typically about 100 variables with 30 or more balance edits. These interrelated systems of linear relations that the values must satisfy provide much information about possible errors and missing values.

Since balance edits describe relations between many variables, they are multivariate edits. Moreover, since they are often connected by common variables, they should be treated as a system of linear equations. It is convenient to express such a system in matrix notation. Denoting the five variables in (1.3) and (1.4) by x_1 (*Profit*), x_2 (*Turnover*), x_3 (*Total costs*), x_4 (*Employee costs*), and x_5 (*Other costs*), the system can be written as

$$\begin{pmatrix} 1 & -1 & 1 & 0 & 0 \\ 0 & 0 & -1 & 1 & 1 \end{pmatrix} \begin{pmatrix} x_1 \\ x_2 \\ x_3 \\ x_4 \\ x_5 \end{pmatrix} = \begin{pmatrix} 0 \\ 0 \end{pmatrix},$$

or $\mathbf{Ax} = \mathbf{0}$. The admissible values of a vector $\mathbf{x}$ subject to a system of balance edits, defined by a restriction matrix $\mathbf{A}$, can then be written as

$$S_{\mathbf{x}} = \{\mathbf{x} \mid \mathbf{Ax} = \mathbf{0}\}.$$

1.4 Basic Methods for Statistical Data Editing and Imputation

In this section we have a first brief look at various methods that can be used to edit and impute data. Before we sketch these methods, we first look back to see why these methods were developed.

Computers have been used in the editing process for a long time [see, e.g., Nordbotten (1963)]. In the early years their role was, however, restricted to checking which edits were violated. Subject-matter specialists entered data into a mainframe computer. Subsequently, the computer checked whether these data satisfied all specified edits. For each record all violated edits were listed. Subject-matter specialists then used these lists to correct the records. That is, they retrieved all paper questionnaires that did not pass all edits and corrected these questionnaires. After they had corrected the data, these data were again entered into the mainframe computer, and the computer again checked whether the data satisfied all edits. This iterative process was continued until (nearly) all records passed all edits.

A major problem of this approach was that during the manual correction process the records were not checked for consistency. As a result, a record that was "corrected" could still fail one or more specified edits. Such a record hence required more correction. It was not exceptional that some records had to be corrected several times. It is therefore not surprising that editing in this way was very costly, both in terms of money as well as in terms of time. In the literature it was estimated that 25% to 40% of the total budget was spent on editing [see e.g. Federal Committee on Statistical Methodology (1990) and Granquist and Kovar (1997)].

1.4.1 EDITING DURING THE DATA COLLECTION PHASE

The most efficient editing technique of all is no editing at all, but instead ensuring that correct data are obtained during the data collection phase. In this section we briefly discuss ways to obtain data with no or only few errors at data collection.

When one aims to collect correct data at data collection, one generally uses a computer to record the data. This is called computer-assisted data collection. With computer-assisted data collection the computer can immediately check the recorded data for violations of edits. Below we discuss four modes for computer-assisted data collection: CAPI (Computer-Assisted Personal Interviewing), CATI (Computer-Assisted Telephone Interviewing), CASI (Computer-Assisted Self-Interviewing), and CAWI (Computer-Assisted Web Interviewing). For more information on computer-assisted data collection in general, we refer to Couper et al. (1998).

When CAPI is used to collect the data, an interviewer visits the respondent and enters the answers directly into a laptop. When CATI is used to collect the data, the interview is carried out during a telephone call. When CASI or CAWI is used to collect the data, the respondent fills in an electronic questionnaire himself. The difference between these two modes is that for CAWI an electronic questionnaire on the Internet has to be filled in, whereas for CASI an off-line electronic questionnaire is used. When an invalid response is given to a question or an inconsistency between two or more answers is noted during any of these data collection modes, this can be immediately reported by the computer that is used for data entry. The error can then be resolved by asking the respondent the relevant question(s) again. For CASI and CAWI, generally not all edits that could be specified are actually implemented, since the respondent may get annoyed and might refuse to complete the questionnaire when the edits keep on reporting that his/her answers are inconsistent.

Computer-assisted data collection removes the need for data entry by typists, since the data arrive at the statistical office already in digital form. This eliminates one source of potential errors. In many cases, data collected by means of CAPI, CATI, CASI or CAWI also contain fewer errors than data collected by means of paper questionnaires because random errors that affect paper questionnaires are detected and avoided at collection. For face-to-face interviewing CAPI has in fact become the standard. CAPI, CATI, CASI, and CAWI may hence seem to be ideal ways to collect data, but—unfortunately—they too have their disadvantages.

A first disadvantage of CATI and CAPI is that CATI and, especially, CAPI are very expensive. A second disadvantage of CATI and CAPI is that a prerequisite for these two data collection modes is that the respondent is able to answer the questions during the interview. For a survey on persons and households, this is often the case. The respondent often knows (good proxies of) the answers to the questions, or is able to retrieve the answers quickly. For a survey on enterprises the situation is quite different. Often it is impossible to retrieve the correct answers quickly, and often the answers are not even known by one person or one department of an enterprise. Finally, even in the exceptional case that one person knew all answers to the questions, the NSI would generally not know the identity of this person. For the above-mentioned reasons, many NSIs frequently use CAPI and CATI to collect data on persons and households but only rarely for data on enterprises.

Pilot studies and actual applications have revealed that CASI and CAWI are indeed viable data collection modes, but also that several problems arise when these modes are used. Besides IT problems, such as that the software—and the Internet connection—should be fast and reliable and the security of the transmitted data should be guaranteed, there are a number of practical and statistical problems. We have already mentioned the practical problem that if the edits keep on reporting that the answers are inconsistent, the respondent may refuse to fill in the rest of the questionnaire. An example of a statistical problem is that the group of people responding to a web survey may be selective, since Internet usage is not uniformly distributed over the population [see, e.g., Bethlehem (2007)].

Another important problem for CAWI and CASI is that data collected by either of these data collection modes may appear to be of higher statistical quality than data collected by means of paper questionnaires, but in fact are not. When data are collected by means of CASI and CAWI, one can enforce that the respondents supply data that satisfy built-in edits, or one can avoid balance edits by automatically calculating total amounts from the reported components. Because less edits are failed by the collected data, these data may appear to be of higher statistical quality. This need not be the case, however. Each edit that is built into the electronic questionnaire will be automatically satisfied by the collected data, and hence cannot be used to check for errors later on. Therefore, the collected data may appear to contain only few errors, but this might be due to a lack of relevant edits. There are indications that respondents can be less accurate when filling in an electronic questionnaire, especially if totals are computed automatically [see Børke (2008) and Hoogland and Smit (2008)].

NSIs seem to be moving toward the use of mixed-mode data collection, where data are collected by a mix of several data collection modes. This obviously has consequences for statistical data editing. Some of the potential consequences have been examined by Børke (2008), Hoogland and Smit (2008), and Van der Loo (2008).

1.4.2 MODERN EDITING METHODS

Below we briefly mention editing methods that are used in modern practice. The editing techniques are examined in detail in other chapters of this book.

Interactive Editing. Subject-matter specialists have extensive knowledge on their area of expertise. This knowledge should be used as well as possible. This aim can be achieved by providing subject-matter specialists with efficient and effective interactive data editing tools. Most interactive data editing tools applied at NSIs allow one to check the specified edits during or after data entry, and—if necessary—to correct erroneous data immediately. This is referred to as interactive or computer-assisted editing. To correct erroneous data, several approaches can be followed: The respondent can be recontacted, the respondent's data can be compared to his data from previous years, the respondent's data can be compared to data from similar respondents, and subject-matter knowledge of the human editor can be used. Interactive editing is nowadays a standard way to edit data. It can be used to edit both categorical and numerical data. The number of variables, edits, and records may, in principle, be high. Generally, the quality of data editing in a computer-assisted manner is considered high. Interactive editing is examined in more detail in Section 6.5.

Selective Editing. Selective editing is an umbrella term for several methods to identify the influential errors (i.e., the errors that have a substantial impact on the publication figures) and outliers (i.e., values that do not fit a model of the data well). Selective editing techniques aim to apply interactive editing to a well-chosen subset of the records, such that the limited time and resources available for interactive editing are allocated to those records where it has the most effect on the quality of the final estimates of publication figures. Selective editing techniques try to achieve this aim by splitting the data into two streams: the critical stream and the noncritical stream. The critical stream consists of records that are the most likely ones to contain influential errors; the noncritical stream consists of records that are unlikely to contain influential errors. The records in the critical stream, the critical records, are edited in a traditional interactive manner. The records in the noncritical stream, the noncritical records, are not edited in a computer-assisted manner. They may later be edited automatically. Selective editing is examined in Chapter 6.

Macro-editing. We distinguish between two forms of macro-editing. The first form is sometimes called the aggregation method [see, e.g., Granquist (1990)]. It formalizes and systematizes what every statistical agency does before publication: verifying whether figures to be published seem plausible. This is accomplished by comparing quantities in publication tables with the same quantities in previous publications. Only if an unusual value is observed, a micro-editing procedure is applied to the individual records and fields contributing to the suspicious quantity. A second form of macro-editing is the distribution method. The available data are used to characterize the distribution of the variables. Then, all individual values are compared with the distribution. Typically, measures of location and spread are computed. Records containing values that could be considered uncommon (given the distribution) are candidates for further inspection and possibly for editing. Macro-editing, in particular the aggregation method, has always been applied at statistical offices. Macro-editing is again examined further in Chapter 6.

Automatic Editing. When automatic editing is applied, records are edited by a computer without human intervention. In this sense, automatic editing is the opposite of the traditional approach to the editing problem, where each record is edited manually. Automatic editing has already been used in the 1960s and 1970s [see, e.g., Nordbotten (1963)]. Nevertheless, it has never become very popular. For this we point out two reasons. First, in former days, computers were too slow to edit data automatically. Second, development of a system for automatic editing was often considered too complicated and too costly by many statistical offices. In the last two decades, however, a lot of progress has been made with respect to automatic editing: Computers have become faster and algorithms have been simplified and have also become more efficient. For these reasons we pay more attention to automatic editing than to the other editing techniques in this book. Automatic editing of systematic errors is examined in Chapter 2, and automatic editing of random errors in Chapters 3 to 5.

1.4.3 IMPUTATION METHODS

To estimate a missing value, or a value that was identified as being erroneous during statistical data editing, two main approaches can be used. The first approach is manual imputation or correction, where the corresponding respondent is recontacted or subject-matter knowledge is used to obtain an estimate for the missing or erroneous value. The second approach is automated imputation, which is based on statistical estimation techniques, such as regression models. In this book, we only treat the latter approach.

In imputation, predictions from parametric or nonparametric models are derived for values that are missing or flagged as erroneous. An imputation model predicts a missing value using a function of auxiliary variables, the predictors. The auxiliary variables may be obtained from the current survey or from other sources such as historical information (the value of the missing variable in a previous period) or, increasingly important, administrative data. The most common types of imputation models are variants of regression models with parameters estimated from the observed correct data. However, especially for categorical variables, donor methods are also frequently used. Donor methods replace missing values in a record with the corresponding values from a nearby complete and valid record. Often a donor record is chosen such that it resembles as much as possible the record with missing values. Imputation methods are treated in Chapters 7 to 9 of this book.

1.5 An Edit and Imputation Strategy

Data editing is usually performed as a sequence of different detection and/or correction process steps. In this section we give a global description of an editing strategy. This description is general enough to include the situation for many data sets as special cases, and most editing strategies applied in practice will at least include a number of the elements and principles described here. The process steps can be characterized from different points of view—for instance, by the type of errors they try to detect or resolve or by the methods that are used for detection or correction. Another important distinction is between automatic methods that can be executed without human intervention and interactive editing that is performed by editors.

The global editing strategy as depicted in Figure 1.1 consists of the following five steps that are clarified below.

1. *Treatment of Systematic Errors.* Identify and eliminate errors that are evident and easy to treat with sufficient reliability.

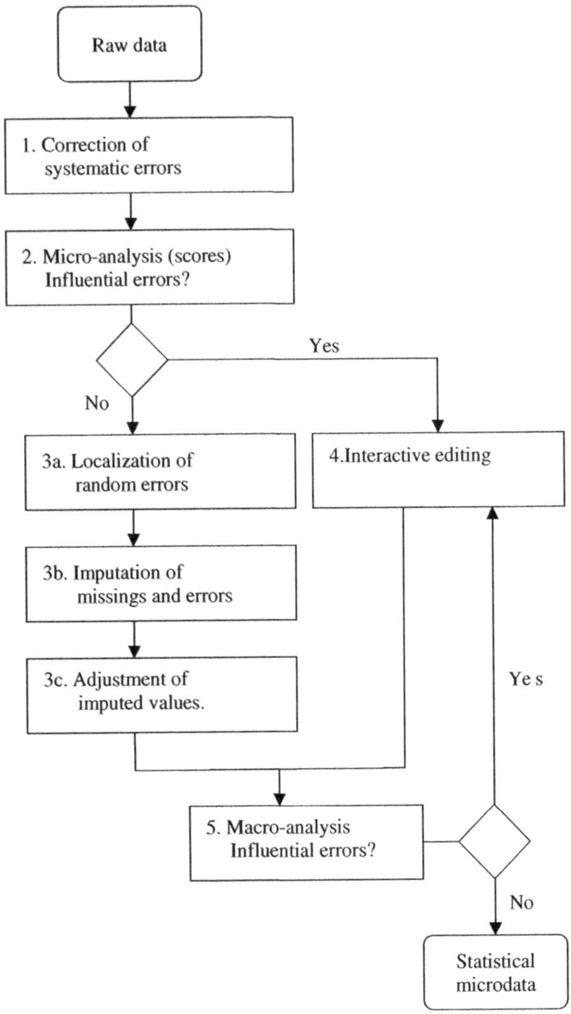

FIGURE 1.1 Example of a process flow.

2. *Micro-selection.* Select records for interactive treatment that contain influential errors that cannot be treated automatically with sufficient reliability.
3. *Automatic Editing.* Apply all relevant automatic error detection and correction procedures to the (many) records that are not selected for interactive editing in step 2.
4. *Interactive Editing.* Apply interactive editing to the minority of the records with influential errors.
5. *Macro-selection.* Select records with influential errors by using methods based on outlier detection techniques and other procedures that make use of all or a large fraction of the response.

We distinguish two kinds of process steps: those that localize or treat errors and those that direct the records through the different stages of the process. The processes in step 2 and 5 are of the latter kind; they are "selectors" that do not actually treat errors, but select records for specific kinds of further processing.

Step 1. *Correction of Systematic Errors.* Detection and correction of systematic errors is an important first step in an editing process. It can be done automatically and reliably at virtually no costs and hence will improve both the efficiency and the quality of the editing process. It is in fact a very efficient and probably often underused correction approach. Systematic and other evident errors and algorithms that can automatically resolve these errors are described in Chapter 2 of this book.

Step 2. *Micro-selection.* Errors that cannot be resolved in the previous step will be taken care of either manually (by subject-matter specialists) or automatically (by specialized edit and imputation algorithms). In this step, the data are split into a critical stream and a noncritical stream, using selective editing techniques as mentioned in Section 1.4.2. The extent to which a record potentially contains influential errors can be measured by a score function [cf. Latouche and Berthelot (1992), Lawrence and McKenzie (2000), and Farwell and Rain (2000)]. This function is constructed such that records with high scores likely contain errors that have substantial effects on estimates of target parameters. For this selection step, a threshold value for the score has been set and all records with scores above this threshold are directed to manual reviewers whereas records with scores below the threshold are treated automatically. More details can be found in Chapter 6.

Apart from the score function, which looks at influential errors, another important selection criterion is imputability. For some variables, very accurate imputation models can be developed. If such a variable fails an edit, the erroneous value can safely be replaced by an imputed value, even if it is an influential value. Note that the correction of systematic errors in the previous step can also be an example of automatic treatment of influential errors, if the systematic error is an influential one.

Step 3a. *Localization of erroneous values (random errors).* The next three steps are automatic detection and correction procedures. In principle, they are designed to solve hard edit failures, including missing values, but they can be applied to soft edits if the soft edit is treated as a hard one. These three steps together represent the vast majority of all edit and imputation methodology. The other chapters of this book are devoted to this methodology for automatic detection and correction of erroneous and missing values.

The first step in the automatic treatment of errors is the localization of errors. Since systematic errors have already been removed, the remaining errors at this stage are random errors. Once the (hard) edits are defined and implemented, it is straightforward to check whether the values in a record are inconsistent in the sense that some of these edits are violated. It is, however, not so obvious how to decide which variables in an inconsistent record are in error. The designation of erroneous values in an inconsistent record is called the error localization problem, which is treated in Chapters 3 to 5 of this book.

Step 3b. *Imputation.* In this step, missing data are imputed in an automatic manner. The imputation method that is best suited for a particular situation will depend on the characteristics of the data set and the research goals. In Chapters 7 to 9 we examine imputation methods in detail.

Step 3c. *Consistency Adjustment of Imputed Values.* In most cases, the edits are not taken into account by the imputation methods; some exceptions are examined in Chapter 9. As a consequence, the imputed records are in general inconsistent with the edits. This problem can be solved by the introduction of an adjustment step in which adjustments are made to the imputed values such that the record satisfies all edits and the adjustments are as small as possible. This problem can be formulated as a linear or a quadratic programming problem and is treated in Chapter 10.

Step 4. *Interactive Editing.* Substantial mistakes by somewhat larger enterprises that have an appreciable influence on publication aggregates and for which no accurate imputation model exists are not considered suitable for the generic procedures of automatic editing. These records are treated by subject-matter specialists in a process step called interactive editing; see Section 1.4.2 above and Chapter 6.

Step 5. *Macro-selection.* The steps considered so far all use micro-editing methods—that is, methods that use the data of a single record and related auxiliary information to check and correct it. Micro-editing processes can be conducted from the start of the data collection phase, as soon as records become available. In contrast, macro-selection techniques use information from other records and can only be applied if a substantial part of the data is collected or has been imputed. Macro-selection techniques are also selective editing techniques in the sense that they aim to direct the attention only to possibly influential erroneous values. Macro-editing is treated in Chapter 6 of this book.

The process flow suggested in Figure 1.1 is just one possibility. Depending on the type of survey and the available resources and auxiliary information, the process flow can be different. Not all steps are always carried out, the order of steps may be different, and the particular methods used in each step can differ between types of surveys. For social surveys, for instance, selective editing is not very important because the contributions of individuals to a publication total are not so much different, unlike the contributions of small and large enterprises in business surveys. Consequently, there is less need for manual editing of influential records, and step 4 need not be performed. Often, in social surveys, due to a lack of hard edits, the main type of detectable error is the missing value, and process steps 3a and 3c are not performed either. For administrative data the collection of all records, or a large part of it, is often available at once. This is different from the situation for surveys where the data are collected over a period of time. For administrative data it is therefore possible to form preliminary estimates immediately and to start with macro-editing as a tool for selective editing, and a process could start with step 1, followed by step 5 and possibly by step 4 and/or step 3.

Although automatic procedures are frequently used for relatively unimportant errors, choosing the most suitable error detection and/or imputation methods is still important. If nonappropriate methods are used, especially for large amounts of random errors and/or missing values, additional bias may be introduced. Furthermore, as the quality of the automatic error localization and imputation methods and models gets better, more records can be entrusted to the automatic treatment in step 3 and less records have to be selected for the time-consuming and costly interactive editing step.

REFERENCES

Barnett, V., and T. Lewis (1994), *Outliers in Statistical Data.* John Wiley & Sons, New York.

Bethlehem, J. (2007), *Reducing the Bias of Web Survey Based Estimates.* Discussion paper 07001, Statistics Netherlands, Voorburg (see also www.cbs.nl).

Børke, S. (2008), *Using "Traditional" Control (Editing) Systems to Reveal Changes when Introducing New Data Collection Instruments.* Working Paper No. 6, UN/ECE Work Session on Statistical Data Editing, Vienna.

Chambers, R., A. Hentges, and X. Zhao (2004), Robust Automatic Methods for Outlier and Error Detection. *Journal of the Royal Statistical Society A 167*, pp. 323–339.

Couper, M. P., R. P. Baker, J. Bethlehem, C. Z. F. Clark, J. Martin, W. L. Nichols II, and J. M. O'Reilly (eds.) (1998), *Computer Assisted Survey Information Collection.* John Wiley & Sons, New York.

Farwell, K., and M. Rain (2000), Some Current Approaches to Editing in the ABS. *Proceedings of the Second International Conference on Establishment Surveys*, Buffalo, pp. 529–538.

Federal Committee on Statistical Methodology (1990), *Data Editing in Federal Statistical Agencies.* Statistical Policy Working Paper 18, U.S. Office of Management and Budget, Washington, D.C.

Granquist, L. (1984), *Data Editing and its Impact on the Further Processing of Statistical Data.* Workshop on Statistical Computing, Budapest.

Granquist, L. (1990), A Review of Some Macro-Editing Methods for Rationalizing the Editing Process. *Proceedings of the Statistics Canada Symposium*, pp. 225–234.

Granquist, L. (1995), Improving the Traditional Editing Process. In: *Business Survey Methods*, B.G. Cox, D.A. Binder, B.N. Chinnappa, A. Christianson, M.J. Colledge, and P.S. Kott, eds. John Wiley & Sons, New York, pp. 385–401.

Granquist, L. (1997), The New View on Editing. *International Statistical Review 65*, pp. 381–387.

Granquist, L. and J. Kovar (1997), Editing of Survey Data: How Much Is Enough? In: *Survey Measurement and Process Quality*, L.E. Lyberg, P. Biemer, M. Collins, E.D. De Leeuw, C. Dippo, N. Schwartz, and D. Trewin, eds. John Wiley & Sons, New York, pp. 415–435.

Hoogland, J., and R. Smit (2008), *Selective Automatic Editing of Mixed Mode Questionnaires for Structural Business Statistics.* Working Paper No. 2, UN/ECE Work Session on Statistical Data Editing, Vienna.

Latouche, M., and J. M. Berthelot (1992), Use of a Score Function to Prioritize and Limit Recontacts in Editing Business Surveys. *Journal of Official Statistics 8*, pp. 389–400.

Lawrence, D., and R. McKenzie (2000), The General Application of Significance Editing. *Journal of Official Statistics 16*, pp. 243–253.

Little, R. J. A., and D. B. Rubin (2002), *Statistical Analysis with Missing Data*, second edition. John Wiley & Sons, New York.

Nordbotten, S. (1955), Measuring the Error of Editing the Questionnaires in a Census. *Journal of the American Statistical Association 50*, pp. 364–369.

Nordbotten, S. (1963), Automatic Editing of Individual Statistical Observations. In: *Conference of European Statisticians Statistical Standards and Studies No. 2*, United Nations, New York.

Rocke, D. M., and D. L. Woodruff (1996), Identification of Outliers in Multivariate Data. *Journal of the American Statistical Association 91*, pp. 1047–1061.

Rousseeuw, P. J., and M. L. Leroy (1987), *Robust Regression & Outlier Detection*. John Wiley & Sons, New York.

Rubin, D. B. (1987), *Multiple Imputation for Non-Response in Surveys*. John Wiley & Sons, New York.

Schafer, J. L. (1997), *Analysis of Incomplete Multivariate Data*. Chapman & Hall, London.

Todorov, V., M. Templ, and P. Filzmoser (2009), *Outlier Detection in Survey Data Using Robust Methods*. Working Paper No. 40, UN/ECE Work Session on Statistical Data Editing, Neuchâtel.

Van der Loo, M. P. J. (2008), *An Analysis of Editing Strategies for Mixed-Mode Establishment Surveys*. Discussion paper 08004, Statistics Netherlands (see also www.cbs.nl).

Wallgren, A., and B. Wallgren (2007), *Register-Based Statistics—Administrative Data for Statistical Purposes*. John Wiley & Sons, Chichester.

Willeboordse, A. (ed.) (1998), *Handbook on the Design and Implementation of Business Surveys*. Office for Official Publications of the European Communities, Luxembourg.

CHAPTER TWO

Methods for Deductive Correction

2.1 Introduction

In the theory of editing, a distinction is often made between systematic errors and random errors, where an error is called systematic if it is reported consistently over time by different respondents; see, for example, EDIMBUS (2007) and Section 1.3 of this book. This type of error typically occurs when a respondent misunderstands or misreads a survey question—for example, by reporting amounts in units rather than multiples of 1000 units. A fault in the data processing system might also introduce a systematic error. Since it is reported consistently by different respondents, an undiscovered systematic error leads to biased estimates. Random errors, on the other hand, occur by accident—for example, when a "1" on a survey form is read as a "7" during data processing. In this chapter, we focus on the detection and correction of systematic errors. General methodology for detecting random errors will be the subject of Chapters 3 to 5.

Detecting and correcting systematic errors at the beginning of the editing process, before other forms of editing are applied, is particularly useful. These errors can often be detected automatically in a straightforward manner, in contrast to the complex methods that are required for the automatic localization of random errors. Moreover, after detection, the correction of a systematic error is trivial, because the underlying error mechanism is assumed to be known. This type of correction is often called "deductive" or "logical." When selective editing is used (see Chapter 6), performing deductive corrections in a separate step increases the efficiency of the editing process, because less manual editing is needed. In addition, solving the error localization problem for random errors

Handbook of Statistical Data Editing and Imputation, by T. de Waal, J. Pannekoek, and S. Scholtus

then becomes easier, because the number of violated edits becomes smaller. This means that more records are eligible for automatic editing.

This chapter is organized as follows. Section 2.2 explains some general principles for the detection and correction of systematic errors. This section also contains a few simple examples. However, different types of systematic errors occur for different surveys, and no generic recipe can be given for the detection of all systematic errors. Instead, to illustrate the possibilities, Section 2.3 works out deductive correction methods for four particular errors in some detail.

We remark that, due to item nonresponse, unedited data typically contain a substantial number of missing values. It is assumed throughout this chapter that, for numerical data, these missing values have been temporarily replaced by zeros. This is merely a precondition for determining which edits are violated and which are satisfied, and it should not be considered a full imputation. When all deductive corrections have been performed, all imputed zeros should be replaced by missing values again, to be re-imputed by a valid method later.

2.2 Theory and Applications

2.2.1 CORRECTING INCONSISTENCIES DEDUCTIVELY

Generally speaking, an inconsistency in the data may be corrected deductively if its occurrence can only be explained in one way, based on logical reasoning and/or subject-matter knowledge. In practice, some assumption always has to be made to exclude other possible explanations. If the assumption is reasonable, then deductive correction may be applied. It is important to test the validity of the underlying assumption, because deductive correction may lead to biased estimates if it is based on an invalid assumption.

EXAMPLE 2.1

A questionnaire asks respondents to state their sex and whether they are currently pregnant. Obviously, the following edit should hold:

(2.1) IF *Sex* = "Male", THEN *Currently Pregnant* = "No".

Suppose that certain records with *Sex* reported as "Male" violate this edit, either because *Currently Pregnant* is reported as "Yes" or because it is missing. Depending on the survey context, it may be reasonable to assume that *Sex* is reported without error. In particular, this is the case if this variable is known from a well-maintained population register. Under this assumption, violations of edit (2.1) may be corrected deductively by imputing the value "No" for *Currently Pregnant*. If this assumption should be untenable, however, the number of pregnancies in the population will be underestimated after editing. ■

EXAMPLE 2.2

A business survey contains the variables T for *Turnover*, C for *Costs*, and P for *Profit*. By definition, we have the following edit:

$$T - C = P. \tag{2.2}$$

The first column of Table 2.1 shows a record that violates (2.2). As a first assumption, we may postulate that only one of the values is reported erroneously, since it suffices to change one variable to obtain consistency. If only one variable is incorrect, its true value can be computed from (2.2) by plugging in the two other observed values. The remaining columns of Table 2.1 show the resulting consistent versions of the record; the adjusted value is displayed in boldface.

TABLE 2.1 An Inconsistent Record with Three Potential Consistent Versions

	Record	Correction 1	Correction 2	Correction 3
T	353	**398**	353	353
C	283	283	**238**	283
P	115	115	115	**70**

Intuitively, using C to resolve the inconsistency is most appealing, because it requires the minor adjustment of "283" to "238". Somewhere during reporting, collecting, and processing, it seems far more likely that the true value "238" was accidentally changed to "283" than that "398" was changed to "353" or "115" was changed to "70". Hence, the following deductive correction method may be suggested: If a record violates (2.2) and can be made consistent by interchanging two digits in one of the reported values, then the record should be corrected this way. Alternatively, if Fellegi–Holt-based automatic editing is used (see Chapter 3), this information could be incorporated in the reliability weights by lowering the weight of the offending variable [cf. Van de Pol, Bakker, and De Waal (1997)].

Interchanging two digits is an example of a "simple typing error." A general method for correcting simple typing errors deductively is discussed in Section 2.3.3. ■

Example 2.2 illustrates an idea that is often used in formulating deductive correction methods: If a certain inconsistency can be resolved by making only a minor adjustment to the data (and other possible corrections involve changes that are more drastic), then this adjustment is likely to return the true data.

We remark that deductive correction is always possible if the underlying error mechanism is known or, rather, can be guessed with reasonable accuracy. Thus, in some cases it may be attractive to correct certain random errors deductively, as we did in the two examples above. However, in practice the approach is mainly suited for the correction of systematic errors, because more information about the error mechanism tends to be available for errors of this type.

2.2.2 DETECTING NEW SYSTEMATIC ERRORS

New systematic errors can be discovered through a thorough analysis of edit violations in a data file. If a certain edit is frequently violated in a similar manner, this may indicate the presence of a systematic error. For a better understanding of the cause of the error, the original questionnaire should be consulted. Once a systematic error has been identified, it is usually not difficult to construct an algorithm for the automatic detection and correction of the error.

This analysis requires that a large part of the data have been collected. Its conclusions can be used to make improvements for future versions of the survey. It is important to realize that these improvements do not have to be restricted to the editing process. The presence of a systematic error may indicate that many respondents have difficulties with a certain feature of the questionnaire. Sometimes the occurrence of a particular systematic error can be prevented by rewording or redesigning part of the questionnaire. If this is so, then preventing the error at the source is, in principle, preferable to detecting and correcting the error later on. Nevertheless, automatic correction methods can be useful, because some errors appear to be impossible to prevent (e.g., sign errors; see Section 2.3.2 below). Moreover, since constantly redesigning the questionnaire is impractical and may have unwanted side effects when time series are published, it is often convenient to adopt an automatic detection algorithm at first and to accumulate knowledge on the prevention of systematic errors over time until a major revision of the survey process is due. Then all suggested improvements to the design of the questionnaire can be made simultaneously.

EXAMPLE 2.3

The following example of the discovery of a new systematic error is taken from the annual survey of structural business statistics at Statistics Netherlands. A certain balance edit of the form

$$x_1 + x_2 + x_3 + x_4 = x_5 \tag{2.3}$$

should hold. Table 2.2 displays four examples of inconsistent records.

In these records, it holds that $x_2 + x_3 + x_4 = x_5$, which suggests that respondents ignored the value of x_1 when they computed the value of x_5. The underlying problem is revealed when we look at the questionnaire. Figure 2.1 shows the design of the answer boxes for the variables $x_1, \ldots, x_5$ on the survey form. There is a gap between the box corresponding to x_1 and the other boxes. As a result, it is not clear from the design whether x_1 should contribute to the sum or not.

TABLE 2.2 **Examples of Records that Violate Edit (2.3)**

	Record 1	Record 2	Record 3	Record 4
x_1	1,100	364	1,135	901
x_2	88	46	196	134
x_3	40	34	68	0
x_4	42	0	42	0
x_5	170	80	306	134

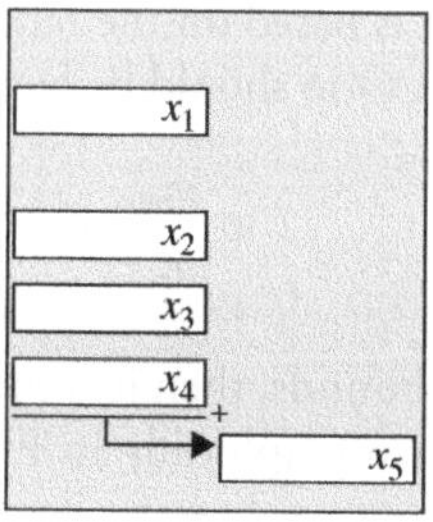

FIGURE 2.1 **Design of the survey form for edit (2.3).**

It is not difficult to construct an automatic method for detecting and correcting this systematic error. However, in this case there is an obvious way of preventing the error by improving the design of the survey form. In fact, a new questionnaire has since been implemented at Statistics Netherlands, which has all answer boxes evenly spaced. ■

2.3 Examples

In this section, we describe methods for the deductive correction of four types of errors. Section 2.3.1 deals with the infamous unity measure error. The other examples illustrate that deductive correction methods can go beyond simple if–then rules. Section 2.3.2 introduces the concept of a sign error. Section 2.3.3 discusses simple typing errors. Finally, Section 2.3.4 describes one possible way of handling rounding errors—that is, very small inconsistencies with respect to balance edits. The efficiency of the editing process can be increased by correcting rounding errors deductively, and any resulting bias in estimates will be negligible, because the required adaptations are very small and tend to cancel out on an aggregated level.

2.3.1 THE UNITY MEASURE ERROR

Respondents are often asked to round off reported numerical values to a certain base—for instance, to multiples of 1000 units. Some respondents ignore this instruction and consequently report amounts that are consistently too high by a certain factor. This is called a "unity measure error." It is important to detect and correct this error during editing, because otherwise publication figures of many items would be overestimated. Since the error tends to be made consistently throughout the questionnaire, it typically leads to few edit violations.

Unity measure errors are usually detected by comparing the raw survey data with reference data. Examples of reference data are: edited data from the same respondent in a previous survey, data from similar respondents in a previous survey or the current survey (by taking a robust measure such as the median), and auxiliary data on the respondent from another source (e.g., a register).

A widely used method is based on the ratio of the raw amount x_{raw} and the reference amount x_{ref}. This ratio should be between certain bounds,

$$l < \frac{x_{raw}}{x_{ref}} < u. \tag{2.4}$$

If the ratio is outside these bounds, then it is concluded that x_{raw} contains a unity measure error. This is shown graphically in Figure 2.2: A point that lies in the shaded region corresponds to a value of x_{raw} outside the bounds given by ratio edit (2.4), which is therefore considered erroneous.

Once a unity measure error has been detected, it is corrected by dividing x_{raw} by the appropriate factor. Note that this method assumes that x_{ref} is not affected by unity measure errors.

Good values for the lower and upper bounds in (2.4) have to be chosen in practice. There is a trade-off, because specifying a wide interval leads to a higher number of missed errors (i.e., observations that are considered correct but actually contain a unity measure error), while specifying a narrow interval leads to a higher number of false hits (i.e., observations that supposedly contain a unity measure error but are actually correct). If previously edited data are available, a simulation study can be conducted to experiment with different bounds. During the study, the number of missed errors and the number of

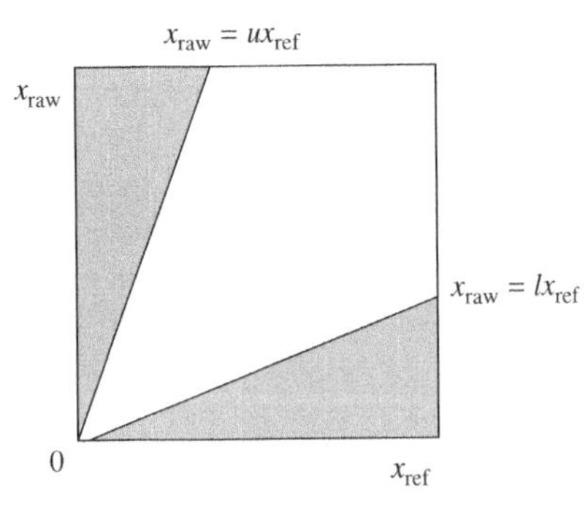

FIGURE 2.2 Graphical representation of the ratio edit method for detecting unity measure errors.

false hits are determined for each choice of bounds. The bounds that yield the smallest number of misclassifications may then be used in the editing process for localizing unity measure errors in raw data. A practical example of such a simulation study is discussed in Section 11.3.5.

EXAMPLE 2.4

The editing process for the Dutch monthly short-term statistics contains a step that detects and corrects so-called "thousand-errors" in the reported *Turnover*. A thousand-error is a unity measure error, where the respondent reports amounts in euros instead of the requested multiples of 1000 euros. For the Dutch short-term statistics, the error can be detected in two possible ways. If the respondent has reported at least once during the previous six months, then the raw value of *Turnover* is compared with the last known value. Absolute values are taken to avoid working with negative values of *Turnover*. A record is flagged to contain a thousand-error, if the ratio of the two absolute values is larger than 300; that is, (2.4) is used with $l = -300$ and $u = 300$.

If the respondent has not reported recently, the raw value of *Turnover* is compared instead with its within-stratum median in the edited data from the previous month, using a stratification based on *Economic Activity* and *Number of Employees*. In this case, a narrower interval is used for detection: $l = -100$ and $u = 100$ in (2.4). ■

Al-Hamad, Lewis, and Silva (2008) describe an alternative method for detecting unity measure errors, by looking at the difference between the number of digits in the raw value and the reference value:

$$diff = |\lceil \log_{10} x_{\text{raw}} \rceil - \lceil \log_{10} x_{\text{ref}} \rceil|, \tag{2.5}$$

where $\lceil a \rceil$ denotes the smallest integer larger than or equal to a. Different types of unity measure errors may be detected by flagging records that yield a certain value of *diff*. For instance, a thousand-error corresponds to $diff = 3$. By taking the absolute value in (2.5), this method also detects unity measure errors that remain in the reference data.

A more complex method for detecting unity measure errors, based on explicit modeling of the error, is suggested by Di Zio, Guarnera, and Luzi (2005).

2.3.2 SIGN ERRORS

The unity measure error is widely known and well-documented in the literature. The remaining examples in this section—sign errors, simple typing errors, and rounding errors—are less well known. These three errors are associated with violations of balance edits in numerical data; hence, they are a common problem in business surveys, where respondents have to do a lot of accounting when filling in a questionnaire, but not so much in social surveys. The main references for these three examples are Scholtus (2008, 2009); the Dutch structural business statistics are the motivating example for the research in these papers.

Sign errors occur frequently in surveys where respondents have to perform calculations involving both subtraction and addition. In particular, this is the case for the so-called profit-and-loss account, which is a part of the questionnaire used for the Dutch structural business statistics. The profit-and-loss account consists of a number of balance variables, which we denote by $x_0, x_1, \ldots, x_{n-1}$. A final balance variable x_n is found by adding up the other balance variables. Thus, the data should conform to the following edit:

$$x_0 + x_1 + \cdots + x_{n-1} = x_n. \tag{2.6}$$

Edit (2.6) is sometimes referred to as the external sum. A balance variable is defined as the difference between a returns item and a costs item. If these items are also asked in the questionnaire, the following edit should hold:

$$x_{k,r} - x_{k,c} = x_k, \tag{2.7}$$

where $x_{k,r}$ denotes the returns item and $x_{k,c}$ the costs item. Edits of the form (2.7) are referred to as internal sums. Note that the statistical office may decide not to ask both the balance variable and the underlying returns and costs items in a questionnaire—for instance, to reduce the response burden.

We use a general description of the profit-and-loss account, in which the returns and costs are not necessarily asked for every balance variable in the survey. To keep the notation simple, we assume that the balance variables are arranged such that only $x_0, x_1, \ldots, x_m$ are split into returns and costs, for some $m \in \{0, 1, \ldots, n-1\}$. Thus, the following set of edits is used:

$$\begin{aligned} x_0 &= x_{0,r} - x_{0,c}, \\ &\vdots \\ x_m &= x_{m,r} - x_{m,c}, \\ x_n &= x_0 + x_1 + \cdots + x_{n-1}. \end{aligned} \tag{2.8}$$

In this notation the 0th balance variable x_0 stands for *operating results*, and $x_{0,r}$ and $x_{0,c}$ represent *operating returns* and *operating costs*, respectively.

Table 2.3 displays the structure of the profit-and-loss account from the structural business statistics questionnaire that was used at Statistics Netherlands until 2005. The associated edits are given by (2.8), with $n = 4$ and $m = n - 1 = 3$. Table 2.3 also displays three examples of inconsistent records.

In example (a), two edits are violated: the external sum and the internal sum with $k = 1$. Interestingly, the profit-and-loss account can be made fully consistent with respect to all edits by changing the value of x_1 from 10 to -10 (see Table 2.4). This is the natural way to obtain a consistent profit-and-loss

TABLE 2.3 Structure of the Profit-and-Loss Account, with Three Example Records

Variable	Full Name	(a)	(b)	(c)
$x_{0,r}$	Operating returns	2100	5100	3250
$x_{0,c}$	Operating costs	1950	4650	3550
x_0	Operating results	150	450	300
$x_{1,r}$	Financial revenues	0	0	110
$x_{1,c}$	Financial expenditure	10	130	10
x_1	Operating surplus	10	130	100
$x_{2,r}$	Provisions rescinded	20	20	50
$x_{2,c}$	Provisions added	5	0	90
x_2	Balance of provisions	15	20	40
$x_{3,r}$	Exceptional income	50	15	30
$x_{3,c}$	Exceptional expenses	10	25	10
x_3	Exceptional result	40	10	20
x_4	Pretax results	195	610	−140

TABLE 2.4 Corrected Versions of the Example Records from Table 2.3[a]

Variable	Full Name	(a)	(b)	(c)
$x_{0,r}$	Operating returns	2100	5100	3250
$x_{0,c}$	Operating costs	1950	4650	3550
x_0	Operating results	150	450	**−300**
$x_{1,r}$	Financial revenues	0	**130**	110
$x_{1,c}$	Financial expenditure	10	**0**	10
x_1	Operating surplus	**−10**	130	100
$x_{2,r}$	Provisions rescinded	20	20	**90**
$x_{2,c}$	Provisions added	5	0	**50**
x_2	Balance of provisions	15	20	40
$x_{3,r}$	Exceptional income	50	**25**	30
$x_{3,c}$	Exceptional expenses	10	**15**	10
x_3	Exceptional result	40	10	20
x_4	Pretax results	195	610	−140

[a]Changes are in boldface.

account here, since any other explanation would require more variables to be changed. Moreover, it is quite conceivable that the minus sign in x_1 was left out by the respondent or "lost" during data processing.

Two internal sums are violated in example (b), but the external sum holds. The natural way to obtain a consistent profit-and-loss account here, is by interchanging the values of $x_{1,r}$ and $x_{1,c}$, and also of $x_{3,r}$ and $x_{3,c}$ (see Table 2.4). By treating the inconsistencies this way, full use is made of the amounts actually filled in by the respondent and no imputation of synthetic values is necessary.

The two types of errors found in examples (a) and (b) are quite common. We will refer to them as "sign errors" and "interchanged returns and costs," respectively. For the sake of brevity, we also use the term "sign error" to refer to both types. Sign errors and interchanged returns and costs are closely related and should therefore be searched for by one detection algorithm. We now formulate such an algorithm, working from the assumption that if an inconsistent record can be made to satisfy all edits in (2.8) by only changing signs of balance variables and/or interchanging returns items and costs items, this is indeed the way the record should be corrected.

It should be noted that the 0th returns and costs items differ from the other variables in the profit-and-loss account in the sense that they are also present in other edits, connecting them to items from other parts of the survey. For instance, *operating costs* should equal the sum of *total labor costs*, *total machine costs*, and so on. If $x_{0,r}$ and $x_{0,c}$ were interchanged to suit the 0th internal sum, other edits might become violated. When detecting sign errors, we therefore introduce the constraint that we are not allowed to interchange $x_{0,r}$ and $x_{0,c}$. Because of the way the questionnaire is designed, it seems highly unlikely that any respondent would mix up these two amounts anyway.

As stated above, a record contains a sign error if it satisfies the following two conditions:

- At least one edit in (2.8) is violated.
- It is possible to satisfy (2.8) by only changing the signs of balance amounts and/or interchanging returns and costs items other than $x_{0,r}$ and $x_{0,c}$.

An equivalent way of formulating this is to say that an inconsistent record contains a sign error if the following set of equations has a solution:

$$\begin{aligned}
x_0 s_0 &= x_{0,r} - x_{0,c},\\
x_1 s_1 &= \left(x_{1,r} - x_{1,c}\right) t_1,\\
&\vdots\\
x_m s_m &= \left(x_{m,r} - x_{m,c}\right) t_m,\\
x_n s_n &= x_0 s_0 + x_1 s_1 + \cdots + x_{n-1} s_{n-1},\\
(s_0, \ldots, s_n;\ t_1, \ldots, t_m) &\in \{-1, 1\}^{n+m+1}.
\end{aligned} \tag{2.9}$$

Note that in (2.9) the x's are used as known constants rather than unknown variables. Thus, a different set of equations in $(s_0, \ldots, s_n; t_1, \ldots, t_m)$ is found for each record. Once a solution to (2.9) has been found, it immediately tells us how to correct the sign error: If $s_j = -1$, then the sign of x_j must be changed; and if $t_k = -1$, then the values of $x_{k,r}$ and $x_{k,c}$ must be interchanged. It is easy to see that the resulting record satisfies all edits (2.8). Since $x_{0,r}$ and $x_{0,c}$ should not be interchanged, no variable t_0 is present in (2.9).[1]

[1] The absence of t_0 in (2.9) also has a technical reason: The value of s_0 is now fixed by the first equation in the system. Fixing one variable is necessary to obtain a unique solution to (2.9), because otherwise any solution could be transformed into a new solution by multiplying each variable by -1.

EXAMPLE 2.5

We set up system (2.9) for example (c) from Table 2.3:

$$
\begin{aligned}
300s_0 &= -300,\\
100s_1 &= 100t_1,\\
40s_2 &= -40t_2,\\
20s_3 &= 20t_3,\\
-140s_4 &= 300s_0 + 100s_1 + 40s_2 + 20s_3,\\
(s_0, s_1, s_2, s_3, s_4; t_1, t_2, t_3) &\in \{-1, 1\}^8 .
\end{aligned}
$$

Solving this system yields

$$
\begin{aligned}
&s_0 = -1, \quad s_1 = 1, \quad s_2 = 1, \quad s_3 = 1, \quad s_4 = 1;\\
&t_1 = 1, \quad t_2 = -1, \quad t_3 = 1.
\end{aligned}
$$

This solution tells us that the value of x_0 should be changed from 300 to -300 and that the values of $x_{2,r}$ and $x_{2,c}$ should be interchanged. This correction indeed yields a fully consistent profit-and-loss account with respect to (2.8), as can be seen in Table 2.4. ■

An important question is: Does system (2.9) always have a unique solution? Scholtus (2008) establishes the following sufficient condition for uniqueness: If $x_0 \neq 0$ and $x_n \neq 0$ and if the equation

$$\lambda_0 x_0 + \lambda_1 x_1 + \cdots + \lambda_{n-1} x_{n-1} = 0 \tag{2.10}$$

does not have any solution $\lambda_0, \lambda_1, \ldots, \lambda_{n-1} \in \{-1, 0, 1\}$ for which at least one term $\lambda_j x_j \neq 0$, then *if* the inconsistency in the record can be resolved by changing signs and/or interchanging returns and costs, it can be done so in a unique way. Translated roughly, this means that the inconsistency can be resolved uniquely unless there exists some simple linear relation between the balance amounts $x_0, x_1, \ldots, x_{n-1}$—for example, if two balance amounts happen to be equal. Since no linear relation of the form (2.10) exists by design, for the great majority of inconsistent records containing sign errors the inconsistency can indeed be resolved uniquely.

For instance, in the data of the Dutch wholesale structural business statistics of 2001, it was found that over 95% of all records satisfied this condition for uniqueness (Scholtus, 2008). In particular, this condition shows that the solution found in Example 2.5 is unique. We emphasize that the condition is sufficient for uniqueness, but not necessary. Thus, an instance of (2.9) that does not satisfy this condition may still have a unique solution; see Scholtus (2008) for an example.

Detecting a sign error in a given record is equivalent to solving the corresponding system (2.9). Therefore all that is needed to implement the detection of sign errors, is a systematic method to solve this system. The least sophisticated way of finding a solution to (2.9) would be to simply try all possible combinations of s_j and t_k. Since m and n are small in this situation, the number of possibilities is not very large and this approach is actually quite feasible. However, it is also possible to reformulate the problem as a so-called binary linear programming problem. This has the advantage that standard software may be used to implement the method.

To reformulate the problem, we introduce the following binary variables:

$$\sigma_j = \frac{1 - s_j}{2}, \qquad j \in \{0, 1, \ldots, n\},$$

$$\tau_k = \frac{1 - t_k}{2}, \qquad k \in \{1, \ldots, m\}.$$

Note that $\sigma_j = 0$ if $s_j = 1$, and $\sigma_j = 1$ if $s_j = -1$, and the same holds for τ_k and t_k.

Finding a solution to (2.9) may now be restated as follows:

$$\text{Minimize} \sum_{j=0}^{n} \sigma_j + \sum_{k=1}^{m} \tau_k$$

such that

$$
\begin{aligned}
x_0(1 - 2\sigma_0) &= x_{0,r} - x_{0,c} \\
x_1(1 - 2\sigma_1) &= \left(x_{1,r} - x_{1,c}\right)(1 - 2\tau_1) \\
&\vdots \\
x_m(1 - 2\sigma_m) &= \left(x_{m,r} - x_{m,c}\right)(1 - 2\tau_m) \\
x_n(1 - 2\sigma_n) &= x_0(1 - 2\sigma_0) + \cdots + x_{n-1}(1 - 2\sigma_{n-1}) \\
(\sigma_0, \ldots, \sigma_n; \tau_1, \ldots, \tau_m) &\in \{0, 1\}^{n+m+1}
\end{aligned}
\tag{2.11}
$$

Observe that in this formulation the number of variables s_j and t_k that are equal to -1 is minimized; that is, the solution is searched for that results in the smallest number of changes being made in the record. Obviously, if a unique solution to (2.9) exists, then this is also the solution to (2.11). The binary linear programming problem may be solved by applying a standard branch and bound algorithm. Since n and m are small, very little computation time is needed to find the solution.

The following plan summarizes the correction method for sign errors and interchanged returns and costs. The input consists of a record that does not satisfy (2.8).

ALGORITHM 2.1

Deductive correction of sign errors and interchanged returns and costs:

Step 1. Set up the binary linear programming problem (2.11) for the values in the input record.

Step 2. Solve (2.11). If the problem is infeasible, then the record does not contain a sign error. If a solution is found, continue.

Step 3. Replace x_j by $-x_j$ for every $\sigma_j = 1$ and interchange $x_{k,r}$ and $x_{k,c}$ for every $\tau_k = 1$.

2.3.3 SIMPLE TYPING ERRORS

In Example 2.2 in Section 2.2.1 we encountered an inconsistent record that could be corrected deductively by assuming that the respondent had interchanged two digits in a numerical value. Interchanging two adjacent digits is an example of a "simple typing error." Other examples include:

- Adding a digit—for example, writing "46297" instead of "4627"
- Omitting a digit—for example, writing "427" instead of "4627"
- Replacing a digit—for example, writing "4687" instead of "4627"

A common feature of these four simple typing errors is that they always affect one variable at a time. This is not true of errors in general. Another common feature of these types of errors is that they result in an observed erroneous value, which is related to the unobserved correct value in an easily recognizable way. Again, the same cannot be said of errors in general.

Formally, a simple typing error can be seen as a function $f : \mathbb{Z} \to \mathbb{Z}$ acting on the true value x. Due to the error, the value $f(x)$ is observed instead of x. It is not difficult to write down explicit expressions for the functions that describe the four simple typing errors mentioned above; such expressions can be found in Scholtus (2009).

In Example 2.2, a simple typing error was detected by using the fact that the variables should satisfy one balance edit. In this section, we describe a method for detecting and correcting simple typing errors in numerical data that should satisfy a number of balance edits, and possibly also other edits. The added difficulty in this situation stems from the fact that in general the balance edits are interrelated, so variables should satisfy different edits simultaneously.

Analyzing Violated and Satisfied Edits. For now, we assume that the variables $\mathbf{x} = (x_1, \ldots, x_p)^T$ have to satisfy only balance edits. Suppose that the balance edits are given by

$$a_{k1}x_1 + \cdots + a_{kp}x_p = 0, \qquad k = 1, \ldots, r,$$

where all coefficients a_{kj} are integers. Together, these edits can be written as $\mathbf{Ax} = \mathbf{0}$, where $\mathbf{A} = (a_{kj})$ is an $r \times p$ matrix of coefficients and $\mathbf{0}$ is the r vector of zeros. We discuss an extension of the method that also handles other types of edits at the end of this section.

Each edit defines a three-way partition of $\{1, \ldots, p\}$:

$$
\begin{aligned}
J_1^{(k)} &= \{j : a_{kj} > 0\}, \\
J_2^{(k)} &= \{j : a_{kj} < 0\}, \\
J_3^{(k)} &= \{j : a_{kj} = 0\}
\end{aligned}
\tag{2.12}
$$

for $k = 1, \ldots, r$. The kth edit can be written as

$$
\sum_{j \in J_1^{(k)}} a_{kj} x_j = - \sum_{j \in J_2^{(k)}} a_{kj} x_j. \tag{2.13}
$$

When $j \in J_3^{(k)}$, we say that x_j is not involved in the kth edit. The complement $\bar{J}_3^{(k)} = J_1^{(k)} \cup J_2^{(k)}$ contains the indices of all variables involved in the kth edit.

Similarly, each variable defines a partition of $\{1, \ldots, r\}$:

$$
\begin{aligned}
K_1^{(j)} &= \{k : a_{kj} > 0\}, \\
K_2^{(j)} &= \{k : a_{kj} < 0\}, \\
K_3^{(j)} &= \{k : a_{kj} = 0\}
\end{aligned}
\tag{2.14}
$$

for $j = 1, \ldots, p$. The complement $\bar{K}_3^{(j)} = K_1^{(j)} \cup K_2^{(j)}$ contains the indices of all edits that involve x_j. We assume throughout that each variable is involved in at least one edit (i.e., $\bar{K}_3^{(j)} \neq \emptyset$ for all j), since a variable that is not involved in any edits can be ignored during editing.

Given an observed record $\mathbf{x}$, it is possible to compute, for each edit, two partial sums:

$$
s_1^{(k)} = \sum_{j \in J_1^{(k)}} a_{kj} x_j, \quad s_2^{(k)} = - \sum_{j \in J_2^{(k)}} a_{kj} x_j, \qquad k = 1, \ldots, r.
$$

The record violates the kth edit, and we write $\phi(k) = 1$, if $s_1^{(k)} \neq s_2^{(k)}$ (see (2.13)). Otherwise, the record satisfies the kth edit, and we write $\phi(k) = 0$. Thus, the set of edits is split into two groups:

$$
E_1 = \{k : \phi(k) = 1\}, \qquad E_2 = \{k : \phi(k) = 0\}.
$$

The edits with indices in E_1 are violated by the current record, whereas the edits with indices in E_2 are satisfied.

Finally, we define the following subset of the variables:

$$J_0 = \bigcap_{k \in E_2} J_3^{(k)} = \left\{ j \,:\, E_2 \subseteq K_3^{(j)} \right\}. \tag{2.15}$$

This subset has the following interpretation: It is the index set of variables that are not involved in any edit that is satisfied by the current record. In other words, all edits that involve a variable from J_0 are violated by the current record. When searching for simple typing errors, we only want to perform corrections that increase the number of satisfied edits, without causing previously satisfied edits to become violated. This provision implies that the only variables we can safely change are those in J_0. The equivalence between the two definitions in (2.15) is trivial.

Generating Automatic Corrections. As we saw in Example 2.2, a record can be made to satisfy a violated balance edit by changing one of the variables involved in that edit. In particular, if $j \in \bar{J}_3^{(k)}$ and the kth edit is currently violated, then the edit becomes satisfied if we change the value of x_j to

$$\tilde{x}_j^{(k)} = \frac{1}{a_{kj}} \left(s_2^{(k)} - s_1^{(k)} + a_{kj} x_j \right). \tag{2.16}$$

Namely, if $j \in J_1^{(k)}$, then this operation changes the value of $s_1^{(k)}$ to

$$\tilde{s}_1^{(k)} = s_1^{(k)} - a_{kj} x_j + a_{kj} \tilde{x}_j^{(k)} = s_2^{(k)};$$

and if $j \in J_2^{(k)}$, then this operation changes the value of $s_2^{(k)}$ to

$$\tilde{s}_2^{(k)} = s_2^{(k)} + a_{kj} x_j - a_{kj} \tilde{x}_j^{(k)} = s_1^{(k)}.$$

In both cases, the edit is no longer violated.[2]

For each $j \in J_0$, a list of values $\tilde{x}_j^{(k)}$ can be generated by computing (2.16) for all $k \in \bar{K}_3^{(j)}$. Next, we check, for each value on the list, whether a simple typing error could have produced the observed value x_j if the true value were $\tilde{x}_j^{(k)}$. This is the case if

$$x_j = f(\tilde{x}_j^{(k)}) \tag{2.17}$$

for one of the functions corresponding to a simple typing error. If a function can be found such that (2.17) holds, it seems plausible that a simple typing error has changed the true value $\tilde{x}_j^{(k)}$ to the observed value x_j. Before drawing any conclusions, however, it is important to examine all other possible corrections.

[2] In the case that $|a_{kj}| > 1$, it is possible that formula (2.16) yields a noninteger value. As an example, consider a record with $x_2 = 4$ and $x_3 = 11$, where we want to find the value of x_1 such that $2x_1 + x_2 = x_3$ holds. Using (2.16), we obtain $x_1 = 7/2$. For our present purpose, a noninteger $\tilde{x}_j^{(k)}$ can be immediately discarded, because it is never explained by a simple typing error.

For now, we keep the value $\tilde{x}_j^{(k)}$ on the list. On the other hand, if no function is found such that (2.17) holds, then $\tilde{x}_j^{(k)}$ is removed from the list, because no simple typing error could have changed this value into the observed value x_j.

After discarding some of the values from the list for x_j, it is possible that only an empty list remains. In that case, we do not consider this variable anymore. On the other hand, the reduced list may contain duplicate values, if the same value of x_j can be used to satisfy more than one edit. We denote the unique values that occur on the reduced list by $\tilde{x}_{j,1}, \ldots, \tilde{x}_{j,T_j}$, and we denote the number of times that value $\tilde{x}_{j,t}$ occurs by $\nu_{j,t}$. If $T_j = 1$, we drop the second index and simply write $\tilde{x}_j$ and ν_j. We remark that $\nu_{j,t}$ represents the number of currently violated edits that become satisfied when x_j is changed to $\tilde{x}_{j,t}$. By construction, it holds that $\nu_{j,t} \geq 1$.

The above procedure is performed for each $j \in J_0$. For each variable, we find a (possibly empty) list of potential changes that can be explained by simple typing errors and that, when considered separately, cause one or more violated edits to become satisfied. The question now remains how to make an optimal selection from these potential changes. Ideally, the optimal selection should return the true values of all variables affected by simple typing errors. Since we do not know the true values, a more pragmatic solution is to select the changes that together lead to a maximal number of satisfied edits. In the simple case that exactly one potential change is found for exactly one variable, the choice is straightforward. If more than one potential change is found and/or if more than one variable can be changed, the choice requires more thought, because clearly we cannot change the same variable twice and we should not change two variables involved in the same edit. On the other hand, a record might contain several independent typing errors, and we do want to resolve as many of these errors as possible.

The selection problem from the previous paragraph can be formulated as a mathematical optimization problem:

$$\begin{aligned}
&\text{Maximize } \textstyle\sum_{j\in J_0} \sum_{t=1}^{T_j} \nu_{j,t}\delta_{j,t} \\
&\text{such that} \\
&\textstyle\sum_{j\in \bar{J}_3^{(k)} \cap J_0} \sum_{t=1}^{T_j} \delta_{j,t} \leq 1 \qquad \text{for } k \in E_1 \\
&\delta_{j,t} \in \{0,1\} \qquad \text{for } j \in J_0 \text{ and } t \in \left\{1, \ldots, T_j\right\}.
\end{aligned} \tag{2.18}$$

The binary variable $\delta_{j,t}$ equals 1 if we choose to replace x_j with the value $\tilde{x}_{j,t}$, and 0 otherwise. Note that the criterion function in (2.18) counts the number of resolved edit violations. We seek values for $\delta_{j,t}$ that maximize this number, under the inequality constraints in (2.18). These constraints state that at most one change is allowed for each $j \in J_0$ and that at most one variable may be changed per violated edit. Here, the assumption is used that each variable is involved in at least one edit.

To solve problem (2.18), a standard branch and bound algorithm may be applied, constructing a binary tree to enumerate all choices of $\delta_{j,t}$. Branches of the binary tree may be pruned if they do not lead to a feasible solution with respect to the inequality constraints. Note that in this case many branches can be pruned because the constraints are quite strict: Once we set $\delta_{j,t} = 1$ for a particular combination (j, t), all other δ-values that occur in the same constraint must be set to zero. This helps to speed up the branch and bound algorithm.

Once a solution to (2.18) has been found, the value of x_j is changed to $\tilde{x}_{j,t}$ if $\delta_{j,t} = 1$. If $\delta_{j,t} = 0$ for all $t = 1, \ldots, T_j$, then the value of x_j is not changed. Formally, for each $j \in J_0$ the new value of x_j is given by

$$\hat{x}_j = \sum_{t=1}^{T_j} \tilde{x}_{j,t}\delta_{j,t} + x_j \left(1 - \sum_{t=1}^{T_j} \delta_{j,t}\right). \tag{2.19}$$

EXAMPLE 2.6

Suppose that the unedited data consist of records with $p = 11$ numerical variables that should conform to $r = 5$ balance edits:

$$\begin{aligned} x_1 + x_2 &= x_3, \\ x_2 &= x_4, \\ x_5 + x_6 + x_7 &= x_8, \\ x_3 + x_8 &= x_9, \\ x_9 - x_{10} &= x_{11}. \end{aligned}$$

The corresponding partitions (2.12) and (2.14) are displayed in Table 2.5.

TABLE 2.5 Partitions of Variables and Edits for Example 2.6

(a) Partition of Variables According to (2.12)

k	$J_1^{(k)}$	$J_2^{(k)}$	$J_3^{(k)}$
1	$\{1, 2\}$	$\{3\}$	$\{4, 5, 6, 7, 8, 9, 10, 11\}$
2	$\{2\}$	$\{4\}$	$\{1, 3, 5, 6, 7, 8, 9, 10, 11\}$
3	$\{5, 6, 7\}$	$\{8\}$	$\{1, 2, 3, 4, 9, 10, 11\}$
4	$\{3, 8\}$	$\{9\}$	$\{1, 2, 4, 5, 6, 7, 10, 11\}$
5	$\{9\}$	$\{10, 11\}$	$\{1, 2, 3, 4, 5, 6, 7, 8\}$

(b) Partition of Edits According to (2.14)

j	$K_1^{(j)}$	$K_2^{(j)}$	$K_3^{(j)}$
1	$\{1\}$	$\emptyset$	$\{2, 3, 4, 5\}$
2	$\{1, 2\}$	$\emptyset$	$\{3, 4, 5\}$
3	$\{4\}$	$\{1\}$	$\{2, 3, 5\}$
4	$\emptyset$	$\{2\}$	$\{1, 3, 4, 5\}$
5	$\{3\}$	$\emptyset$	$\{1, 2, 4, 5\}$
6	$\{3\}$	$\emptyset$	$\{1, 2, 4, 5\}$
7	$\{3\}$	$\emptyset$	$\{1, 2, 4, 5\}$
8	$\{4\}$	$\{3\}$	$\{1, 2, 5\}$
9	$\{5\}$	$\{4\}$	$\{1, 2, 3\}$
10	$\emptyset$	$\{5\}$	$\{1, 2, 3, 4\}$
11	$\emptyset$	$\{5\}$	$\{1, 2, 3, 4\}$

Moreover, suppose that we are given the following observed record:

x_1	x_2	x_3	x_4	x_5	x_6	x_7	x_8	x_9	x_{10}	x_{11}
1452	116	1568	161	323	76	12	411	19,979	1842	137

This record violates the second, fourth, and fifth edits. Thus $E_1 = \{2, 4, 5\}$ and $E_2 = \{1, 3\}$. Using (2.15), we find that $J_0 = J_3^{(1)} \cap J_3^{(3)} = \{4, 9, 10, 11\}$; the variables x_4, x_9, x_{10}, and x_{11} are only involved in violated edits. Therefore, we only consider these four variables.

Since x_4 is only involved in the second edit, formula (2.16) yields one possible value: $\tilde{x}_4 = 116$. From this value, the observed value $x_4 = 161$ can be explained by a simple typing error, namely the interchanging of two adjacent digits in the true value. Choosing this value only changes the status of the second edit from violated to satisfied, so $v_4 = 1$. Variable x_9 is involved in the fourth and fifth edits. According to (2.16),

$$\tilde{x}_9^{(4)} = -(19{,}979 - 1979 - 19{,}979) = 1979$$

and

$$\tilde{x}_9^{(5)} = 1979 - 19{,}979 + 19{,}979 = 1979.$$

so both edits become satisfied by the same choice of $\tilde{x}_9$. Moreover, the observed value can be explained by a simple typing error: adding a digit in the true value. Thus, we find $\tilde{x}_9 = 1979$ with $v_9 = 2$. Variables x_{10} and x_{11} are only involved in the fifth edit, and we find

$$\tilde{x}_{10}^{(5)} = -(1979 - 19{,}979 - 1842) = 19{,}842$$

and

$$\tilde{x}_{11}^{(5)} = -(1979 - 19{,}979 - 137) = 18{,}137.$$

Changing 19,842 to 1842 can be explained by a simple typing error (omitting a digit from the true value), so $\tilde{x}_{10} = 19{,}842$ with $v_{10} = 1$. Changing 18,137 to 137 requires multiple typing errors, so we do not consider variable x_{11} anymore.

Since several potential changes have been found, we set up problem (2.18) to determine the optimal choice. We obtain:

$$\begin{aligned}
&\text{Maximize } \delta_4 + 2\delta_9 + \delta_{10} \\
&\text{such that} \\
&\delta_4 \leq 1, \\
&\delta_9 \leq 1, \\
&\delta_9 + \delta_{10} \leq 1, \\
&\delta_4, \delta_9, \delta_{10} \in \{0, 1\}.
\end{aligned}$$

It is easy to see that the optimal solution is $\{\delta_4 = 1, \delta_9 = 1, \delta_{10} = 0\}$. This solution yields the following adapted record:

x_1	x_2	x_3	$\hat{x}_4$	x_5	x_6	x_7	x_8	$\hat{x}_9$	x_{10}	x_{11}
1452	116	1568	116	323	76	12	411	1979	1842	137

This adapted record satisfies all edits. ■

Extension to Other Types of Edits. We have assumed until now that only balance edits are specified. In practice, numerical data often also have to satisfy inequalities, such as

$$\textit{Number of Employees (in persons)} \geq \textit{Number of Employees (in FTE)},$$

and conditional edits, such as

$$\text{IF } \textit{Number of Employees} > 0, \text{ THEN } \textit{Wages} > 0.$$

There is an obvious way to extend the method to this more general situation. First, all nonbalance edits are ignored and a list of possible corrected values is constructed using formula (2.16), as before. Now, when reducing the list to $\tilde{x}_{j,1}, \ldots, \tilde{x}_{j,T_j}$, we use an additional criterion: A potential correction should not introduce any new edit violations in the set of inequalities and conditional edits. If a potential correction does lead to new edit violations, it is removed from the list. The rest of the method remains the same.

The following plan summarizes the correction method for simple typing errors. The input consists of a record that violates at least one balance edit.

ALGORITHM 2.2

Deductive correction of simple typing errors:

Step 1. Determine the index set J_0 according to (2.15).

Step 2. For each $j \in J_0$, construct a list of values $\tilde{x}_j^{(k)}$ according to (2.16), for all $k \in \bar{K}_3^{(j)}$.

Step 3. For each $j \in J_0$, reduce the list to all values $\tilde{x}_{j,1}, \ldots, \tilde{x}_{j,T_j}$ that can be explained by simple typing errors and do not cause additional violations of nonbalance edits. Let $v_{j,t}$ denote the number of balance edits that become satisfied by choosing $\tilde{x}_{j,t}$.

Step 4. Set up and solve problem (2.18). For each $j \in J_0$, the corrected value $\hat{x}_j$ is given by (2.19).

2.3.4 ROUNDING ERRORS

In this section, we look at very small inconsistencies with respect to balance edits; for example, the situation that a total value is just one unit smaller or larger than the sum of the component items. We call such inconsistencies "rounding errors," because they may be caused by values being rounded off to a common base, such as multiples of 1000. It is not straightforward to obtain a so-called "consistent rounding"—that is, to make sure that the rounded values satisfy the same restrictions as the original values. For example, if the terms of the sum $2.7 + 7.6 = 10.3$ are rounded off to natural numbers the ordinary way, the additivity is destroyed: $3 + 8 \neq 10$. Several algorithms for consistent rounding are available from the literature; see, for example, Salazar-González et al. (2004). Obviously, very few respondents are aware of these methods or indeed inclined to use them while filling in a questionnaire.

By their nature, rounding errors have virtually no influence on aggregates, and in this sense the choice of method to correct them is unimportant. However, as we shall see in Chapters 3 to 5, the complexity of the automatic error localization problem for random errors increases rapidly as the number of violated edits becomes larger, irrespective of the magnitude of these violations. Thus, a record containing many rounding errors and very few "real" errors might not be suitable for automatic editing and might have to be edited manually. This is clearly a waste of resources. When selective editing is used, it is therefore advantageous to resolve all rounding errors in the early stages of the editing process.

In the remainder of this section, we describe a heuristic method to resolve rounding errors in numerical survey data. We call this method a heuristic method because it does not return a solution that is "optimal" in some sense—for example, that the number of changed variables or the total change in values is minimized. The rationale of using a heuristic method is that the adaptations needed to resolve rounding errors are very small and that it is therefore not necessary to use a sophisticated and potentially time-consuming optimization algorithm.

When the vector $\mathbf{x} = \left(x_1, \ldots, x_p\right)^T$ contains the survey variables, the balance edits can be written as a linear system

$$\mathbf{Ax} = \mathbf{b}, \tag{2.20}$$

where each row of the $r \times p$ matrix $\mathbf{A}$ defines an edit and each column corresponds to a survey variable. The vector $\mathbf{b} = (b_1, \ldots, b_r)^T$ contains any constant terms that occur in the edits. Denoting the kth row of $\mathbf{A}$ by $\mathbf{a}_k^T$, the kth balance edit is violated if $|\mathbf{a}_k^T\mathbf{x} - b_k| > 0$. The inconsistency is called a rounding error when $0 < |\mathbf{a}_k^T\mathbf{x} - b_k| \leq \delta$, where $\delta > 0$ is small.

Similarly, the edits that take the form of a linear inequality can be written as

$$\mathbf{Bx} \geq \mathbf{c}, \tag{2.21}$$

where each edit is defined by a row of the $q \times p$ matrix $\mathbf{B}$ together with a constant from $\mathbf{c} = (c_1, \ldots, c_q)^T$. Note that in (2.21) the $\geq$ sign should be taken elementwise. We assume throughout this section that all edits can be formulated as either a linear equality or a linear inequality.

EXAMPLE 2.7

The following small-scale example will be used to illustrate the heuristic method. In this example, records of 11 variables $x_1, \ldots, x_{11}$ should conform to the following five balance edits:

$$\begin{aligned} x_1 + x_2 &= x_3, \\ x_2 &= x_4, \\ x_5 + x_6 + x_7 &= x_8, \\ x_3 + x_8 &= x_9, \\ x_9 - x_{10} &= x_{11}. \end{aligned} \tag{2.22}$$

These edits may be written as $\mathbf{Ax} = \mathbf{0}$, with $\mathbf{x} = (x_1, \ldots, x_{11})^T$ and

$$\mathbf{A} = \begin{pmatrix} 1 & 1 & -1 & 0 & 0 & 0 & 0 & 0 & 0 & 0 & 0 \\ 0 & 1 & 0 & -1 & 0 & 0 & 0 & 0 & 0 & 0 & 0 \\ 0 & 0 & 0 & 0 & 1 & 1 & 1 & -1 & 0 & 0 & 0 \\ 0 & 0 & 1 & 0 & 0 & 0 & 0 & 1 & -1 & 0 & 0 \\ 0 & 0 & 0 & 0 & 0 & 0 & 0 & 0 & 1 & -1 & -1 \end{pmatrix}. \tag{2.23}$$

This is an instance of (2.20) with $\mathbf{b} = \mathbf{0}$. Moreover, suppose that we are given the following inconsistent record:

x_1	x_2	x_3	x_4	x_5	x_6	x_7	x_8	x_9	x_{10}	x_{11}
12	4	15	4	3	1	8	11	27	41	−13

This record violates all edits, except for $x_2 = x_4$. If we take $\delta \geq 1$, the violations are all small enough to qualify as rounding errors. ■

Although the idea behind our heuristic method is quite simple, we need some results on matrix algebra to explain why it works. Therefore, we first provide a brief summary of the necessary background. We then describe the heuristic method for the case that only balance edits have to be satisfied. Finally, the method is extended to include linear inequalities.

Theoretical Background on Matrices. We recall that Cramer's Rule is a result on linear systems named after the Swiss mathematician Gabriel Cramer (1704–1752), which states the following.

LEMMA 2.1 Cramer's Rule.

Let $\mathbf{C} = (c_{kj})$ be an invertible $r \times r$-matrix. The unique solution $\mathbf{v} = (v_1, \ldots, v_r)^T$ to the system $\mathbf{Cv} = \mathbf{w}$ is given by

$$v_j = \frac{\det \mathbf{C}_j}{\det \mathbf{C}}, \qquad j = 1, \ldots, r,$$

where $\mathbf{C}_j$ denotes the matrix found by replacing the jth column of $\mathbf{C}$ by $\mathbf{w}$.

An alternative way of stating this result, is that for every invertible matrix $\mathbf{C}$,

$$\mathbf{C}^{-1} = \frac{1}{\det \mathbf{C}} \mathbf{C}^{\dagger}. \tag{2.24}$$

where $\mathbf{C}^{\dagger}$ denotes the adjoint matrix of $\mathbf{C}$. The adjoint matrix is found by transposing the matrix of cofactors; that is, the (j, k)th element of $\mathbf{C}^{\dagger}$ equals

$$(-1)^{k+j} \det \mathbf{C}_{-k,-j},$$

where $\mathbf{C}_{-k,-j}$ denotes the matrix $\mathbf{C}$ with the kth row and the jth column removed. For a proof of (2.24), and thus of Cramer's Rule, see, for example, Section 13.5 of Harville (1997).

A square matrix is called "unimodular" if its determinant is equal to 1 or -1. The following lemma is an immediate consequence of Cramer's Rule.

LEMMA 2.2

If $\mathbf{C}$ is an integer-valued unimodular matrix and $\mathbf{w}$ is an integer-valued vector, then the solution to the system $\mathbf{Cv} = \mathbf{w}$ is also integer-valued.

A matrix for which the determinant of every square submatrix equals 0, 1, or -1 is called "totally unimodular." That is to say, every square submatrix of a totally unimodular matrix is either singular or unimodular. Clearly, in order to be totally unimodular, a matrix must have all elements equal to 0, 1 or -1. We stress that a totally unimodular matrix itself need not be square.

A stronger version of Lemma 2.2 holds for the submatrices of a totally unimodular matrix.

LEMMA 2.3

Let $\mathbf{C}$ be a square submatrix of a totally unimodular matrix. If $\mathbf{C}$ is invertible, all elements of $\mathbf{C}^{-1}$ are in $\{-1, 0, 1\}$.

Proof. We use the adjoint matrix $\mathbf{C}^{\dagger}$. Since $|\det \mathbf{C}| = 1$ and all cofactors are equal to 0, 1 or -1, the property follows immediately from (2.24). ■

Clearly, it is infeasible to test whether a given matrix is totally unimodular just by applying the definition, unless the matrix happens to be very small. The number of determinants that need to be computed becomes prohibitively large even for matrices of moderate size. General results that can be used to establish the total unimodularity of a given matrix in practice are discussed by, among others, Heller and Tompkins (1956), Tamir (1976), and Raghavachari (1976).

In addition, the following "reduction method" can be useful if the matrix under scrutiny happens to be sparse, which is usually the case for restriction matrices that arise in practice. We begin by making an observation.

LEMMA 2.4

Let $\mathbf{C}$ be a matrix containing only elements from $\{-1, 0, 1\}$ that has the following form, possibly after a permutation of columns:

$$\mathbf{C} = \begin{pmatrix} \mathbf{C}_1 & \mathbf{C}_2 \end{pmatrix},$$

where each column of $\mathbf{C}_1$ contains at most one nonzero element. Then $\mathbf{C}$ is totally unimodular if and only if $\mathbf{C}_2$ is totally unimodular.

Proof. We prove that if $\mathbf{C}_2$ is totally unimodular, it follows that $\mathbf{C}$ is totally unimodular, the other implication being trivial. To do this we must show that the determinant of every square submatrix of $\mathbf{C}$ lies in $\{-1, 0, 1\}$. The proof works by induction on the order of the submatrix. The statement is clearly true for all 1×1 submatrices. Suppose that the statement holds for all $(s-1) \times (s-1)$ submatrices of $\mathbf{C}$ (with $s > 1$) and let $\mathbf{C}^{(s)}$ be any $s \times s$ submatrix of $\mathbf{C}$.

We may assume that $\mathbf{C}^{(s)}$ is invertible and also that it contains at least one column from $\mathbf{C}_1$, since otherwise there is nothing to prove. Suppose that the jth column of $\mathbf{C}^{(s)}$ comes from $\mathbf{C}_1$. Since $\mathbf{C}^{(s)}$ is invertible, this column must contain a nonzero element, say in the kth row. If we denote this nonzero element by c_{kj}, it follows by expanding the determinant of $\mathbf{C}^{(s)}$ on the jth column that

$$\det \mathbf{C}^{(s)} = (-1)^{k+j} c_{kj} \det \mathbf{C}^{(s-1)},$$

where $\mathbf{C}^{(s-1)}$ is the $(s-1) \times (s-1)$ matrix found by removing the kth row and the jth column of $\mathbf{C}^{(s)}$. Since $|c_{kj}| = 1$, this means that $|\det \mathbf{C}^{(s)}| = |\det \mathbf{C}^{(s-1)}|$. It follows from the invertibility of $\mathbf{C}^{(s)}$ and from the induction hypothesis that $|\det \mathbf{C}^{(s)}| = 1$. ■

Since $\mathbf{C}$ is totally unimodular if and only if $\mathbf{C}^T$ is totally unimodular, the next result is equivalent to Lemma 2.4.

LEMMA 2.5

Let $\mathbf{C}$ be a matrix containing only elements from $\{-1, 0, 1\}$ that has the following form, possibly after a permutation of rows:

$$\mathbf{C} = \begin{pmatrix} \mathbf{C}_1 \\ \mathbf{C}_2 \end{pmatrix},$$

where each row of $\mathbf{C}_1$ contains at most one nonzero element. Then $\mathbf{C}$ is totally unimodular if and only if $\mathbf{C}_2$ is totally unimodular.

These two results sometimes allow us to determine whether a given matrix $\mathbf{C}$ is totally unimodular by considering a much smaller matrix. Instead of $\mathbf{C}$, it suffices, by Lemma 2.4, to consider the submatrix $\mathbf{C}_2$ that consists of all columns of $\mathbf{C}$ containing two or more nonzero elements. Similarly, instead of $\mathbf{C}_2$, it suffices, by Lemma 2.5, to consider the submatrix $\mathbf{C}_{22}$ that consists of all rows of $\mathbf{C}_2$ containing two or more nonzero elements. Next, it can happen that some columns of $\mathbf{C}_{22}$ contain less than two nonzero elements, so we may again apply Lemma 2.4 and consider the submatrix $\mathbf{C}_{222}$ found by deleting these columns from $\mathbf{C}_{22}$. This iterative process may be continued until we either come across a matrix of which all columns and all rows contain at least two nonzero elements, or a matrix that is clearly totally unimodular (or clearly not).

EXAMPLE 2.7 (*continued*)

As an illustration, we apply the reduction method to the 5×11 matrix $\mathbf{A}$ defined in (2.23). By iteratively applying Lemma 2.4 and Lemma 2.5, we obtain the following matrices:

$$\mathbf{A} \xrightarrow{\substack{\text{Lemma}\\ 2.4}} \begin{pmatrix} 1 & -1 & 0 & 0 \\ 1 & 0 & 0 & 0 \\ 0 & 0 & -1 & 0 \\ 0 & 1 & 1 & -1 \\ 0 & 0 & 0 & 1 \end{pmatrix} \xrightarrow{\substack{\text{Lemma}\\ 2.5}} \begin{pmatrix} 1 & -1 & 0 & 0 \\ 0 & 1 & 1 & -1 \end{pmatrix} \xrightarrow{\substack{\text{Lemma}\\ 2.4}} \begin{pmatrix} -1 \\ 1 \end{pmatrix}.$$

The final 2×1 matrix is clearly totally unimodular, and we immediately know that the original matrix is also totally unimodular. Note that we have obtained this result without computing a single determinant. ■

The Basic Scapegoat Algorithm. The idea behind the heuristic method is as follows. For each record containing rounding errors, a set of variables is selected beforehand. Next, the rounding errors are resolved by only adjusting the values of these selected variables. Hence the name "scapegoat algorithm" seems appropriate. In fact, the algorithm performs the selection in such a way that exactly one choice of values exists for the selected variables such that all rounding errors are resolved. Different variables are selected for each record to minimize the effect of the adaptations on aggregates.

It is assumed that the $r \times p$ matrix $\mathbf{A}$ satisfies $r \leq p$ and rank($\mathbf{A}$) $= r$; that is, the number of variables should be at least as large as the number of restrictions and no redundant restrictions may be present. Clearly, these are very mild assumptions. In addition, the scapegoat algorithm becomes simpler if $\mathbf{A}$ is a totally unimodular matrix. So far, we have found that matrices of balance edits used for structural business statistics at Statistics Netherlands are always of this type. A similar observation is made in Section 3.4.1 of De Waal (2002).

Suppose we have an inconsistent record $\mathbf{x}$, possibly containing both rounding errors and other errors. In the first step of the scapegoat algorithm, all rows of $\mathbf{A}$ for which $|\mathbf{a}_k^T\mathbf{x} - b_k| > \delta$ are removed from the matrix and the associated constants are removed from $\mathbf{b}$. We denote the resulting $r_0 \times p$ matrix by $\mathbf{A}_0$ and the resulting r_0 vector of constants by $\mathbf{b}_0$. It may happen that the record satisfies the remaining balance edit rules $\mathbf{A}_0\mathbf{x} = \mathbf{b}_0$, because it does not contain any rounding errors. In that case the algorithm stops here.

It is easy to see that if $\mathbf{A}$ satisfies the assumptions above, then so does $\mathbf{A}_0$. Hence rank($\mathbf{A}_0$) $= r_0$ and $\mathbf{A}_0$ has r_0 linearly independent columns. The r_0 leftmost linearly independent columns may be found by putting the matrix in row echelon form through Gaussian elimination, as described in Section 2.2 of Fraleigh and Beauregard (1995), or alternatively by performing a *QR* decomposition with column pivoting, as discussed in Section 4.4 of Golub and Van Loan (1996). (How these methods work is irrelevant for our present purpose.) Since we want the choice of scapegoat variables and hence of columns to vary between records, a random permutation of columns is performed beforehand, yielding $\tilde{\mathbf{A}}_0$. The variables of $\mathbf{x}$ are permuted accordingly to yield $\tilde{\mathbf{x}}$.

Next, $\tilde{\mathbf{A}}_0$ is partitioned into two submatrices $\mathbf{A}_1$ and $\mathbf{A}_2$. The first of these is an $r_0 \times r_0$ matrix that contains the first r_0 linearly independent columns of $\tilde{\mathbf{A}}_0$, and the second is an $r_0 \times (p - r_0)$ matrix containing all other columns of $\tilde{\mathbf{A}}_0$. The vector $\tilde{\mathbf{x}}$ is also partitioned into subvectors $\mathbf{x}_1$ and $\mathbf{x}_2$, containing the variables associated with the columns of $\mathbf{A}_1$ and $\mathbf{A}_2$, respectively. Thus

$$\tilde{\mathbf{A}}_0\tilde{\mathbf{x}} = \mathbf{b}_0$$

becomes

$$\mathbf{A}_1\mathbf{x}_1 + \mathbf{A}_2\mathbf{x}_2 = \mathbf{b}_0.$$

At this point, the variables from $\mathbf{x}_1$ are selected as scapegoat variables and the variables from $\mathbf{x}_2$ remain fixed. Therefore the values of $\mathbf{x}_2$ are filled in from the original record and we are left with the system

$$\mathbf{A}_1\mathbf{x}_1 = \mathbf{b}_0 - \mathbf{A}_2\mathbf{x}_2 \equiv \mathbf{b}^*, \tag{2.25}$$

where $\mathbf{b}^*$ is a vector of known constants.

By construction the square matrix $\mathbf{A}_1$ is of full rank and therefore invertible. Thus (2.25) has the unique solution $\hat{\mathbf{x}}_1 = \mathbf{A}_1^{-1}\mathbf{b}^*$. In general, this solution might contain fractional values, whereas most business survey variables are restricted to be integer-valued. If this is the case, a controlled rounding algorithm similar to the one described by Salazar-González et al. (2004) can be applied to the values of $\left(\hat{\mathbf{x}}_1^T, \mathbf{x}_2^T\right)^T$ to obtain an integer-valued solution to $\mathbf{A}_0\mathbf{x} = \mathbf{b}_0$. Note, however, that this is not possible without slightly changing the value of at least one variable from $\mathbf{x}_2$ too.

If $\mathbf{A}$ happens to be a totally unimodular matrix, this problem does not occur. In that case $\det \mathbf{A}_1 = \pm 1$, and we know from Lemma 2.2 that $\hat{\mathbf{x}}_1$ is always integer-valued. In the remainder of this discussion, we assume that $\mathbf{A}$ is indeed totally unimodular.

EXAMPLE 2.7 (*continued*)

For $\mathbf{A}$ defined in (2.23), it is easily established that rank$(\mathbf{A}) = 5$. Moreover, we have demonstrated that $\mathbf{A}$ is totally unimodular. Since all violations qualify as rounding errors, $\mathbf{A}_0$ is identical to $\mathbf{A}$ in this example.

A random permutation is applied to the elements of $\mathbf{x}$ and the columns of $\mathbf{A}$. Suppose that the permutation is given by $1 \to 11$, $2 \to 8$, $3 \to 2$, $4 \to 5, 5 \to 10, 6 \to 9, 7 \to 7, 8 \to 1, 9 \to 4, 10 \to 3, 11 \to 6$. This yields the following result:

$$\tilde{\mathbf{A}} = \begin{pmatrix} 0 & -1 & 0 & 0 & 0 & 0 & 0 & 1 & 0 & 0 & 1 \\ 0 & 0 & 0 & 0 & -1 & 0 & 0 & 1 & 0 & 0 & 0 \\ -1 & 0 & 0 & 0 & 0 & 0 & 1 & 0 & 1 & 1 & 0 \\ 1 & 1 & 0 & -1 & 0 & 0 & 0 & 0 & 0 & 0 & 0 \\ 0 & 0 & -1 & 1 & 0 & -1 & 0 & 0 & 0 & 0 & 0 \end{pmatrix}.$$

The first five columns of $\tilde{\mathbf{A}}$ happen to be linearly independent, so these together form the invertible matrix $\mathbf{A}_1$, while $\mathbf{A}_2$ consists of the other six columns. The scapegoat variables are those that correspond with the columns of $\mathbf{A}_1$—that is, x_8, x_3, x_{10}, x_9, and x_4. For the remaining variables, we fill in the original values from the record to calculate the constant vector $\mathbf{b}^*$:

$$\mathbf{b}^* = -\mathbf{A}_2 \begin{pmatrix} x_{11} \\ x_7 \\ x_2 \\ x_6 \\ x_5 \\ x_1 \end{pmatrix} = - \begin{pmatrix} 0 & 0 & 1 & 0 & 0 & 1 \\ 0 & 0 & 1 & 0 & 0 & 0 \\ 0 & 1 & 0 & 1 & 1 & 0 \\ 0 & 0 & 0 & 0 & 0 & 0 \\ -1 & 0 & 0 & 0 & 0 & 0 \end{pmatrix} \begin{pmatrix} -13 \\ 8 \\ 4 \\ 1 \\ 3 \\ 12 \end{pmatrix} = \begin{pmatrix} -16 \\ -4 \\ -12 \\ 0 \\ -13 \end{pmatrix}.$$

We obtain the following system in $\mathbf{x}_1$:

$$\mathbf{A}_1\mathbf{x}_1 = \begin{pmatrix} 0 & -1 & 0 & 0 & 0 \\ 0 & 0 & 0 & 0 & -1 \\ -1 & 0 & 0 & 0 & 0 \\ 1 & 1 & 0 & -1 & 0 \\ 0 & 0 & -1 & 1 & 0 \end{pmatrix} \begin{pmatrix} x_8 \\ x_3 \\ x_{10} \\ x_9 \\ x_4 \end{pmatrix} = \begin{pmatrix} -16 \\ -4 \\ -12 \\ 0 \\ -13 \end{pmatrix} = \mathbf{b}^*.$$

Solving this system yields $\hat{x}_3 = 16$, $\hat{x}_4 = 4$, $\hat{x}_8 = 12$, $\hat{x}_9 = 28$ and $\hat{x}_{10} = 41$.

When the original values of the variables in $\mathbf{x}_1$ are replaced by these new values, the record becomes consistent with respect to (2.22):

x_1	x_2	$\hat{x}_3$	$\hat{x}_4$	x_5	x_6	x_7	$\hat{x}_8$	$\hat{x}_9$	$\hat{x}_{10}$	x_{11}
12	4	16	4	3	1	8	12	28	41	−13

We remark that in this example it was not necessary to change the value of every scapegoat variable. In particular, x_4 and x_{10} have retained their original values. ■

The solution vector $\hat{\mathbf{x}}_1$ is constructed by the scapegoat algorithm without any explicit use of the original vector $\mathbf{x}_1$. Therefore, it is not completely trivial that the adjusted values remain close to the original values, which is obviously what we would hope for. We now derive an upper bound on the size of the adjustments, under the assumption that $\mathbf{A}$ is totally unimodular.

Recall that the maximum norm of a vector $\mathbf{v} = (v_1, \ldots, v_p)^T$ is defined as

$$|\mathbf{v}|_\infty = \max_{j=1,\ldots,p} |v_j|.$$

The associated matrix norm is [cf. Section 4.4 of Stoer and Bulirsch (2002)]

$$||\mathbf{C}||_\infty = \max_{k=1,\ldots,r} \sum_{j=1}^{p} |c_{kj}|,$$

with $\mathbf{C} = \left(c_{kj}\right)$ any $r \times p$ matrix. It is easily shown that

$$|\mathbf{C}\mathbf{v}|_\infty \leq ||\mathbf{C}||_\infty |\mathbf{v}|_\infty, \tag{2.26}$$

for every $r \times p$ matrix $\mathbf{C}$ and every p-vector $\mathbf{v}$.

We now turn to the scapegoat algorithm. By construction $\hat{\mathbf{x}}_1$ satisfies $\mathbf{A}_1\hat{\mathbf{x}}_1 = \mathbf{b}^*$. The original vector $\mathbf{x}_1$ satisfies $\mathbf{A}_1\mathbf{x}_1 = \mathbf{b}'$, with $\mathbf{b}' \neq \mathbf{b}^*$. Thus

$$\hat{\mathbf{x}}_1 - \mathbf{x}_1 = \mathbf{A}_1^{-1}\left(\mathbf{b}^* - \mathbf{b}'\right). \tag{2.27}$$

It follows from (2.26) and (2.27) that

$$|\hat{\mathbf{x}}_1 - \mathbf{x}_1|_\infty \leq ||\mathbf{A}_1^{-1}||_\infty |\mathbf{b}^* - \mathbf{b}'|_\infty \leq r_0 |\mathbf{b}^* - \mathbf{b}'|_\infty, \tag{2.28}$$

where for the last inequality we observe that Lemma 2.3 implies $||\mathbf{A}_1^{-1}||_\infty \leq r_0$.

We write $\hat{\mathbf{x}} = \left(\hat{\mathbf{x}}_1^T, \mathbf{x}_2^T\right)^T$. By

$$\begin{aligned}
\mathbf{b}^* - \mathbf{b}' &= \mathbf{A}_1\hat{\mathbf{x}}_1 - \mathbf{A}_1\mathbf{x}_1 \\
&= \mathbf{A}_1\hat{\mathbf{x}}_1 + \mathbf{A}_2\mathbf{x}_2 - \mathbf{b}_0 - (\mathbf{A}_1\mathbf{x}_1 + \mathbf{A}_2\mathbf{x}_2 - \mathbf{b}_0) \\
&= \tilde{\mathbf{A}}_0\hat{\mathbf{x}} - \mathbf{b}_0 - (\mathbf{A}_0\mathbf{x} - \mathbf{b}_0) \\
&= -(\mathbf{A}_0\mathbf{x} - \mathbf{b}_0),
\end{aligned}$$

we see that $|\mathbf{b}^* - \mathbf{b}'|_\infty = |\mathbf{A}_0\mathbf{x} - \mathbf{b}_0|_\infty = \delta_{\max}$, where $\delta_{\max} \leq \delta$ is the magnitude of the largest rounding error that occurs for this particular record. Plugging this into (2.28), we find

$$|\hat{\mathbf{x}}_1 - \mathbf{x}_1|_\infty \leq r_0\delta_{\max}. \tag{2.29}$$

Inequality (2.29) shows that the solution found by the scapegoat algorithm cannot be arbitrarily far from the original record. The fact that the upper bound on the absolute difference between elements of $\hat{\mathbf{x}}_1$ and $\mathbf{x}_1$ is proportional to the order of $\mathbf{A}_1$ suggests that we should expect ever larger adjustments as the number of balance edits increases, which is somewhat worrying. In practice we find much smaller adjustments than $r_0\delta_{\max}$, though. For instance, in Example 2.7 the maximal absolute difference according to (2.29) equals 5, but actually no value was changed more than one unit. Nevertheless, Scholtus (2008) shows that it is possible to construct a pathological example for which this upper bound is achieved.

A more interesting view on the size of the adjustments may be provided by the quantity

$$\frac{1}{r_0}\sum_{j=1}^{r_0} |\hat{x}_{1j} - x_{1j}|,$$

which measures the *average* size of the adjustments, rather than the maximum. Starting from (2.27), we see that

$$|\hat{x}_{1j} - x_{1j}| = \left| \sum_{k=1}^{r_0} \left(\mathbf{A}_1^{-1}\right)_{jk} \left(b_k^* - b_k'\right) \right| \leq \sum_{k=1}^{r_0} \left| \left(\mathbf{A}_1^{-1}\right)_{jk} \right| \cdot |b_k^* - b_k'|.$$

Using again that $|b_k^* - b_k'| \leq |\mathbf{b}^* - \mathbf{b}'|_\infty = \delta_{\max}$, we find

$$\frac{1}{r_0} \sum_{j=1}^{r_0} |\hat{x}_{1j} - x_{1j}| \leq \frac{\delta_{\max}}{r_0} \sum_{j=1}^{r_0} \sum_{k=1}^{r_0} \left| \left(\mathbf{A}_1^{-1}\right)_{jk} \right| \equiv \gamma(\mathbf{A}_1)\delta_{\max}, \tag{2.30}$$

where $\gamma(\mathbf{A}_1) = \frac{1}{r_0} \sum_{j=1}^{r_0} \sum_{k=1}^{r_0} |(\mathbf{A}_1^{-1})_{jk}|$. This gives an upper bound on the average adjustment size that can be evaluated beforehand.

Suppose that a set of balance edits (2.20) is given. Restricting ourselves to the case $r_0 = r$, we can compute $\gamma(\mathbf{A}_1)$ for various invertible $r \times r$ submatrices of $\mathbf{A}$ to assess the magnitude of the upper bound in (2.30). It can be shown that $\mathbf{A}$ has exactly det $(\mathbf{A}\mathbf{A}^T)$ invertible $r \times r$ submatrices [cf. Scholtus (2008)]. In practice, this number is very large and it is infeasible to compute $\gamma(\mathbf{A}_1)$ for all of them. In that case, we can take a random sample of reasonable size, by repeatedly performing the part of the scapegoat algorithm that constructs an invertible submatrix.

EXAMPLE 2.7 (*continued*)

For the 5×11 matrix in (2.23), det $(\mathbf{A}\mathbf{A}^T) = 121$, so $\mathbf{A}$ has 121 invertible 5×5-submatrices. Since this number is not too large, we have evaluated $\gamma(\mathbf{A}_1)$ for all these matrices. The mean value of $\gamma(\mathbf{A}_1)$ turns out to be 1.68, with a standard deviation of 0.39. Thus, since $\delta_{\max} = 1$, according to (2.30) the average adjustment size is bounded on average by 1.68. ■

At the end of this section, we look at the adjustments that occur in a real-world example. These turn out to be quite small.

The Scapegoat Algorithm with Inequalities. In addition to balance edits, numerical survey variables usually have to satisfy a large number of edits that take the form of linear inequalities. For instance, it is very common that most variables are restricted to be nonnegative. The scapegoat algorithm as described above does not consider this and might therefore change a nonnegative variable from 0 to −1, resulting in a new violation of an edit. We now extend the algorithm to prevent this.

Suppose that in addition to balance edits (2.20), the data also have to satisfy inequalities (2.21). For a given record, we call a variable "critical" if it occurs in an inequality that (almost) becomes an exact equality when the current values of the survey variables are filled in. That is to say,

$$(2.31)\qquad x_j \text{ is critical } \Leftrightarrow b_{kj} \neq 0 \text{ and } 0 \leq \mathbf{b}_k^T\mathbf{x} - c_k \leq \varepsilon_k \qquad \text{for some } k,$$

where $\mathbf{b}_k^T$ denotes the kth row of $\mathbf{B}$ and $\varepsilon_k \geq 0$ marks the margin we choose for the kth restriction. As a particular case, x_j is called critical if it must be nonnegative and currently has a value between 0 and $\varepsilon_{k(j)}$, with $k(j)$ the index of the row in $\mathbf{B}$ corresponding to the nonnegativity constraint for x_j. To prevent the violation of edits in (2.21), no critical variable should be selected for change during the execution of the scapegoat algorithm.

A way to achieve this works as follows: Rather than randomly permuting all variables (and all columns of $\mathbf{A}_0$), two separate permutations should be performed for the noncritical and the critical variables. The permuted columns associated with the noncritical variables are then placed to the left of the columns associated with the critical variables. This ensures that linearly independent columns are found among those that are associated with noncritical variables, provided that the record contains a sufficient number of noncritical variables. If the number of survey variables is much larger than the number of balance edits (as is typically the case), this should not be a problem.

If a record contains many critical variables, some of these might still be selected as scapegoat variables. In itself, this is not a problem, as long as no inequality edits become violated.[3] We can build in a check at the end of the algorithm that rejects the solution if a new violation of an edit from (2.21) is detected. It then seems advantageous to let the record be processed again, because a different permutation of columns may yield a feasible solution after all. To prevent the algorithm from getting stuck, the number of attempts should be maximized by a preset constant M. If no feasible solution has been found after M attempts, the record remains untreated.

Good values of ε_k and M have to be determined in practice. However, in our opinion, not too much effort should be put into this, because these parameters only affect a limited number of records. In an alternative version of the real-world application discussed below (Example 2.8), we found only a handful of infeasible solutions when the inequality edits were not taken into account by the scapegoat algorithm.

The following plan summarizes the scapegoat algorithm. The input consists of an inconsistent record $\mathbf{x}$ (p variables), a set of r balance edits $\mathbf{Ax} = \mathbf{b}$, a set of q inequalities $\mathbf{Bx} \geq \mathbf{c}$, and parameters δ, $\varepsilon_k (k = 1, \ldots, q)$ and M. It is assumed that $\mathbf{A}$ is a totally unimodular matrix.

ALGORITHM 2.3

A heuristic method for resolving rounding errors:

Step 1. Remove all edits for which $|\mathbf{a}_k^T\mathbf{x} - b_k| > \delta$. The remaining system is written as $\mathbf{A}_0\mathbf{x} = \mathbf{b}_0$. The number of rows in $\mathbf{A}_0$

[3] This is in fact the reason why we also permute the critical variables in the previous paragraph: The algorithm may yield a useful solution even if critical variables are selected as scapegoat variables.

is called r_0. If $\mathbf{A}_0\mathbf{x} = \mathbf{b}_0$ holds: stop. Otherwise: Determine the critical variables according to (2.31).

Step 2. **(a)** Perform random permutations of the critical and noncritical variables separately. Then permute the corresponding columns of $\mathbf{A}_0$ the same way. Put the noncritical variables and their columns before the critical variables and their columns.

(b) Determine the r_0 leftmost linearly independent columns in the permuted matrix $\tilde{\mathbf{A}}_0$. Together, these columns are a unimodular matrix $\mathbf{A}_1$ and the associated variables form a vector $\mathbf{x}_1$ of scapegoat variables. The remaining columns are a matrix $\mathbf{A}_2$ and the associated variables form a vector $\mathbf{x}_2$.

(c) Fix the values of $\mathbf{x}_2$ from the record and compute $\mathbf{b}^* = \mathbf{b}_0 - \mathbf{A}_2\mathbf{x}_2$.

Step 3. Solve the system $\mathbf{A}_1\mathbf{x}_1 = \mathbf{b}^*$.

Step 4. Replace the values of $\mathbf{x}_1$ by the solution just found. If the resulting record does not violate any new edits from $\mathbf{Bx} \geq \mathbf{c}$, we are done. If it does, return to step 2a, unless this has been the Mth attempt. In that case the record is not adjusted.

EXAMPLE 2.8

The scapegoat algorithm has been tested at Statistics Netherlands, using data from the Dutch wholesale structural business statistics of 2001. There are 4725 records containing 97 variables each. These variables should conform to a set of 26 balance edits and 120 inequalities, of which 92 represent nonnegativity constraints. The resulting 26×97 matrix $\mathbf{A}$ is totally unimodular, as can be determined very quickly using the reduction method described above. (Note that it would be practically impossible to determine whether a matrix of this size is totally unimodular just by computing all the relevant determinants.)

We used an implementation of the algorithm in *S-Plus* to treat the data. The parameters used were $\delta = 2$, $\varepsilon_k = 2$ $(k = 1, \ldots, 120)$, and $M = 10$. The total computation time on an ordinary desktop PC was less than three minutes.

Table 2.6 summarizes the results of applying the scapegoat algorithm. No new violations of inequalities were found. In fact, the adjusted data happen to satisfy four additional inequalities.

According to (2.29), the size of the adjustments made by the algorithm is theoretically bounded by $26 \times 2 = 52$, which is rather high. A random sample of $10,000$ invertible 26×26 submatrices of $\mathbf{A}$ was drawn to evaluate (2.30). The sample mean of $\gamma(\mathbf{A}_1)$ is 1.89, with a standard

TABLE 2.6 Results of Applying the Scapegoat Algorithm to the Wholesale Data

Number of records	4,725
Number of variables per record	97
Number of adjusted records	3,176
Number of adjusted variables	13,531
Number of violated edit rules (before)	34,379
Balance edit rules	26,791
Inequalities	7,588
Number of violated edit rules (after)	23,054
Balance edit rules	15,470
Inequalities	7,584

deviation of 0.27. Thus, the average adjustment size is bounded on average by $1.89 \times 2 \approx 3.8$. We remark that this value of $\gamma(\mathbf{A}_1)$ is only marginally higher than the one obtained for the much smaller restriction matrix from Example 2.7.

Table 2.7 displays the adjustment sizes that were actually found for the wholesale data. These turn out to be very reasonable. The average adjustment size equals 1.13, which shows that the second theoretical bound given above is still too high.

TABLE 2.7 Distribution of the Adjustments (in Absolute Value)

Magnitude	Frequency
1	11,953
2	1,426
3	134
4	12
5	4
6	2

■

Finally, we remark that when searching for other types of errors, it is convenient to ignore rounding errors—that is, treat edits that are violated by a small amount as though they were satisfied. As described in Sections 2.3.2 and 2.3.3, the correction methods for sign errors and simple typing errors do not take into account that the data may contain rounding errors. Fortunately, they can be easily modified to ignore rounding errors. For sign errors, problem (2.11) should be replaced by

$$
\begin{aligned}
&\text{Minimize } \textstyle\sum_{j=0}^{n} \sigma_j + \sum_{k=1}^{m} \tau_k \\
&\text{such that} \\
&-\delta \leq x_0(1-2\sigma_0) - \left(x_{0,r} - x_{0,c}\right) \leq \delta, \\
&-\delta \leq x_1(1-2\sigma_1) - \left(x_{1,r} - x_{1,c}\right)(1-2\tau_1) \leq \delta, \\
&\quad\vdots \\
&-\delta \leq x_m(1-2\sigma_m) - \left(x_{m,r} - x_{m,c}\right)(1-2\tau_m) \leq \delta, \\
&-\delta \leq x_0(1-2\sigma_0) + \cdots + x_{n-1}(1-2\sigma_{n-1}) - x_n(1-2\sigma_n) \leq \delta, \\
&(\sigma_0, \ldots, \sigma_n; \tau_1, \ldots, \tau_m) \in \{0,1\}^{n+m+1}.
\end{aligned}
$$

See Scholtus (2008) for more details.

For simple typing errors, it suffices to redefine E_1 and E_2 as follows:

$$
\begin{aligned}
E_1 &= \{1, \ldots, r\} \setminus E_2, \\
E_2 &= \left\{k \,:\, -\delta \leq s_1^{(k)} - s_2^{(k)} \leq \delta\right\}.
\end{aligned}
$$

See Scholtus (2009) for more details.

2.4 Summary

In this chapter, we discussed methods for correcting errors deductively, when the underlying error mechanism is assumed to be known. Deductive corrections are, in a sense, the best possible corrections, because they yield the true values with certainty if our assumptions are valid. Systematic errors in particular lend themselves to deductive correction. However, no generic recipe for the detection of systematic errors can be given, and different errors occur for different surveys.

REFERENCES

Al-Hamad, A., D. Lewis, and P. L. N. Silva (2008), *Assessing the Performance of the Thousand Pounds Automatic Editing Procedure at the ONS and the Need for an Alternative Approach.* Working Paper No. 21, UN/ECE Work Session on Statistical Data Editing, Vienna.

De Waal, T. (2002), *Algorithms for Automatic Error Localisation and Modification.* Paper prepared for the DATACLEAN 2002 conference, Jyväskylä.

Di Zio, M., U. Guarnera, and O. Luzi (2005), Editing Systematic Unity Measure Errors through Mixture Modelling. *Survey Methodology 31*, pp. 53–63.

EDIMBUS (2007), *Recommended Practices for Editing and Imputation in Cross-Sectional Business Surveys.* Manual prepared by ISTAT, Statistics Netherlands and SFSO.

Fraleigh, J. B., and R. A. Beauregard (1995), *Linear Algebra*, third edition. Addison-Wesley, Reading, MA.

Golub, G. H., and C. F. Van Loan (1996), *Matrix Computations*, third edition. The Johns Hopkins University Press, Baltimore.

Harville, D. A. (1997), *Matrix Algebra from a Statistician's Perspective*. Springer, New York.

Heller, I., and C. B. Tompkins (1956), An Extension of a Theorem of Dantzig's. In: *Linear Inequalities and Related Systems*, H. W. Kuhn and A. W. Tucker, eds. Princeton University Press, Princeton, NJ, pp. 247–254.

Raghavachari, M. (1976), A Constructive Method to Recognize the Total Unimodularity of a Matrix. *Zeitschrift für Operations Research 20*, pp. 59–61.

Salazar-González, J. J., P. Lowthian, C. Young, G. Merola, S. Bond, and D. Brown (2004), Getting the Best Results in Controlled Rounding with the Least Effort. In: *Privacy in Statistical Databases*, J. Domingo-Ferrer and V. Torra, eds. Springer, Berlin, pp. 58–72.

Scholtus, S. (2008), *Algorithms for Correcting Some Obvious Inconsistencies and Rounding Errors in Business Survey Data*. Discussion paper 08015, Statistics Netherlands, The Hague (see also www.cbs.nl).

Scholtus, S. (2009), *Automatic Correction of Simple Typing Errors in Numerical Data with Balance Edits*. Discussion paper 09046, Statistics Netherlands, The Hague (see also www.cbs.nl).

Stoer, J., and R. Bulirsch (2002), *Introduction to Numerical Analysis*, third edition. Springer, New York.

Tamir, A. (1976), On Totally Unimodular Matrices. *Networks 6*, pp. 373–382.

Van de Pol, F., F. Bakker, and T. de Waal (1997), *On Principles for Automatic Editing of Numerical Data with Equality Checks*. Report 7141-97-RSM, Statistics Netherlands, Voorburg.

CHAPTER THREE

Automatic Editing of Continuous Data

3.1 Introduction

This chapter discusses the detection of random errors in an automatic manner. When automatic editing is applied, records are edited by a computer without human intervention. Automatic editing is the opposite of the traditional approach to the editing problem, where each record is edited manually. Automatic editing can be applied to both categorical and continuous data, and even to a mix of categorical, continuous, and integer-valued data. The main advantage of automatic editing is that it allows one to correct a large number of records quickly and at low costs. Automatic editing has been applied since the early 1960s [see Nordbotten (1963)].

The present chapter is the first in a sequence of three chapters. This chapter focuses on automatically detecting random errors in continuous data. The next two chapters describe extensions to categorical, and integer-valued data, respectively.

To automate the statistical data editing process, one often divides this process into two steps. In the first step, the error localization step, the errors in the data are detected. Often, during this step, edits are used to determine whether a record is consistent or not. Inconsistent records are considered to contain errors, while consistent records are considered error-free. If a record contains errors, the erroneous fields in this record are also identified in the error localization step. In the second step, the imputation step, erroneous data are replaced by more accurate data and missing data are imputed. The error localization step only

Handbook of Statistical Data Editing and Imputation, by T. de Waal, J. Pannekoek, and S. Scholtus

determines which fields are considered erroneous; the imputation step determines the actual values of these fields. To automate statistical data editing both the error localization step and the imputation step need to be automated. In this chapter we restrict ourselves to discussing the former step. The imputation step is discussed in Chapter 7 and further.

Generally speaking, we can subdivide the methods for automatic error localization of random errors into methods based on statistical models, methods based on deterministic checking rules, and methods based on solving a mathematical optimization problem. Methods based on statistical models can be further subdivided into methods based on outlier detection techniques and neural networks. In Section 3.2 we briefly describe these methods in general.

Outlier detection techniques and neural networks are extensively discussed in the literature. Treating the error localization problem as a mathematical optimization problem has been relatively neglected so far. In this chapter we aim to fill up this gap in the literature. In particular, we give an overview of algorithms based on the so-called (generalized) Fellegi–Holt paradigm for solving the error localization problem for continuous data. This paradigm says that the data of a record should be made to satisfy all edits by changing the fewest possible (weighted) number of fields. Here each variable in a record is given a nonnegative weight, the so-called reliability weight of this variable. A reliability weight is a measure of "confidence" in the value of this variable. Variables that are generally correctly observed should be given a high reliability weight; variables that are often incorrectly observed should be given a low reliability weight. A reliability weight of a variable corresponds to the error probability of this variable; that is, the probability that its observed value is erroneous. The higher the reliability weight of a variable, the lower its error probability. In the original version of the Fellegi–Holt paradigm, all reliability weights were set to 1. In Section 3.4.1 we give a mathematical formulation of the error localization problem for continuous data and linear edits, using the (generalized) Fellegi–Holt paradigm.

In this chapter we focus on giving an overview of various algorithms for solving the error localization problem for continuous data based on the Fellegi–Holt paradigm. Because continuous data mostly occur in business surveys, the algorithms described in the present chapter are mainly suitable for business surveys and less so for social surveys.

In Section 3.4.2 we discuss a naive approach for solving the error localization problem, and we show that this approach does not work. In most of our algorithms, Fourier–Motzkin elimination plays an important role. This technique is described in Section 3.4.3. In Section 3.4.4 we sketch the method for solving the error localization problem proposed by Fellegi and Holt (1976). Sections 3.4.5 to 3.4.8 describe various alternatives to this method, namely approaches based on solving an integer programming problem, on vertex generation, on branch-and-bound, and on cutting planes. Section 3.5 concludes the chapter by giving a summary.

To a substantial part, this chapter is based on De Waal and Coutinho (2005). That article can be seen as an update of the overview article by Liepins, Garfinkel, and Kunnathur (1982).

3.2 Automatic Error Localization of Random Errors

In this section we give a general description of methods for automatic error localization. In principle, most of these of methods can also be used for selective editing (see Chapter 6). When used for selective editing, the "errors"—or better: outlying or suspicious values, identified by these methods—are later examined manually. When used for automatic editing, the identified "errors" are later "corrected" by means of automatic imputation without any human interference.

The first two classes of methods, automatic error localization based on outlier detection techniques (see Section 3.2.1) and automatic error localization based on neural networks (see Section 3.2.2), do not explicitly depend on user-specified edits. These edits are implicitly taken into account in the process of identifying the data values that are "inconsistent" with the fitted statistical model(s). The third and fourth class of methods, automatic error localization based on deterministic checking rules (see Section 3.2.3) and automatic error localization based on solving an optimization problem (see Section 3.2.4), respectively, explicitly depend on user-specified edits. In the former class of methods the edits are used to detect errors; subsequently a deterministic rule is used to correct these errors. In the latter class of methods, edits are used as the constraints of the mathematical optimization problem.

In the EUREDIT project (see http://www.cs.york.ac.uk/euredit), automatic error localization methods from three of the four classes mentioned above were developed and evaluated, namely automatic error localization based on outlier detection techniques, automatic error localization based on neural networks, and automatic error localization based on solving an optimization problem. The EUREDIT project was a large international research and development project on statistical data editing and imputation involving 12 institutes from seven different countries. The project lasted from March 2000 until March 2003. Important aims were (a) the evaluation of current "in-use" methods for data editing and imputation and (b) the development and evaluation of a selected range of new or recent techniques for data editing and imputation. Sections 3.2.1 and 3.2.2 focus on the work carried out under the EUREDIT project. For more information on the methods examined in the EUREDIT project, we refer to Chambers (2004).

3.2.1 AUTOMATIC ERROR LOCALIZATION BASED ON OUTLIER DETECTION TECHNIQUES

To detect erroneous data values by means of outlier detection techniques one, implicitly or explicitly, constructs a statistical model for the variables under consideration. Often, such models are regression models [see, e.g., Little and Smith (1987) and Ghosh-Dastidar and Schafer (2003)]. If an outlier detection technique is used for automatic error localization, data values that are outliers under the fitted model are identified as being erroneous ones.

To measure the "outlyingness" of a set of data values, a metric needs to be defined. An abundance of such metrics is described in the statistical literature [see, e.g., Barnett and Lewis (1994) and Rousseeuw and Leroy (1987)]. For continuous data, a well-known metric is the Mahalanobis distance. The basic problem with the application of this metric is that it depends on the unknown underlying mean and variance of the data generating process. Obtaining reliable estimates for the mean and variance is a complicated problem. Several iterative procedures have been proposed in the statistical literature to overcome this problem [see, e.g., Rocke and Woodruff (1993) and Woodruff and Rocke (1994)]. These iterative procedures tend to be extremely time-consuming, especially for the large data sets that are typical for National Statistical Institutes (NSIs). In the EUREDIT project, attention was therefore focused on less time-consuming outlier detection techniques. Below we briefly list these techniques.

Transformed Rank Correlations. This method uses bivariate Spearman rank correlations and a transformation into the space of the principal axes to construct a positive definite covariance matrix. This positive definite covariance matrix is subsequently used to measure distances between data values and a center of the data [see Béguin and Hulliger (2004)].

Epidemic Algorithm. This method is based on the idea of spreading a simulated epidemic, starting at the spatial median of the data set. From that spatial median the epidemic reaches other data points. Data values that are infected very late in the process, or are not infected after a fixed time, are identified as outliers [see Béguin and Hulliger (2004)].

Forward Search Algorithms. These methods start with the selection of an outlier-free subset of the data set, the so-called *clean* data [cf. Hadi and Simonoff (1993), Atkinson (1994), Rocke and Woodruff (1996), Kosinski (1999), Billor, Hadi, and Velleman (2000), and Riani and Atkinson, (2000)]. Based on this clean subset, a model for the variable(s) of interest is estimated by means of standard multivariate statistical techniques. The estimated parameters of the model, in particular the mean and the variance, are then used to calculate distances of the sample data values to the center of the data—for instance, by means of the Mahalanobis distance. Next, the clean data set is redefined by declaring all observations with a distance to the center of the clean subset below a certain threshold to be outlier-free. This iterative procedure is repeated until convergence—that is, until all observations outside the clean subset are identified as outliers with respect to the distances generated by the model fitted to the clean subset.

Robust Tree Modeling. A regression tree model [cf. Breiman et al. (1984)] classifies the data in terms of the values of a set of predictor variables. The original data set is sequentially divided into subgroups, or nodes, that are increasingly more homogeneous with respect to the values of a numerical response variable. This process of sequentially dividing the data set into subsets continues until a stopping criterion is met. In the outlier-robust regression tree models that have been evaluated under the EUREDIT project, outliers are down-weighted when calculating the measure of within node heterogeneity, with the weights based on outlier-robust influence functions. These weights reflect the distance from a robust estimate of the mean for the values in the node. Once the tree has been generated, the outliers are determined. An outlier is defined as an observation with an average weight over all node splits less than a specified threshold [see Chambers, Hentges, and Zhao (2004)].

3.2.2 AUTOMATIC ERROR LOCALIZATION USING NEURAL NETWORKS

Neural networks can be seen as a nonparametric class of regression models. A neural network consists of elementary units, called neurons, linked by weighted connections [see, e.g., Bishop (1995) and Fine (1999)]. The neurons are arranged in layers. There is an input layer representing the input fields, one or more hidden layers, and an output layer representing the output field(s). Similarly to automatic error localization based on outlier detection techniques, data values that do not fit the model constructed by the neural network are considered erroneous. In the EUREDIT project, several kinds of neural networks were developed and evaluated for the purpose of automatic error localization. Below we briefly describe these neural networks.

Multilayer Perceptron (MLP). An MLP is a network with at least one hidden layer. Like all neural networks, an MLP learns through training. During the iterative training phase the network examines individual records, generates a prediction for each record, and makes adjustments to the connection weights whenever the network makes an incorrect prediction. Error localization can be carried out by defining the target variable for the network as the zero–one indicator denoting presence/absence of errors. Alternatively, one can define the variable under consideration as the target variable. In this case, if the predicted value differs substantially from the observed value, the observed value is considered an outlier, and hence as (possibly) erroneous [see Larsen and Madsen (1999)].

Correlation Matrix Memory (CMM). CMM is a type of neural network that is trained to associate pairs of patterns, comprising an input pattern and an output pattern [see Austin and Lees (2000)]. Once the associations have been learned, input patterns are fed into the CMM. For each input pattern, the learned association is recalled, leading to recovery of the originally paired output pattern. The CMM is used as a kind of filter to remove patterns that do not match closely with the input pattern, and thus to find a subset of near-matching records for each input record. For outlier detection, the subset of near-matching records is processed to determine the distance from the record under consideration to its Kth neighbor record (for a suitable preset value of K, with records ranked according to distance). Records with a larger distance, and the variables in such records which contribute most to this larger distance, are judged more likely to be outlying, and hence to be in error.

Tree Structured Self-Organizing Maps. The Self-Organizing Map (SOM) is a multivariate algorithm that models the joint distribution of data. It constructs a lower-dimensional latent surface of the data set. The implementation of the SOM uses a discrete set of nodes (i.e., neurons) to construct the surface. These nodes can be interpreted as data clusters that are smoothed along the SOM lattice. In the EUREDIT project a tree-structured version of the SOM, called TS-SOM [cf. Koikkalainen and Oja (1990)], was used. As usual for statistical modeling techniques, outlier detection was carried out by comparing each observation to its fitted value under the SOM model. If the model did not explain the observation well enough, it was considered to be an outlier.

3.2.3 AUTOMATIC ERROR LOCALIZATION BASED ON DETERMINISTIC CHECKING RULES

Deterministic checking rules state which variables are considered erroneous when the edits in a certain record are violated. An example of such a rule is: If component variables do not sum up to the corresponding total variable, the total variable is considered to be erroneous. Advantages of this approach are its transparency and its simplicity. A drawback of this approach is that many detailed checking rules have to be specified, which can be time- and resource-consuming. Another drawback is that maintaining and checking the validity of a high number of detailed checking rules can be complex. Moreover, in some cases it may be impossible to develop deterministic checking rules that are powerful enough to identify errors in a reliable manner. A final disadvantage is the fact that bias may be introduced as one aims to correct random errors in a systematic manner. In our opinion, this latter observation makes the use of deterministic checking rules more suited for correction of systematic errors than for random errors.

Automatic error localization based on deterministic checking rules is applied in practice at NSIs, such as Statistics Netherlands [see, e.g., Aelen and Smit (2009)].

3.2.4 AUTOMATIC ERROR LOCALIZATION BASED ON SOLVING AN OPTIMIZATION PROBLEM

To formulate the error localization problem as a mathematical optimization problem, a guiding principle (i.e., an objective function) for identifying the erroneous fields in a record is needed. Freund and Hartley (1967) were among the first to propose such a guiding principle. It is based on minimizing the sum of the distance between the observed data and the "corrected" data and a measure for the violation of the edits. That paradigm never became popular, however, possibly because a "corrected" record may still fail to satisfy the specified edits. A similar guiding principle, based on minimizing a quadratic function measuring the distance between the observed data and the "corrected" data subject to the constraint that the "corrected" data satisfy all edits, was later proposed by Casado Valera et al. (1996). A third guiding principle is based on first imputing missing data and potentially erroneous data for records failing edits by means of donor imputation and then selecting an imputed record that satisfies all edits and that is "closest" to the original record. This paradigm forms the basis of NIM (Nearest-neighbor Imputation Methodology). Thus far, NIM has mainly been used for demographic data, sometimes in combination with methodology based on the Fellegi–Holt paradigm [see Manzari (2004)]. For details on NIM we refer to Bankier (1999), Bankier et al. (2000), and Section 4.5 of this book.

The best-known and most-used guiding principle is the (generalized) paradigm of Fellegi and Holt (1976), which we already described in Section 3.1. Using this guiding principle, as we do in the remainder of the present chapter, the error localization problem can be formulated as a mathematical optimization problem (see Section 3.4.1). The (generalized) Fellegi–Holt paradigm can be applied to numerical data as well as to categorical (discrete) data. In this chapter we restrict ourselves to numerical—and in particular continuous—data. Extensions to categorical and integer-valued data are discussed in Chapters 4 and 5, respectively.

3.3 Aspects of the Fellegi–Holt Paradigm

The Fellegi–Holt paradigm can often be used successfully in the case of nonsystematic, random errors. Its direct aim is the construction of internally consistent records, not the construction of a data set that possesses certain (distributional) properties. Assuming that only few errors are made, the Fellegi–Holt paradigm obviously is a sensible one. Provided that the set of edits used is sufficiently powerful, application of this paradigm generally results in data of higher statistical quality, especially when used in combination with other editing techniques. This is confirmed by various evaluation studies such as Hoogland and Van der Pijll (2003).

A strong aspect of using the Fellegi–Holt paradigm is its flexibility. For instance, one can let the edits that have to be satisfied depend on the record under consideration. One could start with a basic set of edits. Depending on certain statistical characteristics of the record under consideration, one can add some additional edits to that basic set or remove some edits from that set. One can also let the reliability weights depend on the record under consideration. Van de Pol, Bakker, and De Waal (1997) describe a method to dynamically adjust the reliability weights in order to take the occurrence of potential typing errors into account. If violated edits can be satisfied by, for example, swapping two digits of a certain value, the reliability weight of the corresponding variable is dynamically lowered.[1] This idea can be extended further. One can, for example, also compare the value of a variable to a "normal" value for this variable. If there is a substantial difference, the reliability weight of this variable may be lowered. For all algorithms described in the rest of this chapter, except for the algorithm of

[1] A better approach would be to detect and correct swapped digits by means of the techniques examined in Chapter 2.

Section 3.4.4, using a set of edits or a set of reliability weights conditional on the record under consideration does not pose any extra computational challenges.

Using the Fellegi–Holt paradigm has a number of drawbacks. A first drawback is that the class of edits that can be handled is restricted to hard edits. That is, a system based on the Fellegi–Holt paradigm considers all edits specified for a certain record as hard ones. Especially in the case of automatic editing, one should be careful not to specify too many "soft" edits in order to avoid over-editing [see Di Zio, Guarnera, and Luzi (2005)]—that is, to avoid that too much effort and time is spent on correcting errors that do not have a noticeable impact on the ultimately published figures, or—even worse—that unlikely, but correct, data are "corrected". Such unjustified alterations can be detrimental for data quality.

A second drawback is that the class of errors that can safely be treated by means of the Fellegi–Holt paradigm is limited to random errors that are not caused by a systematic reason, but by accident. An example is an observed value where a respondent by mistake typed in a digit too many. Systematic errors—that is, errors that are reported consistently between (some of the) responding units—cannot be handled by a system based on the Fellegi–Holt paradigm. Such errors can, for instance, be caused by the consistent misunderstanding of a question on the survey form by (some of) the respondents. Examples are financial figures reported in euros instead of the requested thousands of euros. Since such errors occur in groups of related variables such as all variables related to purchases, they often do not violate edits. For methods for detecting and correcting systematic errors, we refer to Chapter 2.

A final drawback is that automatic editing alone is generally not enough to obtain data of sufficient statistical quality. Very suspicious or highly influential records should be edited interactively, because adjustments to such records should be as accurate as possible. At the moment, human beings are still better than machines when it comes to making accurate and often subtle changes to influential records. In some cases they may even be able to re-contact the provider of the data, although statistical offices generally try to avoid this as much as possible. Only the less suspicious and less influential records can be edited automatically—for instance by using the Fellegi–Holt paradigm.

Several systems for automatic editing are based on the Fellegi–Holt paradigm. Examples for continuous data are: GEIS [Generalized Edit and Imputation System; see Kovar and Whitridge (1990) and Statistics Canada (1998)] and Banff (Banff Support Team, 2008), both by Statistics Canada, SPEER [Structured Programs for Economic Editing and Referrals; see Winkler and Draper (1997)] by the U.S. Bureau of the Census; AGGIES [Agricultural Generalized Imputation and Edit System; see Todaro (1999)] by the National Agricultural Statistics Service, a SAS program developed by the Central Statistical Office of Ireland [cf. Central Statistical Office (2000)]; and CherryPi [cf. De Waal (1996)] and SLICE (De Waal, 2001), both by Statistics Netherlands. Examples for categorical data are SCIA [Sistema Controllo e Imputazione Automatici; see Barcaroli et al. (1995)] by Istituto Nazionale di Statistica and DISCRETE [cf. Winkler and Petkunas (1997)] by the U.S. Bureau of the Census.

3.4 Algorithms Based on the Fellegi–Holt Paradigm

3.4.1 THE ERROR LOCALIZATION PROBLEM FOR CONTINUOUS DATA USING THE FELLEGI–HOLT PARADIGM

In this section we describe the error localization problem as a mathematical optimization problem, using the (generalized) Fellegi–Holt paradigm. This mathematical optimization problem is solved for each record separately. Because we do not want to complicate the notation unnecessarily, we will not use an index to indicate the record we are dealing with. We denote the continuous variables in a certain record by x_j ($j = 1, \ldots, p$). The record itself is denoted by the vector $(x_1, \ldots, x_p)$. We assume that edit k ($k = 1, \ldots, K$) is written in either of the two following forms:

(3.1) $$a_{k1}x_1 + \cdots + a_{kp}x_p + b_k = 0$$

or

(3.2) $$a_{k1}x_1 + \ldots + a_{kp}x_p + b_k \geq 0.$$

An example of an edit of type (3.1) is

(3.3) $$T = P + C,$$

where T is the turnover of an enterprise, P its profit, and C its costs. Edit (3.3) expresses that the profit and the costs of an enterprise should sum up to its turnover. This is a typical example of a hard edit. A record not satisfying this edit is obviously incorrect and has to be corrected.

An example of an edit of type (3.2) is

(3.4) $$P \leq 0.5T,$$

expressing that the profit of an enterprise should be at most 50% of its turnover. Edit (3.4) does not hold true for all possible enterprises and is an example of a soft edit. Soft edits are treated as if they were hard ones and therefore have to be satisfied by all records.

Edit k ($k = 1, \ldots, K$) is satisfied by a record $(x_1, \ldots, x_p)$ if (3.1), respectively (3.2) holds. A variable x_j is said to *enter*, or to be *involved in*, edit k given by (3.1) or (3.2) if $a_{kj} \neq 0$. That edit is then said to *involve* this variable. All edits given by (3.1) or (3.2) have to be satisfied simultaneously. We assume that the edits can indeed be satisfied simultaneously. Any field for which the value is missing is considered to be erroneous. Edits in which a variable with a missing value is involved are considered to be violated.

For each record $(x_1, \ldots, x_p)$ in the data set that is to be edited automatically, we wish to determine—or, more precisely, ensure the existence of—a synthetic record $(\check{x}_1, \ldots, \check{x}_p)$ such that $(\check{x}_1, \ldots, \check{x}_p)$ satisfies all edits k given by (3.1) and (3.2) $(k = 1, \ldots, K)$, none of the $\check{x}_j$ $(j = 1, \ldots, p)$ is missing, and

$$\sum_{j=1}^{p} w_j y_j \tag{3.5}$$

is minimized, where the variables y_j $(j = 1, \ldots, p)$ are defined by

$$y_j = \begin{cases} 1 & \text{if } \check{x}_j \neq x_j \text{ or } x_j \text{ is missing,} \\ 0 & \text{otherwise.} \end{cases} \tag{3.6}$$

In (3.5), w_j is the reliability weight of variable x_j $(j = 1, \ldots, p)$. The variables of which the values in the synthetic record differ from the original values, together with the variables of which the values were originally missing, form an optimal solution to the error localization problem. The above formulation is a mathematical formulation of the generalized Fellegi–Holt paradigm. As we mentioned before, in the original Fellegi–Holt paradigm, all reliability weights w_j $(j = 1, \ldots, p)$ were set to 1 in (3.5). A solution to the error localization problem is basically just a list of all variables that need to be changed—that is, the variables x_j for which the corresponding $y_j = 1$ need to be changed.

In practice, records that contain a lot of errors are not suited for automatic edit and imputation in our opinion. Automatic edit and imputation of those records simply cannot lead to a trustworthy record. We feel that such records should therefore be treated in a different way—for instance, by traditional manual editing—or should be discarded completely. For the error localization problem, this implies that only records containing less than a certain number of errors N_{max} should be edited automatically. If one detects during the error localization process that a record cannot be made to satisfy all edits by changing at most N_{max} variables, this record can be discarded as far as automatic edit and imputation is concerned.

Unless otherwise noted, whenever we refer to the error localization problem in this chapter, we will mean the mathematical optimization problem described in this section.

The error localization problem may have several optimal solutions for a record. In contrast to what is common in the literature on the error localization problem where usually only one optimal solution is found, most of the algorithms we describe aim to find all optimal solutions. The reason for generating all optimal solutions is that the problem of obtaining records of sufficiently high quality, the ultimate goal of statistical data editing, is more comprehensive than the error localization problem as it is formulated in the literature. To obtain records of sufficiently high quality, statistical aspects, such as the distribution of the corrected data, should also be taken into account. By generating all optimal solutions to the error localization problem, we gain the option to later use a secondary, statistical criterion to select one optimal solution that is best from a statistical point of view. The variables involved in the selected solution are set to missing and are subsequently imputed during the imputation step by any statistical imputation method that one prefers, such as regression imputation or donor imputation [for an overview of imputation methods, see Kalton and Kasprzyk (1986), Kovar and Whitridge (1995), Schafer (1997), Little and Rubin (2002), and Chapter 7 and further of the present book]. In the present chapter we will not explore the process of selecting one optimal solution from several optimal solutions, nor will we explore the imputation step. One way to select one optimal solution from several optimal solutions is discussed in Chapter 11 on application studies.

3.4.2 A NAIVE APPROACH

It is clear that at least one value per violated edit should be changed. A naive approach would be to assume that it is sufficient to change any value per violated edit. In that case the error localization problem reduces to an associated set-covering problem [for more on set-covering problems in general, see, e.g., Nemhauser and Wolsey (1988)], given by: minimize (3.5) subject to the condition that in each violated edit at least one variable should be changed.

We define

$$c_{kj} = \begin{cases} 1 & \text{if variable } x_j \text{ is involved in edit } k, \\ 0 & \text{otherwise.} \end{cases} \tag{3.7}$$

Then the condition that in each violated edit at least one variable should be changed can be written as

$$\sum_{j=1}^{p} c_{kj} y_j \geq 1 \quad \text{for } k \in \Omega_V \tag{3.8}$$

where Ω_V is the set of edits violated by the record under consideration.

Unfortunately, our assumption that any value per violated edit can be changed generally does not hold. The solution to the associated set-covering problem is usually not a solution to the corresponding error localization problem. This is illustrated by the following example.

EXAMPLE 3.1

Suppose the explicitly specified edits are given by

$$T = P + C, \tag{3.9}$$

$$0.5T \leq C, \tag{3.10}$$

$$C \leq 1.1T, \tag{3.11}$$

$$T \leq 550N, \tag{3.12}$$

$$T \geq 0, \tag{3.13}$$

$$C \geq 0, \tag{3.14}$$

$$N \geq 0, \tag{3.15}$$

where T again denotes the turnover of an enterprise, P its profit, C its costs, and N the number of employees. The turnover, profit, and costs are given in thousands of euros. Edits (3.10) and (3.11) give bounds for the costs of an enterprise in terms of the turnover, edit (3.12) gives a bound for the turnover in terms of the number of employees, and edits (3.13) to (3.15) say that the turnover, the costs of an enterprise, and the number of employees are nonnegative.

Let us consider a specific record with values $T = 100$, $P = 40{,}000$, $C = 60{,}000$, and $N = 5$. Edits (3.10) and (3.12) to (3.15) are satisfied, whereas edits (3.9) and (3.11) are violated. We assume that the reliability weights of variables T, P, and C equal 1, and the reliability weight of variable N equals 2. That is, variable N is considered more reliable than the financial variables T, P, and C.

The set-covering problem associated to the error localization problem has the optimal solution that the value of T should be changed, because this variable *covers* the violated edits (i.e., each violated edit involves T), and has a minimal reliability weight.[2] The optimal value of the objective function (3.5) of the set-covering problem equals 1. However, to satisfy edit (3.9) by changing the value of T the value of T should be set to 100,000, but in that case edit (3.12) would be violated. The optimal solution of the set-covering problem is not a feasible solution to the error localization problem, because variable T cannot be imputed consistently, i.e. such that all edits become satisfied. In fact, it can be shown that this particular instance of the error localization problem has the optimal solution that variables P and C should both be changed. The optimal value of the objective function (3.5) to the error localization problem equals 2. This is larger than the optimal value of the objective function of the associated set-covering problem. Possible values for variables P and C are $P = 40$ and $C = 60$. The resulting record passes all edits. Note that in this example the respondent probably forgot that the values of P and C should be given in thousands of euros. ■

A feasible solution to the error localization problem is always a feasible solution to the associated set-covering problem, but not vice versa. Hence, the value of

[2] In general a set of variables S is said to cover the violated edits, if in each violated edit at least one variable from S is involved.

the optimal solution to the error localization problem is at least equal to the value of the optimal solution to the associated set-covering problem. In Sections 3.4.4 and 3.4.8 we will examine how we can strengthen the set-covering problem in such a way that a solution to the set-covering problem does correspond to a solution to the error localization problem.

3.4.3 FOURIER–MOTZKIN ELIMINATION

Fourier–Motzkin elimination [see, e.g., Duffin (1974)] extended to include equalities, plays a fundamental role in several algorithms for solving the error localization problem described in this chapter. This technique can be used to eliminate a variable from a set of (in)equalities. If we want to eliminate a variable x_r from the set of current edits by means of Fourier–Motzkin elimination, we start by copying all edits not involving this variable from the set of current edits to the new set of edits Ψ.

If variable x_r occurs in an equation, we express x_r in terms of the other variables. Say x_r occurs in edit s of type (3.1), we then write x_r as

$$x_r = -\frac{1}{a_{sr}}\left(b_s + \sum_{j \neq r} a_{sj}x_j\right) \tag{3.16}$$

Expression (3.16) is used to eliminate x_r from the other edits involving x_r. If x_r is involved in several equations, we select any of these equations to express x_r in terms of other variables. These other edits are hereby transformed into new edits, not involving x_r, that are logically implied by the old ones. These new edits are added to our new set of edits Ψ. Note that if the original edits are consistent—that is, can be satisfied by certain values u_j $(j = 1, \ldots, p)$—then the new edits are also consistent because they can be satisfied by u_j $(j = 1, \ldots, p; j \neq r)$. Conversely, note that if the new edits are consistent, say they can be satisfied by the values u_j $(j = 1, \ldots, p; j \neq r)$, then the original edits are also consistent because they can be satisfied by the values u_j $(j = 1, \ldots, p)$ where u_r is defined by filling u_j $(j = 1, \ldots, p; j \neq r)$ into (3.16).

If x_r does not occur in an equality but only in inequalities, we consider all pairs of edits of type (3.2) involving x_r. Suppose we consider the pair consisting of edit s and edit t. We first check whether the coefficients of x_r in those inequalities have opposite signs; that is, we check whether $a_{sr} \times a_{tr} < 0$. If that is not the case, we do not consider this particular combination (s,t) of edits anymore. If the coefficients of x_r do have opposite signs, one of the edits, say edit s, can be written as an upper bound on x_r—that is, as

$$x_r \leq -\frac{1}{a_{sr}}\left(b_s + \sum_{j \neq r} a_{sj}x_j\right) \tag{3.17}$$

and the other edit, edit t, as a lower bound on x_r—that is, as

$$x_r \geq -\frac{1}{a_{tr}}\left(b_t + \sum_{j \neq r} a_{tj}x_j\right). \tag{3.18}$$

Edits (3.17) and (3.18) can be combined into

$$-\frac{1}{a_{tr}}\left(b_t + \sum_{j \neq r} a_{tj}x_j\right) \leq x_r \leq -\frac{1}{a_{sr}}\left(b_s + \sum_{j \neq r} a_{sj}x_j\right),$$

which yields an implicit edit not involving x_r given by

$$-\frac{1}{a_{tr}}\left(b_t + \sum_{j \neq r} a_{tj}x_j\right) \leq -\frac{1}{a_{sr}}\left(b_s + \sum_{j \neq r} a_{sj}x_j\right). \tag{3.19}$$

After all possible pairs of edits involving x_r have been considered and all implicit edits given by (3.19) have been generated and added to our new set of edits Ψ, we delete the original edits involving x_r that we started with. In this way we obtain a new set of edits Ψ not involving variable x_r. This set of edits Ψ may be empty. This occurs when all current edits involving x_r are inequalities and the coefficients of x_r in all those inequalities have the same sign. Note that—similar to the case where x_r is involved in an equation (see above)—if the original edits are consistent, say they can be satisfied by certain values u_j ($j = 1, \ldots, p$), then the new edits are also consistent as they can be satisfied by u_j ($j = 1, \ldots, p; j \neq r$). This is by definition also true if the new set of edits is empty. Conversely, note that if the new edits are consistent, say they can be satisfied by certain values u_j ($j = 1, \ldots, p; j \neq r$), then the minimum of the right-hand sides of (3.19) for the u_j ($j = 1, \ldots, p; j \neq r$) is larger than, or equal to, the maximum of the left-hand sides of (3.19) for the u_j ($j = 1, \ldots, p; j \neq r$), since *all* pairs of edits (3.2) are considered. So, we can find a value u_r in between the maximum of the left-hand sides of (3.19) and the minimum of the right-hand sides of (3.19). This implies that u_r satisfies

$$-\frac{1}{a_{tr}}\left(b_t + \sum_{j \neq r} a_{tj}u_j\right) \leq u_r \leq -\frac{1}{a_{sr}}\left(b_s + \sum_{j \neq r} a_{sj}u_j\right) \quad \text{for all pairs } s \text{ and } t,$$

which in turn implies that the original edits are consistent. We have demonstrated the main property of Fourier–Motzkin elimination: A set of edits is consistent if and only if the set of edits after elimination of a variable is consistent. Note that as one only has to consider pairs of edits, the number of implied edits is obviously finite.

We illustrate Fourier–Motzkin elimination by means of a simple example.

EXAMPLE 3.2

Suppose the edits are given by

(3.20) $$T = P + C,$$

(3.21) $$P \leq 0.5T,$$

(3.22) $$-0.1T \leq P,$$

(3.23) $$T \geq 0,$$

(3.24) $$T \leq 550N.$$

To illustrate how to eliminate a variable involved in a balance edit, we eliminate variable P from (3.20) to (3.24). We obtain

(3.25) $$T - C \leq 0.5T \quad \text{(equivalently: } 0.5T \leq C\text{)},$$

(3.26) $$-0.1T \leq T - C \quad \text{(equivalently: } C \leq 1.1T\text{)},$$

(3.23), and (3.24). Edits (3.23) to (3.26) are the edits that have to be satisfied by T, C, and N. The main property of Fourier–Motzkin elimination says that if T, C, and N satisfy (3.23) to (3.26), a value for P exists such that (3.20) to (3.24) are satisfied.

To illustrate how to eliminate a variable that is only involved in inequality edits, we eliminate variable C from (3.23) to (3.26). We obtain

(3.27) $$0.5T \leq 1.1T,$$

(3.23), and (3.24). Edit (3.27) is equivalent to edit (3.23). Edits (3.23) and (3.24) are hence the edits that have to be satisfied by T and N. The main property of Fourier–Motzkin elimination implies that if T and N satisfy (3.23) and (3.24), a value for C exists such that edits (3.23) to (3.26) can be satisfied. Combining the two results we have found, we conclude if T and N satisfy (3.23) and (3.24), values for P and C exist such that (3.20) to (3.24) are satisfied. ■

We would like to end this section with a brief remark on the history of Fourier–Motzkin elimination. In the early nineteenth century, Jean-Baptiste Joseph Fourier became interested in systems of linear inequalities. In particular, Fourier was interested in the problem of whether a feasible solution to a specified set of linear inequalities exists. Fourier developed an interesting method for solving this problem based on successively eliminating variables from the system of inequalities. It was published for the first time in 1826 [see Fourier (1826) and Kohler (1973)]. Fourier himself was not too impressed by his discovery. In his book on determinate equations, *Analyse des equations déterminées*, which was published after his death in 1831, this method for eliminating variables was omitted.

Fourier's method eliminates variables from a system of linear inequalities and equalities. The number of linear inequalities and equalities may change while eliminating variables. This may be a problem for large systems. For large systems the number of linear (in)equalities often becomes so high that Fourier–Motzkin elimination becomes impractical; that is, too much computing time and computer memory is required to apply Fourier–Motzkin elimination.

There is a so-called dual version of Fourier–Motzkin elimination. Here linear (in)equalities are eliminated rather than variables. The number of variables may vary while eliminating constraints. The dual version of Fourier–Motzkin elimination suffers from the same problem as the primal version: For large systems the number of variables often becomes so high that the method is impractical.

For many years Fourier's method was forgotten. In the 20th century it was rediscovered several times. For instance, Motzkin (1936) rediscovered the method. For this reason the method is usually referred to as Fourier–Motzkin elimination. Other people who have rediscovered the method are Dines (1927) and Chernikova (1964, 1965). Chernikova in fact rediscovered the dual version of Fourier–Motzkin elimination (see also Section 3.4.6).

In 1976 Fellegi and Holt again rediscovered Fourier–Motzkin elimination. Fellegi and Holt, however, not only rediscovered Fourier–Motzkin elimination for numerical data, they also extended the method to categorical variables (see the next section for a brief description of the Fellegi–Holt method for continuous data, and see Section 4.3 for a description of the Fellegi–Holt method for categorical data and also for a mix of categorical and continuous data). This extension is similar to the so-called resolution technique known from machine learning and mathematical logic [see, for example, Robinson (1965, 1968), Russell and Norvig (1995), Williams and Brailsford (1996), Marriott and Stuckey (1998), Warners (1999), and Ben-Ari (2001)].

In this chapter and the next two chapters Fourier–Motzkin elimination and its extensions to categorical and integer-valued data play a major role.

3.4.4 THE FELLEGI–HOLT METHOD

Fellegi and Holt (1976) describe a method for solving the error localization problem automatically. In this section we sketch their method, for details we refer to the article by Fellegi and Holt (1976). The method is based on generating so-called *implicit*, or *implied*, edits. Such implicit edits are logically implied by the explicit edits—that is, the edits specified by the subject-matter specialists. Implicit edits are redundant. They can, however, reveal important information about the feasible region defined by the explicit edits. This information is already contained in the explicitly defined edits, but there the information may be rather hidden.

The method proposed by Fellegi and Holt starts by generating a well-defined set of implicit and explicit edits that is referred to as the *complete set of edits* [see Fellegi and Holt (1976) for a precise definition of this set; see also Section 4.3 of this book; in the present section we will restrict ourselves to giving an example later on]. It is called the *complete* set of edits not because all possible implicit edits are generated, but because this set of (implicit and explicit) edits suffices to translate the error localization problem into a set-covering problem. The complete set of edits comprises the explicit edits and the so-called *essentially new* implicit ones [cf. Fellegi and Holt (1976)]. Once the complete set of edits has been generated one only needs to find a set of variables S that covers the violated (explicit and implicit) edits of the complete set of edits with a minimum sum of reliability weights in order to solve the error localization problem.

The complete set of edits is generated by repeatedly selecting a variable, which Fellegi and Holt (1976) refer to as the generating field. Subsequently, all pairs of edits (explicit or implicit ones) are considered, and it is checked whether (essentially new) implicit edits can be obtained by eliminating the selected generating field from these pairs of edits by means of Fourier–Motzkin elimination. This process continues until no (essentially new) implicit edits can be generated anymore, whatever generating field is selected. The complete set of edits has then been determined.

Example 3.3 illustrates the Fellegi–Holt method. This example is basically an example provided by Fellegi and Holt themselves, except for the fact that in their article the edits indicate a condition of edit failure (i.e., if a condition holds true, the edit is violated), whereas here the edits indicate the opposite condition of edit consistency (i.e., if a condition holds true, the edit is satisfied).

EXAMPLE 3.3

Suppose we have four variables x_j ($j = 1, \ldots, 4$). The explicit edits are given by

(3.28)
$$x_1 - x_2 + x_3 + x_4 \geq 0$$

and

(3.29)
$$-x_1 + 2x_2 - 3x_3 \geq 0.$$

The (essentially new) implicit edits are then given by

(3.30)
$$x_2 - 2x_3 + x_4 \geq 0,$$

(3.31)
$$x_1 - x_3 + 2x_4 \geq 0,$$

$$2x_1 - x_2 + 3x_4 \geq 0.$$

For instance, edit (3.30) is obtained by selecting x_1 as the generating field and eliminating this variable from edits (3.28) and (3.29). The above five edits form the complete set of edits because no (essentially new) implicit edit can be generated anymore.

An example of an implicit edit that is not an essentially new one is

$$-x_1 + 3x_2 - 5x_3 + x_4 \geq 0,$$

which is obtained by multiplying edit (3.29) by 2 and adding the result to edit (3.28). This is not an essentially new implicit edit, because no variable has been eliminated from the edits (3.28) and (3.29).

Now, suppose we are editing a record with values (3,4,6,1), and suppose that the reliability weights are all equal to 1. Examining the explicit edits, we see that edit (3.28) is satisfied, whereas edit (3.29) is violated. From the explicit edits, it is not clear which of the fields should be changed. If we also examine the implicit edits, however, we see that edits (3.29), (3.30) and (3.31) fail. Variable x_3 occurs in all three violated edits. So, we can satisfy all edits by changing x_3. For example, x_3 could be made equal to 1. Changing x_3 is the only optimal solution to the error localization problem in this example. ■

An important practical drawback of the Fellegi–Holt method for numerical data is that the number of required implicit edits may be very high, and in particular that the generation of all these implicit edits may be very time-consuming. For most real-life problems the computing time to generate all required implicit edits becomes extremely high even for a small to moderate number of explicit edits, as some early experiments at Statistics Netherlands confirmed. At Statistics Netherlands, we therefore quickly abandoned this path based on the Fellegi–Holt method and decided not to investigate it any further in our attempt to develop software for automatic editing of random errors. An exception to the rule that the number of (essentially new) implicit edits becomes extremely high is formed by ratio edits—that is, ratios of two variables that are bounded by constant lower and upper bounds. For ratio edits the number of (essentially new) implicit edits is low, and the Fellegi–Holt method is exceedingly fast in practice [cf. Winkler and Draper (1997)].

The Fellegi–Holt method has been improved upon since it was first proposed. The improvements have mainly been developed for categorical data, but may possibly be extended to numerical data. Garfinkel, Kunnathur, and Liepins (1986) propose improvements to the edit generation process of the Fellegi and Holt method [see also Boskovitz (2008)]. Winkler (1998) proposes further improvements on the edit generation process of the Fellegi–Holt method. Some of these improvements are discussed in Section 4.3.3.

3.4.5 USING STANDARD SOLVERS FOR INTEGER PROGRAMMING PROBLEMS

In this section we describe how standard solvers for mixed integer programming (MIP) problems can be used to solve the error localization problem. To apply such solvers, we make the assumption that the values of the variables x_j $(j = 1, \ldots, p)$ are bounded. That is, we assume that for variable x_j $(j = 1, \ldots, p)$, constants α_j and β_j exist such that

$$\alpha_j \leq x_j \leq \beta_j$$

for all consistent records. In practice, such values α_j and β_j always exist although they may be very large, because numerical variables that occur in data of statistical offices are by nature bounded. The values of α_j and β_j may be negative.

The problem of minimizing (3.5) so that all edits (3.1) and (3.2) become satisfied can be formulated as a MIP problem as follows [see Riera-Ledesma and Salazar-González (2003)]: Minimize (3.5) subject to (3.1), (3.2),

$$x_j^0 - (x_j^0 - \alpha_j)y_j \leq x_j \leq x_j^0 + (\beta_j - x_j^0)y_j$$

and

$$y_j \in \{0, 1\},$$

where x_j^0 denotes the original value of variable x_j $(j = 1, \ldots, p)$.

If the value of x_j $(j = 1, \ldots, p)$ is missing, we fill in the value zero and remove the variable y_j from the objective function (3.5)—that is, we set the weight w_j in (3.5) equal to zero. We also set the corresponding y_j equal to one.

The above MIP problem can be solved by applying commercially available solvers. McKeown (1984) also formulates the error localization problem for continuous data as a standard MIP problem. Schaffer (1987) and De Waal (2003a) give extended IP formulations that include categorical data.

3.4.6 THE VERTEX GENERATION APPROACH

In this section we examine a well-known and often-applied approach for solving the error localization problem. This approach is based on the observation that the error localization problem can, in principle, be solved by generating the vertices of a certain polyhedron. Since this is a well-known approach that has been implemented in several computer packages, such as GEIS (Kovar and Whitridge, 1990), Banff (Banff Support Team, 2008), CherryPi (De Waal, 1996), AGGIES

(Todaro, 1999), and a SAS program developed by the Central Statistical Office of Ireland [cf. Central Statistical Office (2000)], we examine this approach in a bit more detail than most of the other approaches described in this chapter.

If the set of edits (3.1) and (3.2) is not satisfied by a record $(x_1^0, \ldots, x_p^0)$, where x_j^0 $(j = 1, \ldots, p)$ denotes the observed value of variable x_j, then we seek values $x_j^+ \geq 0$ and $x_j^- \geq 0$ $(j = 1, \ldots, p)$ corresponding to positive and negative changes, respectively, to x_j^0 $(j = 1, \ldots, p)$ such that all edits (3.1) and (3.2) become satisfied. The objective function (3.5) is to be minimized subject to the constraint that the new, synthetic record

$$(x_1^0 + x_1^+ - x_1^-, \ldots, x_p^0 + x_p^+ - x_p^-)$$

satisfies all edits (3.1) and (3.2). That is, the x_j^+ and x_j^- $(j = 1, \ldots, p)$ have to satisfy

$$a_{k1}(x_1^0 + x_1^+ - x_1^-) + \ldots + a_{kp}(x_p^0 + x_p^+ - x_p^-) + b_k = 0 \tag{3.32}$$

and

$$a_{k1}(x_1^0 + x_1^+ - x_1^-) + \ldots + a_{kp}(x_p^0 + x_p^+ - x_p^-) + b_k \geq 0 \tag{3.33}$$

for each edit k $(k = 1, \ldots, K)$ of type (3.1) and type (3.2), respectively.

As in Section 3.4.5, if the value of x_j $(j = 1, \ldots, p)$ is missing, we fill in the value zero and remove the variable y_j from the objective function (3.5). We also remember that the value of x_j has to be modified.

The set of constraints given by (3.32) and (3.33) defines a so-called polyhedron for the unknowns x_j^+ and x_j^- $(j = 1, \ldots, p)$. The vertex generation approach for solving the error localization problem is based on Theorem 3.1 below.

THEOREM 3.1

An optimal solution to the error localization problem corresponds to a vertex of the polyhedron defined by the set of constraints (3.32) and (3.33).

Proof. Note that in an optimal solution to the error localization problem either $x_j^+ = 0$ or $x_j^- = 0$ $(j = 1, \ldots, p)$, because we can replace x_j^+ and x_j^- by $x_j^+ - \min(x_j^+, x_j^-)$ and $x_j^- - \min(x_j^+, x_j^-)$ $(j = 1, \ldots, p)$, respectively, and still retain the same optimal solution. Now, suppose that an optimal solution to the error localization problem is attained in a point satisfying

(a) $x_j^+ = 0$ for $j \in I^+$, $x_j^+ \neq 0$ otherwise,

(b) $x_j^- = 0$ for $j \in I^-$, $x_j^- \neq 0$ otherwise, for certain index sets I^+ and I^-. We then consider the problem of minimizing the linear function

$$\sum_{j \in I^+} x_j^+ + \sum_{j \in I^-} x_j^- \tag{3.34}$$

subject to the constraints given by (3.32) and (3.33). The function (3.34) is nonnegative, because $x_j^+ \geq 0$ and $x_j^- \geq 0$ ($j = 1, \ldots, p$). Moreover, (3.34) equals zero only for a point satisfying (a) and (b) above. In other words, our optimal solution to the error localization problem is also a minimum of (3.34) subject to the constraints given by (3.32) and (3.33), and, conversely, a minimum to (3.34) subject to the constraints given by (3.32) and (3.33) is also an optimum to the error localization problem. When a linear target function, such as (3.34), subject to a set of linear constraints, such as (3.32) and (3.33), is minimized, there are three options: There are no feasible solutions at all, the value of the target function is unbounded, or the minimum of the target function is bounded. In the latter case, the target function attains its minimum in a vertex of the feasible polyhedron described by the set of linear constraints [see, e.g., Nemhauser and Wolsey (1988)]. In our case the minimum of the target function is obviously bounded, and hence there is a vertex in which this minimum is attained.

We hence conclude that an arbitrary optimal solution to the error localization problem corresponds to a vertex of the polyhedron defined by the constraints given by (3.32) and (3.33). ■

The above theorem implies that one can find all optimal solutions to the error localization problem by generating the vertices of the polyhedron defined by the constraints (3.32) and (3.33) and then selecting the ones with the lowest objective value (3.5).

We illustrate the idea of the vertex generation approach by means of Example 3.4 below.

EXAMPLE 3.4

Suppose two edits involving three continuous variables have been specified. These edits are given by

$$2x_1 + x_2 \leq 5$$

and

$$-x_1 + x_3 \leq 1.$$

Suppose furthermore that an erroneous record given by the vector (1,4,3) is to be edited automatically. The x_j^+ and x_j^- ($j = 1, 2, 3$) then have to satisfy

(3.35) $$-2x_1^+ - x_2^+ + 2x_1^- + x_2^- \geq 1,$$

(3.36) $$x_1^+ - x_3^+ - x_1^- + x_3^- \geq 1,$$

(3.37) $$x_j^+ \geq 0 \quad \text{for } j = 1, 2, 3,$$

and

(3.38) $$x_j^- \geq 0 \quad \text{for } j = 1, 2, 3.$$

The theory developed so far in this section now says that an optimal solution to the error localization problem can be found in a vertex of the polyhedron for the x_j^+ and x_j^- defined by (3.35) to (3.38). ■

Unfortunately, the number of vertices of the polyhedron defined by (3.32) and (3.33) is often too high in practice to generate all of them. Instead, one should therefore generate a suitable subset of the vertices only. In the remainder of this section we focus on this aspect of the vertex generation approach.

There are a number of vertex generation algorithms that efficiently generate such a suitable subset of vertices of a polyhedron. An example of such a vertex generation algorithm is an algorithm proposed by Chernikova (1964, 1965). Probably most computer systems for automatic edit and imputation of numerical data are based on adapted versions of this algorithm. The best-known example of such a system is GEIS (Kovar and Whitridge, 1990; Statistics Canada, 1998) and its successor Banff (Banff Support Team, 2008). Other examples are CherryPi (De Waal, 1996), AGGIES (Todaro, 1999), and a SAS program developed by the Central Statistical Office of Ireland [cf. Central Statistical Office (2000)]. The original algorithm of Chernikova is rather slow for solving the error localization problem. It has been accelerated by various modifications [cf. Rubin (1975, 1977), Sande (1978), Schiopu-Kratina and Kovar (1989), Fillion and Schiopu-Kratina (1993), and De Waal (2003b)].

Only the last four of these papers focus on the error localization problem itself. Sande (1978) discusses the error localization problems for numerical data, categorical data, and mixed data. The discussion of the error localization problem in mixed data is very brief, however. Schiopu-Kratina and Kovar (1989) and Fillion and Schiopu-Kratina (1993) propose a number of improvements on Sande's method for solving the error localization problem for numerical data. They do not consider the error localization problems for categorical or mixed data. In the remainder of this section on the vertex generation approach, we sketch the algorithm by Chernikova and some later improvements. This is quite a technical description and may be skipped, unless one wants to apply Chernikova's algorithm oneself. Readers not interested in applying this algorithm, may continue with Section 3.4.7.

Because in an optimal solution to the error localization problem either $x_j^+ = 0$ or $x_j^- = 0$ $(j = 1, \ldots, p)$, we will use, instead of the objective function (3.5), the alternative objective function

$$\sum_{j=1}^{p} w_j \left(\delta(x_j^+) + \delta(x_j^-)\right), \tag{3.39}$$

where $\delta(x) = 1$ if $x \neq 0$ and $\delta(x) = 0$ otherwise, in the rest of the section.

Chernikova's algorithm (Chernikova, 1964, 1965) was in fact designed to generate the edges of a system of linear inequalities given by

$$\mathbf{Cx} \geq \mathbf{0} \tag{3.40}$$

and

$$\mathbf{x} \geq \mathbf{0}, \tag{3.41}$$

where $\mathbf{C}$ is a constant $n_r \times n_c$ matrix and $\mathbf{x}$ an n_c-dimensional vector of unknowns. The algorithm is described in the Appendix to this chapter. It can be used to find the vertices of a system of linear inequalities because of the following lemma [cf. Rubin (1975, 1977)].

LEMMA 3.1

The vector $\mathbf{x}^0$ is a vertex of the system of linear inequalities

$$\mathbf{Ax} \leq \mathbf{b} \tag{3.42}$$

and

$$\mathbf{x} \geq \mathbf{0} \tag{3.43}$$

if and only if $\left\{(\lambda\mathbf{x}^0|\lambda)^T, \lambda \geq 0\right\}$ is an edge of the cone described by

$$(-\mathbf{A}|\mathbf{b}) \begin{pmatrix} \mathbf{x} \\ \xi \end{pmatrix} \geq \mathbf{0} \tag{3.44}$$

and

$$\begin{pmatrix} \mathbf{x} \\ \xi \end{pmatrix} \geq \mathbf{0}. \tag{3.45}$$

Here $\mathbf{A}$ is an $n_r \times n_v$ matrix, $\mathbf{b}$ is an n_r vector, $\mathbf{x}$ is an n_v-vector, and ξ and λ are scalar variables.

For notational convenience we write

$$n_c = n_v + 1$$

throughout this section. The matrix in (3.44) is then an $n_r \times n_c$ matrix just like in (3.40), so we can use the same notation as in Rubin's formulation of Chernikova's algorithm (see also the Appendix).

If Chernikova's algorithm is used to determine the edges of (3.44) and (3.45), then after the termination of the algorithm the vertices of (3.42) and (3.43) correspond to those columns j_0 of $\mathbf{L}^{n_r}$ (cf. Appendix) for which $l^{n_r}_{n_c j_0} \neq 0$. The entries of such a vertex $\mathbf{x}'$ are given by

$$x'_{i_0} = l^{n_r}_{i_0 j_0} / l^{n_r}_{n_c j_0} \quad \text{for } i_0 = 1, \ldots, n_v. \tag{3.46}$$

We illustrate the use of Chernikova's algorithm (see the Appendix for details) by means of an example.

EXAMPLE 3.5

Suppose we want to use Chernikova's algorithm to determine the vertices of the following system:

$$x_1 + x_2 - 2x_3 \leq 1, \tag{3.47}$$

$$5x_1 - 2x_2 + 2x_3 \leq 3, \tag{3.48}$$

$$3x_1 + x_2 - 4x_3 \leq 2, \tag{3.49}$$

$$-x_1 - x_2 + 2x_3 \leq -1, \tag{3.50}$$

where the x_j ($j = 1, 2, 3$) are nonnegative.

The matrix $\mathbf{Y}^0$ is then given by

$$\mathbf{Y}^0 = \begin{pmatrix} -1 & -1 & 2 & 1 \\ -5 & 2 & -2 & 3 \\ -3 & -1 & 4 & 2 \\ 1 & 1 & -2 & -1 \\ \hline 1 & 0 & 0 & 0 \\ 0 & 1 & 0 & 0 \\ 0 & 0 & 1 & 0 \\ 0 & 0 & 0 & 1 \end{pmatrix}$$

The horizontal line separates the upper matrix $\mathbf{U}^0$ from the lower matrix $\mathbf{L}^0$. We process the first row and obtain the following result.

$$\mathbf{Y}^1 = \left(\begin{array}{cccccc} 2 & 1 & 0 & 0 & 0 & 0 \\ -2 & 3 & -12 & -2 & 2 & 5 \\ 4 & 2 & -2 & -1 & 2 & 1 \\ -2 & -1 & 0 & 0 & 0 & 0 \\ \hline 0 & 0 & 2 & 1 & 0 & 0 \\ 0 & 0 & 0 & 0 & 2 & 1 \\ 1 & 0 & 1 & 0 & 1 & 0 \\ 0 & 1 & 0 & 1 & 0 & 1 \end{array}\right)$$

For instance, the first two columns of $\mathbf{Y}^1$ are obtained by copying the last two columns of $\mathbf{Y}^0$. The third column of $\mathbf{Y}^1$ is obtained by adding the third column of $\mathbf{Y}^0$ to two times the first column of $\mathbf{Y}^0$. In a similar way the other columns of $\mathbf{Y}^1$ are obtained.

We now process the fourth row. The result is

$$\mathbf{Y}^2 = \left(\begin{array}{cccc} 0 & 0 & 0 & 0 \\ -12 & -2 & 2 & 5 \\ -2 & -1 & 2 & 1 \\ 0 & 0 & 0 & 0 \\ \hline 2 & 1 & 0 & 0 \\ 0 & 0 & 2 & 1 \\ 1 & 0 & 1 & 0 \\ 0 & 1 & 0 & 1 \end{array}\right).$$

Processing the third row yields

$$\mathbf{Y}^3 = \left(\begin{array}{cccc} 0 & 0 & 0 & 0 \\ 2 & 5 & -10 & 3 \\ 2 & 1 & 0 & 0 \\ 0 & 0 & 0 & 0 \\ \hline 0 & 0 & 2 & 1 \\ 2 & 1 & 2 & 1 \\ 1 & 0 & 2 & 0 \\ 0 & 1 & 0 & 2 \end{array}\right).$$

Note that columns 1 and 4 of $\mathbf{Y}^2$ are not combined into a column of $\mathbf{Y}^3$ (see Step 9 of Chernikova's algorithm). Likewise, columns 2 and 3 of $\mathbf{Y}^2$ are not combined into a column of $\mathbf{Y}^3$.

Finally, we process the second row and obtain the final matrix:

$$\mathbf{Y}^4 = \left(\begin{array}{ccccc} 0 & 0 & 0 & 0 & 0 \\ 2 & 5 & 3 & 0 & 0 \\ 2 & 1 & 0 & 10 & 0 \\ 0 & 0 & 0 & 0 & 0 \\ \hline 0 & 0 & 1 & 2 & 16 \\ 2 & 1 & 1 & 12 & 16 \\ 1 & 0 & 0 & 7 & 6 \\ 0 & 1 & 2 & 0 & 20 \end{array}\right).$$

Note that columns 2 and 3 of $\mathbf{Y}^3$ are not combined into a column of $\mathbf{Y}^4$.

In matrix $\mathbf{Y}^4$ we can see that the vertices of system (3.47) to (3.50) are given by (0,1,0), (0.5,0.5,0) and (0.8,0.8,0.3) [see (3.46)]. ■

Now, we explain how Chernikova's algorithm can be used to solve the error localization problem. The set of constraints given by (3.1) and (3.2) can be written in the form (3.42) and (3.43). We can find the vertices of the polyhedron corresponding to this set of constraints by applying Chernikova's algorithm to (3.44) and (3.45). Vertices of the polyhedron defined by the edits (3.1) and (3.2) are given by columns $\mathbf{y}_{*s}^{n_r}$ for which $u_{is}^{n_r} \geq 0$ for all i and $l_{n_c s}^{n_r} > 0$, where n_c is the number of rows of the final matrix $\mathbf{L}^{n_r}$ (cf. Appendix). In our case, n_c equals the total number of variables x_j^+ and x_j^- plus one [corresponding to ξ in (3.44) and (3.45)], that is, $n_c = 2p + 1$. The values of the variables x_j^+ and x_j^- in such a vertex are given by the corresponding values $l_{j0s}^{n_r}/l_{n_c s}^{n_r}$.

A minor technical problem is that we cannot use the objective function (3.39) directly when applying Chernikova's algorithm, because the values of x_j^+ and x_j^- are not known during the execution of this algorithm. Therefore, we introduce a new objective function that associates a value to each column of the matrix $\mathbf{Y}^t$ (cf. Appendix). Let us assume that the first p entries of a column $\mathbf{l}_{*s}^t$ of $\mathbf{L}^t$ correspond to the x_j^+ variables and the subsequent p entries to the x_j^- variables. We define the following objective function:

$$\sum_{j=1}^{p} w_j^r \left(\delta(l_{j,s}^t) + \delta(l_{p+j,s}^t)\right) \tag{3.51}$$

EXAMPLE 3.6

In Example 3.4 the set of constraints (3.35) to (3.38) were defined for the x_j^+ and x_j^+ ($j = 1, 2, 3$). Application of Chernikova's algorithm yields three vertices given by

1. $x_1^+ = 1$, $x_2^- = 3$, $x_1^- = x_2^+ = x_3^+ = x_3^- = 0$;
2. $x_1^- = 0.5$, $x_3^- = 1.5$, $x_1^+ = x_2^+ = x_2^- = x_3^+ = 0$;
3. $x_2^- = 1$, $x_3^- = 1$, $x_1^+ = x_1^- = x_2^+ = x_3^+ = 0$.

Now, suppose that the reliability weights are given by the vector $\mathbf{w} = (1, 2, 3)$. It is then easy to check that the value of the objective function (3.5) of these three vertices equals 3, 4, and 5, respectively. The optimal solution of the error localization problem is hence given by: Change the values of x_1 and x_2. ■

To accelerate Chernikova's algorithm, we aim to limit the number of vertices that are generated as much as possible. Once we have found a (possibly suboptimal) solution to the error localization problem for which the objective value (3.51) equals η, say, we from then on look only for vertices corresponding to solutions with an objective value at most equal to η.

Besides using the current best value of the objective function (3.51) to limit the number of vertices generated, there are a few other tricks that can be used to avoid the generation of unnecessary vertices. These tricks are examined below.

A *correction pattern* associated with column $\mathbf{y}_{*s}^t$ in an intermediate matrix $\mathbf{Y}^t$, where $\mathbf{Y}^t$ can be split into an upper matrix $\mathbf{U}^t$ and lower matrix $\mathbf{L}^t$ with n_r and n_c rows, respectively (cf. Appendix), is defined as the n_c-dimensional vector with entries $\delta(y_{js}^t)$ for $n_r < j \leq n_r + n_c$. For each x_j^+ and x_j^- a correction pattern contains an entry with a value in {0,1}. Sande (1978) notes that once a vertex has been found, all columns with correction patterns with ones on the same places as in the correction pattern of this vertex can be removed.

The concept of correction patterns has been improved upon by Fillion and Schiopu-Kratina (1993), who note that it is not important how the value of a variable is changed, but only whether the value of a variable is changed or not. A *generalized correction pattern* associated with column $\mathbf{y}_{*s}^t$ in an intermediate matrix $\mathbf{Y}^t$ is defined as the p-dimensional vector of which the jth entry equals 1 if and only if the entry corresponding to the jth variable in column $\mathbf{y}_{*s}^t$ is different from 0, and 0 otherwise. For each variable involved in the error localization problem, a generalized correction pattern contains an entry with a value in {0,1}. Again, once a vertex has been found, all columns with generalized correction patterns with ones on the same places as in the generalized correction pattern of this vertex can be deleted.

Fillion and Schiopu-Kratina (1993) define a *failed row* as a row that contains at least one negative entry placed on a column of which the last entry is nonzero. They note that in order to solve the error localization problem, we can already terminate Chernikova's algorithm as soon as all failed rows have been processed.

The final adaptation of Fillion and Schiopu-Kratina (1993) to Chernikova's algorithm is a method to speed-up the algorithm in case of missing values. Suppose the error localization problem has to be solved for a record with missing values. For each numerical variable of which the value is missing, we first fill in an arbitrary value, say zero. Next, only the entries corresponding to variables with nonmissing values are taken into account when calculating the value of function (3.51) for a column. The solution to the error localization problem is given by the variables corresponding to the determined optimal generalized correction patterns plus the variables with missing values. In this way, unnecessary generalized correction patterns according to which many variables with nonmissing values should be changed are discarded earlier than in the standard algorithm.

At Statistics Netherlands, the modified algorithm has been implemented in a computer program called CherryPi. CherryPi has been used for several years to automatically edit data in the uniform production system for annual structural business surveys at Statistics Netherlands [cf. De Jong (2002)]. Owing to memory and speed restrictions, a heuristic rule was used in CherryPi in order to reduce the computational burden for some records. As a consequence of this pragmatic heuristic rule, for some records only suboptimal solutions were found, and for some other records no solutions at all were found. Another effect of this rule is that the optimality of the found solutions is not guaranteed. There is a trade-off between the quality of the solutions found by CherryPi and its computing time. For more information on the implemented pragmatic rule we refer to De Waal (2003a, 2003b).

An interesting aspect of Chernikova's algorithm in the light of the other algorithms described in this chapter is—as we already mentioned—that it is a dual version of Fourier–Motzkin elimination: Whereas in Fourier–Motzkin elimination the variables are eliminated at the possible expense of creating additional constraints, in Chernikova's algorithm the constraints are eliminated at the possible expense of creating additional variables. For a more detailed summary of several papers on Chernikova's algorithm and an extension to categorical data we refer to De Waal (2003a, 2003b).

3.4.7 A BRANCH-AND-BOUND ALGORITHM

De Waal and Quere (2003) proposed a branch-and-bound algorithm for solving the error localization problem. In this chapter we briefly discuss the method for continuous data. In the next chapters we extend the method to categorical data and to integer data.

The basic idea of the algorithm is that for each record a binary tree is constructed. In our case, we use a binary tree to split up the process of searching for solutions to the error localization problem. We need some terminology with respect to binary trees before we can explain our algorithm. Following Cormen, Leiserson, and Rivest (1990), we recursively define a binary tree as a structure on a finite set of nodes that either contains no nodes or comprises three disjoint sets of nodes: a root node, a left (binary) subtree, and a right (binary) subtree. If the left subtree is nonempty, its root node is called the left child node of the root node of the entire tree, which is then called the parent node of the left child node. Similarly, if the right subtree is nonempty, its root node is called the right child node of the root node of the entire tree, which is then called the parent node of

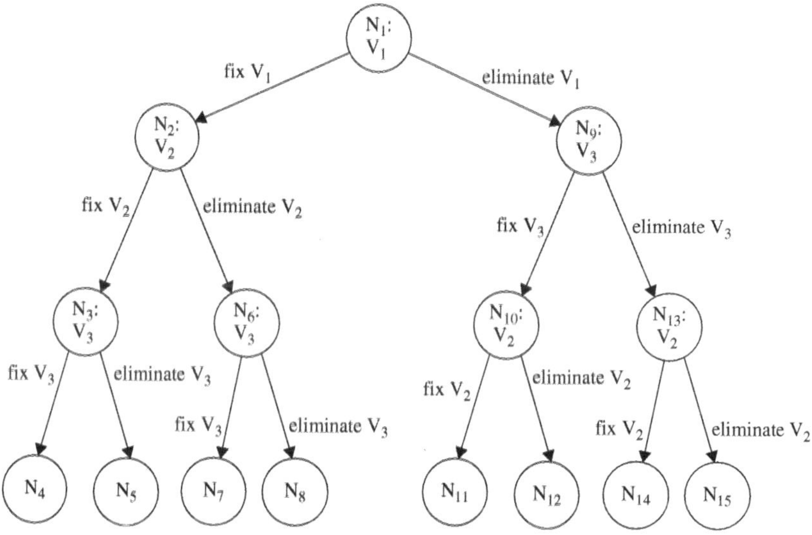

FIGURE 3.1 A binary tree.

the right child node. All nodes except the root node in a binary tree have exactly one parent node. Each node in a binary tree can have at most two (nonempty) child nodes. A node in a binary tree that has only empty subtrees as its child nodes is called a terminal node, or also a leaf. A nonleaf node is called an internal node. In Figure 3.1 we have drawn a small binary tree involving 15 nodes.

Node N_1 is the root node of the entire binary tree. Its child nodes are N_2 and N_9. The terminal nodes are nodes N_4, N_5, N_7, N_8, N_{11}, N_{12}, N_{14}, and N_{15}.

In each internal node of the binary tree generated by our algorithm a variable is selected that has not yet been selected in any predecessor node. If all variables have already been selected in a predecessor node, we have reached a terminal node of the tree.

We first assume that no values are missing. After the selection of a variable, two branches (i.e., subtrees) are constructed: in one branch we assume that the observed value of the selected variable is correct, in the other branch we assume that the observed value is incorrect. For instance, in Figure 3.1 variable V_1 is selected in node N_1. In the left-hand branch, variable V_1 is fixed to its original value; in the right-hand branch, V_1 is eliminated. By constructing a binary tree, we can, in principle, examine all possible error patterns and search for the best solution to the error localization problem.

In the branch in which we assume that the observed value is correct, the variable is fixed to its original value in the set of edits. In the branch in which we assume that the observed value is incorrect, the selected variable is eliminated from the set of edits. A variable that has either been fixed or eliminated is said to have been treated (for the corresponding branch of the tree). To each node in the tree, we have an associated set of edits for the variables that have not yet been treated in that node. The set of edits corresponding to the root node of our tree is the original set of edits.

Eliminating a variable is nontrivial because removing a variable from a set of edits may imply additional edits for the remaining variables. To illustrate why edits may need to be generated, we give a very simple example. Suppose we have three variables x_1, x_2, and x_3, and two edits: $x_1 \leq x_2$ and $x_2 \leq x_3$. If we want to eliminate variable x_2 from these edits, we cannot simply delete this variable and the two edits, but have to generate the new edit $x_1 \leq x_3$ implied by the two old ones because otherwise we could have that $x_1 > x_3$ and the original set of edits cannot be satisfied. See also Example 3.1 for a further explanation of why implied edits may be needed.

To ensure that the original set of edits can be satisfied, Fourier–Motzkin elimination is used. For inequalities, Fourier–Motzkin elimination basically consists of using the variable to be eliminated to combine these inequalities pairwise (if possible), as we did in the above example (see also Section 3.4.3). If the variable to be eliminated is involved in a balance edit, we use this equation to express this variable in terms of the other variables and then use this expression to eliminate the variable from the other edits.

In each branch the set of current edits is updated. Updating the set of current edits is the most important step of the algorithm. How the set of edits has to be updated depends on whether the selected variable is fixed or eliminated. Fixing a variable to its original value is done by substituting this value in all current edits, failing as well as nonfailing ones. Conditional on fixing the selected variable to its original value, the new set of current edits is a set of implied edits for the remaining variables in the tree. That is, conditional on the fact that the selected variable has been fixed to its original value, the remaining variables have to satisfy the new set of edits. As a result of fixing the selected variable to its original value, some edits may become tautologies—that is, may become satisfied by definition. An example of a tautology is "$1 \geq 0$". Such a tautology may, for instance, arise if a variable x has to satisfy the edit $x \geq 0$, the original value of x equals 1, and x is fixed to its original value. These tautologies may be discarded from the new set of edits. Conversely, some edits may become self-contradicting. An example of a self-contradicting relation is "$0 = 1$". If self-contradicting edits are generated, this particular branch of the binary tree cannot result in a solution to the error localization problem.

Eliminating a variable by means of Fourier–Motzkin elimination amounts to generating a set of implied edits that do not involve this variable. This set of implied edits has to be satisfied by the remaining variables. In the generation process we need to consider both the failing edits and the nonfailing ones in the set of current edits. The generated set of implied edits becomes the set of edits corresponding to the new node of the tree.

If values are missing in the original record, the corresponding variables only have to be eliminated (and not fixed) from the set of edits, because these variables always have to be imputed.

After all variables have been treated, we are left with a set of relations involving no unknowns. If and only if this set of relations contains no self-contradicting relations, the variables that have been eliminated in order to reach the corresponding terminal node of the tree can be imputed consistently—that is, such that all original edits can be satisfied (cf. Theorems 4.3 and 4.4 in Chapter 4). The set of relations involving no unknowns may be the empty set, in which case it obviously does not contain any self-contradicting relations. In the algorithm we check for each terminal node of the tree whether the variables that have been eliminated in order to reach this node can be imputed consistently. Of all the sets of variables that can be imputed consistently, we select the ones with the lowest sum of reliability weights. In this way we find all optimal solutions to the error localization problem (cf. Theorem 4.5 in Chapter 4).

We illustrate the branch-and-bound approach by means of an example.

EXAMPLE 3.7

Suppose the explicit edits are given by

(3.52) $$T = P + C,$$

(3.53) $$P \leq 0.5T,$$

(3.54) $$-0.1T \leq P,$$

(3.55) $$T \geq 0,$$

(3.56) $$T \leq 550N,$$

where T denotes the turnover of an enterprise, P its profit, C its costs, and N the number of employees. Let us consider a specific erroneous record with values $T = 100$, $P = 40{,}000$, $C = 60{,}000$, and $N = 5$. Edits (3.54), (3.55), and (3.56) are satisfied, whereas edits (3.52) and (3.53) are violated. The reliability weights of the variables T, P, and C equal 1, and the reliability weight of variable N equals 2. As edits (3.52) and (3.53) are violated, the record contains errors.

We select a variable, say T, and construct two branches: one where T is eliminated and one where T is fixed to its original value. We consider the first branch and eliminate T from the set of edits. We obtain the following new edits.

(3.57) $P \leq 0.5(P + C)$ [combination of (3.52) and (3.53)],

(3.58) $-0.1(P + C) \leq P$ [combination of (3.52) and (3.54)],

(3.59) $P + C \geq 0$ [combination of (3.52) and (3.55)],

(3.60) $P + C \leq 550N$ [combination of (3.52) and (3.56)].

Edits (3.57) to (3.59) are satisfied, edit (3.60) is violated. Because edit (3.60) is violated, changing T is not a solution to the error localization problem. If we were to continue examining the branch where T is eliminated by eliminating and fixing more variables, we would find that the best solution in this branch has an objective value equal to 3. We now consider the other branch where T is fixed to its original value. We fill in the original value of T in edits (3.52) to (3.56) and obtain (after removing any tautology that might arise) the following edits:

$$100 = P + C, \tag{3.61}$$

$$P \leq 50, \tag{3.62}$$

$$-10 \leq P, \tag{3.63}$$

$$100 \leq 550N. \tag{3.64}$$

Edits (3.63) and (3.64) are satisfied, edits (3.61) and (3.62) are violated. We select another variable, say P, and again construct two branches: one where P is eliminated and one where P is fixed to its original value. Here, we only examine the former branch and obtain the following edits:

$$100 - C \leq 50,$$

$$-10 \leq 100 - C, \tag{3.65}$$

$$100 \leq 550N. \tag{3.66}$$

Only edit (3.65) is violated. We select variable C and again construct two branches: one where C is eliminated and another one where C is fixed to its original value. In this example, we only examine the former branch and obtain edit (3.66) as the only implied edit. Because this edit is satisfied by the original value of N, changing P and C is a solution to the error localization problem. By examining all branches of the tree, including the ones that we have skipped here, we find that this is the only optimal solution to this record. ■

3.4.8 A CUTTING PLANE APPROACH

In 1986, Garfinkel, Kunnathur, and Liepins proposed a cutting plane algorithm for solving the error localization problem for categorical data, and in 1988 a similar algorithm for continuous data. In these algorithms a set-covering problem is solved to optimality in order to determine a potential solution to the error localization problem. If the potential solution turns out to be infeasible for the error localization problem, additional constraints, so-called cuts or cutting planes, are generated and subsequently added to the set-covering problem. Next, the new set-covering problem is solved. This iterative process goes on until an optimal solution to the error localization problem has been found. To check the feasibility of a potential solution to the error localization problem for continuous data, the first phase of the simplex algorithm [see Chvátal (1983) for more information on the simplex algorithm] could be used. To generate additional cuts for continuous data, a linear programming problem could be solved. Ragsdale and McKeown (1996) propose to solve a *modified* set-covering problem instead of solving an ordinary set-covering problem as an improvement on the algorithm by Garfinkel, Kunnathur, and Liepins (1988) for continuous data. The results by Ragsdale and McKeown (1996) indicate that their adaptation indeed leads to a reduced computing time.

In the present section we describe an algorithm for solving the error localization problem for continuous data similar to the ones proposed by Garfinkel, Kunnathur, and Liepins (1988) and Ragsdale and McKeown (1996). In our algorithm, again potential solutions to the error localization problem are determined that are subsequently checked for feasibility. However, instead of executing the first phase of the simplex algorithm, we eliminate all variables involved in a potential solution by means of Fourier–Motzkin elimination. By eliminating variables involved in a potential solution to the error localization problem, we can check the feasibility of a potential solution and simultaneously generate additional cuts in the case that the potential solution turns out to be infeasible. In other words, the former two steps of checking the feasibility of a potential solution and generating additional constraints are combined into a single step. In contrast to the algorithms by Garfinkel, Kunnathur, and Liepins (1986, 1988) and Ragsdale and McKeown (1996), which aim to find only one optimal solution for each record, our algorithm aims to find all optimal solutions for each record.

We start by describing the (modified) set-covering problem associated to an instance of the error localization problem. After that, we will be ready to describe our algorithm.

The Associated (Modified) Set-Covering Problem. As we have already seen in Section 3.4.2, to each instance of the error localization problem we can associate a set-covering problem. This set-covering problem is based on the observation that the value of at least one variable involved in each violated edit should be changed. A solution to the set-covering problem corresponds to a potential solution to the error localization problem. In order to give a mathematical formulation of the set-covering problem, we use the 0–1 variables y_j ($j = 1, \ldots, p$) defined in Section 3.4.1 [see (3.6)]. For a general set of edits of type (3.1) and (3.2) the set-covering problem is given by: Minimize the objective function given by (3.5) subject to the constraint that in each violated edit at least one variable should be changed. We use the c_{kj} defined by (3.7), that is,

$$c_{kj} = \begin{cases} 1 & \text{if variable } x_j \text{ is involved in edit } k, \\ 0 & \text{otherwise.} \end{cases}$$

We recall that the constraint that in each violated edit at least one variable should be changed can be written as

$$\sum_{j=1}^{p} c_{kj} y_j \geq 1 \quad \text{for } k \in \Omega_V,$$

where Ω_V is the set of edits violated by the record under consideration [see also (3.8)].

As we have seen in Section 3.4.2 minimizing (3.5) subject to the above constraint is, however, insufficient to solve the error localization problem. Ragsdale and McKeown (1996) strengthen the above formulation by noting that the set-covering problem does not take the direction of change of the variables into account. The above formulation only says that at least one variable involved in each violated edit should be changed. It does not consider pairs of edits. However, in order to satisfy a particular violated edit by changing a certain variable, it may be necessary to increase this variable's value; whereas to satisfy another violated edit, it may be necessary to decrease its value. Obviously, it is impossible to satisfy both edits simultaneously by changing the value of this variable only. Ragsdale and McKeown (1996) replace each variable x_j ($j = 1, \ldots, p$) in the violated edits by $x_j^0 + x_j^+ - x_j^-$, where x_j^0 denotes the observed value in the data set, $x_j^+ \geq 0$ a positive change in value, and $x_j^- \geq 0$ a negative change in value. This yields a set of constraints for the x_j^+ and the x_j^-. Next, they specify a modified set-covering problem involving 0–1 variables y_j^+ and y_j^- defined by

$$y_j^+ = \begin{cases} 1 & \text{if } x_j^+ > 0, \\ 0 & \text{if } x_j^+ = 0 \end{cases}$$

and

$$y_j^- = \begin{cases} 1 & \text{if } x_j^- > 0, \\ 0 & \text{if } x_j^- = 0. \end{cases}$$

In the modified set-covering problem, the objective function (3.5) is replaced by

$$\sum_{j=1}^{p} w_j (y_j^+ + y_j^-).$$

The modified set-covering problem can be solved by operation research techniques [see, e.g., Ragsdale and McKeown (1996)]. We illustrate the set-covering problem and its modified version by means of an example, which is taken from Ragsdale and McKeown (1996).

EXAMPLE 3.8

Suppose the same two edits involving three continuous variables as in Example 3.4 and Example 3.6 have been specified:

$$2x_1 + x_2 \leq 5$$

and

$$-x_1 + x_3 \leq 1.$$

As in Examples 3.4 and 3.6, we also assume that an erroneous record given by the vector (1,4,3) is to be edited automatically. Moreover, as in Example 3.6 we again assume that the reliability weights are given by the vector $\mathbf{w} = (1, 2, 3)$. Both edits are violated. The associated set-covering problem is then given by

$$\text{Minimize } y_1 + 2y_2 + 3y_3$$

subject to

$$y_1 + y_2 \geq 1,$$

$$y_1 + y_3 \geq 1,$$

where $y_1, y_2, y_3 \in \{0, 1\}$. An optimal solution to this set-covering problem is given by the vector $\mathbf{y} = (1, 0, 0)$. However, the corresponding potential solution to the error localization problem, change x_1, is not feasible because there is no value x_1 such that the vector $\mathbf{x} = (x_1, 4, 3)$ satisfies both edits.

To obtain the modified set-covering problem, we replace x_j in the violated edits by $x_j^0 + x_j^+ - x_j^-$. This yields the following two constraints for the x_j^+ and the x_j^-:

$$-2x_1^+ - x_2^+ + 2x_1^- + x_2^- \geq 1 \tag{3.67}$$

and

$$x_1^+ - x_3^+ - x_1^- + x_3^- \geq 1. \tag{3.68}$$

To satisfy (3.67) at least one of the variables x_1^- and x_2^- must assume a positive value. To satisfy (3.68), at least one of the variables x_1^+ and x_3^- must assume a positive value. This leads to the following modified set-covering problem:

$$\text{Minimize } y_1^+ + y_1^- + 2y_2^+ + 2y_2^- + 3y_3^+ + 3y_3^-$$

subject to

$$y_1^- + y_2^- \geq 1, \tag{3.69}$$

$$y_1^+ + y_3^- \geq 1, \tag{3.70}$$

$$y_1^+ + y_1^- \leq 1, \tag{3.71}$$

where $y_j^+, y_j^- \in \{0, 1\}$ for $j = 1, 2, 3$. The constraints (3.69) and (3.70) are the familiar set-covering constraints induced by (3.67) and (3.68) and the fact that the variables involved in those latter two constraints are nonnegative. Constraint (3.71) expresses that x_1 cannot be increased and decreased simultaneously. The solution to the above modified set-covering problem is $y_1^+ = y_2^- = 1$ and $y_1^- = y_2^+ = y_3^- = y_3^+ = 0$. The corresponding potential solution to the error localization problem, change x_1 and x_2 (more precisely: increase x_1 and decrease x_2), turns out to be indeed feasible. ■

If the value of x_j is missing, we obviously cannot replace x_j by $x_j^0 + x_j^+ - x_j^-$ as the observed value x_j^0 of x_j does not exist. As in Sections 3.4.5 and 3.4.6, if the value x_j^0 of x_j ($j = 1, \ldots, p$) is missing, we fill in the value zero for x_j^0, and remove the variable x_j from the objective function (3.5). We remember that the value of x_j has to be modified.

A solution to an instance of the error localization problem provides a feasible solution to the associated (modified) set-covering problem. Unfortunately, the converse does not hold. Our algorithm therefore strengthens the original set of modified set-covering constraints by iteratively adding constraints, cuts, to this set.

A Cutting Plane Algorithm. In our algorithm we denote, for $s > 0$, the set of implicit edits that have been generated during the sth iteration by Ω_s and the set of modified set-covering constraints after the sth iteration by Γ_s.

0. We denote the set of explicit edits that has to be satisfied by Ω_0. We define $\Gamma_0 = \emptyset$, and set $s = 1$.
1. Determine the edits in Ω_{s-1} that are violated by the record under consideration. If no edits are violated, the record is considered correct and the algorithm terminates. Otherwise, go to Step 2.
2. For each violated edit in Ω_{s-1}, we construct the corresponding constraints for the modified set-covering problem (see the above section on the modified set-covering problem). Let $\Gamma_s := \Gamma_{s-1} \cup \{\text{new set-covering constraints}\}$. Go to Step 3.
3. Determine all optimal solutions to the modified set-covering problem where the constraints are given by the elements of Γ_s. That is, determine all covers with a minimal sum of reliability weights. These optimal covers are denoted by the vectors $\check{\mathbf{y}}^t$ ($t = 1, \ldots, T_s$), where T_s is the number of optimal covers to the modified set-covering problem defined by Γ_s, and the index set of the tth optimal cover is denoted by $\check{I}^t$, that is, $\check{I}^t = \{j \mid \check{y}_j^t = 1\}$. If the sum of the

reliability weights of these solutions exceeds N_{max}, the maximum (weighted) number of fields that may be modified (see Section 3.4.1), we stop: The record is considered too erroneous for automatic editing and is discarded. Otherwise, go to Step 4.

4. For each optimal cover $\check{\mathbf{y}}^t$ ($t = 1, \ldots, T_s$), eliminate all variables in $\check{I}^t$ from Ω_0 by means of Fourier–Motzkin elimination (see Section 3.4.3). If the resulting set of edits, denoted by Ω_s^t, does not contain any edit violated by the values of the remaining variables, the variables in $\check{I}^t$ form an optimal solution to the error localization problem. If any optimal solution to the error localization problem has been found in Step 4, we output all found optimal solutions and stop. If no optimal solution to the error localization problem has been found, we go to Step 5.
5. Set $\Omega_s = \bigcup_{t=1}^{T_s} \Omega_s^t$, and let $s := s + 1$. Go to Step 1.

The algorithm is illustrated in Figure 3.2.

In words, our cutting plane algorithm works as follows. In Step 0, the initialization step, we set Ω_0 equal to the set of explicit, original edits. No

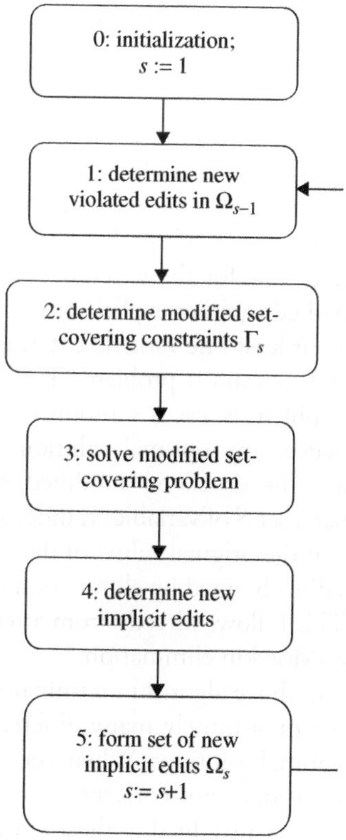

FIGURE 3.2 A graphical representation of our cutting plane algorithm.

modified set-covering constraints have been generated at that moment, and we therefore set Γ_0 equal to the empty set. If none of the edits are violated, we exit the algorithm. After Step 0 the iterative part of the algorithm begins. In Step 1 of iteration s, we determine all edits that are violated in Ω_{s-1}, the set of implicit edits that have been generated during the $(s-1)$th iteration. In Step 2, we generate the corresponding modified set-covering constraints and add these to the modified set-covering constraints and thereby obtain Γ_s. In Step 3, we then determine all potential optimal solutions to the error localization problem by solving the associated modified set-covering problem defined by Γ_s. In Step 4, we eliminate the variables involved in each potential optimal solution to the error localization problem from the original edits and thus generate one or more sets of implicit edits Ω_s^t. Finally, in Step 5 we collect all implicit edits that have been generated in iteration s into one set Ω_s.

We have the following theorem.

THEOREM 3.2

After termination of the above algorithm, all optimal solutions to the error localization problem have been determined. Moreover, the algorithm is guaranteed to terminate after a finite number of iterations.

Proof. A necessary condition for a set S of variables to be a feasible solution to the error localization problem is that after elimination of these variables the original values of the remaining variables satisfy the edits (explicit ones and implied ones obtained by elimination of variables in S) involving only these latter variables. So, none of these edits for the latter variables may be violated. This means that for each violated edit (either explicit or implied) at least one variable should be changed; that is, at least one variable entering this edit should be part of a solution to the error localization problem. This shows that any solution to the error localization problem is also a solution to the associated (modified) set-covering problem. Hence, any optimal solution to the error localization problem is also a solution to the associated (modified) set-covering problem.

It remains to show that a set S of variables is indeed a feasible solution to the error localization problem if the original values of the remaining variables satisfy the explicit and implicit edits obtained by elimination of the variables in S from the original set of edits. This follows directly from a repeated application of the main property of Fourier–Motzkin elimination.

It is easy to see that the above algorithm terminates after a finite number of iterations because there are only finitely many different set-covering problems, each with finitely many optimal solutions. Whenever an optimal solution to the current set-covering problem does not correspond to a feasible solution to the error localization problem, cuts are added to the set-covering problem (see Steps 2 and 5) in order to make this optimal set-covering solution infeasible for all subsequent set-covering problems. ■

EXAMPLE 3.9

We consider the same problem as in Example 3.7. As the edits (3.52) and (3.53) are violated, the associated modified set-covering problem is given by

$$\text{(3.72)} \qquad \text{Minimize } y_T^+ + y_T^- + y_P^+ + y_P^- + y_C^+ + y_C^- + 2y_N^+ + 2y_N^-$$

subject to the constraints

$$\text{(3.73)} \qquad y_T^+ + y_P^- + y_C^- \geq 1,$$

$$\text{(3.74)} \qquad y_T^+ + y_P^- \geq 1,$$

$$\text{(3.75)} \qquad y_j^+ + y_j^- \leq 1 \quad \text{for } j = T, P, C, N,$$

where $y_T^+, y_T^-, y_P^+, y_P^-, y_C^+, y_C^-, y_N^+, y_N^- \in \{0, 1\}$. For instance, to obtain constraint (3.73) we first substitute $100 + x_T^+ - x_T^-$ for T, $40{,}000 + x_P^+ - x_P^-$ for P, and $60{,}000 + x_C^+ - x_C^-$ for C in (3.52), where x_T^+, x_T^-, x_P^+, x_P^-, x_C^+, and x_C^- are nonnegative variables. This results in

$$x_T^+ + x_P^- + x_C^- - x_T^- - x_P^+ - x_C^+ = 99{,}900.$$

So, either the value of T should be increased, the value of P should be decreased, or the value of C should be decreased in order to satisfy (3.52). Constraint (3.73) expresses this in terms of the 0–1 variables y_T^+, y_P^-, and y_C^-. Constraint (3.75) expresses that a variable cannot be increased and decreased simultaneously.

The optimal solutions to the above modified set-covering problem are

(a) $y_T^+ = 1$ and $y_T^- = y_P^+ = y_P^- = y_C^+ = y_C^- = y_N^+ = y_N^- = 0$ (corresponding to: increase T);

(b) $y_P^- = 1$ and $y_T^+ = y_T^- = y_P^+ = y_C^+ = y_C^- = y_N^+ = y_N^- = 0$ (corresponding to: decrease P).

We first eliminate the variable T from the edits (3.52) to (3.56) by means of Fourier–Motzkin elimination to check whether changing the value of T is indeed a solution to the error localization problem. We obtain the edits given by (3.57) to (3.60). Edits (3.57) to (3.59) are satisfied, but edit (3.60) is violated. Because edit (3.60) is violated, changing the value of T is not a solution to the error localization problem.

To check whether changing the value of P is a solution to the error localization problem, we eliminate the variable P from the edits (3.52) to (3.56). We obtain four edits given by

$$\text{(3.76)} \qquad T - C \leq 0.5T \qquad \text{[combination of (3.52) and (3.53)]},$$

$$\text{(3.77)} \qquad -0.1T \leq T - C \qquad \text{[combination of (3.52) and (3.54)]},$$

and (3.55) and (3.56). Edits (3.55), (3.56), and (3.76) are satisfied, but edit (3.77) is violated. Therefore, changing the value of P is not a solution to the error localization problem.

The modified set-covering constraints corresponding to violated edits (3.60) and (3.77) are given by

$$y_P^- + y_C^- + y_N^+ \geq 1$$

and

$$y_T^+ + y_C^- \geq 1,$$

respectively.

We add these modified set-covering constraints to the constraints (3.73) and (3.74), and we minimize the objective function (3.72) subject to the updated system of modified set-covering constraints. The optimal solutions to the new modified set-covering problem are:

(a) $y_P^- = y_C^- = 1$ and $y_T^+ = y_T^- = y_P^+ = y_C^+ = y_N^+ = y_N^- = 0$ (corresponding to: decrease both P and C);

(b) $y_T^+ = y_C^- = 1$ and $y_T^- = y_P^+ = y_P^- = y_C^+ = y_N^+ = y_N^- = 0$ (corresponding to: increase T and decrease C);

(c) $y_P^- = y_T^+ = 1$ and $y_T^- = y_P^+ = y_C^+ = y_C^- = y_N^+ = y_N^- = 0$ (corresponding to: decrease P and increase T).

We check the feasibility of the first potential solution to the error localization problem by eliminating the variables P and C from the explicit edits (3.52) to (3.56). We first eliminate the variable P, and again obtain the constraints (3.55), (3.56), (3.76), and (3.77). Now, we eliminate the variable C from these latter constraints. We obtain (3.55), (3.56), and

$$0.5T \leq 1.1T \qquad \text{[combination of (3.76) and (3.77)]}.$$

This set of edits is satisfied by the values of the remaining variables. Hence, an optimal solution to the error localization problem is: change (decrease) the values of P and C. By checking the other two potential optimal solutions to the error localization problem, we find that this is the only optimal solution. ■

3.4.9 COMPUTATIONAL ASPECTS

In this section we provide computational results for (prototype) computer programs based on the methods of Sections 3.4.5 to 3.4.8 implemented at Statistics Netherlands on six realistic data sets. In the literature some evaluation studies for numerical data are already described [see Garfinkel, Kunnathur, and

Liepins (1988), Kovar and Winkler (1996), and Ragsdale and McKeown (1996)]. It is difficult to compare our results to the results described in the literature. First, because in most cases the actual computing speeds of the computer systems used in those studies are difficult to retrieve, and hence difficult to compare to the computing speed of present-day PCs. Second, because Garfinkel, Kunnathur, and Liepins (1988) and Ragsdale and McKeown (1996) use randomly generated data whereas we use realistic data. We feel that realistic data rather than randomly generated data should be used for evaluation studies, because the properties of realistic data are different from those of randomly generated data. Kovar and Winkler (1996) do use realistic data, but the data set used in their evaluation study is not generally available.

We are aware of the fact that the reader cannot check how well the computational comparisons presented in this chapter have been carried out. We note, however, that this is basically always the case for such computational comparisons, even if the source code of the used algorithms is described in full detail. Unless the reader is an expert himself, he is generally unable to determine the quality of the source code. An interesting aspect of our evaluation study is that we are competing against ourselves. The aim of Statistics Netherlands was to develop software for automatic editing based on the (generalized) Fellegi–Holt paradigm that is both sufficiently fast for practical use and easy to develop and maintain. To achieve this aim, we experimented with a number of algorithms over the course of several years. In 1995 we first experimented with the original Fellegi–Holt method. Considering the poor results we obtained with this approach, we then quickly switched to the vertex generation approach of Section 3.4.6. In 1999 we developed the branch-and-bound approach of Section 3.4.7. In 2002 we experimented with standard solvers for MIP problems for solving the error localization problem (see Section 3.4.5), and we developed the cutting plane method of Section 3.4.8.

The six evaluation data sets come from a wide range of business surveys, namely a survey on labor costs (data set A), a survey on enterprises in the photographic sector (data set B), a survey on enterprises in the building and construction industry (data set C), a survey on the retail sector (data set D), and a survey on environmental expenditures (data set E). Owing to confidentiality reasons, the branch of industry to which the businesses in data set F belong has not been made public. To the best of our knowledge, these six data sets are representative for other data sets from business surveys. A good performance on the six data sets hence suggests that the performance on other data sets arising in practice will also be acceptable.

In Table 3.1 we give a summary of the characteristics of the six data sets. In this table the number of variables, the number of nonnegativity edits, the number of balance edits, the number of inequality edits (excluding the nonnegativity edits), the total number of records, the number of inconsistent records (i.e., records failing edits or containing missing values), and the total number of missing values are listed. In addition, we present the number of records with more than six erroneous fields or missing values. Finally, we list the average number of errors per inconsistent record (excluding the missing values) and the average number

TABLE 3.1 **Characteristics of the Data Sets**

	A	B	C	D	E	F
Number of variables	90	76	53	51	54	26
Number of nonnegativity edits	90	70	36	49	54	22
Number of balance edits	0	18	20	8	21	3
Number of inequality edits	8	2	16	7	0	15
Total number of records	4,347	274	1480	4217	1039	1425
Number of inconsistent records	4,347	157	1404	2152	378	1141
Total number of missing values	259,838	0	0	0	2,230	195
Number of records with more than six errors or missing values	4,346	7	117	16	136	8
Errors per inconsistent record	0.2	2.5	2.6	1.6	5.8	3.0
Number of optimal solutions per inconsistent record	6.1	12.0	6.9	23.3	1.2	11.6

of optimal solutions per inconsistent record. The numbers in the last two rows of Table 3.1 have been determined by carefully comparing the number of fields involved in the optimal solutions, respectively the number of optimal solutions, of the four algorithms we have evaluated. The number of fields involved in the optimal solutions is assumed to be equal to the actual number of errors.

The first algorithm we consider is based on a standard MIP formulation for the error localization problem for continuous data (see Section 3.4.5). This algorithm has been implemented in Visual C++ 6.0; and it calls routines of CPLEX, a well-known commercial MIP-solver, to actually solve the MIP problems involved. We refer to this program as ERR_CPLEX in the remainder of this chapter. ERR_CPLEX finds only one optimal solution to each instance of the error localization problem. To find all optimal solutions, we could—once an optimal solution to the current MIP problem has been determined—iteratively add an additional constraint that basically states that the present optimal solution is excluded but other optimal solutions to the current MIP problem remain feasible, and we could solve the new MIP problem. This process of determining an optimal solution to the current MIP problem and adding an additional constraint to obtain a new MIP problem goes on until all optimal solutions to the error localization problem have been found. We have not implemented this option, however. Resolving the problem from scratch for each optimal solution would be very time-consuming. The alternative is to use a so-called hot restart, where information generated to obtain an optimal solution to a MIP problem is utilized to obtain an optimal solution to a slightly modified MIP problem. A problem with this possibility is that experiences at Statistics Netherlands with CPLEX so far, on linear programming problems arising in statistical disclosure control, show that CPLEX becomes numerically unstable if too many hot restarts in a row are applied. Our results of ERR_CPLEX therefore are only indicative.

If the algorithms we have developed ourselves would be clearly outperformed by ERR_CPLEX, this would suggest that standard MIP-solvers might give better results than our own algorithms.

The second algorithm is based on vertex generation (see Section 3.4.6). This algorithm has been implemented in a program, CherryPi, using the programming language Delphi. The implemented algorithm uses a matrix to solve the error localization problem. The number of rows of this matrix is implied by the number of edits and the number of variables. The actual number of columns is determined dynamically. Due to memory and speed restrictions, a maximum for the allowed number of columns is set in CherryPi. If the actual number of columns exceeds the allowed maximum, certain columns are deleted. This influences the solutions that are found by CherryPi. Due to this pragmatic rule in some cases, only nonoptimal solutions may be found, and in some other cases no solutions at all may be found. Another effect of this pragmatic rule is that if columns have been deleted to arrive at solutions to an instance of the error localization problem, the optimality of the found solutions is not guaranteed. The higher the allowed number of columns, the better the quality of the solutions found by CherryPi, but also the slower the speed of the program. Practical experience has taught us that in many instances, setting the allowed number of columns to 4000 gives an acceptable trade-off between the quality of the found solutions and the computing time of the program. In the version of CherryPi that was used for the comparison study, the allowed number of columns was therefore set to 4000.

The third algorithm is based on the branch-and-bound approach proposed in Section 3.4.7. The algorithm has been implemented in a prototype program, Leo, using the Delphi programming language. The program requires that a maximum cardinality N_{max} for the optimal solutions must be specified beforehand. Only if a record can be corrected by N_{max} changes or less, optimal solutions to the error localization problem are determined by Leo. In Leo the following rule to select a branching variable has been implemented: First select the variables that are involved in at least one failed edit and a minimum number of satisfied edits, then select the variable from this set of variables that occurs most often in the failed edits. In the case there are several "best" variables to branch on, one of them is chosen randomly. Leo has later been re-implemented as the error localization module of the SLICE package (see De Waal, 2001).

The fourth algorithm is based on the cutting plane approach proposed in Section 3.4.8. The algorithm has been implemented, using the programming language Delphi, in a prototype computer program that we will refer to as CUTTING. A fundamental part of the program is a solver for modified set-covering problems. Using well-known ideas from the literature, we have developed this solver, which is based on a recursive branch-and-bound algorithm, ourselves. CUTTING can determine all optimal solutions up to a user-specified maximum cardinality N_{max}. If such a maximum cardinality N_{max} is specified, records requiring more changes than N_{max} are rejected for automatic editing by CUTTING. The program can also work without such a maximum cardinality.

TABLE 3.2 Computational Results of the Error Localization Algorithms for the Data Sets (in seconds)

	A	B	C	D	E	F
ERR_CPLEX	216	9	86	100	12	32
CherryPi	570	96	540	498	622	79
Leo	18	308	531	21	59	7
Leo ($N_{max} = 6$)	7	51	94	19	4	8
CUTTING	601	513	1913	1101	90	94
CUTTING ($N_{max} = 6$)	156	395	695	1036	50	92

For Leo and CUTTING we have performed two kinds of experiment per data set. In the first kind of experiments we have set the maximum cardinality N_{max} to six. In the second kind of experiment for Leo, we have set N_{max} as high as possible without encountering memory problems for many (i.e., 20 or more) records. To be precise: in the second kind of experiment for Leo, we have set N_{max} equal to 90 for data set A, to 8 for data sets B and C, and to 12 for data sets D, E, and F. In the second kind of experiment for CUTTING, we have removed a maximum cardinality all together. For ERR_CPLEX and CherryPi, we have only performed experiments without a specified maximum cardinality.

All programs suffer from some numerical problems. These problems arise because in (erroneous) records the largest values may be a factor 10^9 or more larger than the smallest values. For instance, due to these numerical problems, ERR_CPLEX occasionally generates suboptimal solutions containing too many variables. Owing to numerical and memory problems, the programs could not determine solutions for all records. The number of records for which no solutions could be found was in general (very) low [see also De Waal and Coutinho (2005)]. For all data sets, Leo with $N_{max} = 6$ and CUTTING with $N_{max} = 6$ found all optimal solutions for all records requiring six or less changes.

The experiments have been performed on a PC that was connected to a local area network. Computing times may therefore be influenced by the amount of data that were transmitted through the network at the time of the experiments. To reduce and to estimate this influence, we have performed five experiments per data set at various moments during the day. In Table 3.2 we have mentioned the average computing times of these experiments (in seconds) for the six data sets over the corresponding five experiments. In all experiments, the reliability weight of each variable was set to 1. Note that some programs have a random aspect that influences the computing time. For instance, in Leo the selection of the variable to branch on is partly a random process.

For data set F the computing times of Leo and CUTTING are almost equal to the computing times of Leo with $N_{max} = 6$, respectively CUTTING with $N_{max} = 6$, because data set F contains only eight records with more than six errors (see Table 3.1). Owing to the stochastic variability in the computing times, Leo even outperformed Leo with $N_{max} = 6$ in our experiments. Taking the standard deviations of the experiments into account (not reported here), Leo and Leo with $N_{max} = 6$ are about equally fast for this data set.

Examining the results of Table 3.2, we can conclude that as far as computing speed is concerned, ERR_CPLEX and Leo (either with $N_{max} = 6$ or with $N_{max} > 6$) are the best programs. We note at the same time, however, that this conclusion is not completely justified because ERR_CPLEX determines only one optimal solution whereas the other programs (aim to) determine all optimal solutions. In Table 3.2 we also see that the use of a maximal cardinality for CUTTING and Leo clearly reduces the computing time.

Besides computing speed, other aspects are, of course, important too. We already noted that all programs, even the commercially available CPLEX, occasionally suffered from numerical problems. Leo in addition sometimes suffered from some memory problems. Due to its matrix with a fixed maximum number of columns, CherryPi does not always determine optimal solutions, but less good, suboptimal solutions. Summarizing, it is hard to give a verdict on the quality of the solutions found by the programs as the programs suffer from a diversity of problems. In a pragmatic sense, this also is good news: There is no strong preference for any of the examined algorithms, so one can select the algorithm that one prefers best from a theoretical or practical point of view.

3.5 Summary

In this chapter we have provided some historical background information on automatic error localization of random errors in continuous data. We have briefly mentioned and sketched the possibilities of statistical outlier detection techniques and the use of approaches such as neural networks in this regard. We then focused on automatic error localization of random errors based on solving a mathematical optimization problem, in particular using the so-called Fellegi–Holt paradigm. We have described a number of algorithms for solving the mathematical optimization problem arising from the use of the Fellegi–Holt paradigm, namely the method originally proposed by Fellegi and Holt (1976), a method based on applying standard solvers for integer programming problems, a vertex generation approach, a branch-and-bound approach and a cutting plane algorithm.

Winkler (1999) identified speed improvement as the main area of research for automatic error localization based on the Fellegi–Holt paradigm. This conclusion is not confirmed by the evaluation study of Section 3.4.9 [see also De Waal and Coutinho (2005)]. The conclusion that can be drawn from that study is that all algorithms examined in Sections 3.4.5 to 3.4.8 appear to be sufficiently fast for use in practice at Statistics Netherlands. In a relatively small country as the Netherlands, typically at most a few hundred thousand records of business surveys have to be edited automatically per year. With any of the evaluated algorithms, this can be achieved within a week on a moderate PC. The different conclusions by Winkler (1999) and De Waal and Coutinho (2005) may be caused by the fact that Winkler mainly concentrates on categorical data and the Fellegi–Holt method (see Section 3.4.4), whereas De Waal and Coutinho focus on continuous data and on other kinds of algorithms.

As we already mentioned in Section 3.4.6, a computer package based on the adapted version of Chernikova's algorithm (see Section 3.4.6) has been used for several years in the day-to-day routine for structural business surveys at Statistics Netherlands. Our evaluation results show that the computing speed of that program was acceptable in comparison to other algorithms. They also show, however, that this program was outperformed by the program based on the branch-and-bound algorithm. Further improvements to the adapted version of Chernikova's algorithm—for example, better selection criteria for the row to be processed and better ways to handle missing values—may reduce its computing time. However, such improvements would at the same time increase the complexity of the algorithm, thereby making it virtually impossible to maintain by software engineers at Statistics Netherlands.

Based on the evaluation study, we considered, and still consider, the branch-and-bound algorithm to be a very promising method for solving the error localization problem. The main reason for our choice is the excellent performance of Leo for records with up to six errors. For such records it determines all optimal solutions very fast. We admit that for records with more than six errors the results of Leo become less good, just like the other algorithms. The program begins to suffer from memory problems, and the computing time increases. However, as we argued before, we feel that records with many errors should not be edited in an automatic manner, but in a manual manner. Given this point of view, Leo seems to be an excellent choice. Another reason for selecting the branch-and-bound algorithm as our preferred algorithm is that it can be relatively easily extended to categorical and integer-valued data (see Chapters 4 and 5).

Besides being faster than our adapted version of Chernikova's algorithm, the branch-and-bound algorithm is considerably less complex, and hence easier to maintain, than the adapted version of Chernikova's algorithm. We therefore decided to switch to the branch-and-bound algorithm instead of the adapted version of Chernikova's algorithm for our production software called SLICE [see De Waal (2001)].

We conclude this chapter by noting that research on new algorithms for solving the error localization problem based on the Fellegi–Holt paradigm is still ongoing. Riera-Ledesma and Salazar-González (2003) report good computational results for numerical data using an approach based on Benders' decomposition [cf. Nemhauser and Wolsey (1988)]. A promising new direction is based on the application of techniques from mathematical logic, such as using solvers for the so-called satisfiability problem and resolution methods [see Bruni, Reale, and Torelli (2001), Bruni and Sassano (2001), Boskovitz, Goré, and Hegland (2003), Boskovitz (2008)]. Thus far, these techniques have only been applied to the error localization problem for categorical data. Extension of the techniques to numerical data seems an interesting research topic for the coming years.

3.A Appendix: Chernikova's Algorithm

Rubin's formulation (Rubin, 1975, 1977) of Chernikova's algorithm is as follows:

1. Construct the $(n_r + n_c) \times n_c$ matrix

$$\mathbf{Y}^0 = \begin{pmatrix} \mathbf{U}^0 \\ \mathbf{L}^0 \end{pmatrix},$$

where $\mathbf{U}^0 = \mathbf{C}$ and $\mathbf{L}^0 = \mathbf{I}_{n_c}$: the $n_c \times n_c$ identity matrix. The j_0th column of $\mathbf{Y}^0$, $\mathbf{y}^0_{*j_0}$, will also be denoted as

$$\mathbf{y}^0_{*j_0} = \begin{pmatrix} \mathbf{u}^0_{*j_0} \\ \mathbf{l}^0_{*j_0} \end{pmatrix},$$

where $\mathbf{u}^0_{*j_0}$ and $\mathbf{l}^0_{*j_0}$ are the j_0th columns of $\mathbf{U}^0$ and $\mathbf{L}^0$, respectively.

2. $t := 0$.
3. If any row of $\mathbf{U}^t$ has all components negative, $\mathbf{x} = \mathbf{0}$ is the only point satisfying (3.40) and (3.41). We set $t := n_r$ and the algorithm terminates.
4. If all the elements of $\mathbf{U}^t$ are nonnegative, the columns of $\mathbf{L}^t$ are the edges of the cone described by (3.40) and (3.41). We set $t := n_r$ and the algorithm terminates.
5. If neither 3 nor 4 holds: Choose a row of $\mathbf{U}^t$, say row r, with at least one negative entry.
6. Let $R = \{j \mid y^k_{rj} \geq 0\}$. Let v be the number of elements in R. Then the first v columns of the new matrix $\mathbf{Y}^{t+1}$ are all the columns $\mathbf{y}^t_{*j}$ of $\mathbf{Y}^t$ for $j \in R$.
7. Examine the matrix $\mathbf{Y}^t$.
   (a) If $\mathbf{Y}^t$ has only two columns and $y^t_{r1} \times y^t_{r2} < 0$, then choose $\mu_1, \mu_2 > 0$ such that $\mu_1 y^t_{r1} + \mu_2 y^t_{r2} = 0$. Adjoin the column $\mu_1 \mathbf{y}^t_{*1} + \mu_2 \mathbf{y}^t_{*2}$ to $\mathbf{Y}^{t+1}$. Go to Step 10.
   (b) If $\mathbf{Y}^t$ has more than two columns, then let $S = \{(s, u) \mid y^t_{rs} \times y^t_{ru} < 0 \text{ and } u > s\}$; that is, let S be the set of all pairs of columns of $\mathbf{Y}^t$ whose elements in row r have opposite signs. Let I_0 be the index set of all nonnegative rows of $\mathbf{Y}^t$—that is, all rows of $\mathbf{Y}^t$ with only nonnegative entries. For each $(s, u) \in S$, find all $i_0 \in I_0$ such that $y^t_{i_0 s} = y^t_{i_0 u} = 0$. Call this set $I_1(s, u)$.
8. If $I_1(s, u) = \emptyset$, then $\mathbf{y}^t_{*s}$ and $\mathbf{y}^t_{*u}$ do not contribute another column to the new matrix.
9. If $I_1(s, u) \neq \emptyset$, check to see if there is a w not equal to s or t such that $y^t_{i_0 w} = 0$ for all $i_0 \in I_1(s, u)$. If such a w exists, then $\mathbf{y}^t_{*s}$ and $\mathbf{y}^t_{*u}$ do not contribute a column to the new matrix. If no such w exists, then choose $\mu_1, \mu_2 > 0$ such that $\mu_1 y^t_{rs} + \mu_2 y^t_{ru} = 0$. Adjoin the column $\mu_1 \mathbf{y}^t_{*s} + \mu_2 \mathbf{y}^t_{*u}$ to $\mathbf{Y}^{t+1}$.

10. When all pairs in S have been examined and the additional columns (if any) have been added, we say that row r has been processed. We then define matrices $\mathbf{U}^{t+1}$ and $\mathbf{L}^{t+1}$ by

$$\mathbf{Y}^{t+1} = \begin{pmatrix} \mathbf{U}^{t+1} \\ \mathbf{L}^{t+1} \end{pmatrix},$$

where $\mathbf{U}^{t+1}$ is a matrix with n_r rows and $\mathbf{L}^{t+1}$ a matrix with n_c rows. The j_0th column of $\mathbf{Y}^{t+1}$, $\mathbf{y}_{*j_0}^{t+1}$, will also be denoted as

$$\mathbf{y}_{*j_0}^{t+1} = \begin{pmatrix} \mathbf{u}_{*j_0}^{t+1} \\ \mathbf{l}_{*j_0}^{t+1} \end{pmatrix},$$

where $\mathbf{u}_{*j_0}^{t+1}$ and $\mathbf{l}_{*j_0}^{t+1}$ are the j_0th columns of $\mathbf{U}^{t+1}$ and $\mathbf{L}^{t+1}$, respectively.

11. $t := t + 1$. If $t \leq n_r$ we go to Step 3, else the algorithm terminates.

Chernikova's algorithm can be modified in order to handle equalities more efficiently than by treating them as two inequalities. Steps 3, 5, and 6 should be replaced by the following:

3. If any row of $\mathbf{U}^t$ corresponding to an inequality or equality has all components negative or if any row of $\mathbf{U}^t$ corresponding to an equality has all components positive, $\mathbf{x} = \mathbf{0}$ is the only point satisfying (3.40) and (3.41). We set $t := n_r$ and the algorithm terminates.

5. If neither 3 nor 4 holds: Choose a row of $\mathbf{U}^t$, say row r, with at least one negative entry if the row corresponds to an inequality, and with at least one nonzero entry if the row corresponds to an equality.

6. If row r corresponds to an inequality, then apply Step 6 of the standard algorithm. If row r corresponds to an equality, then let $R = \{j_0 \mid y_{rj_0}^t = 0\}$. Let v be the number of elements in R. Then the first v columns of the new matrix $\mathbf{Y}^{t+1}$ are all the columns $\mathbf{y}_{*j_0}^t$ of $\mathbf{Y}^t$ for $j_0 \in R$.

In Step 5 of Chernikova's algorithm a failed row has to be chosen. Rubin (1975) proposes the following simple rule. Suppose a failed row has z_0 entries equal to zero, p_0 positive entries, and q_0 negative ones. We then calculate for each failed row the value $Z_{\max} = z_0 + p_0 + p_0 q_0$ if the row corresponds to an inequality and the value $Z_{\max} = z_0 + p_0 q_0$ if the row corresponds to an equality, and choose a failed row with the lowest value of $Z_{\max}$.

REFERENCES

Aelen, F., and R. Smit (2009), *Towards an Efficient Data Editing Strategy for Economic Statistics at Statistics Netherlands*. European Establishment Statistics Workshop, Stockholm.

Atkinson, A. C. (1994), Fast Very Robust Methods for the Detection of Multiple Outliers. *Journal of the American Statistical Association 89*, pp. 1329–1339.

Austin, J., and K. Lees (2000), A Search Engine Based on Neural Correlation Matrix Memories. *Neurocomputing 35*, pp. 55–72.

Banff Support Team (2008), *Functional Description of the Banff System for Edit and Imputation.* Technical Report, Statistics Canada.

Bankier, M. (1999), *Experience with the New Imputation Methodology Used in the 1996 Canadian Census with Extensions for Future Censuses.* Working Paper No. 24, UN/ECE Work Session on Statistical Data Editing, Rome.

Bankier, M., P. Poirier, M. Lachance, and P. Mason (2000), A Generic Implementation of the Nearest-Neighbour Imputation Methodology (NIM). *Proceedings of the Second International Conference on Establishment Surveys*, Buffalo, pp. 571–578.

Barcaroli, G., C. Ceccarelli, O. Luzi, A. Manzari, E. Riccini, and F. Silvestri (1995), *The Methodology of Editing and Imputation of Qualitative Variables Implemented in SCIA.* Internal Report, Istituto Nazionale di Statistica, Rome.

Barnett, V., and T. Lewis (1994), *Outliers in Statistical Data.* John Wiley & Sons, New York.

Béguin, C., and B. Hulliger (2004), Multivariate Outlier Detection in Incomplete Survey Data: The Epidemic Algorithm and Transformed Rank Correlation. *Journal of the Royal Statistical Society A 167*, pp. 275–294.

Ben-Ari, M. (2001), *Mathematical Logic for Computer Science*, second edition. Springer-Verlag, London.

Billor, N., A. S. Hadi, and P. F. Velleman (2000), BACON: Blocked Adaptive Computationally Efficient Outlier Nominators. *Computational Statistics and Data Analysis 34*, pp. 279–298.

Bishop, M. C. (1995), *Neural Networks for Pattern Recognition.* Clarendon Press, Oxford.

Boskovitz, A. (2008), *Data Editing and Logic: The Covering Set Method from the Perspective of Logic.* Doctorate thesis, Australian National University.

Boskovitz, A., R. Goré, and M. Hegland (2003), *A Logical Formalisation of the Fellegi–Holt Method of Data Cleaning.* Report, Research School of Information Sciences and Engineering, Australian National University, Canberra.

Breiman, L., J. H. Friedman, R. A. Olshen, and C. J. Stone (1984), *Classification and Regression Trees.* Wadsworth, Pacific Grove.

Bruni, R., A. Reale, and R. Torelli (2001), Optimization Techniques for Edit Validation and Data Imputation. *Proceedings of Statistics Canada Symposium 2001 "Achieving Data Quality in a Statistical Agency: a Methodological Perspective" XVIII-th International Symposium on Methodological Issues.*

Bruni, R., and A. Sassano (2001), *Logic and Optimization Techniques for an Error Free Data Collecting.* Report, University of Rome "La Sapienza".

Casado Valero, C., F. Del Castillo Cuervo-Arango, J. Mateo Ayerra, and A. De Santos Ballesteros (1996), *Quantitative Data Editing: Quadratic Programming Method.* Presented at the COMPSTAT 1996 Conference, Barcelona.

Central Statistical Office (2000), *Editing and Calibration in Survey Processing.* Report SMD-37, Ireland.

Chambers, R. (2004), Evaluation Criteria for Statistical Editing and Imputation. In: *Methods and Experimental Results from the EUREDIT Project*, (J. R. H. Charlton, ed. (http://www.cs.york.ac.uk/euredit/).

Chambers, R., A. Hentges, and X. Zhao (2004), Robust Automatic Methods for Outlier and Error Detection. *Journal of the Royal Statistical Society A 167*, pp. 323–339.

Chernikova, N. V. (1964), Algorithm for Finding a General Formula for the Non-Negative Solutions of a System of Linear Equations. *USSR Computational Mathematics and Mathematical Physics 4*, pp. 151–158.

Chernikova, N. V. (1965), Algorithm for Finding a General Formula for the Non-Negative Solutions of a System of Linear Inequalities. *USSR Computational Mathematics and Mathematical Physics 5*, pp. 228–233.

Chvátal, V. (1983), *Linear Programming*. W. H. Freeman and Company, New York.

Cormen, T. H., C. E. Leiserson and R. L. Rivest (1990), *Introduction to Algorithms*. The MIT Press/McGraw-Hill Book Company, Cambridge, MA.

De Jong, A. (2002), *Uni-Edit: Standardized Processing of Structural Business Statistics in the Netherlands*. Working Paper No. 27, UN/ECE Work Session on Statistical Data Editing, Helsinki.

De Waal, T. (1996), *CherryPi: A Computer Program for Automatic Edit and Imputation*. UN/ECE Work Session on Statistical Data Editing, Voorburg.

De Waal, T. (2001), SLICE: Generalised Software for Statistical Data Editing. In: *Proceedings in Computational Statistics*, J. G. Bethlehem and P. G. M. Van der Heijden, eds. Physica-Verlag, New York, pp. 277–282.

De Waal, T. (2003a), *Processing of Erroneous and Unsafe Data*. Ph.D. Thesis, Erasmus University, Rotterdam (see also www.cbs.nl).

De Waal, T. (2003b), Solving the Error Localization Problem by Means of Vertex Generation. *Survey Methodology 29*, pp. 71–79.

De Waal, T., and W. Coutinho (2005), Automatic Editing for Business Surveys: an Assessment for Selected Algorithms. *International Statistical Review 73*, pp. 73–102.

De Waal, T., and R. Quere (2003), A Fast and Simple Algorithm for Automatic Editing of Mixed Data. *Journal of Official Statistics 19*, pp. 383–402.

Dines, L. L. (1927), On Positive Solutions of a System of Linear Equations. *Annals of Mathematics 28*, pp. 386–392.

Di Zio, M., U. Guarnera, and O. Luzi (2005), *Improving the Effectiveness of a Probabilistic Editing Strategy for Business Data*. ISTAT, Rome.

Duffin, R. J. (1974), On Fourier's Analysis of Linear Inequality Systems. *Mathematical Programming Studies 1*, pp. 71–95.

Fellegi, I. P., and D. Holt (1976), A Systematic Approach to Automatic Edit and Imputation. *Journal of the American Statistical Association 71*, pp. 17–35.

Fillion, J. M., and I. Schiopu-Kratina (1993), *On the Use of Chernikova's Algorithm for Error Localization*. Report, Statistics Canada.

Fine, T. L. (1999), *Feedforward Neural Network Methodology*. Springer-Verlag, New York.

Fourier, J. B. J. (1826), Solution d'une Question Particulière du Calcul des Inégalités. In: *Oeuvres II*, Paris.

Freund, R. J., and H. O. Hartley (1967), A Procedure for Automatic Data Editing. *Journal of the American Statistical Association 62*, pp. 341–352.

Garfinkel, R. S., A. S. Kunnathur, and G. E. Liepins (1986), Optimal Imputation of Erroneous Data: Categorical Data, General Edits. *Operations Research 34*, pp. 744–751.

Garfinkel, R. S., A. S. Kunnathur, and G. E. Liepins (1988), Error Localization for Erroneous Data: Continuous Data, Linear Constraints. *SIAM Journal on Scientific and Statistical Computing 9*, pp. 922–931.

Ghosh-Dastidar, B., and J. L. Schafer (2003), Multiple Edit/Multiple Imputation for Multivariate Continuous Data. *Journal of the American Statistical Association 98*, pp. 807–817.

Hadi, A. S., and J. F. Simonoff (1993), Procedures for the Identification of Multiple Outliers in Linear Models. *Journal of the Royal Statistical Society B 56*, pp. 393–396.

Hoogland, J., and E. Van der Pijll (2003), *Evaluation of Automatic versus Manual Editing of Production Statistics 2000 Trade and Transport*. Working Paper No. 4, UN/ECE Work Session on Statistical Data Editing, Madrid.

Kalton, G., and D. Kasprzyk (1986), The Treatment of Missing Survey Data. *Survey Methodology 12*, pp. 1–16.

Kohler, D. A. (1973), Translation of a Report by Fourier on His Work on Linear Inequalities. *Opsearch 10*, pp. 38–42.

Koikkalainen, P., and E. Oja (1990), Self-Organizing Hierarchical Feature Maps. In: *Proceedings of the International Joint Conference on Neural Networks II*, pp. 279–285, IEEE Press, Piscataway, NJ.

Kosinski, A. S. (1999), A Procedure for the Detection of Multivariate Outliers. *Computational Statistics & Data Analysis 29*, pp. 145–161.

Kovar, J., and P. Whitridge (1990), Generalized Edit and Imputation System; Overview and Applications. *Revista Brasileira de Estadistica 51*, pp. 85–100.

Kovar, J., and P. Whitridge (1995), Imputation of Business Survey Data. In: *Business Survey Methods*, B.G. Cox, D.A. Binder, B.N. Chinnappa, A. Christianson, M.J. Colledge, and P.S. Kott, eds. John Wiley & Sons, New York, pp. 403–423.

Kovar, J. and W. E. Winkler (1996), *Editing Economic Data.* UN/ECE Work Session on Statistical Data Editing, Voorburg.

Larsen, B. S., and B. Madsen (1999), *Error Identification and Imputations with Neural Networks.* Working Paper No. 26, UN/ECE Work Session on Statistical Data Editing, Rome.

Liepins, G. E., R. S. Garfinkel, and A. S. Kunnathur (1982), Error Localization for Erroneous Data: A Survey. *TIMS/Studies in the Management Sciences 19*, pp. 205–219.

Little, R. J. A., and D. B. Rubin (2002), *Statistical Analysis with Missing Data.* John Wiley & Sons, New York.

Little, R. J. A., and P. J. Smith (1987), Editing and Imputation of Quantitative Survey Data. *Journal of the American Statistical Association 82*, pp. 58–68.

Manzari, A. (2004), Combining Editing and Imputation Methods: An Experimental Application on Population Census Data. *Journal of the Royal Statistical Society A 167*, pp. 295–307.

Marriott, K., and P. J. Stuckey (1998), *Programming with Constraints—An Introduction.* MIT Press, Cambridge, MA.

McKeown, P. G. (1984), A Mathematical Programming Approach to Editing of Continuous Survey Data. *SIAM Journal on Scientific and Statistical Computing 5*, pp. 784–797.

Motzkin, T. S. (1936), *Contributions to the Theory of Linear Inequalities* (in German). Dissertation, University of Basel.

Nemhauser, G. L., and L. A. Wolsey (1988), *Integer and Combinatorial Optimization.* John Wiley & Sons, New York.

Nordbotten, S. (1963), Automatic Editing of Individual Statistical Observations. In: *Conference of European Statisticians Statistical Standards and Studies No. 2,* United Nations, New York.

Ragsdale, C. T., and P. G. McKeown (1996), On Solving the Continuous Data Editing Problem. *Computers & Operations Research 23,* pp. 263–273.

Riani, M., and A. C. Atkinson (2000), Robust Diagnostic Data Analysis: Transformations in Regression. *Technometrics 42,* pp. 384–398.

Riera-Ledesma, J., and J. J. Salazar-González (2003), *New Algorithms for the Editing and Imputation Problem.* Working Paper No. 5, UN/ECE Work Session on Statistical Data Editing, Madrid.

Robinson, J. A. (1965), A Machine-Oriented Logic Based on the Resolution Principle. *Journal of the Association of Computing Machinery 12,* pp. 23–41.

Robinson, J. A. (1968), The Generalized Resolution Principle. In: *Machine Intelligence 3,* E. Dale and D. Michie, eds. Oliver and Boyd, Edinburgh, pp. 7 7–93.

Rocke, D. M., and D. L. Woodruff (1993), Computation of Robust Estimates of Multivariate Location and Shape. *Statistica Neerlandica 47,* pp. 27–42.

Rocke, D. M., and D. L. Woodruff (1996), Identification of Outliers in Multivariate Data. *Journal of the American Statistical Association 91,* pp. 1047–1061.

Rousseeuw P. J., and M. L. Leroy (1987), *Robust Regression & Outlier Detection.* John Wiley & Sons, New York.

Rubin, D. S. (1975), Vertex Generation and Cardinality Constrained Linear Programs. *Operations Research 23,* pp. 555–565.

Rubin, D. S. (1977), Vertex Generation Methods for Problems with Logical Constraints. *Annals of Discrete Mathematics 1,* pp. 457–466.

Russell, S., and P. Norvig (1995), *Artificial Intelligence, a Modern Approach.* Prentice-Hall, Englewood Cliffs, NJ.

Sande, G. (1978), *An Algorithm for the Fields to Impute Problems of Numerical and Coded Data.* Technical report, Statistics Canada.

Schafer, J. L. (1997), *Analysis of Incomplete Multivariate Data.* Chapman & Hall, London.

Schaffer, J. (1987), Procedure for Solving the Data-Editing Problem with Both Continuous and Discrete Data Types. *Naval Research Logistics 34,* pp. 879–890.

Schiopu-Kratina, I., and J. G. Kovar (1989), *Use of Chernikova's Algorithm in the Generalized Edit and Imputation System.* Methodology Branch Working Paper BSMD 89-001E, Statistics Canada.

Statistics Canada (1998), *GEIS: Functional Description of the Generalized Edit and Imputation System.* Report, Statistics Canada.

Todaro, T. A. (1999), *Overview and Evaluation of the AGGIES Automated Edit and Imputation System.* Working Paper No. 19, UN/ECE Work Session on Statistical Data Editing, Rome.

Van de Pol, F., F. Bakker, and T. De Waal (1997), *On Principles for Automatic Editing of Numerical Data with Equality Checks.* Report 7141-97-RSM, Statistics Netherlands, Voorburg.

Warners, J. P. (1999), *Non-Linear Approaches to Satisfiability Problems.* Ph.D. Thesis, Eindhoven University of Technology.

Williams, H. P., and S. C. Brailsford (1996), Computational Logic and Integer Programming. In: *Advances in Linear and Integer Programming*, J. E. Beasley, ed. Clarendon Press, Oxford, pp. 249–281.

Winkler, W. E. (1998), *Set-Covering and Editing Discrete Data.* Statistical Research Division Report 98/01, U.S. Bureau of the Census, Washington, D.C.

Winkler, W. E. (1999), *State of Statistical Data Editing and Current Research Problems.* Working Paper No. 29, UN/ECE Work Session on Statistical Data Editing, Rome.

Winkler, W. E., and L. A. Draper (1997), The SPEER Edit System. In: *Statistical Data Editing*, Volume 2: *Methods and Techniques*, United Nations, Geneva.

Winkler, W. E., and T. F. Petkunas (1997), The DISCRETE Edit System. In: *Statistical Data Editing*, Volume 2: *Methods and Techniques*. United Nations, Geneva.

Woodruff, D. L., and D. M. Rocke (1994), Computable Robust Estimation of Multivariate Location and Shape in High Dimension Using Compound Estimators. *Journal of the American Statistical Association 89*, pp. 888–896.

CHAPTER FOUR

Automatic Editing: Extensions to Categorical Data

4.1 Introduction

This chapter focuses on solving the error localization problem for categorical data, and, especially, for a mix of categorical and continuous data. Data of these types are commonly encountered in social surveys. Section 4.2 provides a mathematical formulation for the error localization problem based on the Fellegi–Holt paradigm (see also Chapter 3) for mixed data. The edits we consider are illustrated by several examples in Section 4.2.1. Sections 4.3 to 4.5 describe three algorithms for solving the error localization problem for categorical and mixed data. The algorithms we consider are the Fellegi–Holt method (see Section 4.3 and also Section 3.4.4), an extension of the branch-and-bound approach of Section 3.4.7 to mixed data (see Section 4.4), and the Nearest-neighbor Imputation Methodology (NIM) (see Section 4.5).

In contrast to the algorithms of Sections 4.3 and 4.4, the NIM describes a method that is not based on the Fellegi–Holt paradigm. That is, NIM is not based on first identifying the smallest possible number of values in an inconsistent record as being erroneous and then finding suitable imputations for these values. Instead, NIM is based on first checking whether sets of possible imputations, which are taken from other records, result in a consistent record. Of all sets of possible imputations, the best set according to a certain objective function is selected and used to correct the record.

Handbook of Statistical Data Editing and Imputation, by T. de Waal, J. Pannekoek, and S. Scholtus

Other approaches described in Chapter 3 for continuous data can also be extended to a mix for continuous and categorical data, such as the vertex generation approach described in Section 3.4.6 [see De Waal (2001a, 2003) for this extension] and the cutting plane approach decribed in Section 3.4.8 [see De Waal (2003)]. These approaches are not described in the present book because they are quite technical and we feel that better and easier to understand approaches, in particular the branch-and-bound approach described in Section 4.4, are available.

4.2 The Error Localization Problem for Mixed Data

We denote the categorical variables by v_j ($j = 1, \dots, m$), the numerical variables by $x_j (j = 1, \dots, p)$, and the number of edits by K. For categorical data we denote the domain, i.e. the set of possible values, of variable v_j by D_j. The edits that we consider in this chapter are of the following type:

$$\begin{aligned} &\text{IF} \quad v_j \in F_j^k \qquad \text{for all } j = 1, \dots, m, \\ &\text{THEN} \quad (x_1, \dots, x_p) \in \{\mathbf{x} | a_{k1}x_1 + \cdots + a_{kp}x_p + b_k \geq 0\} \end{aligned} \tag{4.1}$$

or

$$\begin{aligned} &\text{IF} \quad v_j \in F_j^k \qquad \text{for all } j = 1, \dots, m, \\ &\text{THEN} \quad (x_1, \dots, x_p) \in \{\mathbf{x} | a_{k1}x_1 + \cdots + a_{kp}x_p + b_k = 0\}, \end{aligned} \tag{4.2}$$

where $F_j^k \subset D_j (k = 1, \dots, K)$. Numerical variables may attain negative values. For nonnegative numerical variables, an edit of type (4.1) needs to be introduced in order to ensure nonnegativity. All edits E^k ($k = 1, \dots, K$) given by (4.1) and (4.2) have to be satisfied simultaneously. We assume that the edits can indeed be satisfied simultaneously.

A variable such as *Age* can either be considered to be a categorical variable or a continuous one, depending on the kind of edits involving *Age*. However, a variable cannot be considered to be a categorical variable in one edit and a continuous variable in another.

A record that satisfies all edits is called a *consistent* record. The condition after the IF statement—that is, "$v_j \in F_j^k$ for all $j = 1, \dots, m$"—is called the IF condition of edit E^k ($k = 1, \dots, K$). The condition after the THEN-statement is called the THEN condition. If the IF condition does not hold true, the edit is always satisfied, irrespective of the values of the numerical variables.

A categorical variable v_j is said to *enter* an edit E^k given by (4.1) or (4.2) if $F_j^k \subset D_j$ and $F_j^k \neq D_j$—that is, if F_j^k is strictly contained in the domain of variable v_j. That edit is then said to *involve* this categorical variable. A continuous variable x_j is said to *enter* the THEN condition of edit E^k given by (4.1) or (4.2) if $a_{kj} \neq 0$. That THEN condition is then said to *involve* this continuous variable. If the set in the THEN condition of (4.1) or (4.2) is the entire p-dimensional real vector space, then the edit is always satisfied. Such an edit may be discarded. If the set in the THEN condition of (4.1) or (4.2) is empty, then the edit is failed by any record for which the IF condition holds true. Such an edit for which the THEN condition of (4.1) or (4.2) is empty is basically an edit involving only categorical variables. If F_j^k in (4.1) or (4.2) is the empty set, the edit is always satisfied and may be discarded.

The Fellegi–Holt paradigm [see Fellegi and Holt (1976) and Chapter 3] says that for each record $(v_1^0, \ldots, v_m^0, x_1^0, \ldots, x_p^0)$ in the data set to be edited automatically we have to determine a synthetic record $(\check{v}_1, \ldots, \check{v}_m, \check{x}_1, \ldots, \check{x}_p)$ such that (4.1) and (4.2) becomes satisfied for all edits E^k $(k = 1, \ldots, K)$ and such that

$$\sum_{j=1}^{m} w_j^c \delta(v_j^0, \check{v}_j) + \sum_{j=1}^{p} w_j^r \delta(x_j^0, \check{x}_j) \tag{4.3}$$

is minimized. Here w_j^c is the nonnegative reliability weight of categorical variable v_j $(j = 1, \ldots, m)$, w_j^r the nonnegative reliability weight of numerical variable x_j $(j = 1, \ldots, p)$, $\delta(y^0, y) = 1$ if y^0 is missing or $y^0 \neq y$, and $\delta(y^0, y) = 0$ if $y^0 = y$. The variables of which the values in the synthetic record differ from the original values plus the variables for which the original values were missing together form an optimal solution to the error localization problem.

Note that there may be several optimal solutions to a specific instance of the error localization problem. As in Chapter 3, our aim is to find all these optimal solutions, for the same reasons as stated in Chapter 3.

In many practical cases, certain kinds of missing values are acceptable—for example, when the corresponding questions are not applicable to a particular respondent. We assume that for categorical variables such acceptable missing values are coded by special values—for example, the value "NA" (abbreviation for "nonapplicable")—in their domains. Nonacceptable missing values of categorical variables are not coded. The above optimization problem will identify these missing values as being erroneous. We also assume that numerical THEN conditions can only be triggered if the value of none of the numerical variables involved may be missing. Hence, if—for a certain record—a THEN condition involving a numerical variable of which the value is missing is triggered by the categorical values, then either this value is incorrectly missing or at least one of the categorical values is erroneous.

4.2.1 EXAMPLES OF EDITS

Below we illustrate what kind of edits can be expressed in the form (4.1) or (4.2) by means of a number of examples.

1. Simple numerical edit.

$$\textit{Turnover} - \textit{Profit} \geq 0.$$

This is an example of a purely numerical edit. For every combination of categorical values the edit should be satisfied. The edit can be formulated in our standard form as

$$\begin{aligned} &\text{IF} \quad v_j \in D_j \qquad \text{for all } j = 1, \ldots, m, \\ &\text{THEN} \quad (\textit{Profit}, \textit{Turnover}) \in \{(\textit{Profit}, \textit{Turnover}) | \textit{Turnover} - \textit{Profit} \geq 0\}. \end{aligned}$$

In the remaining examples we will be slightly less formal with our notation. In particular, we will omit the terms "$v_j \in D_j$" from the edits.

2. Simple categorical edit.

$$\text{IF } (\textit{Gender} = \text{"Male"}), \text{ THEN } (\textit{Pregnant} = \text{"No"}).$$

This is an example of a purely categorical edit. The edit can be formulated in our standard form as

$$\text{IF } (\textit{Gender} = \text{"Male"}) \text{ AND } (\textit{Pregnant} = \text{"Yes"}), \text{ THEN } \emptyset.$$

3. Simple mixed edit.

$$\begin{aligned} &\text{IF} \quad (\textit{Activity} \in \{\text{"Chemical Industry"}, \text{"Car Industry"}\}), \\ &\text{THEN} \quad (\textit{Turnover} \geq 1{,}000{,}000 \text{ euros}). \end{aligned}$$

This is a typical example of a mixed edit. Given certain values for the categorical variables, a certain numerical constraint has to be satisfied. The edit expresses that an enterprise in the chemical industry or in the car industry should have a turnover of at least one million euros.

4. Complex mixed edit.

$$\begin{aligned} &\text{IF} \quad (\textit{Occupation} = \text{"Statistician"}) \text{ OR } (\textit{Education} = \text{"University"}), \\ &\text{THEN} \quad (\textit{Income} \geq 1000 \text{ euros}). \end{aligned}$$

This edit can be split into two edits given by

$$\begin{aligned} &\text{IF} \quad (\textit{Occupation} = \text{"Statistician"}), \\ &\text{THEN} \quad (\textit{Income} \geq 1000 \text{ euros}) \end{aligned}$$

and

$$\begin{aligned} &\text{IF} \quad (\textit{Education} = \text{"University"}), \\ &\text{THEN} \quad (\textit{Income} \geq 1000 \text{ euros}). \end{aligned}$$

5. Very complicated numerical edit.

$$\text{(4.4)} \qquad \begin{array}{ll} \text{IF} & (\textit{Tax on Wages} > 0), \\ \text{THEN} & (\textit{Number of Employees} \geq 1). \end{array}$$

Edit (4.4) is not in standard form (4.1), because the IF condition involves a numerical variable. To handle this edit, one can carry out a preprocessing step to introduce an auxiliary categorical variable *TaxCond* with domain {"*False*", "*True*"}. Initially, *TaxCond* is given the value "*True*" if *Tax on Wages* > 0 in the unedited record, and the value "*False*" otherwise. The reliability weight for *TaxCond* is set to zero. We can now replace (4.4) by the following three edits of type (4.1):

$$\begin{array}{ll} \text{IF} & (\textit{TaxCond} = \text{"False"}), \\ \text{THEN} & (\textit{Tax on Wages} \leq 0), \\ \text{IF} & (\textit{TaxCond} = \text{"True"}), \\ \text{THEN} & (\textit{Tax on Wages} \geq \varepsilon), \\ \text{IF} & (\textit{TaxCond} = \text{"True"}), \\ \text{THEN} & (\textit{Number of Employees} \geq 1), \end{array}$$

where ε is a sufficiently small positive number.

4.3 The Fellegi–Holt Approach

4.3.1 INTRODUCTION

In Section 3.4.4 we have already sketched the Fellegi and Holt method for continuous data. Fellegi and Holt [see Fellegi and Holt (1976)] also proposed a similar method for categorical data. The Fellegi–Holt method can in fact also be used for a mix of categorical and continuous data. In this section we first examine the Fellegi–Holt method for categorical data, and later we examine the extension to a mix of categorical and continous data.

This section on the Fellegi–Holt method is organized in the following way. In the first three subsections we restrict ourselves to categorical data. In the current subsection we illustrate the basic idea of the Fellegi–Holt approach without any mathematical rigor. Mathematical details are sketched in Section 4.3.2. Proofs are not provided. The interested reader is referred to the original paper by Fellegi and Holt (1976) for these proofs. Improvements on the method proposed by Fellegi and Holt are examined in Section 4.3.3. Numerical and mixed data are discussed in Section 4.3.4.

The method developed by Fellegi and Holt is based on generating so-called *implicit*, or *implied*, edits. Such implicit edits are logically implied by the explicitly specified edits. Implicit edits can be defined for numerical as well as categorical data.

Although implicit edits are redundant, they can reveal important information about the feasible region defined by the explicitly defined edits. This information is, of course, already contained in the explicitly defined edits, but there that information may be rather hidden. Implicit edits sometimes allow one to see relations between variables more clearly. We illustrate this point by means of a simple example, which is taken from Daalmans (2000). We will refer to this example more often in this section.

EXAMPLE 4.1

In a small survey respondents are asked to choose one of the possible alternatives for the following three questions:

1. What is the most important reason for you to buy sugar?
2. Do you drink coffee with sugar?
3. What is the average amount of sugar you consume in one cup of coffee?

The alternatives for the first question are:

- I consume sugar in my coffee.
- I use sugar to bake cherry pie.
- I never buy sugar.
- Other reason.

The alternatives for the second question are:

- Yes.
- No.

The alternatives for the last question are:

- 0 grams.
- More than 0 grams but less than 10.
- More than 10 grams.

The following (explicit) edits have been defined:

1. For someone who does not drink coffee with sugar, the main reason to buy sugar is not to consume it with coffee.
2. The average amount of sugar consumed in one cup of coffee by someone who drinks coffee with sugar is not equal to 0 grams.

3. Someone who never buys sugar does not consume more than 0 grams of sugar in his coffee on average.

   An example of an implicit edit is:

4. Someone who never buys sugar does not consume sugar in his coffee.

This edit is implied by the second and third explicit edit, since the second explicit edit implies that somebody who takes sugar in his/her coffee must consume more than 0 grams of sugar per cup of coffee on average and the third explicit edit says that somebody who consumes more than 0 grams of sugar per cup of coffee on average sometimes buys sugar.

Edit 4 is by definition a redundant edit, because this information is already present in the second and third explicit edit. However, this edit makes the relation between consuming sugar in coffee and buying sugar more clear. This relation is less clear if one only looks at the second and third explicit edit separately. The benefits of generating implicit edits become more apparent later when we continue this example.

A trivial example of another implicit edit is the following:

- Edit 2 or edit 3 has to hold true.

This edit is satisfied if either edit 2 or edit 3 (or both) is satisfied. It is an implicit edit because it follows logically from edit 2 and edit 3, but it is rather useless because it does not make any relation between the variables clearer. ■

Note that for categorical data many "useless" implicit edits can be derived like the second implicit edit in Example 4.1 above. For numerical data the set of edits that are logically implied by the explicitly specified edits contains infinitely many elements. A simple example is given below.

EXAMPLE 4.2

If $x \geq 1$ is an explicit edit, then $\lambda x \geq \lambda$ is an implied edit for all $\lambda \geq 0$. ■

Generating all implicit edits is out of the question for numerical data and is a waste of time and memory for categorical data.

The method proposed by Fellegi and Holt starts by generating a well-defined sufficiently large set of implicit and explicit edits. This set of edits is referred to as the *complete set of edits*. It is referred to as the *complete* set of edits not because all possible implicit edits are generated, but because this is the set of (implicit and explicit) edits that is sufficient and necessary to translate the error localization problem into a so-called set-covering problem [see Nemhauser and Wolsey (1988) for more on the set-covering problem]. The mathematical details on how to generate the complete set of edits are provided in Section 4.3.2.

In particular, once a complete set of edits has been generated, it suffices to find a set of variables S that covers the violated (explicit and implicit) edits; that is, in each violated edit at least one variable of S should be involved.

For categorical data a formal definition of implicit edits and the complete set of edits will be given later. Here we restrict ourselves to giving the complete set of edits for Example 4.1.

EXAMPLE 4.1 (*continued*)

Suppose the explicit edits are given again by the explicit edits of Example 4.1. The complete set of edits is then given by edits 1 to 4, and

5. The average amount of sugar consumed per cup of coffee by someone whose main reason to buy sugar is to consume it with coffee is not equal to 0 grams. ■

Why the complete set of edits is important is precisely explained in mathematical terms in the next subsection. Here we illustrate the idea by means of an example.

EXAMPLE 4.1 (*continued*)

Suppose the explicit edits are once again given by the explicit edits of Example 4.1. Suppose also that the answers recorded for one of the respondents are:

1. What is the most important reason for buying sugar: I never buy sugar;
2. Do you drink coffee with sugar: Yes;
3. What is the average amount of sugar per cup of coffee: 0 grams.

Note that this record does not satisfy the second explicit edit. Obviously either the answer to the second question or the answer to the third question has to be changed. Note that changing the answer to the first question alone cannot result in a consistent record.

A simple approach to see which value can be changed is trial and error. One possibility is to change the third answer to "more than 0 but less than 10 grams." As a consequence, the second explicit edit will become satisfied, but unfortunately the third explicit edit will become failed. So, changing the third answer to "more than 0 but less than 10 grams" is not such a good idea.

Let us try to change the third answer to "more than 10 grams." This is not a good idea either, because the second explicit edit will become satisfied through this change, but again the third explicit edit will become failed.

Now, suppose the answer to the second question is changed to "No," while the third answer is fixed to its original value. Now all edits have become satisfied and a solution to the error localization problem has been found. In this small example a solution is found after a few steps using the trial and error approach. However, for large problems the trial and error approach is not so efficient. In the worst case all $\prod_{j=1}^{m} |D_j|$ possible records have to be checked in order to find all optimal solutions. Here the implicit edits show their importance as we illustrate in the paragraph below.

Consider (implicit) edit 4 of Example 4.1, that is, the edit that says: "Someone who never buys sugar does not consume sugar in his coffee." Note that to determine whether this edit is satisfied or not we only have to consider the answers to the first two questions. That is, whether the edit is satisfied or not does not depend on the answer to the third question. Note also that this edit is failed by the record under consideration. Changing the answer to the third question cannot make this edit satisfied. So, we do not have to consider changing only the value of the third question.

This edit is implied by the second and third explicit edits, so it is obviously redundant. However, as we see it does contain useful information that helps us to identify the most implausible values. ■

4.3.2 FELLEGI–HOLT APPROACH FOR CATEGORICAL DATA: MATHEMATICAL DETAILS

For purely categorical data, the edits that we consider in this book are given by

$$\begin{aligned}&\text{IF} \quad v_j \in F_j^k \qquad \text{for all } j = 1, \ldots, m,\\ &\text{THEN } \emptyset\end{aligned} \tag{4.5}$$

An edit given by (4.5) is violated if $v_j \in F_j^k$ for all $j = 1, \ldots, m$. Otherwise, the edit is satisfied. Alternatively, we will write a categorical edit, E^k, given by (4.5) as

$$P(E^k) = \prod_{j=1}^{m} F_j^k, \tag{4.6}$$

where $\prod$ denotes the Cartesian product. That is, edit E^k is failed if and only if the values v_j ($j = 1, \ldots, m$) of the record under consideration lie in the space given by the right-hand side of (4.6). Fellegi and Holt (1976) refer to (4.6) as the *normal form* of (categorical) edits. Fellegi and Holt show that any system of categorical edits can be expressed in normal form.

A set of edits is satisfied if all edits given by (4.6) are satisfied; that is, a set of edits is failed if at least one edit is failed. If we denote the set of edits E^k $(k = 1, \ldots, K)$ by $\bar{E}$, then a record $\mathbf{v}$ fails this set of edits if and only if

$$\mathbf{v} \in P(\bar{E}),$$

where

$$P(\bar{E}) = \bigcup_{k=1}^{K} P(E^k).$$

A simple example, illustrating the introduced concepts, is given below.

EXAMPLE 4.3

Suppose there are three variables: *Age*, *Marital Status*, and *Sex*. The variable *Age* assumes three values: 1, 2, and 3 (i.e., $D_1 = \{1, 2, 3\}$), representing respectively: "*Age* = 0–14," "*Age* = 15–80," and "*Age* > 80." The variable *Marital Status* only has two possible values: 1 and 2 (i.e., $D_2 = \{1, 2\}$), representing, respectively, "Married" and "Not married." The variable *Sex* assumes two possible values: 1 and 2 (i.e., $D_3 = \{1, 2\}$), representing respectively "Male" and "Female".

The statement

IF (*Age* < 15), THEN (*Marital Status* = "Not Married")

is identical to the statement that a failure occurs if both (*Age* = 0–14) and (*Marital Status* = "Married") hold true, irrespective of the value of variable *Sex*. In more mathematical notation: $F_1^1 = \{1\}$, $F_2^1 = \{1\}$ and $F_3^1 = D_3 = \{1, 2\}$, and in normal form the edit is given by $P(E^1) = \{1\} \times \{1\} \times \{1, 2\}$.

Age and *Marital Status* enter this edit because F_1^1 is a proper subset of D_1 and F_2^1 is a proper subset of D_2, but the edit does not involve *Sex* since F_3^1 is not a proper subset of D_3. ■

Below we continue again with Example 4.1.

EXAMPLE 4.1 (*continued*)

In normal form the explicit edits of Example 4.1 are given by

$$P(E^1) = \{1\} \times \{2\} \times \{1,2,3\},$$
$$P(E^2) = \{1,2,3,4\} \times \{1\} \times \{1\},$$
$$P(E^3) = \{3\} \times \{1,2\} \times \{2,3\}.$$

■

In the algorithm of Fellegi and Holt a subset of all possible explicit and implicit edits, the so-called *complete set of edits*, has to be generated. The following lemma of Fellegi and Holt is the basis of generating implicit edits.

LEMMA 4.1

Let $\bar{E}_g$ be an arbitrary set of edits, explicit or already generated implicit ones, that involve a field g ($g \in \{1, \ldots, m\}$). Let E^* be the edit defined by

$$F_j^* = \bigcap_{k, E^k \in \bar{E}_g} F_j^k, \qquad j \in \{1, \ldots, m\}, j \neq g$$

$$F_g^* = \bigcup_{k, E^k \in \bar{E}_g} F_j^k$$

If $F_j^* \neq \emptyset$ for every $j \in \{1, \ldots, m\}$, then E^* is an implicit edit in normal form.

Fellegi and Holt (1976) prove that all implicit edits required for their complete set of edits can be generated via repeated application of this lemma. Note that this lemma implies that in general implicit edits can be deduced from one, two, three, four, or even more explicit or implicit edits.

The subset $\bar{E}_g$ is called the *contributing set*, it contains the *contributing edits*, or the edits that *imply* E^*. Field g is called the *generating field* of edit E^*. In the example below we again return to Example 4.1.

EXAMPLE 4.1 (*continued*)

In Example 4.1 the (implicit) edit 4, denoted as E^4, with $P(E^4) = \{3\} \times \{1\} \times \{1,2,3\}$, can be obtained using Lemma 4.1 with the third field as the generating field and contributing set $\bar{E}_g = \{E^2, E^3\}$, i.e. the second and third edits.

Besides E^4 there are two other implicit edits: E^5 with $P(E^5) = \{1\} \times \{1,2\} \times \{1\}$ and E^6, with $P(E^6) = \{1,3\} \times \{2\} \times \{2,3\}$. Edit E^5 can be obtained by using Lemma 4.1 with contributing set $\{E^1, E^2\}$ and generating field 2. Edit E^6 can be obtained by using Lemma 4.1 with generating field 1 and contributing set $\{E^1, E^3\}$. In fact E^6 and E^4 could also be combined, using variable 2 as generating field but the resulting implied edit would be identical to E^3. ■

Fellegi and Holt show that it is not necessary to generate all implicit edits by means of Lemma 4.1 in order to solve the error localization problem. To this end they introduce the concept of an essentially new implicit edit. An *essentially new* implicit edit E^* is an implicit edit that does not involve its generating field v_g, i.e. $F_g^* = D_g$. In other words, an implicit edit E^* with generating field v_g is essentially new if

$$\bigcup_{k \in T} F_g^k = D_g \tag{4.7}$$

and

$$\bigcap_{k \in T} F_j^k \neq \emptyset \qquad \text{for all } k = 1, \ldots, g-1, g+1, \ldots, m, \tag{4.8}$$

where T is a certain set of edits. The set of explicit edits together with the set of all essentially new implicit edits is called the complete set of edits.

EXAMPLE 4.1 (*continued*)

It is easy to see that E^4 and E^5 are both essentially new implicit edits, but E^6 is not. The complete set of edits $\bar{E}_C$ is given by $\{E^1, E^2, E^3, E^4, E^5\}$. ■

An essentially new implicit edit E^* can be interpreted as the projection of the edits $\bar{E}_g$ on their entering fields except generating field g. That is, it defines a relation that has to hold for these entering fields except field g. By generating essentially new implicit edits, hidden relations between various variables can be clarified. Once the hidden relations have been made explicit, solving the error localization problem is relatively straightforward.

After generation of the complete set of edits, all failed edits are selected for each record. According to Theorem 4.1—that is, Corollary 2 to Theorem 1 of Fellegi and Holt (1976), mentioned below—sets of variables S have to be found that cover the set of failed edits (i.e., at least one variable contained in S is involved in each failed edit).

THEOREM 4.1

If S is any set of variables that covers the complete set of failed edits, then a set of values exists for the variables in S that together with the set of original values for all other variables will result in a consistent record. That is, the variables in S can be imputed consistently.

Theorem 4.1 says that some set of variables S is a (possibly suboptimal) solution to the error localization problem if S covers the set of failed edits in the complete set of edits. Note that in order to obtain a consistent record, at least one of the entering fields of each failed edit has to be imputed. Therefore it also holds true that a variable set S cannot be imputed consistently, if S does not cover the set of failed edits in the complete set of edits. Consequently, all solutions to the error localization problem can be found by finding all covering sets of variables of the failed edits in the complete set of edits. According to the generalized paradigm of Fellegi and Holt, all sets with a minimal sum of reliability weights among all sets of variables that cover the violated edits are the optimal solutions to the error localization problem.

EXAMPLE 4.1 *(continued)*

Suppose that the reliability weights are 1 for the first field, 2 for the second field, and 1 for the third field. Now, again consider the record $\mathbf{v}$ given by (I never buy sugar; Yes; 0 grams), where the first answer corresponds to the question "What is the most important reason for buying sugar?", the second answer corresponds to "Do you drink coffee with sugar?", and the third answer corresponds to "What is the average amount of sugar per cup of coffee?". The edits E^2 and E^4 are violated by this record. Note that the second field enters both failed edits. In other words, field 2 covers the set of failed edits $\{E^2, E^4\}$. This implies that the variable sets $\{v_1, v_2\}$, $\{v_2, v_3\}$ and $\{v_1, v_2, v_3\}$ also cover the complete set of failed edits. However, according to the generalized paradigm of Fellegi and Holt these latter sets of variables should not be imputed.

There is one other variable set that also covers $\{E^2, E^4\}$, namely $\{v_1, v_3\}$. The sum of the reliability weights of the two variables in this set equals two, exactly the same as the reliability weight of the second field. This means that $\{v_2\}$ and $\{v_1, v_3\}$ are the optimal solutions to the error localization problem. ■

Note that Theorem 4.1 implies that if the variables in a subset S cannot be imputed consistently, then there is a failed (explicit or implicit) edit in which none of the variables of S are involved.

To prove Theorem 4.1, Fellegi and Holt apply a property that we will refer to as the *lifting property*. Define Ω_J as that subset of the set of complete edits that involves only fields $1, \ldots, J$. The lifting property—that is, Theorem 1 of Fellegi and Holt (1976)—then states the following.

THEOREM 4.2

If v_j^0 $(j = 1, \ldots, J-1)$ are values for the first $J-1$ variables that satisfy all edits in Ω_{J-1}, then there exists some value v_J^0 such that the values v_j^0 $(j = 1, \ldots, J)$ satisfy all edits in Ω_J.

Theorem 4.2 states that if the values for $J-1$ variables satisfy all corresponding (explicit and essentially new implicit) edits in the complete set of edits, then there exists a value for the Jth variable such that all edits corresponding to the first J variables become satisfied. In other words, the possibility to satisfy the edits involving only $J-1$ variables is *lifted* to J variables. The lifting property is a very important property.

Note that by repeated application of Theorem 4.2 we can show the following corollary, where m as usual denotes the number of categorical variables.

COROLLARY TO THEOREM 4.2

If v_j^0 $(j = 1, \ldots, J-1)$ are values for the first $J-1$ variables that satisfy all edits in Ω_{J-1}, then there exist values v_j^0 $(j = J, \ldots, m)$ such that the values v_j^0 $(j = 1, \ldots, m)$ satisfy all edits in the complete set of edits.

The correctness of Theorem 4.1 follows immediately from the Corollary to Theorem 4.2.

Using implicit edits to solve the error localization problem was an important methodological breakthrough. The concept of implicit edits was, however, not a completely new idea. For instance, the concept of implicit edits is similar to that of surrogate constraints in linear or integer programming. Moreover, the Fellegi–Holt method to generate implicit edits for categorical data can be considered as a special case of the resolution technique that is used in machine learning and mathematical logic [see, for example, Robinson (1965, 1968), Russell and Norvig (1995), Williams and Brailsford (1996), Marriott and Stuckey (1998), Warners (1999), Chandru and Hooker (1999), Hooker (2000), and Ben-Ari (2001)].

4.3.3 IMPROVEMENTS ON THE FELLEGI–HOLT METHOD FOR CATEGORICAL DATA

In the previous section we have described that a subset of the implicit edits is sufficient in order to solve the error localization problem, namely the subset of the essentially new implicit edits. This subset of essentially new implicit edits is much smaller than the set of all implicit edits. Nevertheless, a practical drawback of the method of Fellegi and Holt is that the number of essentially new implicit edits can be high. As a result, the method may be too slow in practice, or even worse, the number of essentially new implicit edits may be too high to be handled by a computer.

Suppose Lemma 4.1 is applied to generate the essentially new implicit edits with field v_j as generating field. Remember that implicit edits can be deduced from at least two edits. So if N_j denotes the number of edits (explicit and already generated implicit ones) that involve field v_j $(j = 1, \ldots, m)$, then an upper bound on the number of essentially new implicit edits that can be generated by applying Lemma 4.1 is

$$\sum_{i=2}^{N_j} \binom{N_j}{i} = 2^{N_j} - N_j - 1. \tag{4.9}$$

So, generation of all essentially new implicit edits grows exponentially with the number of entering fields in the original edits. Winkler (1999) even reports that the amount of computation required to generate the complete set of edits for K explicit edits is of the order $\exp(\exp(K))$ in the worst case. He also reports that the complete set of edits cannot always be generated in practice, due to the computer memory and computations required.

The complete set of edits can be extremely large. There are two reasons for this, of which the first one plays the by far more important role in most practical situations. This first reason is that to generate the complete set of edits one should, in principle, consider all possible subsets of variables. Each of these subsets should be eliminated from the edits in order to obtain the complete set of edits. The total number of subsets of variables from a set of variables is 2^m, where m is the number of variables.

The second reason is that the number of new edits that are obtained after a variable (or subset of variables) has been eliminated may also be large. For instance, suppose there are s edits, where s is even. Suppose, furthermore, that the domain of the variable to be edited, variable v_g, equals $D_g = \{1, 2\}$. If $F_g^k = \{1\}$ for $k = 1, \ldots, s/2$, and $F_g^k = \{2\}$ for $k = s/2 + 1, \ldots, s$, then the total number of new edits may in the worst case equal $s^2/4$. Reducing computing time and required computer memory are therefore the most important aspects in implementing a Fellegi–Holt-based algorithm.

Garfinkel, Kunnathur, and Liepins (1986) made an important contribution to the theory of the categorical error localization problem. In particular, they provide two Fellegi–Holt-based algorithms. Their first algorithm consists of improved rules for the implicit edit generation. In comparison with the original method proposed by Fellegi and Holt, less implicit edits have to be generated, leading to a reduced computing time. Their idea is to generate only a subset of the essentially new implicit edits, namely the *nonredundant* essentially new implicit edits. This set is sufficiently large to solve the error localization problem. The method developed by Garfinkel, Kunnathur, and Liepins has been further improved by Winkler (1995). This method is sketched below.

The second algorithm of Garfinkel, Kunnathur, and Liepins is a cutting plane algorithm. It solves a sequence of small set-covering problems (SCPs) for each failing record. In the algorithm of Fellegi and Holt the complete set of edits has to be generated and for each record an SCP has to be solved, while in this algorithm for each record only a few implicit edits have to be generated and a number of SCPs have to be solved. This algorithm can lead to a significant improvement in computing time in comparison with the algorithm proposed by Fellegi and Holt, especially in cases in which many fields are involved in the edits.

Besides the method described below, there are other approaches to develop a system based on the Fellegi–Holt method. For instance, if the total number of essentially new implicit edits is too high, then one can resort to generating a set of implicit edits for each violated record separately. Barcaroli and Venturi (1996) show that in order to solve the error localization problem for a certain record, it is sufficient to generate the set of (essentially new) implicit edits implied by the violated explicit edits and the explicit edits in which at least one variable is involved that is also involved in a violated explicit edit.

Improved Implicit Edit Generation. Winkler (1995) provides an adaptation of the Fellegi-Holt algorithm for categorical data sets. His result is an alternative for the first algorithm of Garfinkel, Kunnathur and Liepins (1986). Winkler reduces the number of implicit edits needed to solve the categorical error localization problem. For large problems, Winkler's algorithm leads to a large reduction in computing time in comparison with the algorithm of Fellegi and Holt.

One of his observations concerns the so-called *redundant edits*. An edit E^r is called *redundant*, if there exists another edit E^d that *dominates* edit E^r, that is, $P(E^r) \subseteq P(E^d)$. Note that E^d is failed, if E^r is failed. Note also that the entering fields of E^d are covered by the entering fields of E^r. So, if some set of variables S covers one of the entering variables of E^d, then S covers at least one of the entering variables of E^r. This implies that redundant edits are not needed to find sets of variables that cover the complete set of failed edits. Consequently, redundant edits are not needed to find all optimal solutions to the error localization problem. Furthermore, Winkler (1995) observes that if E^d

replaces E^r in a generating set of edits, then any generated edit would necessarily dominate the edit that would have been obtained if E^r had been used. This implies that redundant edits need not be included in contributing sets of edits. Thus we can conclude that redundant edits can be deleted from the set of edits.

Garfinkel, Kunnathur, and Liepins (1986) observe that if one contributing set is a proper subset of another contributing set and if generating on field v_j (using Lemma 4.1) yields essentially new implicit edits for both contributing sets, then the edit generated using the larger contributing set is redundant to the one using the smaller set. Because redundant edits are not useful for error localization purposes, only prime contributing sets—that is, contributing sets for which no proper subset exists that is also a contributing set of an essentially new implicit edit—have to be used in Lemma 4.1. In other words, it suffices to find *minimal* sets of edits T such that (4.7) and (4.8) are obeyed, but none of their proper subsets obey (4.7).

EXAMPLE 4.4

Suppose the number of categorial variables $m = 2$, $D_1 = \{1, 2, 3\}$, $D_2 = \{1, 2, 3\}$ and

$$P(E^1) = \{1, 2\} \times \{1, 2, 3\},$$
$$P(E^2) = \{1, 3\} \times \{1, 3\},$$
$$P(E^3) = \{2, 3\} \times \{2, 3\},$$
$$P(E^4) = \{1\} \times \{2\}.$$

E^4 is redundant, since it is dominated by E^1. Suppose that Lemma 4.1 is used to generate (essentially new) implicit edits on generating field 1. Consider the contributing sets $S_1 = \{E^1, E^3\}$ and $S_2 = \{E^1, E^2, E^3\}$. Note that S_1 is a proper subset of S_2. Generating on field 1 using contributing set S_1 yields the essentially new implicit edit E^5, with $P(E^5) = \{1, 2, 3\} \times \{2, 3\}$. The essentially new implicit edit E^6, with $P(E^6) = \{1, 2, 3\} \times \{3\}$, is obtained by generating on field 1 using contributing set S_2. Obviously, E^5 dominates E^6. ■

4.3.4 THE FELLEGI–HOLT METHOD FOR A MIX OF CONTINUOUS AND CATEGORICAL DATA

For mixed data—that is, a mix of categorical and continuous data—we can, in principle, also use the Fellegi–Holt method to solve the error localization problem. However, the logic to generate the implicit edits becomes quite complex. This logic becomes so complex, because in order to generate the complete set of edits, categorical variables frequently have to be eliminated from edits in which still some numerical variables are involved. The resulting edits are quite complicated ones.

To eliminate a categorical variable v_g from a set of edits given by (4.1) and (4.2), we start by copying all edits not involving this variable to the new set of edits. Next, we determine all minimal index sets T_r such that (4.7) and (4.8) are satisfied.

Given such a minimal index set T_r we construct the implied edit given by

$$\text{IF } v_g \in D_g, v_j \in \bigcap_{k \in T_r} F_j^k \quad \text{for } j = 1, \ldots, g-1, g+1, \ldots, m,$$

$$\text{THEN } (x_1, ..., x_p) \in \bigcup_{k \in T_r} R_k \tag{4.10}$$

where either

$$R_k = \{\mathbf{x} | a_{k1}x_1 + \cdots + a_{kp}x_p + b_k \geq 0\}$$

or

$$R_k = \{\mathbf{x} | a_{k1}x_1 + \cdots + a_{kp}x_p + b_k = 0\},$$

depending on whether the THEN condition of edit E^k is an inequality or an equality. That is, the THEN condition of this implied edit consists of $|T_r|$ elementary numerical conditions given by

$$\{\mathbf{x} | a_{k1}x_1 + \cdots + a_{kp}x_p + b_k \geq 0\}$$

or

$$\{\mathbf{x} | a_{k1}x_1 + \cdots + a_{kp}x_p + b_k = 0\}$$

for $k \in T_r$. An edit of format (4.10) is satisfied if the IF condition is not satisfied, or if the IF condition is satisfied and at least one of the elementary numerical conditions is satisfied. Note that edit (4.10) is indeed an implied edit. It has to be satisfied by the variables that have not yet been treated.

Edits of format (4.10) arise when categorical variables are eliminated before all numerical variables have been eliminated, even if the edits specified by the subject-matter specialists are of format (4.1) and (4.2).

Edits of format (4.10) can, in principle, be handled by splitting the error localization problem into several subproblems. In each subproblem, only one elementary numerical condition of each edit is involved. That is, in each subproblem the edits are of format (4.1) and (4.2). In total $\prod_r |T_r|$ subproblems have to be solved. In a later stage of the algorithm, these subproblems may themselves be split into new subproblems. The best solutions to the final subproblems are the optimal solutions to the overall problem.

Because of the complexity of the Fellegi–Holt approach when applied to a mix of categorical and continuous data, we will not explore this path any further in this book but will instead examine two other approaches for categorical and mixed data in the next sections.

4.4 A Branch-and-Bound Algorithm for Automatic Editing of Mixed Data

4.4.1 INTRODUCTION

In Section 4.4.2 we describe a branch-and-bound algorithm for a mix of continuous and categorical data. This algorithm is an extension of the algorithm given in Section 3.4.7. Section 4.4.3 gives an example illustrating the algorithm. A proof that the algorithm for mixed data finds all optimal solutions to the error localization problem is given in Section 4.4.4. Section 4.4.5 discusses performance issues related to the algorithm, and Section 4.4.6 concludes the section with a short discussion.

4.4.2 A BRANCH-AND-BOUND ALGORITHM

The basic idea of the branch-and-bound algorithm for mixed data is similar to the algorithm for continuous data (see Section 3.4.7). We first assume that no values are missing. After selection of a variable, two branches are then constructed: In one branch the selected variable is fixed to its original value, while in the other branch the selected variable is eliminated from the set of current edits. In each branch the current set of edits is updated. We treat all continuous variables before any categorical variable is selected and treated.

Updating the set of current edits is the most important step in the algorithm. How the set of edits has to be updated depends not only on whether the selected variable is fixed or eliminated, but also on whether this variable is categorical or continuous. Fixing a variable, either continuous or categorical, to its original value is easy. We simply substitute this value in all current edits, failing as well as nonfailing ones. Note that, given that we fix this variable to its original value, the new set of current edits is a set of implied edits for the remaining variables in the tree; that is, the remaining variables have to satisfy the new set of edits. As a result of fixing the selected variable to its original value, some edits may become satisfied—for instance, when a categorical variable is fixed to a value such that the IF condition of an edit can never become true anymore. These edits may be discarded from the new set of edits. Conversely, some edits may become self-contradictory. In such a case this branch of the binary tree can never result in a solution to the error localization problem.

Eliminating a variable is a relatively complicated process. It amounts to generating a set of implied edits that do not involve this variable. In this generation process we need to consider both the failing edits as well as the nonfailing ones in the current set of edits. That set of implied edits becomes the set of edits corresponding to the new node of the tree. If a continuous variable is to be eliminated, we basically apply Fourier–Motzkin elimination [see Duffin (1974) and Section 3.4.3 of this book] to eliminate that variable from the set of edits. Some care has to be taken in order to ensure that the IF conditions of the resulting edits are correctly defined. In particular, if we want to eliminate a continuous variable x_r from the current set of edits, we start by copying all edits not involving this continuous variable from the current set of edits to the new set of edits. Next, we consider all edits in format (4.1) and (4.2) involving x_r pairwise. Suppose we consider a pair of edit E^s and edit E^t. We start by checking whether the intersection of the IF conditions is nonempty—that is, whether the intersections $F_j^s \bigcap F_j^t$ are nonempty for all $j = 1, \ldots, m$. If any of these intersections is empty, we do not have to consider this pair of edits anymore. So, suppose that all intersections are nonempty.

We now construct an implied edit. If the THEN condition of edit E^s is an equality, we use the equality

$$x_r = -\frac{1}{a_{sr}}\left(b_s + \sum_{j \neq r} a_{sj}x_j\right)$$

to eliminate x_r from the THEN condition of edit E^t. Similarly, if the THEN condition of edit E^s is an inequality and the THEN condition of edit E^t is an equality, the equality in edit E^t is used to eliminate x_r. If the THEN conditions of both edit E^s and edit E^t are inequalities, we check whether the coefficients of x_r in those inequalities have opposite signs. That is, we check whether $a_{sr} \times a_{tr} < 0$. If that is not the case, we do not consider this pair of edits anymore. If the coefficients do have opposite signs, we can write one inequality as an upper bound on x_r and the other as a lower bound on x_r. We generate the following THEN condition for our implied edit:

$$(x_1, \ldots, x_p) \in \{\mathbf{x} | \tilde{a}_1 x_1 + \cdots + \tilde{a}_{r-1}x_{r-1} + \tilde{a}_{r+1}x_{r+1} + \cdots + \tilde{a}_p x_p + \tilde{b} \geq 0\}$$

where

$$\tilde{a}_j = |a_{sr}| \times a_{tj} + |a_{tr}| \times a_{sj} \qquad \text{for } j = 1, \ldots, p; j \neq r$$

and

$$\tilde{b} = |a_{sr}| \times b_t + |a_{tr}| \times b_s.$$

Note that x_r indeed does not enter the resulting THEN condition. The IF condition of the implied edit is given by the intersections $F_j^s \bigcap F_j^t$ for all $j = 1, \ldots, m$.

Note that if we eliminate a continuous variable in any of the ways described above, the resulting set of edits is a set of implied edits for the remaining variables in the tree. That is, this resulting set of edits has to be satisfied by the remaining variables in the tree, given that the eliminated variable may in principle take any real value.

In our algorithm, categorical variables are only treated (i.e., fixed or eliminated) once all continuous variables have been treated. This is done in order to keep our algorithm as simple as possible. If categorical variables would be treated before all continuous ones have been treated, we could obtain edits that are more complex than the edits of type (4.1) and (4.2). For an illustration of this phenomenon we refer to Section 4.3.4. So, once the categorical variables may be selected, the edits in the current set of edits all have the following form:

$$\begin{aligned} &\text{IF} \quad v_j \in F_j^k \qquad \text{for } j = 1, \ldots, m, \\ &\text{THEN} \quad (x_1, \ldots, x_p) \in \emptyset. \end{aligned} \tag{4.11}$$

To eliminate categorical variable v_r from a set of edits given by (4.11), we start by copying all edits not involving this variable to the set of implied edits.

Next, we basically apply the method of Fellegi and Holt to the IF conditions to generate the IF conditions of the implied edits. In the terminology of Fellegi and Holt, variable v_r is selected as the generating field. We start by determining all index sets S such that

$$\bigcup_{k \in S} F_r^k = D_r \tag{4.12}$$

and

$$\bigcap_{k \in S} F_j^k \neq \emptyset \text{ for all } j = 1, \ldots, r-1, r+1, \ldots, m \tag{4.13}$$

From these index sets we select the *minimal* ones, i.e. the index sets S that obey (4.12) and (4.13), but none of their subsets obey (4.12).

Given such a minimal index set S, we construct the implied edit given by

$$\begin{aligned} &\text{IF} \quad v_r \in D_r, v_j \in \bigcap_{k \in S} F_j^k \qquad \text{for } j = 1, \ldots, r-1, r+1, \ldots, m, \\ &\text{THEN} \quad (x_1, \ldots, x_p) \in \emptyset. \end{aligned} \tag{4.14}$$

Note that if we eliminate a categorical variable in the way described above, the resulting set of edits is a set of implied edits for the remaining variables in the tree. That is, this resulting set of edits has to be satisfied by the remaining variables in the tree, given that the eliminated variable may in principle take any value in its domain. We have now explained how the current set of edits changes if we fix or eliminate a variable.

If values are missing in the original record, the corresponding variables only have to be eliminated (and not fixed) from the set of edits, because these variables always have to be imputed.

A natural choice is to treat the variables in the following order:

1. Eliminate all continuous variables with missing values.
2. Fix or eliminate the remaining continuous variables.
3. Eliminate all categorical variables with missing values.
4. Fix or eliminate the remaining categorical variables.

After all categorical variables have been treated, we are left with a set of relations involving no unknowns. This set of relations may be the empty set, in which case it obviously does not contain any self-contradicting relations. A self-contradicting relation is given by

$$\begin{aligned} &\text{IF} \quad v_j \in D_j \qquad \text{for } j = 1, \ldots, m, \\ &\text{THEN} \ (x_1, \ldots, x_p) \in \emptyset. \end{aligned}$$

The set of relations contains no self-contradicting relations if and only if the variables that have been eliminated in order to reach the corresponding terminal node of the tree can be imputed consistently—that is, such that all original edits can be satisfied (cf. Theorems 4.3 and 4.4 in Section 4.4.4).

In the algorithm we check for each terminal node of the tree whether the variables that have been eliminated in order to reach this node can be imputed consistently. Of all sets of variables that can be imputed consistently, we select the ones with the lowest sum of reliability weights. In this way we find all optimal solutions to the error localization problem (cf. Theorem 4.5 in Section 4.4.4).

Equalities in THEN conditions can be handled more efficiently than we have described so far. For instance, if the numerical variable to be eliminated is involved in an equality that has to hold irrespective of the values of the categorical variables, that is, is involved in an edit of the type

$$\begin{aligned} &\text{IF} \quad v_j \in D_j \qquad \text{for } j = 1, \ldots, m, \\ &\text{THEN} \ (x_1, \ldots, x_p) \in \{\mathbf{x} | a_{k1}x_1 + \cdots + a_{kp}x_p + b_k = 0\}, \end{aligned} \tag{4.15}$$

then we do not have to consider all edits pairwise in order to eliminate this variable. Instead, we only have to combine (4.15) with all other current edits. So, if there are K current edits, we do not have to consider $K(K-1)$ pairs, but only $K-1$ pairs. Besides, the number of resulting implied edits is generally less than when all pairs of current edits are considered. We refer to this rule as the equality-elimination rule.

The algorithm sketched in this section is a so-called branch-and-bound algorithm. In a branch-and-bound algorithm a tree is constructed and bounds on the objective function are used to cut off branches of the tree. In Section 4.4.5 we explain how branches can be cut off from our tree.

4.4.3 EXAMPLE

In this section we illustrate the idea of the algorithm presented in the previous section by means of an example. We will not build the entire tree, because this would take too much space and would hardly teach us anything. Instead we will only generate one branch of the tree. Suppose we have to edit a data set containing two categorical variables v_j $(j = 1, 2)$ and two numerical variables x_j $(j = 1, 2)$. The domain D_1 of the first categorical variable is {1,2}, and the domain D_2 of the second categorical variable {1,2,3}. The set of explicit edits is given below.

(4.16) IF $(v_1 = 1 \text{ AND } v_2 \in D_2)$, THEN $\emptyset$,

(4.17) IF $(v_1 \in D_1 \text{ AND } v_2 = 1)$, THEN $\emptyset$,

(4.18) IF $(v_1 = 2 \text{ AND } v_2 \in \{1, 3\})$, THEN $\emptyset$,

(4.19) IF $(v_1 \in D_1 \text{ AND } v_2 \in D_2)$, THEN $x_1 - 12 \geq 0$,

(4.20) IF $(v_1 \in D_1 \text{ AND } v_2 \in \{1, 3\})$, THEN $x_2 = 0$,

(4.21) IF $(v_1 \in D_1 \text{ AND } v_2 = 2)$, THEN $x_2 - 1250 \geq 0$,

(4.22) IF $(v_1 \in D_1 \text{ AND } v_2 = 2)$, THEN $-875x_1 + 12x_2 \geq 0$,

IF $(v_1 \in D_1 \text{ AND } v_2 = 2)$, THEN $1250x_1 - 8x_2 \geq 0$.

Note that this is a rather artificial set of edits because edit (4.16) says that v_1 cannot attain the value 1, and edit (4.17) says the same for variable v_2. In reality, the value 1 would not be an element of the domains D_1 and D_2. The above set of edits helps to illustrate the algorithm of Section 4.4.2 without complicating the issue too much.

Suppose that a record with values $v_1 = 1$, $v_2 = 2$, $x_1 = 25$, and $x_2 = 3050$ is to be edited. Edit (4.16) is violated, so this record is inconsistent. We apply the algorithm described in the previous section and start by selecting a numerical variable, say x_1. In the algorithm, two branches are generated: one branch where x_1 is fixed to its original value 25, and one branch where x_1 is eliminated from the current set of edits. Here we only consider the second branch and eliminate x_1 from the current set of edits.

For instance, if we combine (4.19) and (4.22), we first take the intersection of their IF conditions. This intersection is given by "$v_1 \in D_1$ AND $v_2 = 2$." This intersection is nonempty, so we proceed. The THEN condition of the resulting implied edit is given by $12x_2 \geq 875 \times 12$, or equivalently by $x_2 \geq 875$. The resulting implied edit is hence given by

(4.23) IF $(v_1 \in D_1 \text{ AND } v_2 = 2)$, THEN $x_2 - 875 \geq 0$.

The complete set of resulting (implicit) edits is given by (4.23),

$$\text{IF } (v_1 \in D_1 \text{ AND } v_2 = 2), \text{ THEN } x_2 \geq 0 \tag{4.24}$$

and (4.16), (4.17), (4.18), (4.20), and (4.21).

We select the other numerical variable x_2 and again construct two branches: one branch where x_2 is fixed to its original value 3050, and one branch where x_2 is eliminated from the current set of edits. Here we only consider the first branch and fix x_2 to its original value. As a consequence, some of the resulting edits may be satisfied. Those edits can be discarded. In this case, for instance edit (4.23) becomes satisfied after filling in $x_2 = 3050$ and is discarded from the current branch of the tree. Some other edits might be violated. In such a case the current branch of the tree cannot lead to a solution to the error localization problem. In our example, none of the resulting edits are violated. The resulting set of implicit edits obtained by fixing x_2 to its original value is given by

$$\text{IF } (v_1 \in D_1 \text{ AND } v_2 \in \{1, 3\}), \text{ THEN } \emptyset \tag{4.25}$$

and (4.16), (4.17), and (4.18). Edit (4.25) arises from edit (4.20) by substituting 3050 for x_2. The resulting numerical THEN condition is failed.

All numerical variables have now been treated, either by fixing or by eliminating. We see that the current set of edits is given by the purely categorical explicit edits supplemented with categorical edits that have been generated when the numerical variables were treated. We now treat the categorical variables. We select a categorical variable, say v_1, and again split the tree into two branches: a branch where v_1 is fixed to its original value and a branch where it is eliminated. We only consider the branch where v_1 is eliminated. The resulting set of implicit edits is given by

$$\text{IF } (v_2 = 1), \text{ THEN } \emptyset$$

and

$$\text{IF } (v_2 \in \{1, 3\}), \text{ THEN } \emptyset.$$

We select the other categorical variable v_2. Fixing and eliminating this variable again results in two branches. We only consider the branch where v_2 is fixed to its original value, 2. The resulting set of implicit edits is empty. This implies that the set of original, explicit edits can be satisfied by changing the values of x_1 and v_1 and fixing the other variables to their original values. That is, a (possibly suboptimal) solution to the error localization problem for this record is: Change the values of x_1 and v_1. Possible consistent values are $v_1 = 2$ and $x_1 = 40$.

The other branches of the tree, which we have skipped, also need to be examined, because it is possible that they contain a better solution to the error localization problem. By examining all branches of the tree, one can obtain all optimal solutions to the error localization problem for the record under consideration.

4.4.4 AN OPTIMALITY PROOF

In this section we prove that the algorithm described in Section 4.4.2 indeed finds all optimal solutions to the error localization problem. We do this in three steps.

1. Each time we eliminate or fix a variable the current set of edits is transformed into a new set of edits. The new set of edits involves at least one variable less than the current set of edits. We start by showing that the current set of edits can be satisfied if and only if the new set of edits can be satisfied. This is the content of Theorem 4.3 below.
2. Using this result, we show that if and only if the set of relations involving no unknowns in a terminal node do not contain any self-contradicting relations, we can impute the variables that have been eliminated in order to reach this terminal node consistently—that is, such that the original edits become satisfied. This is the content of Theorem 4.4.
3. The final step consists of observing that the terminal nodes correspond to all potential solutions to the error localization problem, and hence that the algorithm indeed determines all optimal solutions to the error localization problem. This is the content of Theorem 4.5.

Steps 2 and 3 are trivial once the first step has been proved. The proof of the first step is partly similar to the proof of Theorem 1 in Fellegi and Holt (1976). The main differences are that the edits considered by Fellegi and Holt differ from the edits considered in the present chapter and that Fellegi and Holt assume that the so-called complete set of (explicit and implied) edits has been generated. We will not make this assumption.

THEOREM 4.3

Suppose the index set of variables in a certain node is given by T_0 and the current set of edits corresponding to that node by Ω_0. Suppose furthermore that to obtain a next node a certain variable r is either fixed or eliminated. Denote the index set of resulting variables by T_1 ($T_1 = T_0 - \{r\}$) and the set of edits corresponding to this next node by Ω_1. Then there exist values u_j for $j \in T_1$ that satisfy the edits in Ω_1 if and only if there exists a value u_r for variable r such that the values u_j for $j \in T_0$ satisfy the edits in Ω_0.

Proof. It is easy to verify that if there exist values u_j for $j \in T_0 = T_1 \cup \{r\}$ that satisfy the edits in Ω_0, then the same values (except the value of the variable that is fixed or eliminated) automatically satisfy the edits Ω_1 of the next node.

It is a bit more work to prove the converse implication. We have to distinguish between several cases. First, let us suppose that the selected variable is fixed. This is a trivial case. It is clear that if there exist values u_j for $j \in T_1$ that satisfy the edits in Ω_1, there exist values u_j for $j \in T_0$ that satisfy the edits in Ω_0.

Namely, for the fixed variable r we set the value u_r^0 equal to the original value of r.

Let us now suppose that a categorical variable r has been eliminated. Note that in our algorithm all continuous variables have then already been either fixed or eliminated. Suppose that there exist values u_j for $j \in T_1$ that satisfy the edits in Ω_1, but there does not exist a value u_r for the selected variable r such that the values for $j \in T_0$ satisfy the edits in Ω_0. Identify a failed edit in Ω_0 for each possible value of variable r. The index set of these failed edits need not be a minimal one. We therefore remove some of the failed edits such that the corresponding index set S becomes minimal. We then construct the implied edit given by (4.14).

Edit (4.14) is an element of Ω_1. Moreover, the values u_j for $j \in T_1$ do not satisfy this edit. This contradicts our assumption that these values satisfy all edits in Ω_1. So, we can conclude that a value u_r for the selected variable r exists such that the values u_j for the variables in T_0 satisfy the edits in Ω_0.

Finally, let us suppose that a continuous variable r has been eliminated. Suppose that there exist values u_j for $j \in T_1$ that satisfy the edits in Ω_1. Each edit in Ω_1 is obtained either from copying the edits in Ω_0 not involving variable r or from two edits in Ω_0 involving variable r that have been combined.

It is clear that if the edits in Ω_1 that have been obtained from copying the edits in Ω_0 not involving variable r are satisfied by the values u_j for $j \in T_1$, these edits in Ω_0 are also satisfied by the same values for $j \in T_0$.

It remains to prove that if the edits in Ω_1 that have been obtained by combining two edits in Ω_0 are satisfied by $j \in T_1$, there exists a value for variable r such that all edits in Ω_0 involving variable r can be satisfied. First, if the equality-elimination rule has been applied to eliminate variable r, using an edit of type (4.15), and the values u_j for $j \in T_1$ satisfy the edits in Ω_1, the value

$$u_r = -\frac{1}{a_{sr}}\left(b_s + \sum_{j \neq r} a_{sj} u_j\right) \tag{4.26}$$

together with the values u_j for $j \in T_1$ obviously satisfy the edits in Ω_0.

Second, if the equality-elimination rule has not been used to eliminate variable r, we substitute the values u_j for $j \in T_1$ into the edits in Ω_0. As a result, we obtain a number of constraints for the value of the selected variable r. Such a constraint can be an equality involving x_r, a lower bound on x_r, or an upper bound on x_r. That is, these constraints are given by

$$x_r = M_k^E, \tag{4.27}$$

$$x_r \geq M_k^L, \tag{4.28}$$

and

$$x_r \leq M_k^U, \tag{4.29}$$

where M_k^E, M_k^L, and M_k^U are certain constants.

A constraint of type (4.27) has been obtained from an edit in Ω_0 of which the THEN condition can be written in the form

$$x_r = \sum_{j \neq r} d'_{kj} x_j + b'_k \tag{4.30}$$

by filling in the values u_j for $j \in T_1$. Similarly, constraints of types (4.28) and (4.29) have been obtained from edits in Ω_0 of which the THEN conditions can be written in the forms

$$x_r \geq \sum_{j \neq r} d'_{kj} x_j + b'_k$$

and

$$x_r \leq \sum_{j \neq r} d'_{kj} x_j + b'_k,$$

respectively, by filling in the values u_j for $j \in T_1$.

If the constraints given by (4.27) to (4.29) do not contradict each other, we can find a value for variable r such that this value plus the values u_j for $j \in T_1$ satisfy the edits in Ω_0.

So, suppose the constraints given by (4.27) to (4.29) contradict each other. These constraints can only contradict each other if there are constraints s and t given by

1. $x_r = M_s^E$ and $x_r = M_t^E$ with $M_s^E \neq M_t^E$,
2. $x_r = M_s^E$ and $x_r \geq M_t^L$ with $M_s^E < M_t^L$,
3. $x_r \leq M_s^U$ and $x_r = M_t^E$ with $M_s^U < M_t^E$,

or

4. $x_r \leq M_s^U$ and $x_r \geq M_t^L$ with $M_s^U < M_t^L$.

In case 1, constraints s and t have been derived from edits in Ω_0 of which the THEN conditions are equalities. The IF conditions of these edits have a nonempty intersection, because both edits are triggered when we fill in the values u_j for the categorical variables in T_1. So, these edits generate an implied edit in Ω_1 if we eliminate variable r. The THEN condition of this implied edit can be written as

$$\sum_{j \neq r} d'_{sj} x_j + b'_s = \sum_{j \neq r} d'_{tj} x_j + b'_t,$$

where we have used (4.30).

Filling in the values u_j for $j \in T_1$ in this implied edit, we find that M_s^E should be equal to M_t^E. In other words, we have constructed an edit in Ω_1 that would be failed, if we were to fill in the values u_j for $j \in T_1$. This contradicts our assumption that these values satisfy all edits in Ω_1, and we conclude that two constraints given by (4.27) (case 1 above) cannot contradict each other.

For cases 2, 3, and 4 we can show in a similar manner that we would be able to construct a failed implied edit in Ω_1. This contradicts our assumption that the values u_j for $j \in T_1$ satisfy all edits in Ω_1, and we conclude that the constraints given by (4.27) to (4.29) cannot contradict each other. In turn, this allows us to conclude that a value for variable r exists such that this value plus the values u_j for $j \in T_1$ satisfy the edits in Ω_0. ■

THEOREM 4.4

The set of edits corresponding to a terminal node—a set of relations involving no unknowns—contains no self-contradicting relations if and only if the variables that have been eliminated in order to reach this terminal node can be imputed in such a way that the original set of edits becomes satisfied.

Proof. This follows directly from a repeated application of Theorem 4.3. ■

THEOREM 4.5

The algorithm determines all optimal solutions to the error localization problem.

Proof. The terminal nodes of the tree correspond to all possible combinations of fixing and eliminating variables. So, according to Theorem 4.4 above, the algorithm checks which of all possible sets of variables can be imputed consistently. The algorithm selects all optimal sets of variables that can be imputed consistently from all possible sets of variables. So, we can conclude that the algorithm finds all optimal solutions to the error localization problem. ■

4.4.5 COMPUTATIONAL ASPECTS

We have demonstrated in Section 4.4.4 that our algorithm determines all optimal solutions to the error localization problem for mixed data. At first sight, however, the developed algorithm may seem rather slow because an extremely large binary tree has to be generated to find all optimal solutions, even for moderately sized problems. Fortunately, the situation is not nearly as bad as it may seem. First of all, if the minimum number of fields that has to be changed in order to make a record pass all edits is (too) large, the record should not be edited automatically in our opinion (see also Section 3.4.1). We consider the quality of such a record to be too low to allow for automatic correction. In our opinion, such a record should either be edited manually or be discarded completely. By specifying an upper bound on the number of fields that may be changed, the size of the tree can drastically be reduced.

The size of the tree can also be reduced during the execution of the algorithm, because it may already become clear in an intermediate node of the tree that the corresponding terminal nodes cannot generate an optimal solution to the problem. For instance, by fixing wrong variables we may make the set of edits infeasible. This may be noticed in an intermediate node.

The value of the objective function can also be used to reduce the size of the tree. This value cannot decrease while going down the tree. So, if the value of the objective function exceeds the value of an already found (possibly suboptimal) solution, we can again conclude that the corresponding terminal nodes cannot generate an optimal solution to the problem. These terminal nodes need not be examined and can be cut off from the tree.

Because the size of the tree, and hence the computing time of the algorithm, can be influenced by the order in which the variables are treated, this ordering is very important in practice. The ordering must not be fixed before the execution of the algorithm because this would lead to a high computing time on the average. Instead the ordering should be determined dynamically—that is, during the execution of the algorithm. Each time a variable is to be treated, the "best" variable, according to a suitable ordering strategy, should be selected.

For computational results of the branch-and-bound algorithm on continuous data, we refer to Section 3.4.9.

4.4.6 DISCUSSION OF THE BRANCH-AND-BOUND ALGORITHM

The proposed branch-and-bound algorithm is not very complex to implement and maintain. One of the reasons for the simplicity of the algorithm is that it is a very "natural" one. For instance, in the algorithm categorical and continuous variables are treated in almost the same manner, only the underlying method to generate implicit edits differs. Moreover, searching for optimal solutions to the error localization problem is also a natural process. All possible solutions are simply checked, and the best solutions found are the optimal ones. Because of the simplicity of the branch-and-bound algorithm discussed in this section, maintaining software based on this algorithm is relatively simple. Not only Operations Research specialists can understand the algorithm in detail, but also the IT specialists who develop and maintain the final computer program based on the mathematical algorithm. The algorithm also gives good computational results on continuous data (see Section 3.4.9).

Statistics Netherlands has therefore implemented this algorithm in a module of version 1.5 of the SLICE system (De Waal, 2001b). This version reads an upper bound on the number of missing values per record as well as a separate upper bound on the number of errors (excluding missing values) per record. The former number is allowed to be quite high, say 50 or more, whereas the latter number is allowed to be moderate, say 10. If the number of missing values or the number of errors (excluding missing values) in a record exceeds either of these upper bounds, this record is rejected for automatic editing. The new module includes the equality-elimination rule. In addition, it contains a heuristic to handle integer data, which will be explained in Chapter 5. The new module hence solves the error localization problem for a mix of categorical, continuous and integer data.

One may argue that some users of SLICE will want to edit records with many erroneous fields automatically despite our arguments against editing such records. These users might then be disappointed, because the new module will not be able to handle such records. To overcome this problem, we propose to use a simple heuristic treatment of these records instead of applying the module of SLICE. For purely numerical data one could, for instance, minimize the sum of the absolute differences between the original values and the final values subject to the condition that all edits become satisfied. The resulting mathematical problem can be formulated as a linear programming problem, and it can be solved quickly by means of, e.g., the simplex algorithm (Chvátal, 1983). For further details on this linear programming approach and a heuristic approach for a mix of categorical and continuous data, we refer to Chapter 10 of the present book.

We are willing to admit that our choice for the branch-and-bound algorithm is to some extent a subjective choice, but we feel that our choice is a justifiable one.

4.5 The Nearest-Neighbor Imputation Methodology

4.5.1 INTRODUCTION

So far in this chapter and the previous one, we have discussed methods for automatic error localization based on the Fellegi–Holt paradigm. This will also be the subject of Chapter 5. A common feature of these methods is that they only locate erroneous fields. New values are imputed for the erroneous fields in a separate step, and for this an imputation method is chosen independently of the error localization method. Commonly used imputation methods will be introduced in Chapters 7 and 8. In this section, we describe an alternative method for automatic editing called the Nearest-neighbor Imputation Methodology (NIM), which uses a different approach. With the NIM, the localization of erroneous fields and the imputation of new values are not achieved in two separate steps, but simultaneously. Instead of the Fellegi–Holt paradigm, a different minimization criterion is used, which also takes the original and imputed values into account. We choose to treat the NIM at this point, because, like the other automatic editing methods in this chapter, the NIM can handle a combination of categorical and numerical data.

The imputation method for the NIM cannot be chosen freely: It has to be a form of hot deck donor imputation. With this imputation method, values from one record (the "donor") are used to replace erroneous and missing values in another record (the "recipient"). The name "hot deck" indicates that the donor and the recipient come from the same data set. Only records that are error-free may be used as donors. This brief introduction to hot deck donor imputation should be sufficient to understand the description of the NIM below. Hot deck donor imputation will be discussed more fully in Section 7.6.

Section 4.5.2 discusses incentives for developing a new editing method that does not use the traditional Fellegi–Holt paradigm, by examining an example from the 1991 Canadian Census. A basic description of the NIM is given in Section 4.5.3. Aspects of generating so-called feasible imputation actions in an efficient manner are discussed in Section 4.5.4. Finally, a brief comparison between Fellegi–Holt-based editing and the NIM follows in Section 4.5.5.

The NIM has been implemented in a software package called CANCEIS (an acronym for CANadian Census Edit and Imputation System). A detailed description of the program can be found in CANCEIS (2006). Based on practical experience, improvements are continuously being made to CANCEIS [see, e.g., Bankier and Crowe (2009).]

4.5.2 HISTORICAL NOTES

The NIM was originally developed at Statistics Canada to improve the edit and imputation process of the population census. Bankier et al. (1994) sketch the background of this development. An assessment of the 1991 Canadian Census, which had been edited according to traditional Fellegi–Holt methodology in combination with hot deck donor imputation, revealed that the edit and imputation process had produced some unwanted effects. According to Bankier et al. (1994), "many individual imputation actions were implausible and small but important groups in the population had their numbers falsely inflated by the imputation actions." The following example, taken from Bankier et al. (1994), illustrates this.

Table 4.1 displays a six-person household with four variables: *Relationship to Person 1*, *Sex*, *Marital Status*, and *Age*. The original household data, shown in panel (a), are inconsistent because of a too small age difference between Person 1 and Person 2, the eldest daughter of Person 1. In fact, this record violates an edit which states that if Person 2 is a son or daughter of Person 1, then the decade of birth of Person 2 cannot be the same as or precede the decade of birth of Person 1. The data can be made consistent by

(i) increasing the value of *Age* for Person 1;

(ii) lowering the value of *Age* for Person 2;

(iii) changing the value of *Relationship to Person 1* for Person 2.

Panel (b) shows the household after editing with the methodology used in the 1991 Canadian Census. Option (ii) above was used to obtain a consistent household. While the resulting household does not fail any explicit edits, it clearly has some curious properties: Person 1 has five daughters, the eldest of which was born when he was 12 years old, Person 1 and Person 2 are married, but not to each other, and both their respective spouses do not belong to the household. Any one of these features is only true for a small fraction of six-person

TABLE 4.1 **Editing a Six Person Household from the 1991 Canadian Census**

	Relationship to Person 1	Sex	Marital Status	Age
	(a) Original Failed Edit Household			
1	Person 1	M	Married	34
2	Son/Daughter	F	Married	32
3	Son/Daughter	F	Single	14
4	Son/Daughter	F	Single	11
5	Son/Daughter	F	Single	6
6	Son/Daughter	F	Single	2
	(b) Household after Editing with Fellegi–Holt Methodology			
1	Person 1	M	Married	34
2	Son/Daughter	F	Married	22
3	Son/Daughter	F	Single	14
4	Son/Daughter	F	Single	11
5	Son/Daughter	F	Single	6
6	Son/Daughter	F	Single	2
	(c) A more plausible edited version of the household			
1	Person 1	M	Married	34
2	Spouse	F	Married	32
3	Son/Daughter	F	Single	14
4	Son/Daughter	F	Single	11
5	Son/Daughter	F	Single	6
6	Son/Daughter	F	Single	2

households in the population; to find them combined in the same household is very unlikely indeed.

Intuitively, a sensible way to obtain consistency in this example is to change the value of *Relationship to Person 1* for Person 2 to "Spouse," since Person 1 and Person 2 are of similar *Age* and opposite *Sex*, they are both married, and they are the only adults in this household. Panel (c) shows the resulting household data. This solution is also empirically more plausible than the solution of panel (b). Bankier et al. (1994) write: "When available donors were investigated (. . .), it was found that there were 97 (person 1/spouse/four child) households for every 3 (person 1/five child) households."

To avoid unwanted inflation of small subpopulations, the adjustments made by the edit and imputation process should reflect the distribution of donors. In this particular example, the imputation action of panel (c) should be performed with a much higher probability than the imputation action of panel (b), in order to avoid unwanted inflation of the number of six-person households with one adult and five children. This is difficult to achieve with a Fellegi–Holt-based editing method, because donors are only used to generate imputations *after* the fields to impute have been selected. The selection of fields to impute itself is done mechanically by solving a minimization problem, without explicit reference to the set of donors.

As mentioned previously, in this example the household can be made consistent by imputing either of three possible fields. Hence, there are three equivalent solutions to the minimization problem (assuming that no reliability weights are used). During the 1991 Canadian Census, if more than one solution to the minimization problem was found, one of the solutions was selected at random. In the example of Table 4.1, the imputed household would be a one-adult/five-child household with probability 2/3 and a two-adult/four-child household with probability 1/3. Thus, in this example, the Fellegi–Holt-based edit and imputation process is actually twice as likely to create the type of household that is rare according to the distribution of donors, instead of the type of household that is common among the available donors. This means that the edit and imputation process achieves the opposite of the desired property mentioned above.

In principle, it is possible to improve the Fellegi–Holt-based editing method used in the 1991 Canadian Census, to try to prevent unwanted inflation of small subpopulations. One possibility is to dynamically assign reliability weights—that is, to choose a different set of weights for each record. Information on the distribution of donors could then be incorporated into these reliability weights. Also, compared to randomly selecting a solution to the Fellegi–Holt minimization problem, a slight improvement may be expected if all solutions are imputed and the most plausible imputed record is selected. However, this will severely increase the amount of computational work needed for the edit and imputation process. The NIM uses a very different approach, which turns out to be less computationally intensive than most Fellegi–Holt-based algorithms.

4.5.3 BASIC DESCRIPTION OF THE NIM

Bankier (1999) mentions the following objectives for an editing and imputation methodology based on hot deck donor imputation:

1. The imputed record should closely resemble the original record.
2. The imputed data should come from a single donor, rather than two or more, whenever this is possible. In addition, the imputed record should closely resemble that single donor record.
3. Equally good imputation actions should have a similar chance of being selected.

As we shall see, the NIM is designed to achieve these three objectives.

In the first step of the NIM, all incoming records are checked for consistency against a set of edits. Records that do not fail any edits are placed in a *donor pool* $\mathcal{D}$—that is, a collection of potential donors for hot deck imputation. It is assumed that the number of potential donors is large, compared to the number of records to impute. Thus, in order to successfully apply the NIM, the quality of incoming data should be high.

Suppose that a data file contains records of p variables. In order to apply the concept of a *nearest neighbor*, a distance measure between records has to be defined. The distance between two records $\mathbf{x}_1 = \left(x_{11}, \ldots, x_{1p}\right)^T$ and $\mathbf{x}_2 = \left(x_{21}, \ldots, x_{2p}\right)^T$ is

$$D(\mathbf{x}_1, \mathbf{x}_2) = \sum_{j=1}^{p} w_j D_j(x_{1j}, x_{2j}), \tag{4.31}$$

where a (nonnegative) weight w_j and a local distance function $D_j(x_{1j}, x_{2j})$ are associated to each variable. It is assumed that $0 \leq D_j(x_{1j}, x_{2j}) \leq 1$ for each distance function and that $D_j(x_{1j}, x_{2j}) = 0$ if $x_{1j} = x_{2j}$. Otherwise, the local distance functions may be chosen freely. A higher value of w_j implies that the jth variable has more influence on the distance measure and that, for instance, it is considered more important that a donor record matches a failed record on the jth variable. A variable can be left out of the distance measure by choosing $w_j = 0$.

The NIM treats the records outside $\mathcal{D}$—that is, the records that fail at least one edit—one at a time. For a given record $\mathbf{x}_f$ that fails at least one edit, a search is conducted for the N_d potential donor records $\mathbf{x}_d \in \mathcal{D}$ with the smallest values of $D(\mathbf{x}_f, \mathbf{x}_d)$. These potential donor records are referred to as nearest neighbors. The number N_d is specified by the user; Bankier, Lachance, and Poirier (2000) suggest taking $N_d = 40$.

Each nearest neighbor is used to generate *imputation actions*. Performing an imputation action means adapting a failed record by copying the values of some fields from a donor record. Formally, an imputation action is described by the triplet $I = (\mathbf{x}_f, \mathbf{x}_d, \boldsymbol{\delta})$, where $\boldsymbol{\delta} = \left(\delta_1, \ldots, \delta_p\right)^T$ is a binary p-vector with $\delta_j = 1$ if the jth variable is imputed from the donor, and $\delta_j = 0$ otherwise. The imputation action yields an adapted record $\mathbf{x}_a$, given by

$$x_{aj} = \delta_j x_{dj} + (1 - \delta_j) x_{fj}, \qquad j = 1, \ldots, p,$$

or equivalently by

$$\mathbf{x}_a = \text{diag}(\boldsymbol{\delta})\mathbf{x}_d + \left(\mathbf{I} - \text{diag}(\boldsymbol{\delta})\right) \mathbf{x}_f, \tag{4.32}$$

where $\mathbf{I}$ is the $p \times p$ identity matrix and diag($\boldsymbol{\delta}$) is the $p \times p$ diagonal matrix with $\boldsymbol{\delta}$ on the main diagonal. Clearly, only fields with $x_{fj} \neq x_{dj}$ can be used to construct useful imputation actions.

An imputation action is called *feasible* if the resulting record $\mathbf{x}_a$ does not fail any edits. All infeasible imputation actions are immediately discarded. We remark that finding a feasible imputation action is trivial: If we set $\delta_1 = \cdots = \delta_p = 1$, by (4.32), the adapted record $\mathbf{x}_a$ is identical to $\mathbf{x}_d$. By definition, the donor $\mathbf{x}_d$ does not fail any edits. However, the objective of the NIM is to find a feasible imputation action that changes the original record as little as possible. To make this more precise, the following size measure is defined for feasible imputation actions:

$$\mu(I) = \alpha D(\mathbf{x}_f, \mathbf{x}_a) + (1 - \alpha) D(\mathbf{x}_a, \mathbf{x}_d), \tag{4.33}$$

with $\alpha \in (1/2, 1]$ a fixed parameter. This size measure is a convex combination of two distance functions of type (4.31): first, the distance between the original record and the adapted record; and second, the distance between the adapted record and the donor record.

Clearly, an imputation action with a small value of $D(\mathbf{x}_f, \mathbf{x}_a)$ has the desirable property that much of the original record is preserved. Minimizing this expression is in line with—albeit not entirely equivalent to—the Fellegi–Holt paradigm.[1] The second term in (4.33) is used as an indicator of the plausibility of the adapted record. Simply minimizing the first term in (4.33) might lead to an artificial record with certain properties that are unlikely to occur in practice; this was illustrated by the example in Section 4.5.2. If $D(\mathbf{x}_a, \mathbf{x}_d)$ is small, the adapted record is plausible because it resembles an unimputed record from the donor pool. Note that minimizing expression (4.33) captures the first two objectives from Bankier (1999) mentioned above.

The choice of α determines the way the two terms in (4.33) are balanced. Using (4.31) and (4.32), it is not difficult to establish that[2]

$$D(\mathbf{x}_f, \mathbf{x}_d) = D(\mathbf{x}_f, \mathbf{x}_a) + D(\mathbf{x}_a, \mathbf{x}_d) \tag{4.34}$$

holds for all imputation actions. From this and (4.33), it follows that

$$\mu(I) = (2\alpha - 1)D(\mathbf{x}_f, \mathbf{x}_a) + (1 - \alpha)D(\mathbf{x}_f, \mathbf{x}_d). \tag{4.35}$$

This explains why only values of $\alpha \in (1/2, 1]$ are allowed: Taking $\alpha < 1/2$ means that a larger value of $D(\mathbf{x}_f, \mathbf{x}_a)$ leads to a smaller value of $\mu(I)$, and taking $\alpha > 1$ has the same effect with $D(\mathbf{x}_f, \mathbf{x}_d)$. In both cases, undesirable imputation actions would have the smallest values of $\mu(I)$. The values $\alpha = 3/4$ and $\alpha = 9/10$ have been used in practice at Statistics Canada [cf. Bankier (2006)].

Next, assume that all feasible imputation actions have been constructed from the N_d nearest neighbors. Note that often multiple feasible imputation actions can be constructed from the same donor. For each feasible imputation action, we evaluate $\mu(I)$. Denote the smallest occurring value by $\mu_{\min}$. A feasible imputation action I is called a *near-minimum change imputation action* (NMCIA) if it satisfies

$$\mu(I) \leq \gamma \mu_{\min}, \tag{4.36}$$

---

[1] In fact, to recover the generalized Fellegi–Holt paradigm we have to define the local distance functions $D_j(x_{1j}, x_{2j}) = 0$ if $x_{1j} = x_{2j}$ and $D_j(x_{1j}, x_{2j}) = 1$ otherwise for all j, assume that $\mathcal{D}$ contains all (possibly infinitely many) feasible records, and take $N_d = |\mathcal{D}|$.

[2] Bankier (2006) shows that, more generally,

$$\sum_{j=1}^{p} w_j D_j^r(x_{fj}, x_{dj}) = \sum_{j=1}^{p} w_j D_j^r(x_{fj}, x_{aj}) + \sum_{j=1}^{p} w_j D_j^r(x_{aj}, x_{dj})$$

holds for all $r \geq 1$. Expression (4.34) follows as a special case with $r = 1$.

where $\gamma \geq 1$ is set by the user. Only NMCIAs are retained, since it is desirable to change as little as possible to the failed record. Since feasible imputation actions with $\mu(I)$ only marginally higher than $\mu_{\min}$ are almost as good as those that exactly achieve the minimum, we can choose to also retain these imputation actions by taking γ slightly above 1. Retaining some nearly optimal imputation actions helps to prevent that the same donor records are used over and over again. The value $\gamma = 11/10$ has been used in practice at Statistics Canada [cf. Bankier (2006)].

Finally, an NMCIA is randomly selected. Bankier (2006) suggests selecting each NMCIA with probability

$$P(I) \propto \left(\frac{\mu_{\min}}{\mu(I)}\right)^t, \tag{4.37}$$

where $t \geq 0$ determines the selection mechanism. Taking $t = 0$ means that all NMCIAs have an equal probability of selection. By letting $t \to \infty$, only imputation actions with $\mu(I) = \mu_{\min}$ have a nonzero probability of selection; but this can be achieved more directly by taking $\gamma = 1$ in the previous step. An intermediate value of t results in a higher selection probability for imputation actions with $\mu(I)$ closer to $\mu_{\min}$. Note that assigning selection probabilities according to (4.37) captures the third objective from Bankier (1999) mentioned above.

After selection of an NMCIA, the record $\mathbf{x}_f$ has been treated. It is replaced by $\mathbf{x}_a$ in the edited data file. In this way, the original failed records are edited and imputed one at a time.

4.5.4 EFFICIENT GENERATION OF NMCIAs

We now examine certain aspects of the search for feasible NMCIAs in more detail. This discussion is mostly based on Bankier (2006).

Edits. Throughout this section, we formulate edits as conflict rules. This means that if a record satisfies all conditions given by the edit, it fails the edit and consequently contains an inconsistency. Clearly, this is just a matter of convention, and any edit can be reformulated as a conflict rule if necessary.

We consider only edits that can be written in the following form:

$$(\Delta_1 \lhd 0) \wedge \cdots \wedge (\Delta_S \lhd 0), \tag{4.38}$$

where each $\Delta_s \lhd 0$ represents a linear proposition of the form

$$\Delta_s = a_{s1}x_1 + \cdots + a_{sp}x_p - b_s, \tag{4.39}$$

and the symbol $\lhd$ stands for one of $<, >, =, \neq, \leq, \geq$. In (4.38), propositions are combined to form a conflict rule by the 'and'-operator, denoted by $\wedge$. This means that a record fails the edit if and only if it satisfies every proposition involved.

Both numerical and categorical variables x_j are allowed. We use natural numbers to represent the values of a categorical variable. It is implicitly understood that arithmetical operations in (4.39) are only performed with numerical variables, where they are meaningful. In other words, if x_j is categorical, it can only occur in propositions of the form $x_j - b_s = 0$ and $x_j - b_s \neq 0$. Here, b_s represents one of the codes from the domain of x_j.

EXAMPLE 4.5

Suppose that records consist of four variables: *Age*, *Income*, *Marital Status* and *Relationship to Head of Household*. The variables *Age* and *Income* are numerical with the set of natural numbers as domain. The other two variables are categorical. *Marital Status* takes values in the domain {Married, Unmarried, Widowed, Divorced}, which are coded as $\{1, 2, 3, 4\}$, and *Relationship to Head of Household* takes values in {Spouse, Son/Daughter, Other}, which are coded as $\{1, 2, 3\}$. The following edits are defined:

1. It is impossible for someone under 18 to be (or have been) married.
2. It is impossible for someone under 12 to earn a positive income.
3. It is impossible for someone who is not married to be the spouse of the head of the household.

We use *A*, *I*, *MS*, and *RHH* as (obvious) abbreviations of the variable names. The edits can be written in the form (4.38) as follows:

1. $(A - 18 < 0) \wedge (MS - 2 \neq 0)$
2. $(A - 12 < 0) \wedge (I > 0)$
3. $(MS - 1 \neq 0) \wedge (RHH - 1 = 0)$ ■

The six possible signs represented by $\triangleleft$ can be reduced to just $<, >, =$, by allowing negations of propositions in (4.38) as well. Namely, if $\Delta_s \leq 0$ is used in an edit, then it can be replaced by the negation $\neg(\Delta_s > 0)$, and similar arguments hold for the other discarded signs.

Suppose that, after this simplification, the distinct propositions occurring in the set of edits are given by $(\Delta_1 \triangleleft 0), \ldots, (\Delta_S \triangleleft 0)$. Now, every edit can be represented by a vector $\mathbf{e}_k = (e_{k1}, \ldots, e_{kS})^T$, with

$$e_{ks} = \begin{cases} 1 & \text{if } (\Delta_s \triangleleft 0) \text{ is part of the } k\text{th edit,} \\ -1 & \text{if } \neg(\Delta_s \triangleleft 0) \text{ is part of the } k\text{th edit,} \\ 0 & \text{otherwise.} \end{cases} \tag{4.40}$$

Moreover, we can evaluate each proposition for a given record $\mathbf{x}$, and write the result as a condition vector $\mathbf{t} = (t_1, \ldots, t_S)^T$, with

$$t_s = \begin{cases} 1 & \text{if } (\Delta_s \lhd 0) \text{ is true for the values in } \mathbf{x}, \\ -1 & \text{if } (\Delta_s \lhd 0) \text{ is false for the values in } \mathbf{x}. \end{cases} \tag{4.41}$$

For a given record, the value t_s is called its *condition result* for the sth proposition. It is easy to see that a record fails the kth edit if and only if $e_{ks} = t_s$ for all $e_{ks} \neq 0$.

EXAMPLE 4.5 (*continued*)

The following propositions occur in the set of edits in our example:

1. $A - 18 < 0$
2. $MS - 2 = 0$
3. $A - 12 < 0$
4. $I > 0$
5. $MS \quad 1 = 0$
6. $RHH - 1 = 0$

In terms of these six propositions, the three edits are represented by the following vectors:

$$\mathbf{e}_1 = (1, -1, 0, 0, 0, 0)^T,$$
$$\mathbf{e}_2 = (0, 0, 1, 1, 0, 0)^T,$$
$$\mathbf{e}_3 = (0, 0, 0, 0, -1, 1)^T.$$

Now, suppose that we are given a record with values $A = 9$, $I = 25{,}000$, $MS = 2$ and $RHH = 1$. By evaluating each proposition, we obtain the following condition vector:

$$\mathbf{t} = (1, 1, 1, 1, -1, 1)^T.$$

By comparing this vector with $\mathbf{e}_1$, $\mathbf{e}_2$, and $\mathbf{e}_3$, we see that this record fails both the second and the third edit. ■

Thinking of edits in this vector representation makes it easy to recognize redundant edits. An edit is redundant if the set of edits contains another edit such that the first edit is only failed if the second edit is failed. In that case, the first edit can be dropped from the set of edits, because the second edit suffices to find all records that fail the first edit. It is easy to see that, in terms of the vector representation, the edit given by $\mathbf{e}_k$ is redundant if there is an edit $\mathbf{e}_l$ such that all nonzero elements of $\mathbf{e}_l$ are identical to the corresponding elements of $\mathbf{e}_k$.

In Example 4.5,

$$(A - 12 < 0) \wedge (I > 0) \wedge \neg(MS - 2 = 0)$$

would be an example of a redundant edit, since any record that fails this edit also fails the second edit. Its vector representation is $\mathbf{e} = (0, -1, 1, 1, 0, 0)^T$. This vector contains the nonzero elements of $\mathbf{e}_2$ as a subset.

This vector representation of edits is used in CANCEIS to run consistency checks for records in an efficient manner; see Bankier (2006) for more details.

Generating Imputation Actions. A record $\mathbf{x}_f$ outside $\mathcal{D}$ is to be imputed, using $\mathbf{x}_d \in \mathcal{D}$ as donor record. Given the failed record and the donor record, an imputation action I is completely determined by the choice of the binary vector $\boldsymbol{\delta}$ in (4.32). It is obvious that only nonmatching fields of $\mathbf{x}_f$ and $\mathbf{x}_d$ are relevant to the imputation action, and we can define $\delta_j = 0$ for all j with $x_{fj} = x_{dj}$. For notational purposes, we assume that the nonmatching fields are those with $j = 1, \ldots, p_0$, for some $p_0 \leq p$.

In this section, we describe elements of a search algorithm to construct imputation actions in a systematic manner, using a binary tree. The concept of a binary tree was introduced in Section 3.4.7. In the root node of the binary tree, no values have been set for $\delta_1, \ldots, \delta_{p_0}$. We select one of the undecided variables, say δ_1, and construct two branches. In the nodes of the first branch, we choose to keep the original value for this field and hence take $\delta_1 = 0$; in the nodes of the second branch, we choose to impute the value from the donor and take $\delta_1 = 1$. In a similar fashion, more branches are constructed by selecting one remaining undecided variable at a time and assigning a value to it. When we have assigned a value to all variables $\delta_1, \ldots, \delta_{p_0}$, we have reached a leaf of the tree. Each leaf corresponds to one of the 2^{p_0} possible imputation actions.

Because the number of leaves to evaluate increases exponentially with the number of nonmatching fields, it is crucial to make the algorithm efficient by pruning the tree as much as possible. Since we are interested in imputation actions that are feasible—that is, the adapted record does not violate any edits—a branch may be pruned from the tree if it becomes clear that it can only lead to infeasible imputation actions. Moreover, the goal is to generate NMCIAs, as defined by (4.36), so we can also prune all branches that lead to imputation actions with a too high value of $\mu(I)$. Finally, we are only interested in imputation actions that are *essential* in the following sense: It is not possible to obtain a feasible imputation action by imputing a true subset of the variables with $\delta_j = 1$. Thus, every time a feasible imputation action is found, we can prune all branches of the binary tree leading to imputation actions that impute a set of variables which contains all the imputed variables of the current imputation action as a subset.

We now make a number of observations.

1. If in a certain node we have assigned values to, say, $\delta_1, \ldots, \delta_j$, but not to $\delta_{j+1}, \ldots, \delta_{p_0}$, then the imputation action with $\delta_{j+1} = \cdots = \delta_{p_0} = 0$ has the smallest value of $\mu(I)$ among all branches that can be constructed from

this node. This is true, because of these imputation actions, the one just mentioned has the smallest value of $D(\mathbf{x}_f, \mathbf{x}_a)$, and it follows from (4.35) that $\mu(I)$ is monotone increasing in $D(\mathbf{x}_f, \mathbf{x}_a)$, given $\mathbf{x}_d$. Bankier (2006) refers to this minimal imputation action as the *branch imputation action* for that node.

2. If the branch imputation action in a certain node is feasible, all other imputation actions that can be reached from this node may be ignored, because they are not essential. Thus, we can prune all branches that can be constructed from this node, except for the one leading to the branch imputation action.
3. We can prune all branches from a certain node, if the branch imputation action has $\mu(I) > \gamma\mu_{\min}$, with $\mu_{\min}$ the smallest value of $\mu(I)$ among feasible imputation actions that were found previously (including those found using other donors). Namely, in that case the branch imputation action is not an NMCIA according to (4.36), and by the first observation neither is any imputation action that can be reached from this node.
4. In addition, if the current node has an infeasible branch imputation action with $\mu(I) \leq \gamma\mu_{\min}$, we can identify undecided variables that cannot be imputed from the donor because the resulting imputation actions would have a too large size measure. These undecided variables must then be given a δ value of 0. In other words, all further branches with a δ value of 1 for these variables are pruned from the tree. To identify these variables, observe that

$$D(\mathbf{x}_f, \mathbf{x}_a) = \sum_{j=1}^{p} w_j \delta_j D_j(x_{fj}, x_{dj}),$$

and hence that, from (4.35),

$$\mu(I) = (2\alpha - 1)\sum_{j=1}^{p} w_j \delta_j D_j(x_{fj}, x_{dj}) + (1 - \alpha)D(\mathbf{x}_f, \mathbf{x}_d).$$

Now, suppose that I_0 is the branch imputation action in the current node and that the jth variable is undecided and satisfies

$$w_j D_j(x_{fj}, x_{dj}) > \frac{\gamma\mu_{\min} - \mu(I_0)}{2\alpha - 1}, \tag{4.42}$$

with $\mu_{\min}$ the smallest size measure among previously found NMCIAs. If we would construct a branch from the current node with $\delta_j = 1$ and denote its branch imputation action by I_0', it is easily seen that

$$\mu(I_0') = \mu(I_0) + (2\alpha - 1)w_j D_j(x_{fj}, x_{dj}) > \gamma\mu_{\min}.$$

So, the branch imputation action I_0' cannot be an NMCIA. By the previous observation, we may prune the entire branch with $\delta_j = 1$ from the tree. Thus, for every undecided variable that satisfies (4.42), we only construct further branches with $\delta_j = 0$.

These simple observations enable us to prune branches of the tree that lead to imputation actions that are not essential or have a too large size measure. Identifying branches that lead to infeasible imputation actions, which we discuss next, is more complicated. In the following discussion, we assume that the branch imputation action for the current node is infeasible, because otherwise we could immediately apply the second observation above.

An imputation action produces an adapted record $\mathbf{x}_a$, given by (4.32). Plugging this formula into (4.39), we can rewrite each proposition in terms of $\boldsymbol{\delta}$:

$$\text{(4.43)}\quad \Delta_s = \sum_{j=1}^{p} a_{sj}x_{aj} - b_s = \sum_{j=1}^{p} a_{sj}\left[\delta_j x_{dj} + (1-\delta_j)x_{fj}\right] - b_s = \sum_{j=1}^{p} a_{sj}^*\delta_j - b_s^*,$$

with $a_{sj}^* = a_{sj}(x_{dj} - x_{fj})$ and $b_s^* = b_s - \sum_{j=1}^{p} a_{sj}x_{fj}$. For a given failed record and donor record, a_{sj}^* and b_s^* are fixed, so expression (4.43) can be used to evaluate each proposition for any imputation action from the binary tree. This information is summarized in a condition vector $\mathbf{t}$, given by (4.41).

Suppose that we are in a node of the binary tree with some variables still undecided. We denote the condition vector of the branch imputation action by $\mathbf{t}^0$. For the s-th proposition, let J_s^+ denote the index set of undecided variables with $a_{sj}^* > 0$, and let J_s^- denote the index set of undecided variables with $a_{sj}^* < 0$. Furthermore, suppose that this proposition takes the form $\Delta_s < 0$. Denote the value of Δ_s for the branch imputation action of the current node by Δ_s^0, and suppose that $\Delta_s^0 < 0$. In this case, we have $t_s^0 = 1$.

We can now answer three questions:

1. Is it possible to construct an imputation action from the current node with $t_s \neq t_s^0$? *Answer:* Yes, if and only if

$$\Delta_s^0 + \sum_{j \in J_s^+} a_{sj}^* \geq 0.$$

2. If the answer to the first question is "Yes," is there an undecided variable that must be imputed to obtain an imputation action with $t_s \neq t_s^0$? *Answer:* If there is a $j_1 \in J_s^+$ such that

$$\Delta_s^0 + \sum_{j \in J_s^+ \setminus \{j_1\}} a_{sj}^* < 0,$$

then we have to take $\delta_{j_1} = 1$ to be able to change the condition result.

3. If the answer to the first question is "Yes," is there an undecided variable that must not be imputed to obtain an imputation action with $t_s \neq t_s^0$? *Answer:* If there is a $j_1 \in J_s^-$ such that

$$\Delta_s^0 + \sum_{j \in J_s^+} a_{sj}^* + a_{sj_1}^* < 0,$$

then we have to take $\delta_{j_1} = 0$ to be able to change the condition result.

We have answered these questions for a proposition of the form $\Delta_s < 0$ with $\Delta_s^0 < 0$. Similar answers can be derived for three other cases: $\Delta_s < 0$ with $\Delta_s^0 \geq 0$, $\Delta_s > 0$ with $\Delta_s^0 > 0$, and $\Delta_s > 0$ with $\Delta_s^0 \leq 0$. Table 4.2 displays the conditions for each of these cases. It is assumed in panels (b) and (c) that the condition from panel (a) holds, because otherwise the condition result for this proposition cannot be changed anyway.

Two more cases remain: $\Delta_s = 0$ with $\Delta_s^0 = 0$ and $\Delta_s = 0$ with $\Delta_s^0 \neq 0$. The former is not difficult: in this case it is possible to change the condition result provided there is at least one undecided variable with $a_{sj}^* \neq 0$ (in other

TABLE 4.2 Establishing the Existence of Imputation Actions with $t_s \neq t_s^0$

	Case	Condition
	(a) Ensuring that the Condition Result Can Be Changed	
1.	$\Delta_s < 0,\ \Delta_s^0 < 0$	$\Delta_s^0 + \sum_{j \in J_s^+} a_{sj}^* \geq 0$
2.	$\Delta_s < 0,\ \Delta_s^0 \geq 0$	$\Delta_s^0 + \sum_{j \in J_s^-} a_{sj}^* < 0$
3.	$\Delta_s > 0,\ \Delta_s^0 > 0$	$\Delta_s^0 + \sum_{j \in J_s^-} a_{sj}^* \leq 0$
4.	$\Delta_s > 0,\ \Delta_s^0 \leq 0$	$\Delta_s^0 + \sum_{j \in J_s^+} a_{sj}^* > 0$
5.	$\Delta_s = 0,\ \Delta_s^0 = 0$	$J_s^+ \cup J_s^- \neq \emptyset$
6.	$\Delta_s = 0,\ \Delta_s^0 \neq 0$	See end of section
	(b) Variable j_1 Must Be Imputed to Change the Condition Result	
1.	$\Delta_s < 0,\ \Delta_s^0 < 0$	$\Delta_s^0 + \sum_{j \in J_s^+ \setminus \{j_1\}} a_{sj}^* < 0$ for $j_1 \in J_s^+$
2.	$\Delta_s < 0,\ \Delta_s^0 \geq 0$	$\Delta_s^0 + \sum_{j \in J_s^- \setminus \{j_1\}} a_{sj}^* \geq 0$ for $j_1 \in J_s^-$
3.	$\Delta_s > 0,\ \Delta_s^0 > 0$	$\Delta_s^0 + \sum_{j \in J_s^- \setminus \{j_1\}} a_{sj}^* > 0$ for $j_1 \in J_s^-$
4.	$\Delta_s > 0,\ \Delta_s^0 \leq 0$	$\Delta_s^0 + \sum_{j \in J_s^+ \setminus \{j_1\}} a_{sj}^* \leq 0$ for $j_1 \in J_s^+$
5.	$\Delta_s = 0,\ \Delta_s^0 = 0$	$J_s^+ \cup J_s^- = \{j_1\}$
6.	$\Delta_s = 0,\ \Delta_s^0 \neq 0$	See end of section
	(c) Variable j_1 Must Not Be Imputed to Change the Condition Result	
1.	$\Delta_s < 0,\ \Delta_s^0 < 0$	$\Delta_s^0 + \sum_{j \in J_s^+} a_{sj}^* + a_{sj_1}^* < 0$ for $j_1 \in J_s^-$
2.	$\Delta_s < 0,\ \Delta_s^0 \geq 0$	$\Delta_s^0 + \sum_{j \in J_s^-} a_{sj}^* + a_{sj_1}^* \geq 0$ for $j_1 \in J_s^+$
3.	$\Delta_s > 0,\ \Delta_s^0 > 0$	$\Delta_s^0 + \sum_{j \in J_s^-} a_{sj}^* + a_{sj_1}^* > 0$ for $j_1 \in J_s^+$
4.	$\Delta_s > 0,\ \Delta_s^0 \leq 0$	$\Delta_s^0 + \sum_{j \in J_s^+} a_{sj}^* + a_{sj_1}^* \leq 0$ for $j_1 \in J_s^-$
5.	$\Delta_s = 0,\ \Delta_s^0 = 0$	None
6.	$\Delta_s = 0,\ \Delta_s^0 \neq 0$	See end of section

words: $J_s^+ \cup J_s^- \neq \emptyset$). If there happens to be exactly one undecided variable with $a_{sj}^* \neq 0$, then this variable must be imputed in order to change the condition result. Otherwise, the answer to the second question is "No." The answer to the third question is always "No" in this case. These results can also be found in Table 4.2.

The final case, $\Delta_s = 0$ with $\Delta_s^0 \neq 0$, is tricky. We defer the discussion of this case to the end of this section and assume for now that the three questions can be answered for this case also.

Consider an edit, written as $\mathbf{e}_k$ in vectorized form (4.40). If the branch imputation action does not fail $\mathbf{e}_k$, we can ignore this edit. So, suppose that the branch imputation action fails edit $\mathbf{e}_k$; that is, $e_{ks} = t_s^0$ for every $e_{ks} \neq 0$. Based on the previous evaluation of propositions, we can determine, for each proposition with $e_{ks} \neq 0$, whether an imputation action with $t_s \neq t_s^0$ can be constructed from the current node. Denote the index set of all propositions with $e_{ks} \neq 0$ for which such an imputation action exists by S_k.

It is possible that $S_k = \emptyset$. This means that all imputation actions that can be reached from this node are infeasible, because they violate edit $\mathbf{e}_k$. Therefore, this node and all branches extending down from it may be pruned. If S_k is nonempty, we check the following:

- Is there a variable j_1 that must be imputed to change the condition result for each proposition in S_k? If the answer is "Yes," this variable is called *essential to impute* for this edit. If j_1 is essential to impute, all feasible imputation actions that can be reached from this node have $\delta_{j_1} = 1$.
- Is there a variable j_1 that must not be imputed to change the condition result for each proposition in S_k? If the answer is "Yes," this variable is called *essential not to impute* for this edit. If j_1 is essential not to impute, all feasible imputation actions that can be reached from this node have $\delta_{j_1} = 0$.

In this fashion, each edit is treated. If we come across an edit with $S_k = \emptyset$, we are done because no feasible imputation actions can be constructed from the current node. Otherwise, we may find a set of undecided variables that are essential to impute or essential not to impute.[3] By choosing the required δ value for these variables, the number of undecided variables becomes smaller. Effectively, we prune all branches extending down from the current node except for the one corresponding to the correct choice for all variables that are essential to impute and essential not to impute.

EXAMPLE 4.5 (*continued*)

It was established previously that the record with values $A = 9$, $I = 25{,}000$, $MS = 2$, and $RHH = 1$ is inconsistent with respect to the edits.

[3] Note that it is impossible to come across a variable that is both essential to impute and essential not to impute. In that case, the current branch leads only to infeasible imputation actions and therefore would have been pruned at an earlier stage.

Suppose that we want to use the NIM to edit this record, by imputing values from a donor record with $A = 8$, $I = 0$, $MS = 2$, and $RHH = 2$. The reader may verify that this record qualifies as a donor, because it does not fail any edits. We write the vector describing the imputation actions as $\boldsymbol{\delta} = (\delta_A, \delta_I, \delta_{MS}, \delta_{RHH})^T$.

We begin by observing that the failed record matches the donor record on the variable MS. As there is no point in imputing this variable, we take $\delta_{MS} = 0$. There are three nonmatching variables, hence there are $2^3 = 8$ possible imputation actions. In this example, there is not much work in checking every possible imputation action. However, for the purpose of illustration, we shall use the method described above to reduce the search for feasible imputation actions.

Expressing the six propositions in terms of $\boldsymbol{\delta}$ by (4.43), we obtain:

1. $-\delta_A - 9 < 0$
2. $0 = 0$
3. $-\delta_A - 3 < 0$
4. $-25000\delta_I + 25000 > 0$
5. $1 = 0$
6. $\delta_{RHH} = 0$

At the root node of the binary tree, three undecided variables remain, namely A, I, and RHH. The branch imputation action for this node does not impute any variables; hence the resulting record is just the original failed record. We have already evaluated its condition vector: $\mathbf{t}^0 = (1, 1, 1, 1, -1, 1)^T$. Thus, the branch imputation action fails edits $\mathbf{e}_2 = (0, 0, 1, 1, 0, 0)^T$ and $\mathbf{e}_3 = (0, 0, 0, 0, -1, 1)^T$. For these edits, we consider the propositions involved.

Edit $\mathbf{e}_2$ involves propositions $(\Delta_3 < 0)$ and $(\Delta_4 > 0)$. Considering $(\Delta_3 < 0)$, we see, either by inspection or by formally using the expression in Table 4.2, that there is no imputation action with $t_3 \neq t_3^0 = 1$. To change the condition result for $(\Delta_4 > 0)$, we must take $\delta_I = 1$. Since these are the only propositions involved, we conclude that any imputation action that does not fail $\mathbf{e}_2$ must have $\delta_I = 1$. Thus, the variable I is essential to impute here. Similarly, for edit $\mathbf{e}_3$, which involves propositions $\neg(\Delta_5 = 0)$ and $(\Delta_6 = 0)$, we find that the variable RHH is essential to impute: any imputation action that does not fail $\mathbf{e}_3$ must have $\delta_{RHH} = 1$.

Thus, we can move directly from the root node of the binary tree to the node corresponding with $\delta_I = 1$, $\delta_{MS} = 0$, $\delta_{RHH} = 1$, and δ_A undecided. In this node, the branch imputation action $\boldsymbol{\delta} = (0, 1, 0, 1)^T$ has the following condition result vector: $\mathbf{t}^0 = (1, 1, 1, -1, -1, -1)^T$. By comparing this vector with $\mathbf{e}_1$, $\mathbf{e}_2$, and $\mathbf{e}_3$, we see that the branch imputation action is feasible. Hence, there is no need to construct more branches of the tree, because the resulting imputation actions will not be essential.

The feasible imputation action we just found imputes the variables I and RHH. The resulting consistent record has $A = 9$, $I = 0$, $MS = 2$ and $RHH = 2$. This is the only essential feasible imputation action that can be obtained from this donor. ■

We have briefly discussed ways of pruning the binary tree, by identifying as soon as possible imputation actions that are infeasible, not essential or not NMCIAs, thus making the construction of imputation actions more efficient. For a more detailed discussion, and a description of how these methods are implemented in the algorithm used by CANCEIS, the reader is referred to Bankier (2006).

We conclude this section by examining case 6 of Table 4.2. This case involves a proposition of the form $\Delta_s = 0$ for which the branch imputation action has $\Delta_s^0 \neq 0$. Thus, either $\Delta_s^0 > 0$ or $\Delta_s^0 < 0$ holds. If $\Delta_s^0 > 0$, we have to impute undecided variables with $a_{sj}^* < 0$ to change the condition result. Therefore,

$$\Delta_s^0 + \sum_{j \in J_s^-} a_{sj}^* \leq 0$$

is a necessary condition for the possibility of changing the condition result. Similarly, if $\Delta_s^0 < 0$, a necessary condition for the possibility of changing the condition result is

$$\Delta_s^0 + \sum_{j \in J_s^+} a_{sj}^* \geq 0.$$

If the relevant condition is not satisfied, the condition result cannot be changed. However, unlike the conditions we found for the other cases, these conditions are not sufficient: It is possible that the condition result cannot be changed even if the relevant condition is satisfied. Thus, the above conditions can be used to detect some infeasible branches of the binary tree, but not all of them.

To determine with certainty whether the condition result can be changed, we have to assess whether there exists an index set of undecided variables $J \subseteq (J_s^+ \cup J_s^-)$ such that

$$\Delta_s^0 + \sum_{j \in J} a_{sj}^* = 0.$$

However, there appears to be no way of assessing this, other than simply checking every possible imputation action. Since this amounts to generating all remaining branches of the binary tree, there is no efficiency to be gained here.

Finding some of the undecided variables that must (not) be imputed to change the condition result is possible in this case. If $\Delta_s^0 > 0$, the conditions from case 3 in panels (b) and (c) of Table 4.2 can be used to identify a subset of these variables. In a similar fashion, the conditions from case 1 can be used if $\Delta_s^0 < 0$. As a result of imputing these variables, the sign of Δ_s^0 may change. If this happens, more undecided variables that must (not) be imputed can be identified in an iterative process. The identification of such variables stops once the sign of Δ_s^0 remains constant. Unless we obtain exactly $\Delta_s^0 = 0$ after imputing these variables, we do not know whether all variables that must (not) be imputed have been identified.

Consequently, it is possible that we only identify a subset of essential (not) to impute variables for a failed edit, if we encounter propositions from case 6. Thus, the search algorithm may involve redundant steps, because some branches of the binary tree that could have been pruned are still constructed. This does not affect the correctness of the algorithm, however.

4.5.5 NIM VERSUS FELLEGI–HOLT

In this section, we compare the NIM with traditional Fellegi–Holt-based editing, from a theoretical point of view. In particular, we identify situations where one of the two might be preferred. For a comparison of the two editing approaches in practice, see, for example, Chen, Thibaudeau, and Winkler (2003).

In comparison with the Fellegi–Holt-based editing methodology, the NIM has a number of advantages:

- The NIM works fast in practice. A limited number of imputation actions is generated, using a limited number of donors, and one of these imputation actions is then selected. Theoretically, the number of imputation actions generated by the NIM can be high, but the computational work is greatly reduced by making use of the methods described in Section 4.5.4. Hence, data can be edited by the NIM at great speed.
- The NIM is able to handle numerical and categorical data simultaneously. At the time the NIM was first developed, all existing Fellegi–Holt-based applications could handle either numerical data or categorical data, but not a combination (cf. Bankier et al., 1994). We have seen in this chapter that a Fellegi–Holt-type error localization problem *can* be solved for mixed data; in particular, a Fellegi–Holt-based algorithm that handles a combination of numerical and categorical data has been implemented in the software package SLICE, developed at Statistics Netherlands. While the ability to handle mixed data is therefore no longer a "unique selling point," the NIM still outperforms Fellegi–Holt-based methods in terms of computational speed when applied to a mix of numerical and categorical data.
- The NIM uses properties of the distribution of donors to identify plausible imputation actions in a natural way. For instance, in the example of Section 4.5.2 the NIM would impute a two adult/four child household with a much higher probability than a one adult/five child household, because households of the first type are much more common among donors than households of the second type. This approach avoids the false inflation of small subpopulations. This objective is more difficult to achieve in

Fellegi–Holt-based editing, because of the strict separation between the localization of erroneous fields and the imputation of new values.[4]

These advantages are particularly relevant in the case of a population census, for which the NIM was originally developed. Here, a very large number of records has to be edited, so computational efficiency is important. A typical census form contains mostly categorical fields (e.g., *Sex*, *Marital Status*), but also a small number of numerical fields (e.g., *Age*). Therefore, the editing methodology should be able to handle mixed data. Finally, since a population census concerns by definition the entire population, there are many small subpopulations on which statistics will be published (e.g., centenarians, households with many children, ethnic minorities). It is important that the editing and imputation methodology does not falsely inflate the size of these subpopulations.

The NIM also has properties that make it unsuited to certain applications:

- As mentioned in Section 4.5.3, the NIM requires a large number of donor records. Use of the NIM is only advised if the majority of incoming records does not require any imputation. If the number of donor records is smaller than or even about the same size as the number of records to impute, the same donor may be used to impute many records. This is likely to produce unwanted effects in subsequent estimates.
- Since hot deck donor imputation forms an integral part of the NIM, the methodology is unsuited to applications where this form of imputation does not work well. For instance, in data collected for structural business statistics, there are many numerical variables that should conform to many interrelated balance edits; it is virtually impossible to find a donor record that produces a feasible imputation action here. In this case, Fellegi–Holt-based editing is to be preferred. More generally, the NIM appears to be better suited to problems with (mostly) categorical data than problems with (mostly) numerical data.
- The NIM uses the distribution of donors as an approximation to the population distribution. This approximation works very well in the case of a population census with little editing required. In the case of a sample survey, some form of weighting is usually necessary for the sample to correctly represent the population. Also, there may be selectivity due to nonresponse. In principle, it is possible to incorporate sample weights in hot deck imputation [see Andridge and Little (2009)], but the current implementation of the NIM in CANCEIS does not have this feature.

Summarizing, we can say that the NIM works particularly well for particular applications, especially census editing. For these applications, it outperforms Fellegi–Holt-based editing. However, Fellegi–Holt-based editing is more widely applicable and can be used in situations where the NIM would not work at all.

[4] In Fellegi–Holt-based editing, this can partly be achieved by dynamically increasing or decreasing reliability weights, depending on the availability of plausible imputations.

REFERENCES

Andridge, R. R., and R. J. Little (2009), The Use of Sample Weights in Hot Deck Imputation. *Journal of Official Statistics 25*, pp. 21–36.

Bankier, M. (1999), *Experience with the New Imputation Methodology Used in the 1996 Canadian Census with Extensions for Future Censuses*. Working Paper No. 24, UN/ECE Work Session on Statistical Data Editing, Rome.

Bankier, M. (2006), *Imputing Numeric and Qualitative Variables Simultaneously*. Memo, Statistics Canada, Social Survey Methods Division.

Bankier, M., and S. Crowe (2009), *Enhancements to the 2011 Canadian Census E&I System*. Working Paper No. 15, UN/ECE Work Session on Statistical Data Editing, Neuchâtel.

Bankier, M., J.-M. Fillion, M. Luc, and C. Nadeau (1994), Imputing Numeric and Qualitative Variables Simultaneously. In: *Proceedings of the Section on Survey Research Methods*, American Statistical Association, pp. 242–247.

Bankier, M., M. Lachance, and P. Poirier (2000), *2001 Canadian Census Minimum Change Donor Imputation Methodology*. Working Paper No. 17, UN/ECE Work Session on Statistical Data Editing, Cardiff.

Barcaroli, G., and M. Venturi (1996), *The Probabilistic Approach to Automatic Edit and Imputation: Improvements of the Fellegi–Holt Methodology*. UN/ECE Work Session on Statistical Data Editing, Voorburg.

Ben-Ari, M. (2001), *Mathematical Logic for Computer Science*, second edition. Springer-Verlag, London.

CANCEIS (2006), *CANCEIS Version 4.5, User's Guide*. Statistics Canada, Social Survey Methods Division.

Chandru, V., and J. N. Hooker (1999), *Optimization Methods for Logical Inference*. John Wiley & Sons, New York.

Chen, B., Y. Thibaudeau, and W. E. Winkler (2003), *A Comparison Study of ACS If-Then-Else, NIM, DISCRETE Edit and Imputation Systems Using ACS Data*. Working Paper No. 7, UN/ECE Work Session on Statistical Data Editing, Madrid.

Chvátal, V. (1983), *Linear Programming*. W. H. Freeman and Company, New York.

Daalmans, J. (2000), *Automatic Error Localization of Categorical Data*. Report (research paper 0024), Statistics Netherlands, Voorburg.

De Waal, T. (2001a), Solving the Error Localization Problem by Means of Vertex Generation. *Survey Methodology 29*, pp. 71–79.

De Waal, T. (2001b), SLICE: Generalised Software for Statistical Data Editing. In: *Proceedings in Computational Statistics*, J. G. Bethlehem and P. G. M. Van der Heijden, eds. Physica-Verlag, New York, pp. 277–282.

De Waal, T. (2003), *Processing of Erroneous and Unsafe Data*. Ph.D. Thesis, Erasmus University, Rotterdam (see also www.cbs.nl).

Duffin, R. J. (1974), On Fourier's Analysis of Linear Inequality Systems. *Mathematical Programming Studies 1*, pp. 71–95.

Fellegi, I. P., and D. Holt (1976), A Systematic Approach to Automatic Edit and Imputation. *Journal of the American Statistical Association 71*, pp. 17–35.

Garfinkel, R. S., A. S. Kunnathur, and G. E. Liepins (1986), Optimal Imputation of Erroneous Data: Categorical Data, General Edits. *Operations Research 34*, pp. 744–751.

Hooker, J. (2000), *Logic-Based Methods for Optimization: Combining Optimization and Constraint Satisfaction.* John Wiley & Sons, New York.

Marriott, K., and P. J. Stuckey (1998), *Programming with Constraints—An Introduction.* MIT Press, Cambridge, MA.

Nemhauser, G. L., and L. A. Wolsey (1988), *Integer and Combinatorial Optimisation.* John Wiley & Sons, New York.

Robinson, J. A. (1965), A Machine-Oriented Logic Based on the Resolution Principle. *Journal Assoc. Comput. Mach. 12*, pp. 23–41.

Robinson, J. A. (1968), The Generalized Resolution Principle. In: *Machine Intelligence 3*, E. Dale and D. Michie, eds. Oliver and Boyd, Edinburgh, pp. 77–93.

Russell, S., and P. Norvig (1995), *Artificial Intelligence, A Modern Approach.* Prentice-Hall, Englewood Cliffs, NJ.

Warners, J. P. (1999), *Non-Linear Approaches to Satisfiability Problems.* Ph.D. Thesis, Eindhoven University of Technology.

Williams, H. P., and S. C. Brailsford (1996), Computational Logic and Integer Programming. In: *Advances in Linear and Integer Programming*, J. E. Beasley, ed. Clarendon Press, Oxford, pp. 249–281.

Winkler, W. E. (1995), *Editing Discrete Data.* UN/ECE Work Session on Statistical Data Editing, Athens.

Winkler, W. E. (1999), *State of Statistical Data Editing and Current Research Problems.* Working Paper No. 29, UN/ECE Work Session on Statistical Data Editing, Rome.

CHAPTER FIVE

Automatic Editing: Extensions to Integer Data

5.1 Introduction

In the present chapter we extend the branch-and-bound algorithm of Chapters 3 and 4 to a mix of categorical, continuous, and integer-valued data. The error localization problem for a mix of categorical, continuous, and integer-valued data is the same as the error localization problem for a mix of categorical and continuous data (see Section 4.2) except that the integer variables have to attain an integer value.

The remainder of this chapter is organized as follows. Section 5.2 sketches the error localization problem for a mix of categorical, continuous, and integer-valued data by means of an example. Section 5.4 extends the branch-and-bound algorithm described in Chapters 3 and 4 to a mix of categorical, continuous, and integer data. Essential in this extended algorithm is Fourier–Motzkin elimination for integer data, which we describe in Section 5.3. This elimination method is due to Pugh [cf. Pugh (1992) and Pugh and Wonnacott (1994)], who applied this technique to develop so-called array data dependence testing algorithms. Section 5.5 discusses a heuristic approach based on the exact algorithm described in Section 5.4. This heuristic procedure is easier to implement and maintain than the exact algorithm. Computational results for this heuristic procedure are given in Section 5.6. We conclude the chapter with a brief discussion in Section 5.7.

This chapter is for a substantial part based on De Waal (2005).

Handbook of Statistical Data Editing and Imputation, by T. de Waal, J. Pannekoek, and S. Scholtus

5.2 An Illustration of the Error Localization Problem for Integer Data

We start by illustrating the error localization problem for a mix of continuous and integer data by means of an example. We also sketch the idea of our solution method for such data, which basically consists of testing whether all integer-valued variables involved in a solution to the corresponding continuous error localization problem—that is, the error localization problem where all numerical variables are assumed to be continuous—can indeed attain integer values. For the continuous error localization problem we use the branch-and-bound algorithm of Sections 3.4.7 and 4.4.

EXAMPLE 5.1

Suppose a set of edits is given by

$$T = P + C, \tag{5.1}$$

$$0.5T \leq C, \tag{5.2}$$

$$C \leq 1.1T, \tag{5.3}$$

$$T \leq 550N, \tag{5.4}$$

$$320N \leq C, \tag{5.5}$$

$$T \geq 0, \tag{5.6}$$

$$C \geq 0, \tag{5.7}$$

$$N \geq 0, \tag{5.8}$$

where T denotes the turnover of an enterprise, P its profit, C its costs, and N the number of employees. The turnover, profit, and costs are continuous variables, the number of employees an integer one.

Let us consider a specific record with values $T = 5060, P = 2020, C = 3040$, and $N = 5$. This record fails edit (5.4). We apply the Fellegi–Holt paradigm with all reliability weights set to 1 (see Sections 3.1 and 3.4.1), and we try to make the record satisfy all edits by changing as few variables as possible. As T and N occur in the failed edit, it might be possible to satisfy all edits by changing the value of one of these variables only. However, if we were to change the value of T, we would also need to change the value of P or C in order not to violate (5.1). We therefore start by considering the option of changing N. We first treat N as a continuous variable. To test then whether N can be changed so that all edits (5.1) to (5.8) become satisfied, we eliminate N by means of Fourier–Motzkin elimination [cf. Duffin (1974), Chvátal (1983), and Schrijver (1986); see also Section 3.4.3 of the present book]. We combine all upper bounds on N [in this case only (5.5)] with all lower bounds on N [in this case (5.4) and (5.8)] to eliminate N from these edits. We obtain a new constraint, given by

(5.9) $320T \leq 550C$ [combination of (5.4) and (5.5)].

The constraints not involving N [i.e., (5.1), (5.2), (5.3), (5.6), (5.7), and (5.9)] are all satisfied by the original values of T, P, and C. A fundamental property of Fourier–Motzkin elimination is that a set of (in)equalities can be satisfied if and only if the set of (in)equalities after the elimination of a variable can be satisfied. This implies that the edits (5.1) to (5.8) can be satisfied by changing the value of N only. That is, if N were continuous, the (only) optimal solution to the above error localization problem would be: change the value of N. However, N is an integer-valued variable. So, we need to test whether a feasible *integer* value for N exists. By filling in the values for T, P, and C in (5.4) and (5.5), we find $9.2 \leq N \leq 9.5$. In other words, a feasible integer value for N does not exist. Changing the value of N is hence not a solution to this error localization problem.

The next best solution to the continuous error localization problem is given by: change the values of T, P, and C (see Section 3.4.7 for an algorithm to obtain this solution). This is obviously also a feasible solution to the error localization problem for continuous and integer data under consideration, as in this solution variable N retains its original value, i.e. 5, which is integer. It is the (only) optimal solution to our problem because this is the best solution to the corresponding continuous error localization problem for which all integer-valued variables can indeed attain integer values. ■

In this example it is quite easy to check whether a solution to the continuous error localization problem is also a solution to the error localization problem for continuous and integer data. In general, this is not the case, however. In Sections 5.3 and 5.4 we describe in detail how to test whether integer variables involved in a solution to the continuous error localization problem can indeed attain feasible integer values.

5.3 Fourier–Motzkin Elimination in Integer Data

An important technique used in the algorithm described in Section 3.4.7 is Fourier–Motzkin elimination for eliminating a continuous variable from a set of linear (in)equalities. Fourier–Motzkin elimination can be extended to integer data in several ways. For example, Dantzig and Eaves (1973) and Williams (1976, 1983) describe extensions of Fourier–Motzkin elimination to integer programming problems. Unfortunately, these methods seem too time-consuming in many practical cases. Pugh (1992) proposes an alternative extension that he refers to as the Omega test. Pugh (1992) and Pugh and Wonnacott (1994) claim a good performance of this test for many practical cases. Below we briefly explain the Omega test. For more details we refer to Pugh (1992), and Pugh and Wonnacott (1994).

The Omega test has been designed to determine whether an integer-valued solution to a set of linear (in)equalities exists. For the moment we assume that all variables x_j are integer-valued ($j = 1, \ldots, p$), where p denotes the number of variables. Suppose linear (in)equality k ($k = 1, \ldots, K$) is given by

$$a_{k1}x_1 + \cdots + a_{kp}x_p + b_k \geq 0$$

or by

$$a_{k1}x_1 + \cdots + a_{kp}x_p + b_k = 0.$$

The $a_{kj}(j = 1, \ldots, p;\ k = 1, \ldots, K)$ are assumed to be rational numbers. To simplify our notation, we define $x_0 = 1$ and $a_{k0} = b_k$ ($k = 1, \ldots, K$) and rewrite the above linear (in)equality as

$$a_{k0}x_0 + a_{k1}x_1 + \cdots + a_{kp}x_p \geq 0 \tag{5.10}$$

or

$$a_{k0}x_0 + a_{k1}x_1 + \cdots + a_{kp}x_p = 0, \tag{5.11}$$

respectively.

Without loss of generality we assume that redundant equalities have been removed and that all (in)equalities are normalized—that is, that all a_{kj} ($j = 0, \ldots, p;\ k = 1, \ldots, K$) are integer and the greatest common divisor of the a_{kj} in each constraint k equals 1. All variables x_j ($j = 0, \ldots, p$) are integer-valued in this section.

We start by "eliminating" all equalities until we arrive at a new problem involving only inequalities. In this context, we say that all equalities have been eliminated once we have transformed the original system of equalities (5.10) and inequalities (5.11) into an equivalent system of (in)equalities of the following type:

$$x'_k = \sum_{j>k} d'_{kj}x'_j \qquad \text{for } k = 0, \ldots, s-1, \tag{5.12}$$

$$\sum_{j\geq s} d'_{kj}x'_j \geq 0 \qquad \text{for } k = s, \ldots, K', \tag{5.13}$$

where s is the number of equalities in the system (5.12), and the d'_{kj} are integer. The x'_j are a permutation of the x_j, possibly supplemented by some additional, auxiliary variables (see Section 5.3.1). We call a set of equalities (5.10) and inequalities (5.11) equivalent to a set of equalities (5.12) and inequalities (5.13) if a solution to the system (5.10) and (5.11) can be extended to a corresponding solution to the system (5.12) and (5.13), and conversely a solution to the system (5.12) and (5.13) is also a solution to the system (5.10) and (5.11) if we disregard the additional variables. In (5.12) and (5.13), the first s x'_j, which are only involved in equalities, are expressed in terms of the remaining variables, which may also be involved in inequalities. Owing to the possible introduction of additional variables, the system (5.12) and (5.13) may have more constraints than the original system (5.10) and (5.11), so $K' \geq K$.

The original system (5.10) and (5.11) has an integer-valued solution if and only if the system (5.13) has an integer-valued solution. Namely, an integer solution for the x'_j $(j \geq s)$ to the system (5.13) yields an integer solution to the system consisting of (5.12) plus (5.13), by applying back-substitution to the x'_j $(j < s)$. In other words, to check whether a system (5.10) and (5.11) has an integer-valued solution, we only need to check whether the inequalities (5.13) of the equivalent system (5.12) and (5.13) have an integer-valued solution. In this sense the equalities of (5.11) have been eliminated once we have transformed a system given by (5.10) and (5.11) into an equivalent system given by (5.12) and (5.13).

5.3.1 ELIMINATING EQUALITIES

We now discuss how to eliminate an equality. As usual we denote the number of numerical—in this section: integer-valued—variables by p. We define the operation $c \,\overline{\text{mod}}\, d$ involving two integers c and d by

$$c \,\overline{\text{mod}}\, d = c - d\lfloor c/d + 1/2 \rfloor, \tag{5.14}$$

where $\lfloor y \rfloor$ denotes the largest integer less than or equal to y. If d is odd, the value of $c \,\overline{\text{mod}}\, d$ lies in $[-(d-1)/2, (d-1)/2]$. If d is even, the value of $c \,\overline{\text{mod}}\, d$ lies in $[-d/2, d/2 - 1]$. If $c/d - \lfloor c/d \rfloor < 1/2$, then $c \,\overline{\text{mod}}\, d = c \bmod d$. If $c/d - \lfloor c/d \rfloor \geq 1/2$, then $c \,\overline{\text{mod}}\, d = -c \bmod d$. Here, the mod d operator assumes values in $[0, d-1]$.

To eliminate an equality s given by

$$\sum_{j=0}^{p} a_{sj} x_j = 0, \tag{5.15}$$

we select an r such that $a_{sr} \neq 0$ and $|a_{sr}|$ has the smallest value among the a_{sj} $(j = 0, \ldots, p)$. If $|a_{sr}| = 1$, we eliminate the equality by using this equality to express x_r in terms of the other variables, and substitute this expression for x_r into the other (in)equalities. Otherwise, we define $\gamma = |a_{sr}| + 1$. Now we introduce a new variable σ defined by

$$\gamma\sigma = \sum_{j=0}^{p} (a_{sj}\ \overline{\mathrm{mod}}\ \gamma)x_j. \tag{5.16}$$

This variable σ is integer-valued. This can be shown as follows.

(5.17)

$$\sum_{j=0}^{p} (a_{sj}\ \overline{\mathrm{mod}}\ \gamma)x_j = \sum_{j=0}^{p} (a_{sj} - \gamma \lfloor a_{sj}/\gamma + 1/2 \rfloor)x_j = -\sum_{j=0}^{p} \gamma \lfloor a_{sj}/\gamma + 1/2 \rfloor x_j,$$

where we have used (5.15). So, σ equals $-\sum_{j=0}^{p} \lfloor a_{sj}/\gamma + 1/2 \rfloor)x_j$, which is integer because the x_j $(j = 0, \ldots, p)$ and their coefficients in (5.17) are integer.

It is easy to see that $a_{sr}\ \overline{\mathrm{mod}}\ \gamma = -\mathrm{sign}(a_{sr})$, where $\mathrm{sign}(y) = 1$ if $y > 0$, $\mathrm{sign}(y) = 0$ if $y = 0$, and $\mathrm{sign}(y) = -1$ if $y < 0$. Now, we use (5.16) to express x_r in terms of the other variables.

$$x_r = -\mathrm{sign}(a_{sr})\gamma\sigma + \sum_{j=0, j\neq r}^{p} \mathrm{sign}(a_{sr})(a_{sj}\ \overline{\mathrm{mod}}\ \gamma)x_j. \tag{5.18}$$

Substituting (5.18) into the original equality (5.15) gives

$$-|a_{sr}|\gamma\sigma + \sum_{j=0, j\neq r}^{p} (a_{sj} + |a_{sr}| a_{sj}\ \overline{\mathrm{mod}}\ \gamma)x_j = 0. \tag{5.19}$$

Because $|a_{sr}| = \gamma - 1$, (5.19) can be written as

$$-|a_{sr}|\gamma\sigma + \sum_{j=0, j\neq r}^{p} (a_{sj} - (a_{sj}\ \overline{\mathrm{mod}}\ \gamma) + \gamma(a_{sj}\ \overline{\mathrm{mod}}\ \gamma))x_j = 0. \tag{5.20}$$

Using (5.14) on (5.20) and dividing by γ gives

$$-|a_{sr}|\sigma + \sum_{j=0, j\neq r}^{p} (\lfloor a_{sj}/\gamma + 1/2 \rfloor + (a_{sj}\ \overline{\mathrm{mod}}\ \gamma))x_j = 0. \tag{5.21}$$

In (5.21), all coefficients are integer-valued.

It is clear that if the coefficient of variable x_j $(j = 0, \ldots, p)$ equals zero in (5.15), the corresponding coefficient in (5.21) also equals zero. It is also clear that the absolute value of the coefficient of σ in (5.21) is equal to the absolute value of the coefficient of x_r in (5.15). However, for all other variables with a nonzero coefficient in (5.15), the absolute value of the corresponding coefficient in (5.21) is smaller than the absolute value of the coefficient in (5.15). To prove this statement, we first rewrite the coefficient of x_j ($j \neq r$) in (5.21) in the following way:

$$\begin{aligned}\lfloor a_{sj}/\gamma + 1/2\rfloor &+ (a_{sj}\ \overline{\text{mod}}\ \gamma)\\ &= \lfloor a_{sj}/\gamma + 1/2\rfloor + a_{sj} - \gamma\lfloor a_{sj}/\gamma + 1/2\rfloor\\ &= -|a_{sr}|\lfloor\frac{a_{sj}}{|a_{sr}|+1} + \frac{1}{2}\rfloor + a_{sj}, \equiv \hat{a}_{sj},\end{aligned}$$

where we have used again that $\gamma = |a_{sr}| + 1$. We now consider the cases where a_{sj} is positive and negative separately. If $a_{sj} > 0$, then $a_{sj} \geq |a_{sr}|$ by our choice of r. Suppose $a_{sj} = \lambda|a_{sr}|$, where $\lambda \geq 1$. We then have

$$\hat{a}_{sj} = a_{sj}\left(-\frac{1}{\lambda}\lfloor\frac{a_{sj}}{|a_{sr}|+1} + \frac{1}{2}\rfloor + 1\right).$$

Using

$$1 \leq \lfloor\frac{a_{sj}}{|a_{sr}|+1} + \frac{1}{2}\rfloor \leq \lambda$$

for all possible values of $|a_{sr}|$, we obtain $0 \leq \hat{a}_{sj} \leq (1 - (1/\lambda))\, a_{sj}$. Hence, we can conclude that $|\hat{a}_{sj}| \leq |a_{sj}|$. In a similar way, one can show that if $a_{sj} < 0$, then too $|\hat{a}_{sj}| \leq |a_{sj}|$. This is left for the reader to verify.

After a repeated application of the above substitution rule, where each time a new variable is introduced and an old variable is eliminated, to the original equality (5.15) and its derived form(s) (5.21), the equality is transformed into an equality in which (at least) one of the coefficients has absolute value 1. The corresponding variable can then be expressed in terms of the other variables. We substitute this expression into the other (in)equalities. The equality has then been eliminated.

This process continues until we have eliminated all equalities and we have obtained a system of the form (5.12) and (5.13). In the next section we explain how integer variables can be eliminated from a set of linear inequalities (5.13), but first we give an example of how equalities are eliminated.

EXAMPLE 5.2

We repeat part of an example given by Pugh (1992). In this example, four constraints have been specified:

(5.22) $$7x + 12y + 31z = 17,$$

(5.23) $$3x + 5y + 14z = 7,$$

(5.24) $$1 \leq x \leq 40.$$

Note that (5.24) stands for two inequalities. We wish to eliminate equality (5.22). Note that $\gamma = 8$, and using (5.16) we introduce a variable σ defined by

$$8\sigma = -x - 4y - z - 1. \tag{5.25}$$

We eliminate x from (5.22) to (5.24). Applying rule (5.21) on constraint (5.22) yields

$$-7\sigma - 2y + 3z = 3, \tag{5.26}$$

and applying rule (5.18) on constraints (5.23) and (5.24) yields

$$-24\sigma - 7y + 11z = 10, \tag{5.27}$$

$$1 \leq -8\sigma - 4y - z - 1 \leq 40. \tag{5.28}$$

The absolute values of the coefficients of y and z in (5.26) are smaller than the absolute values of the corresponding coefficients in (5.22). The system (5.25) to (5.28) is equivalent to the system (5.22) to (5.24). ■

5.3.2 ELIMINATING AN INTEGER VARIABLE FROM A SET OF INEQUALITIES

When an integer variable is eliminated from a set of inequalities involving only integer-valued variables, two different regions are determined. The first region is referred to as the ***real shadow***. This is simply the region described by the set of inequalities that results if we apply the standard form of Fourier–Motzkin elimination. That is, the real shadow results if we treat the integer variable that is being eliminated as continuous. The second region is referred to as the ***dark shadow***. This dark shadow is constructed in such a way that if it contains a feasible (integer) solution, then the existence of a feasible (integer) solution to the original inequalities is guaranteed.

We describe the construction of the dark shadow. Suppose that two inequalities

$$ax \leq \alpha \tag{5.29}$$

and

$$bx \geq \beta \tag{5.30}$$

are combined to eliminate the integer variable x. Here a and b are positive integer constants, and α and β are linear expressions that may involve all variables except x. Each variable involved in α or β is assumed to have an integer coefficient. The real shadow obtained by eliminating x from the pair of inequalities (5.29) and (5.30) is defined by

$$a\beta \le b\alpha. \tag{5.31}$$

We define the real shadow obtained by eliminating a variable x from a set of inequalities S to be the region described by the inequalities in S not involving x, and the inequalities (5.31) generated by all pairs of upper bounds (5.29) on x and lower bounds (5.30) on x in S.

Now, consider the case in which there is an integer value larger than or equal to $a\beta$ and smaller than or equal to $b\alpha$, but there is no integer solution for x to $a\beta \le abx \le b\alpha$. Let $q = \lfloor \beta/b \rfloor$, then by our assumptions we have

$$abq < a\beta \le b\alpha < ab(q+1).$$

We clearly have $a(q+1) - \alpha > 0$. Since the values of a, b, α and β are integer, we have $a(q+1) - \alpha \ge 1$, and hence

$$ab(q+1) - b\alpha \ge b \tag{5.32}$$

Similarly, we obtain

$$a\beta - abq \ge a \tag{5.33}$$

Combining (5.32) and (5.33), we arrive at

$$b\alpha - a\beta \le ab - a - b$$

In other words, if

$$b\alpha - a\beta \ge ab - a - b + 1 = (a-1)(b-1), \tag{5.34}$$

then an integer solution for x necessarily exists.

To be able to satisfy (5.29) and (5.30) by choosing an appropriate integer value for x, it is sufficient that (5.34) holds true. We therefore define the dark shadow obtained by eliminating variable x from the pair of inequalities (5.29) and (5.30) by the region described by (5.34). Note that if (5.34) holds true, there is an integer value larger than or equal to $a\beta$ and smaller than or equal to $b\alpha$. We define the dark shadow obtained by eliminating a variable x from a set of inequalities S to be the region described by the inequalities in S not involving x, along with the inequalities (5.34) generated by all pairs of upper bounds (5.29) on x and lower bounds (5.30) on x in S.

We now consider a set of inequalities S with only integer-valued coefficients and variables. If the real shadow and the dark shadow resulting from the elimination of x from S are identical, we say that the elimination, or projection, is *exact*. In that case, an integer solution exists if and only if an integer solution to the real/dark shadow exists. If the real shadow and the dark shadow are not identical, we have the following possibilities:

1. If the dark shadow has an integer solution, the set of inequalities S has an integer solution.
2. If the real shadow does not contain a feasible (integer) solution, there is no integer solution to the set of inequalities S.
3. In all other cases, it is not yet clear whether an integer solution to the set of inequalities S exists.

In the latter case we know that if an integer solution to the set of inequalities S were to exist, a pair of constraints $ax \leq \alpha$ and $\beta \leq bx$ would exist such that $ab - a - b \geq b\alpha - a\beta$ and $b\alpha \geq abx \geq a\beta$. From this we can conclude that in such a case an integer solution to the set of inequalities S would satisfy $ab - a - b + a\beta \geq abx \geq a\beta$. We can check whether an integer solution to the set of inequalities S exists by examining all possibilities. Namely, we determine the largest coefficient a_{max} of x for all upper bounds (5.29) on x. For each lower bound $\beta \leq bx$ we then test whether an integer solution exists to the original constraints S combined with $bx = \beta + u$ for each integer u satisfying

$$(a_{max}b - a_{max} - b)/a_{max} \geq u \geq 0.$$

That is, in the latter case we examine $\lfloor (a_{max}b - a_{max} - b)/a_{max} \rfloor + 1$ subproblems of the original problem. These subproblems are referred to as *splinters*.

The theory discussed so far shows that if the dark shadow or one of the splinters has an integer solution, then the original set of inequalities S has an integer solution. Conversely, because we examine all possibilities, it also holds true that if the original set of inequalities S has an integer solution, then the dark shadow or one of the splinters has an integer solution. So, we have demonstrated the following theorem.

THEOREM 5.1

> If and only if an integer solution to the dark shadow or one of the splinters exists, then an integer solution to the original set of inequalities S exists.

Note that if the original set of inequalities S involves p integer variables, the dark shadow and the splinters involve only $p - 1$ integer variables (for the splinters the added equality $bx = \beta + u$ first has to be eliminated in order to arrive at a system of inequalities involving $p - 1$ variables). We have now explained how we can check whether a feasible integer value exists for an integer variable involved in a set of linear inequalities by eliminating this variable. In the next section we examine how we can test whether an integer solution exists for several variables simultaneously by eliminating these variables.

5.3.3 ELIMINATING SEVERAL INTEGER VARIABLES FROM A SET OF INEQUALITIES

Suppose we want to determine whether an integer solution exists for a set of linear inequalities involving p variables. We solve this problem by eliminating these p variables. During the elimination process the original problem may split into several subproblems owing to the splinters that may arise. We apply the procedure sketched below. We focus on the idea underlying the procedure; the computational efficiency of the procedure is ignored here.

We construct a list of subproblems. At the start of the procedure the only (sub)problem is the original problem involving all p variables. We treat each subproblem that may arise separately. We now consider one of those subproblems. We eliminate all variables involved in this subproblem by means of standard Fourier–Motzkin elimination; that is, we repeatedly determine the real shadow until all variables have been eliminated. If the final real shadow without any unknowns is inconsistent, the subproblem does not have a continuous solution, let alone an integer solution. In such a case this subproblem can be discarded.

If the final real shadow of a subproblem is consistent and a continuous solution hence exists, we examine the subproblem again and test whether there is an integer solution to this subproblem. For this subproblem we iteratively select a variable from the set of variables that have not yet been eliminated. The selected variable will be eliminated, using the method of Section 5.3.2. In order to keep the number of computations limited we choose the variable so that the elimination will be exact if possible. As a secondary aim we may then also minimize the number of constraints resulting from the combination of upper and lower bounds. If an exact elimination is not possible, we select a variable with coefficients as close as possible to zero. For such a variable the number of splinters will be relatively small. Testing all splinters for integer solutions can be quite time-consuming, so creating splinters and testing them for integer solutions should be avoided as much as possible. For the subproblem under consideration, we determine the dark shadow and the splinters (if any) by eliminating the selected variable, using the method of Section 5.3.2. The dark shadow and the splinters define new subproblems and are added to the list of subproblems. After this, we have dealt with the subproblem under consideration, and it is deleted from the list of subproblems. We continue this process until all variables have been eliminated from all subproblems on the list of subproblems. The final "subproblems"—or better: final sets of relations—involve only numbers and no unknowns. As in the continuous case (see Chapter 3), such a relation can be self-contradicting—for example, "$0 \geq 1$". We have the following theorem.

THEOREM 5.2

If any of the final sets of relations does not contain a self-contradicting relation, the original set of inequalities has an integer solution. Conversely, if all final sets of relations contain a self-contradicting relation, the original set of inequalities does not have an integer solution.

Proof. This follows from a repeated application of Theorem 5.1. ■

5.4 Error Localization in Categorical, Continuous, and Integer Data

In this section we integrate the Omega test described in Section 5.3 with the branch-and-bound approach for solving the error localization problem for categorical and continuous data proposed by De Waal and Quere (2003) (see also Chapter 4). The result of this integration is an algorithm for solving the error localization problem for categorical, continuous, and integer-valued data. The idea of this algorithm is to test whether the integer-valued variables involved in a solution to the continuous error localization problem—that is, the error localization problem where all numerical variables are assumed to be continuous—can attain integer values. This is illustrated in Figure 5.1.

For a given combination of categorical values, our integrality test reduces to the Omega test. In other words, we basically apply the Omega test on each possible combination of categorical values. What complicates the issue is that we do not explicitly enumerate and test all possible combinations of categorical values. Before we describe the algorithm, we first explain in Section 5.4.1 how balance edits involving integer variables can be "eliminated" and in Section 5.4.2 how integer variables can be eliminated from inequality edits. Finally, Section 5.4.3 describes our algorithm for solving the error localization problem for categorical, continuous, and integer data.

As usual, the edits are given by (5.10) and (5.11). For notational convenience, we define $x_0 = 1$ and $a_{k0} = b_k$ for $k = 1, \ldots, K$, where K is the number of edits, like we also did in Section 5.3.

5.4.1 ERROR LOCALIZATION: ELIMINATING BALANCE EDITS INVOLVING INTEGER VARIABLES

In our integrality test (see Section 5.4.3), integer variables are treated after all continuous variables have been treated and before any categorical variable is treated. That is, once the integer variables are treated, all edits involve only categorical and integer variables. If integer variables are involved in balance edits, we then first "eliminate" these edits. We select a balance edit and basically apply the technique explained in Section 5.3.1 to arrive at an equality in which the absolute value of the coefficient of an integer variable equals 1. During this

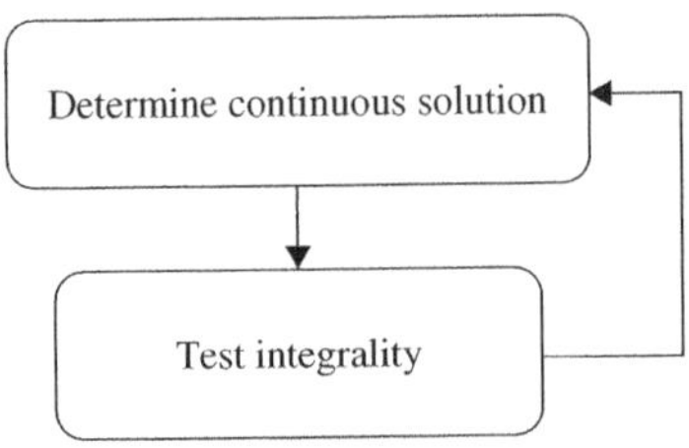

FIGURE 5.1 The basic idea of the error localization algorithm.

process the IF condition of the edit under consideration does not alter. To be more precise, if the selected edit s is given by

$$\text{IF} \quad v_j \in F_j^s \text{ (for } j = 1, \ldots, m),$$

$$\text{THEN } (x_1, \ldots, x_p) \in \{\mathbf{x} \mid \sum_{j=0}^{p} a_{sj}x_j = 0\}) \tag{5.35}$$

with the a_{sj} $(j = 0, \ldots, p)$ integer coefficients and the x_j $(j = 0, \ldots, p)$ integer variables, we transform this edit into

$$\text{IF} \quad v_j \in F_j^s \text{ (for } j = 1, \ldots, m),$$

$$\text{THEN } (\check{x}_1, \ldots, \check{x}_{\check{p}}) \in \{\check{\mathbf{x}} \mid \sum_{j=0}^{\check{p}} \check{a}_{sj}\check{x}_j = 0\} \tag{5.36}$$

where the $\check{a}_{sj}$ $(j = 0, \ldots, \check{p})$ are integer coefficients, and the $\check{x}_j$ $(j = 0, \ldots, \check{p})$ are the transformed integer variables, possibly supplemented by some auxiliary integer variables owing to the elimination of the equality. The total number of variables $\check{x}_j$ is denoted by $\check{p}$ $(\check{p} \geq p)$. In (5.36), at least one integer variable, say $\check{x}_r$, has a coefficient $\check{a}_{sr}$ with $|\check{a}_{sr}| = 1$. Below we describe the procedure to transform (5.35) into (5.36). For notational convenience, we write $\check{p}$ again as p. Likewise, we write the transformed coefficients $\check{a}_{sj}$ $(j = 0, \ldots, \check{p})$ and transformed variables $\check{x}_j$ $(j = 0, \ldots, \check{p})$ again as a_{sj} and x_j. It is important to keep in mind, though, that these coefficients and variables may differ from the original coefficients and variables.

Because auxiliary variables may need to be introduced during the elimination process of a balance edit, we may in fact need to introduce some auxiliary balance edits of which the THEN conditions are given by equations of type (5.16) [or equivalently: of type (5.18)] and the IF conditions are given by the IF condition of the selected edit s. In each of these auxiliary equations, the new auxiliary variable is expressed in terms of the other integer variables x_j $(j = 1, \ldots, p)$—that is, the original integer variables and the already generated auxiliary variables. The other edits are written in terms of the new auxiliary variable by applying the substitution (5.18) to the numerical THEN conditions as far as this is permitted by the IF conditions. The IF conditions of these other edits are changed by the substitution process. In particular, an edit t given by

$$\text{IF} \quad v_j \in F_j^t \text{ (for } j = 1, \ldots, m),$$

$$\text{THEN } (x_1, \ldots, x_p) \in \{\mathbf{x} \mid \sum_{j=0}^{p} a_{tj}x_j \geq 0\} \tag{5.37}$$

involving x_r in its THEN condition gives rise to (at most) two edits given by

$$\begin{aligned} &\text{IF} \quad v_j \in F_j^t \bigcap F_j^s \text{ (for } j = 1, \ldots, m), \\ &\text{THEN } (x_1, \ldots, x_p) \in \{\mathbf{x} \mid -\operatorname{sign}(a_{sr}) a_{tr} \gamma \sigma + \\ &\sum_{j=0, j \neq r}^{p} (a_{tj} + \operatorname{sign}(a_{sr}) a_{tr} (a_{sj} \overline{\operatorname{mod} \gamma})) x_j \geq 0\} \end{aligned} \tag{5.38}$$

and

$$\begin{aligned} &\text{IF} \quad v_j \in F_j^t - F_j^s \text{ (for } j = 1, \ldots, m), \\ &\text{THEN } (x_1, \ldots, x_p) \in \{\mathbf{x} \mid \sum_{j=0}^{p} a_{tj} x_j \geq 0\}. \end{aligned} \tag{5.39}$$

In (5.37) to (5.39) the inequality sign may be replaced by an equality sign. Edits of type (5.38) for which $F_j^t \bigcap F_j^s = \emptyset$ (for some $j = 1, \ldots, m$), as well as edits of type (5.39) for which $F_j^t - F_j^s = \emptyset$ (for some $j = 1, \ldots, m$), may be discarded. Edits given by (5.37) not involving x_r are not modified.

Once we have obtained an edit of type (5.36) with a coefficient a_{sr} such that $|a_{sr}| = 1$, we use the THEN condition of this edit to express the variable x_r in terms of the other variables. That is, we use

$$x_r = -\operatorname{sign}(a_{sr}) \sum_{j=0, j \neq r}^{p} a_{sj} x_j. \tag{5.40}$$

This expression for x_r is then substituted into the THEN conditions of the other edits as far as this is permitted by the IF conditions. The IF conditions of these other edits are changed by the substitution process. In particular, owing to this substitution process an edit given by (5.37) involving x_r in its THEN condition gives rise to (at most) two edits given by (5.39) and

$$\begin{aligned} &\text{IF} \quad v_j \in F_j^t \bigcap F_j^s \text{ (for } j = 1, \ldots, m), \\ &\text{THEN } (x_1, \ldots, x_p) \in \{\mathbf{x} \mid \sum_{j=0, j \neq r}^{p} (a_{tj} - \operatorname{sign}(a_{sr}) a_{tr} a_{sj}) x_j \geq 0\}. \end{aligned} \tag{5.41}$$

In (5.37), (5.39), and (5.41) the inequality sign may be replaced by an equality sign. Edits of type (5.41) for which $F_j^t \bigcap F_j^s = \emptyset$ (for some $j = 1, \ldots, m$), as well as edits of type (5.39) for which $F_j^t - F_j^s = \emptyset$ (for some $j = 1, \ldots, m$), may be discarded. Edits given by (5.37) not involving x_r are not modified.

The new system of edits is equivalent to the original system of edits, in the sense that a solution to the original system of edits corresponds to a solution to the new system, and vice versa. Namely, for the categorical values for which we can use equation (5.18) or (5.40) to eliminate variable x_r, we do this [see (5.38) and (5.41)]. For the categorical values for which we cannot use equation (5.18) or (5.40) to eliminate x_r, we simply leave x_r untouched [see (5.39)]. Note that the IF conditions of an edit of type (5.38) or (5.41) where x_r has been eliminated and an edit still involving x_r have an empty overlap. An edit of type (5.38) or (5.41) and an edit still involving x_r will hence never be combined when eliminating integer variables from inequality edits (see Section 5.4.2 for the elimination of integer variables from inequality edits).

We continue "eliminating" balance edits until for each possible combination of categorical values the associated set of numerical THEN conditions is either the empty set or a system of type (5.12) and (5.13). Note that the balance edits will be eliminated after finitely many steps. Namely, for each possible combination of categorical values we in fact implicitly apply the elimination process of Section 5.3.1, which terminates after a finite number of steps.

After the termination of the above elimination process, we delete all balance edits. We are then left with a set of edits with linear inequalities involving only integer variables as THEN conditions. Because auxiliary variables may have been introduced to eliminate the balance edits, the total number of integer variables in this system of edits may be larger than the original number of integer variables. How we deal with a set of inequality edits involving only integer variables is explained in the next section.

5.4.2 ERROR LOCALIZATION: ELIMINATING INTEGER VARIABLES FROM INEQUALITY EDITS

In this section we assume that each THEN condition is either a linear inequality involving only integer variables or the empty set. When an integer variable is eliminated from a set of inequality edits, a dark shadow and possibly several splinters are generated. Below we describe how this dark shadow and these splinters are defined. We start by selecting an integer variable that we want to eliminate, say x_r. The current edits involving x_r are combined into implicit edits not involving x_r. We consider all edits involving x_r pairwise. Such a pair of edits is given by

$$\text{IF} \quad v_j \in F_j^s \text{ (for } j = 1, \ldots, m),$$

$$\text{THEN } (x_1, \ldots, x_p) \in \{\mathbf{x} \mid \sum_{j=0}^{p} a_{sj} x_j \geq 0\} \tag{5.42}$$

and

$$\text{IF} \quad v_j \in F_j^t \text{ (for } j = 1, \ldots, m),$$

$$\text{THEN } (x_1, \ldots, x_p) \in \{\mathbf{x} \mid \sum_{j=0}^{p} a_{tj} x_j \geq 0\}, \tag{5.43}$$

where all involved numerical variables are integer-valued. We assume that the a_{sj} and the a_{tj} ($j = 0, \ldots, p$), respectively, are normalized.

The real shadow obtained by eliminating x_r from the pair of edits (5.42) and (5.43) is defined only if $a_{sr} \times a_{tr} < 0$. Its THEN condition is then given by

$$\tilde{a}_1 x_1 + \cdots + \tilde{a}_{r-1} x_{r-1} + \tilde{a}_{r+1} x_{r+1} + \cdots + \tilde{a}_p x_p + \tilde{b} \geq 0,$$

where

$$\tilde{a}_j = |a_{sr}| a_{tj} + |a_{tr}| a_{sj} \qquad \text{for } j = 1, \ldots, r-1, r+1, \ldots, p$$

and

$$\tilde{b} = |a_{sr}| b_t + |a_{tr}| b_s$$

and its IF condition is given by

$$v_j \in F_j^s \bigcap F_j^t \qquad \text{for } j = 1, \ldots, m.$$

The dark shadow is also only defined if $a_{sr} \times a_{tr} < 0$. In that case, one coefficient is larger than zero, say $a_{sr} > 0$, and the other coefficient is less than zero, $a_{tr} < 0$. The dark shadow obtained by eliminating x_r from the pair of edits (5.42) and (5.43) is then defined by

$$\text{IF} \quad v_j \in F_j^s \bigcap F_j^t \text{ (for } j = 1, \ldots, m),$$

$$\text{(5.44)} \quad \text{THEN } x \in \{\mathbf{x} \mid \sum_{j=0}^{p} (a_{sr} a_{tj} - a_{tr} a_{sj}) x_j \geq (a_{sr} - 1)(-a_{tr} - 1)\}.$$

If $F_j^t \bigcap F_j^s$ is empty for some $j = 1, \ldots, m$, edit (5.44) is deleted. As for the real shadow, the IF condition of the dark shadow (5.44) is given by the intersections $F_j^s \bigcap F_j^t$ ($j = 1, \ldots, m$), because two numerical THEN conditions can only be combined into an implicit numerical THEN condition for the overlapping parts of their corresponding categorical IF conditions. Note that for this overlapping part the THEN condition of the dark shadow is given by (5.34). The dark shadow obtained by eliminating x_r from a set of inequality edits is by definition given by the edits not involving x_r plus the dark shadows (5.44), assuming they exist, for all pairs of edits (5.42) and (5.43).

Defining the splinters obtained by eliminating x_r from a set of inequality edits is more complicated than in Section 5.3. The reason is that here we want to define splinters for different combinations of categorical values simultaneously, whereas Section 5.3 considers the case without any categorical variables. We describe one possibility to define splinters; for an alternative possibility we refer to De Waal (2003). We write the inequality edits involving x_r as

$$\text{IF} \quad v_j \in F_j^k \text{ (for } j = 1, \ldots, m),$$

$$\text{(5.45)} \qquad \text{THEN } a_{kr} x_r \geq - \sum_{j=0, j \neq r}^{p} a_{kj} x_j.$$

For negative coefficients a_{kr}, the THEN condition of (5.45) provides an upper bound on x_r. For positive coefficients a_{kr}, the THEN condition of (5.45) provides a lower bound on x_r. We start by determining the smallest negative coefficient a_{qr} of x_r for all edits (5.45); that is, a_{qr} is the coefficient of x_r in all upper bounds on x_r with the largest absolute value. For each lower bound on x_r, we then test whether an integer solution exists to the original edits combined with

$$\text{IF} \quad v_j \in F_j^k \ (\text{for } j = 1, \ldots, m)$$

$$\text{THEN} \quad a_{kr}x_r = -\sum_{j \neq r} a_{kj}x_j + u \tag{5.46}$$

for each integer u satisfying $(-a_{qr}a_{kr} + a_{qr} - a_{kr})/(-a_{qr}) \geq u \geq 0$. For each possible combination of categorical values, all splinters required according to the Omega test described in Section 5.3.2 are taken into consideration. For some combinations of categorical values, more splinters than necessary are taken into consideration. These superfluous splinters increase the computing time, but do no harm otherwise. We have the following theorem.

THEOREM 5.3

> The original set of edits with linear inequalities involving only integer variables as THEN conditions has a solution if and only if the dark shadow or a splinter resulting from the elimination of variable x_r has a solution.

Theorem 5.3 follows immediately by noting that for arbitrary, fixed categorical values, it reduces to Theorem 5.1.

We now eliminate all integer-valued variables from the original set of inequality edits. During this process we may have to consider several different sets of edits and corresponding variables owing to the splinters that may arise. We consider each such set of edits (and corresponding variables) separately. If a set of edits contains a balance edit, which happens if this set of edits is a splinter, we apply the technique of Section 5.4.1 to eliminate that equality from this set. For a set of edits involving only inequality edits, we select a variable that has not yet been eliminated and proceed to eliminate this variable using the technique of this section. We continue until all integer variables in all sets of edits have been eliminated, and we are left with one or more sets of edits involving only categorical variables. The theory of Section 5.4.1 and a repeated application of Theorem 5.3 yield the following theorem.

THEOREM 5.4

> A set of edits with THEN conditions involving only integer variables has a solution if and only if any of the sets of edits involving only categorical variables arising after the elimination of all integer-valued variables has a solution.

To check the existence of a solution to a set of edits involving only categorical variables one can use the methodology described in Chapter 4.

5.4.3 ERROR LOCALIZATION: ALGORITHM FOR CATEGORICAL, CONTINUOUS AND INTEGER DATA

After the preparations in the previous subections, we are now able to state our algorithm for solving the error localization problem for a mix of categorical, continuous, and integer data. We denote the error localization problem in categorical, continuous, and integer data under consideration by PI. To solve PI we first apply the branch-and-bound algorithm presented in Chapter 4 without taking into account that some of the variables are integer-valued; that is, we first treat the integer variables as being continuous. We denote the problem where integer variables are treated as continuous ones by PC. PC is the continuous error localization problem.

Just like in Section 4.2 we use the objective function

$$\sum_{j-1}^{m} w_j^c \delta(v_j^0, \check{v}_j) + \sum_{j-1}^{p} w_j^r \delta(x_j^0, \check{x}_j), \tag{5.47}$$

where w_j^c is the nonnegative reliability weight of categorical variable v_j ($j = 1, \ldots, m$), w_j^r the nonnegative reliability weight of numerical variable x_j ($j = 1, \ldots, p$), $\delta(y^0, y) = 1$ if y^0 is missing or $y^0 \neq y$, and $\delta(y^0, y) = 0$ if $y^0 = y$.

Let c_{obj} denote the value of the objective function (5.47) for the best currently found solution to PI, and let S be the set of currently best solutions to PI. We initialize c_{obj} to ∞, and S to $\emptyset$. A solution to PC not involving any integer variables is automatically also a solution to PI. So, whenever we find a solution to PC not involving any integer variables for which (5.47) is less than c_{obj}, we update c_{obj} with that value of (5.47) and set S equal to the current solution to PC. Also, whenever we find a solution to PC not involving any integer variables for which (5.47) is equal to c_{obj}, we add the current solution to PC to S.

Whenever we find a solution to PC involving integer variables for which (5.47) is at most equal to c_{obj}, we consider PI. We test whether the variables involved in the current solution to PC also constitute a solution to PI. The basic idea of this test is illustrated in Figure 5.2.

For the integrality test we first fill in the values of the variables not involved in the current solution to PC into the edits. Subsequently, we eliminate the continuous variables involved in the solution to PC. This yields a system of edits (5.10) and (5.11) in which only the integer-valued and categorical variables involved in the solution to PC occur.

Next, we eliminate all balance edits with integer-valued variables involved in the current solution to PC in the manner described in Section 5.4.1. Subsequently, we eliminate all integer-valued variables involved in the current solution to PC from all inequality edits in the manner described in Section 5.4.2. During this latter elimination process the original problem may be split into several subproblems owing to the splinters that may arise. Finally, we eliminate

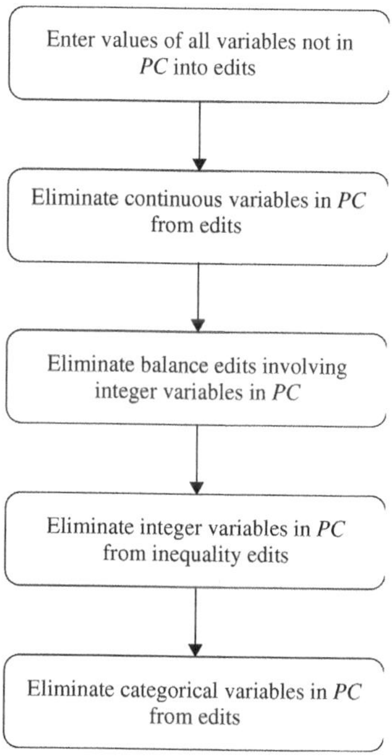

FIGURE 5.2 The basic idea of the integrality test.

all categorical variables from each of these subproblems. For each subproblem we end up with a set of relations not involving any unknowns. Such a set of relations may be empty. If a set of relations we obtain in this way does not contain a self-contradicting relation, which is for instance (by definition) the case if the set of relations is empty, we have found a solution to *PI*. In that case, if the value of (5.47) for the current solution to *PI* is less than c_{obj}, we update c_{obj} accordingly and set *S* equal to the current solution of *PI*, else we add the current solution to *PI* to *S*. If all sets of relations involving no unknowns contain a self-contradicting relation, none of the subproblems leads to a solution to *PI* and the solution to *PC* under consideration is not a solution to *PI*. In that case c_{obj} is not updated, and we continue with finding solutions to *PC*.

Note that in the above approach, the relatively time-consuming integrality test is only invoked once a solution to *PC* with an objective value of c_{obj} or less involving integer-valued variables has been found, so generally only rather infrequently. We have the following theorem.

THEOREM 5.5

The above procedure finds all optimal solutions to *PI*.

Proof. We start by noting that Theorem 4.5, Section 5.4, and Theorem 5.4 show that if and only if any of the final sets of relations involving no unknowns obtained by eliminating all variables involved in a solution to PC does not contain a self-contradicting relation, the original set of edits can be satisfied by modifying the values of the variables involved in this solution. Now, the branch-and-bound algorithm for categorical and continuous data can be used to find all solutions to PC with an objective value (5.47) of c_{obj} or less, for any given value of c_{obj}. For each solution to PC with a value for (5.47) equal to or less than c_{obj}, we test whether it is also a solution to PI. The result of this test is conclusive. We update c_{obj} whenever we have found a better solution to PI than the best one found so far. In other words, all potentially optimal solutions to PI are considered by the procedure, and all optimal solutions to PI are indeed identified as such. ■

We illustrate the algorithm by means of a simple example involving only two integer-valued variables.

EXAMPLE 5.3

We consider a case with only two variables x_1 and x_2, and three edits given by

$$-2x_2 + 5 \geq 0, \quad (5.48)$$

$$5x_1 - x_2 \geq 0,$$

$$-3x_1 + 2x_2 \geq 0.$$

Both variables are integer-valued, and their reliability weights equal one. The original, incorrect record is given by $x_1 = 1$, and $x_2 = 1$. We initialize c_{obj} to ∞, and S to $\emptyset$. We start by solving PC. We select a variable, say x_1, and construct two branches: in the first branch we eliminate x_1 from the set of current edits, in the second branch we fix x_1 to its original value. If we eliminate x_1 from the set of current edits, we obtain (5.48) and $x_2 \geq 0$ as our new set of current edits. This new set of current edits is satisfied by the original value of x_2. Hence, we have found a solution to PC, namely: change x_1. We test whether this is also a solution to PI. To this end, we start by filling in the original value of x_2 into the original set of edits. We obtain the following set of edits involving only x_1.

$$5x_1 - 1 \geq 0, \quad (5.49)$$

$$-3x_1 + 2 \geq 0. \quad (5.50)$$

The dark shadow obtained by eliminating x_1 from (5.49) and (5.50) [see (5.44)] is given by

$$7 \geq 8,$$

which is clearly a self-contradicting relation. We therefore have to consider the splinters. In this simple case there are three splinters. For the first splinter we have to add the constraint

$$5x_1 = 1$$

to (5.49) and (5.50) [see (5.46)], for the second one the constraint

$$5x_1 = 2,$$

and for the third one the constraint

$$5x_1 = 3.$$

It is clear that none of these three splinters has an integer solution for x_1. This would also follow if we were to continue the algorithm by eliminating variable x_1 because we would then obtain only self-contradicting relations. We conclude that although changing x_1 is a solution to PC, it is not a solution to PI.

After this intermezzo during which we tested whether changing the value of only x_1 is a solution to PI, we continue with finding solutions to PC. We now consider the branch where x_1 is fixed to its original value. The corresponding set of current edits is given by (5.48),

$$-x_2 + 5 \geq 0, \tag{5.51}$$

$$2x_2 - 3 \geq 0. \tag{5.52}$$

By eliminating x_2, we see that changing the value of only x_2 is a solution to PC. We check whether this is also a solution to PI. We fill in the original value of x_1 into the original set of edits. We obtain the system (5.48), (5.51), and (5.52). The dark shadow of (5.48) and (5.52) obtained by eliminating x_2 [see (5.44)] is given by

$$4 \geq 1$$

and the dark shadow of (5.51) and (5.52) obtained by eliminating x_2 by

$$7 \geq 0.$$

The above relations are not self-contradicting, so we can conclude that changing the value of x_2 is a solution to PI. As changing the value of x_1 is not a solution to PI, we can even conclude that this is the only optimal solution to PI. A feasible value for x_2 is 2. ■

The method described in this section may appear to be very slow in many cases. Indeed, it is not difficult to design a set of edits for which the method is extremely slow. However, we argue that in practice the situation is not so bad. First, like we already mentioned, the time-consuming algorithm to check potential solutions to *PI* is only invoked once a new solution to *PC* with an objective value less than or equal to the current value of c_{obj} involving integer variables has been found. In practice, the number of times that such a solution to *PC* is found is in most cases rather limited.

Second, whenever we find a solution to *PC* with an objective value less than or equal to the current value of c_{obj}, we only have to test whether the variables involved in this particular solution also form a solution to *PI*. Moreover, often one is only interested in solutions to the error localization problem with a few variables, say 10 or less. We already argued in Section 3.4.1 that if a record requires more than, say, 10 values to be changed, it should not be edited automatically in our opinion, because the statistical quality of the automatically edited record would be too low. This implies that the relatively time-consuming test described in this section involves only a few variables.

Third, the integrality test only becomes really time-consuming when many splinters have to be considered. However, in most edits, either explicit or implicit ones, encountered in practice the coefficients of the integer variables equal -1 or $+1$. This is especially true for balance edits. For an integer variable with coefficient -1 or $+1$ the elimination from inequality edits will be exact; that is, the dark shadow and the real shadow coincide and no splinters have to be generated. For balance edits involving integer variables with coefficients -1 or $+1$, no auxiliary variables have to be introduced in order to eliminate these edits. For such a balance edit the elimination can be performed very fast.

Finally, we can also resort to a heuristic approach based on the exact algorithm. In the next section such a heuristic procedure is described.

5.5 A Heuristic Procedure

At Statistics Netherlands we originally aimed to develop a software package for a mix of categorical and continuous data only. In order to achieve this aim, a number of algorithms were considered. For an assessment of several algorithms on continuous data we refer to Section 3.4.9. As a consequence of our work, the branch-and-bound algorithm described in Chapter 4 has been implemented in SLICE, our general software framework for automatic editing and imputation [cf. De Waal (2001)]. Later the wish to extend the implemented algorithm to include integer-valued data arose. The algorithm described in Section 5.4 was developed to fulfill that wish. However, once this algorithm was developed, we considered it to be too complex to implement and maintain in production software. We therefore decided not to implement the exact algorithm of Section 5.4, but instead to develop a simpler heuristic procedure based on the exact algorithm. That heuristic procedure, which is described below, has been implemented in version 1.5 of SLICE.

Only the integrality test for the integer-valued variables involved in a solution to *PC* (i.e., a potential solution to *PI*) differs for the exact algorithm and the heuristic procedure. In our heuristic procedure, we do not examine splinters, nor do we introduce auxiliary variables in order to eliminate balance edits.

Whenever we have to eliminate an integer-valued variable x_r from a pair of edits s and t in the heuristic checking procedure, we distinguish between two cases. If either edit s or edit t (or both) is a balance edit involving x_r, we examine whether the coefficient of x_r in the corresponding normalized THEN condition (if both edits are balance edits, we examine both normalized THEN conditions) equals $+1$ or -1. If this is not the case, we make the conservative assumption that no feasible integer value for x_r exists, and we reject the potential solution to the *PI*. If both edits s and t are inequality edits, we eliminate x_r from these edits by determining the dark shadow [see (5.44)].

If several integer variables are involved in the solution to *PC* under consideration, we repeatedly apply the above procedure until all these variables have been eliminated. If the resulting set of edits involving only categorical variables has a solution, the solution to *PC* is also a solution to *PI* (see Theorem 5.4). On the other hand, if the resulting set of edits does not have a solution, we make the conservative assumption that the current solution to *PC* is not a solution to *PI*. This assumption is conservative because we do not check the splinters.

The above heuristic procedure is considerably easier to implement and maintain than the exact algorithm of Section 5.4. The price we have to pay for using the heuristic procedure instead of the exact algorithm of Section 5.4 is that we sometimes conclude that an integer solution does not exist, whereas in fact it does.

5.6 Computational Results

In this section we provide some computational results for the heuristic procedure described in Section 5.5. In this section we focus on the number of records that were solved to optimality and the number of records that could not be solved. For more computational results, including exact computing times, we refer to De Waal (2005).

The heuristic procedure has been tested on five realistic data sets. We have used realistic data sets rather than randomly generated synthetic data for our evaluation study, because we feel that the properties of realistic data are completely different than those of randomly generated data. Considering that a production version of SLICE for categorical and continuous data already existed, we decided to implement the heuristic procedure described in Section 5.5 directly in SLICE (version 1.5), without implementing it in prototype software first.

Our experiments have therefore been carried out by means of SLICE 1.5. This software package has been designed for use in the day-to-day routine at Statistics Netherlands. It has been optimized for robustness against misuse and for ease of maintainability. It uses well-tested components that facilitate debugging. Moreover, the software stores different kinds of metadata, such as which fields are identified as being erroneous. SLICE 1.5 has not been optimized for speed. The speed of the software can definitely be improved upon. Compared to the prototype software for categorical and continuous data (called Leo; see Chapter 3) on which this production software is based, the production software is about 16 times or more slower [see De Waal and Quere (2003) and Chapter 3 of the present book, where similar data and edits were used as in the present chapter). The prototype software, however, could handle only a mix of categorical and continuous data, not integer-valued data.

SLICE 1.5 allows the user to specify several parameters, such as a maximum for the number of errors in a record, a maximum for the number of missing values in a record, the maximum computing time per record, the maximum number of (explicit and implicit) edits in a node of the binary search tree, and a maximum for the number of determined solutions. In our evaluation experiments we did not set a limit for the number of missing values in a record. We have set the maximum number of (explicit and implicit) edits in a node to 3000, and we have set the maximum computing time per record to 60 seconds. In our experiments we have varied the maximum number of errors and the maximum for the number of determined solutions.

If a record cannot be made to satisfy all edits by changing at most the specified maximum number of errors, it is discarded by SLICE 1.5. A record is also discarded by SLICE 1.5 if it contains more missing values than the specified maximum. Whenever SLICE 1.5 has found N_{sol} solutions with the lowest value c_{low} for the objective function (5.47) found so far, where N_{sol} is the specified maximum number of determined solutions, it from then on searches only for solutions to the error localization problem for which the value of the objective function (5.47) is strictly less than c_{low}. After SLICE 1.5 has solved the error localization problem for a record, it returns at most N_{sol} solutions with the lowest value for the objective function (5.47). Owing to the use of the heuristic procedure of Section 5.5, these determined solutions may be suboptimal. If the maximum number of edits in a node exceeds 3000 or the maximum computing time per record exceeds 60 seconds, SLICE 1.5 returns the best solutions (if any) it has determined so far. So, even if the maximum number of edits in a node or the maximum computing time per record is exceeded, the heuristic procedure implemented in SLICE 1.5 may return a solution. For some records the heuristic procedure of SLICE 1.5 could not find a solution at all.

In all balance edits corresponding to the five evaluation data sets, the coefficients of the involved variables equal -1 or $+1$. Also, in all inequality edits corresponding to data sets B and D, the coefficients of the involved variables equal -1 or $+1$. In the inequality edits corresponding to data sets C and F, however, many coefficients of the involved variables are not equal to -1 or $+1$.

We have compared the solutions determined by the heuristic procedure implemented in SLICE 1.5 to the optimal solutions. For purely numerical data, the edits reduce to linear constraints, and the error localization problem can easily be formulated as an integer programming problem [see, e.g., Schaffer (1987), Riera-Ledesma and Salazar-González (2003), and Section 3.4.5 of the present book]. We have therefore used a solver for integer programming problems to determine the optimal solutions. For our evaluation study we have used CPLEX [cf. ILOG CPLEX 7.5 Reference Manual (2001)]. Note that although the error localization problem for numerical (either continuous or integer-valued) data can quite easily be solved by a solver for integer programming problems, the error localization problem for a mix of numerical (either continuous or integer-valued) and categorical data quickly becomes very hard to solve for such a solver [see also De Waal (2003)].

In our evaluation study we have used the same data sets as in Chapter 3 with the exception of data set A, which was not used in this particular evaluation study. In Table 5.1 we give the number of inconsistent records for which the heuristic procedure of SLICE 1.5, with the maximum number of errors set to 10, found an optimal solution, the number of inconsistent records for which it found a suboptimal solution, the number of inconsistent records for which it could not find a (possibly suboptimal) solution at all, and the number of inconsistent records for which it exceeded the maximum computing time per record but did find a (possibly suboptimal) solution. The number of inconsistent records for which the heuristic procedure of SLICE 1.5 found an optimal solution plus the number of inconsistent records for which it found a suboptimal solution plus the number of inconsistent records for which it found no solution at all equals the number of inconsistent records given in Table 3.1 of Section 3.4.9.

In our evaluation study the maximum number of edits in a node was never exceeded. Note that records for which the heuristic procedure exceeded the maximum computing time may still be solved to optimality by this procedure.

As described in Sections 5.4 and 5.5, the heuristic procedure of SLICE 1.5 consists of two parts: a branch-and-bound algorithm where all numerical variables are treated as being continuous ones and an integrality test. In order to assess the slowdown of the algorithm owing to the integrality test, we compare the computing times of the heuristic procedure to the computing times if all variables were continuous ones rather than integer-valued ones.

TABLE 5.1 Number of Records that Were Optimally Solved, Could Not Be Solved, and for which the Maximum Computing Time per Record Was Exceeded

	Data Set B	Data Set C	Data Set D	Data Set E	Data Set F
Number of optimally solved records	120	1347	2150	378	1139
Number of suboptimally solved records	33	27	0	0	0
Number of unsolved records	4	30	2	0	2
Number of records for which the maximum computing time (60 seconds) was exceeded	3	14	11	0	0

For data sets B to E, the increase in computing time owing to the integrality test is rather small, namely between 0% and 18%. For data set F, however, the increase in computing time owing to the integrality test is quite large (up to 65%).

The effect of increasing the maximum number of errors on the relative computing time of the integrality test depends on the data set under consideration. For data sets B and E, the relative increase in computing time owing to the integrality test becomes less with increasing maximum numbers of errors. For data set C the relative increase in computing time owing to the integrality test gradually becomes more with increasing maximum numbers of errors. For data set D, this relative increase in computing time is more or less stable for different maximum numbers of errors. Finally, for data set F the relative increase in computing time grows rapidly with increasing maximum numbers of errors.

Determining several solutions instead of one leads to a limited increase in computing time. The largest relative increase in computing time when determining at most 10 solutions instead of only one is for data set B. The computing time increased by approximaltely 6%.

In Table 5.2 we give the total number of erroneous fields according to the heuristic procedure of SLICE 1.5 and the exact algorithm implemented by means of CPLEX for the records that could be solved, possibly in a suboptimal manner, by means of the heuristic procedure.

The number of fields that are unnecessarily identified as being erroneous by the heuristic procedure was very small in our evaluation study. In other words, for the data sets used in our evaluation study, the quality of the solutions determined by the heuristic procedure in terms of the total number of fields identified as erroneous is very good. In the worst case, data set C, the surplus of fields identified as being erroneous by the heuristic procedure in comparison to the number of fields identified as being erroneous by the exact algorithm implemented by means of CPLEX is less than 2% of the latter number of fields.

Finally, we examine the quality of the heuristic procedure in terms of the number of optimal solutions determined. We set both the maximum number of errors and the maximum number of solutions per record to 10. The reason for selecting the latter number is that for records with more than 10 optimal solutions to the error localization problem, it is very hard to later select the correct solution—that is, correctly identify the erroneous fields—anyway. For

TABLE 5.2 Total Number of Erroneous Fields in Solved Records According to the Heuristic Procedure and the Exact Algorithm

	Data Set B	Data Set C	Data Set D	Data Set E	Data Set F
Exact algorithm for integer data (CPLEX)	378	3424	3526	2362	2919
Heuristic procedure (SLICE 1.5)	381	3482	3526	2362	2919

TABLE 5.3 Number of Optimal Solutions of the Heuristic Procedure and the Exact Algorithm

	Data Set B	Data Set C	Data Set D	Data Set E	Data Set F
Exact algorithm for integer data (CPLEX)	701	6609	11404	474	6207
Heuristic procedure (SLICE 1.5)	701 (100%)[a]	6477 (98%)	11404 (100%)	474 (100%)	4828 (78%)

[a]Values in parentheses represent the number of optimal solutions determined by the heuristic procedure in percent of the number of optimal solutions determined by the exact algorithm.

the records for which the heuristic procedure succeeded in determining an optimal solution, we compare the number of optimal solutions determined by the heuristic procedure to the number of optimal solutions determined by the exact algorithm implemented by means of CPLEX. The results are given in Table 5.3. □

For data sets B, D, and E, the heuristic procedure determined the same number of optimal solutions as the exact algorithm. For data sets C and F, the data sets for which the coefficients of the variables involved in the corresponding inequality edits often are unequal to -1 or $+1$, the number of optimal solutions determined by the heuristic procedure is less than the number of optimal solutions determined by the exact algorithm implemented by means of CPLEX. In particular, this is the case for data set F, where the number of optimal solutions determined by the heuristic procedure is only 78% of the number of optimal solutions determined by the exact algorithm. Data set F is the only data set for which the number of inequality edits with coefficients unequal to -1 or $+1$ for the involved variables clearly outnumbers the number of balance edits, which probably explains our result. Note that despite the fact that the number of optimal solutions determined by the heuristic procedure for data set F is clearly less than the actual number of optimal solutions, the heuristic procedure does succeed in solving all records to optimality, except for two records for which it could not find a solution at all.

5.7 Discussion

In this chapter we have developed an exact algorithm for solving the error localization problem for a mix of categorical, continuous, and integer data. This algorithm is quite complex to implement and maintain in a software system, especially in a software system that is meant to be used routinely in practice. Based on this exact algorithm, we have therefore also developed a much simpler heuristic procedure. This heuristic procedure has been implemented in the production software at Statistics Netherlands, SLICE. In this chapter we have also examined the performance of the heuristic procedure.

The exact algorithm and the heuristic procedure described in this chapter have a number of theoretical drawbacks. Both the exact algorithm and the heuristic procedure are extensions to an exact algorithm for continuous and categorical data (see Chapter 4). The computing time of this latter exact algorithm can, theoretically, be exponential in its input parameters, such as the number of variables, the number of edits, and the maximum number of errors. For some data sets in our evaluation study, namely data sets C and E, this exponential increase in the computing time owing to an increase of the maximum number of errors is, unfortunately, also observed in practice. For some practical instances of the error localization problem, this exponential increase in the computing time may be a problem. For such instances, one has to resort to other heuristic approaches, such as setting fields that are likely to be erroneous to "missing" in a preprocessing step (the exact algorithm for continuous and categorical data is generally faster for records with many missing values than for records with many erroneous values), or to an alternative algorithm altogether [see Chapter 3 and De Waal and Coutinho (2005) for references to some papers on alternative approaches].

The computing time of the integrality test of the exact algorithm can, theoretically, also be exponential in the number of variables, the number of edits, and the maximum number of errors. In our evaluation study on the heuristic procedure, the increase in computing time owing to the integrality test is limited for most evaluation data sets. However, for data F the increase in computing time owing to the integrality test grows rapidly when increasing the maximum number of errors. Again, for some practical instances of the error localization problem, this rapid increase in the computing time may be a problem, and one may have to resort to other approaches.

In principle, the number of erroneous fields identified by the heuristic procedure may be (much) higher than the number of erroneous fields identified by an exact algorithm. In our evaluation study, this has, however, not occurred. The number of fields identified as being erroneous by the heuristic procedure is for all evaluation data sets almost equal, and often even precisely equal, to the number of fields identified as being erroneous by an exact algorithm implemented by means of CPLEX.

Another potential drawback of the heuristic procedure is that the number of optimal solutions determined by this procedure can be (much) less than for an exact algorithm. In our evaluation study, this has also not occurred. For most evaluation data sets, the number of optimal solutions determined by the heuristic procedure is equal or almost equal to the number of optimal solutions determined by an exact algorithm implemented by means of CPLEX. The only exception is data set F, where the number of optimal solutions determined by the heuristic procedure drops to about 78% of the number of optimal solutions determined by an exact algorithm implemented by means of CPLEX. Whereas the actual average number of optimal solutions is 5.4 (= 6207/1139, see Tables 5.1 and 5.3) per optimally solved record if the maximum number of optimal solutions determined is set to 10, the heuristic procedure determines only 4.2 optimal solutions on the average. Fortunately, for our purposes at Statistics Netherlands this is an acceptable result.

As mentioned before, at Statistics Netherlands we aimed to implement an algorithm for a mix of categorical, continuous, and integer data. Given the fact we had already implemented the branch-and-bound algorithm for continuous and categorical data described in Chapter 4 in our production software, our main choice to be made was whether we would implement the exact algorithm described in Section 5.4 or the heuristic procedure of Section 5.5 in that production software. Considering the complexity of implementing and maintaining the exact algorithm in production software, we decided to implement the heuristic procedure instead of the exact algorithm. Our, admittedly limited, experience with the heuristic procedure so far suggests that we have made a good choice here. For Statistics Netherlands, the benefits of using the heuristic procedure, in particular a considerable simplification in developing and maintaining the software in comparison to the exact algorithm of Section 5.4, outweigh the disadvantages, possibly worse and less solutions, of using the heuristic procedure instead of the exact algorithm. Despite the earlier mentioned theoretical drawbacks of the heuristic procedure, its computing speed and the quality of its solutions thus far appear to be fully acceptable for application in practice at Statistics Netherlands.

REFERENCES

Chvátal, V. (1983), *Linear Programming*. W. H. Freeman and Company, New York.

Dantzig, G. B. and B. C. Eaves (1973), Fourier–Motzkin Elimination and Its Dual. *Journal of Combinatorial Theory (A) 14*, 288–297.

De Waal, T. (2001), SLICE: Generalised Software for Statistical Data Editing. In: *Proceedings in Computational Statistics*, (J. G. Bethlehem and P. G. M. Van der Heijden, eds. Physica-Verlag, New York, pp. 277–282.

De Waal, T. (2003), *Processing of Erroneous and Unsafe Data*. Ph.D. Thesis, Erasmus University, Rotterdam (see also www.cbs.nl).

De Waal, T. (2005), Automatic Error Localisation for Categorical, Continuous and Integer Data. *Statistics and Operations Research Transactions 29*, pp. 57–99.

De Waal, T. and W. Coutinho (2005), Automatic Editing for Business Surveys: An Assessment of Selected Algorithms. *International Statistical Review 73*, pp. 73–102.

De Waal, T. and R. Quere (2003), A Fast and Simple Algorithm for Automatic Editing of Mixed Data. *Journal of Official Statistics 19*, pp. 383–402.

Duffin, R. J. (1974), On Fouriers Analysis of Linear Inequality Systems. *Mathematical Programming Studies 1*, pp. 71–95.

ILOG (2001), *ILOG CPLEX 7.5 Reference Manual*.

Pugh, W. (1992), The Omega Test: A Fast and Practical Integer Programming Algorithm for Data Dependence Analysis. *Communications of the ACM 35*, pp. 102–114.

Pugh, W. and D. Wonnacott (1994), Experiences with Constraint-Based Array Dependence Analysis. In: *Principles and Practice of Constraint Programming, Second International Workshop*, Lecture Notes in Computer Science 768, Springer-Verlag, Berlin.

Riera-Ledesma, J. and J. J. Salazar-González (2003), *New Algorithms for the Editing and Imputation Problem.* Working Paper No. 5, UN/ECE Work Session on Statistical Data Editing, Madrid.

Schaffer, J. (1987), Procedure for Solving the Data-Editing Problem with Both Continuous and Discrete Data Types. *Naval Research Logistics 34*, pp. 879–890.

Schrijver, A. (1986), *Theory of Linear and Integer Programming*. John Wiley & Sons, New York.

Williams, H. P. (1976), Fourier–Motzkin Elimination Extension to Integer Programming. *Journal of Combinatorial Theory (A) 21*, pp. 118–123.

Williams, H. P. (1983), A Characterisation of All Feasible Solutions to an Integer Program. *Discrete Applied Mathematics 5*, pp. 146–155.

CHAPTER SIX

Selective Editing

6.1 Introduction

As stated in the introductory chapter of this book, the process of improving data quality by detecting and correcting errors encompasses a variety of procedures, both manual and automatic. Traditionally, a large amount of resources have been invested in manually following up edit failures by subject-matter specialists. This manual or interactive editing is very time-consuming and therefore expensive and adversely influencing the timeliness of publications. Moreover, when manual editing involves recontacting the respondents, it will also increase the response burden. This has urged statistical offices to critically investigate the profits of interactive editing. Numerous studies in this area have shown that the number of records that are manually edited can be greatly reduced if the editing effort is focused on the errors with the greatest influence on the estimates of the principal parameters of interest. An editing strategy in which manual editing is limited or prioritized to those errors where this editing has substantial influence on publications figures is called selective or significance editing [cf. Lawrence and McKenzie (2000), Hedlin (2003), Granquist (1995), Granquist and Kovar (1997)].

The diminishing effect of correcting increasingly less important errors on the estimates of totals is illustrated in Figure 6.1. This figure shows the change in the estimate of the total number of employees of small to medium-size firms in the retail trade as a function of the number of edited records, with the records sorted in the order of diminishing influence of corrections on the estimate. The figure shows that correcting the most important errors increases the estimate considerably, but this effect gradually decreases and editing more than 150 of the 350 units hardly changes the outcome. This figure is based on a large number of

Handbook of Statistical Data Editing and Imputation, by T. de Waal, J. Pannekoek, and S. Scholtus

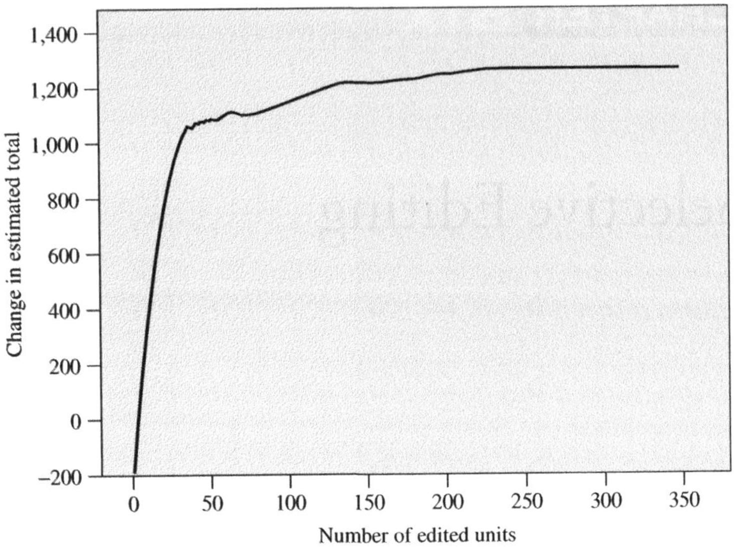

FIGURE 6.1 Effect of editing on estimates.

edited records, and it is only in retrospect that it can be concluded that editing could have been limited to only a fraction of these records.

It is the purpose of selective editing methods to predict beforehand, without manual inspection, which records contain influential errors, so it is worthwhile to spend the time and resources to correct them interactively. Based on these predictions, a selection step can be built into the process flow; this serves to divide the records into those that will be manually treated and those for which automatic treatment is adequate. In Section 1.5 of Chapter 1, a generic process flow was described (see also Figure 1.1). In this process flow, two process steps were identified that did not actually treat errors but served only to make the selection for either manual or automatic further processing of the records. These two selection processes are referred to as *micro-selection* and *macro-selection.*

Many methods for selecting records for interactive editing are specifically designed for use in the early stages of the data collection period. These methods are ***input editing*** or ***micro-selection*** methods and are applied to each incoming record individually; they depend on parameters that are determined before the data collection takes place, often based on previous cycles of the survey, and the values of the target variables in the single record under consideration. The purpose of such methods is to start the time-consuming interactive editing as soon as the first survey returns are received, so that the selection process itself will not delay the further processing of the data. Other methods, referred to as ***output editing***, ***macro-editing***, or ***macro-selection***, are designed to be used when the data collection is (almost) completed. These methods explicitly use the information from all current data to detect suspect and influential values. In this stage preliminary estimates of target parameters can be calculated and the influence of editing outlying values on these parameters can be predicted.

In this chapter we give an overview of the methods that are used to perform the selection of records for interactive editing and we also briefly discuss the interactive editing process itself. In Section 6.2 a brief summary is given of the historical developments that led to the selective editing processes that are in use at NSIs today. A detailed account of the methods used for the selection mechanisms is given in Sections 6.3 and 6.4. Section 6.3 treats the methodology for micro-selection and Section 6.4 the methodology for macro-selection. In Section 6.5 a review of interactive editing is presented and a summary and some conclusions are presented in Section 6.6.

6.2 Historical Notes

Over time the editing process has seen several changes. The first changes were triggered by the increasing power and availability of electronic computers. Statistical institutes have been using electronic computers in the editing process since the 1950s [cf. Nordbotten (1963)]. Using computers as an aid for data processing in general and editing in particular has obvious benefits. Stuart (1966) puts this as follows: "The development of an integrated man–machine screening logic has the effect of designating human roles which require truly human skills, and mechanical processes which require unfeeling machine capabilities. Thus, it can be of advantage to data quality, cost reduction, machine utilization, and perhaps most importantly, the dignity of the human being."

Initially, mainframe computers were used to perform consistency checks only. To this end, data from the paper survey forms were entered into a computer by professional typists. A checking program was run on the mainframe computer, producing detailed lists of edit violations per record. Subject-matter specialists then used printed versions of these lists to make corrections on the original paper questionnaires. The edited data were again entered into the mainframe computer by typing staff, and the checking program was run once more to see if all edit violations had been removed. Usually, this was not the case, because it is very difficult for a human to find values that satisfy a large number of interrelated edit rules simultaneously. Therefore, new lists of edit violations were printed and new corrections were made on the original questionnaires. This iterative cycle of automatic checking and manual correcting was continued until (nearly) all records satisfied all edit rules.

The advent of the microcomputer in the 1980s enabled an improved form of computer-assisted manual editing, called interactive editing. With interactive editing, data are entered into a computer only once. The computer runs consistency checks and displays a list of edit violations per record on the screen. Subject-matter specialists perform manual editing directly on the captured data.

Whenever a possible correction is typed in, the computer immediately reruns the consistency checks to see whether this correction removes edit violations. Each record is edited separately until it does not violate any edit rules. Because the subject-matter specialists get immediate feedback on their actions, this approach is more effective and more efficient than the old one. See Bethlehem (1987) and Pierzchala (1990) for more details.

Studies [cf. Granquist (1995, 1997) and Granquist and Kovar (1997)] have shown that generally not all errors have to be corrected to obtain reliable publication figures. It suffices to remove only the most influential errors. These studies have been confirmed by many years of practical experience at several NSIs. As a result, research has been focused on effective methods to single out the records for which it is likely that interactive editing will lead to a significant improvement in the quality of estimates. These selection methods appear under names as selective editing, significance editing and macro-editing [cf. Hidiroglou and Berthelot (1986), Granquist (1990), Lawrence and McDavitt (1994), Lawrence and McKenzie (2000), De Waal, Renssen, and Van de Pol (2000), and Hedlin, (2003)]. The selection mechanisms have as a common purpose to select the records that likely contain influential errors. These records will then be treated interactively, whereas the other records will either be edited automatically or not edited at all. In this selective editing view, automatic editing is confined to correcting the relatively unimportant errors. The main purpose of automatic editing is then to ensure that the data satisfy fatal edits, such as balance edits, so that obvious inconsistencies cannot occur at any level of aggregation.

During the last decade, new methods and models for automatic editing have been (and are being) developed that aim to broaden the scope of automatic editing to several types of influential errors. For example, new methods for systematic errors have been developed that aim to identify not only the fact that an error has occurred but also the underlying error mechanism [cf. Al-Hamad, Lewis, and Silva (2008), Di Zio, Guarnera, and Luzi (2008), and Scholtus (2008, 2009); see also Chapter 2). For the errors that these methods can identify, reliable correction mechanisms are often available that can be applied whether or not the error is influential. Other methods aim to improve the models with which predictions for the true values can be obtained—for instance, by exploiting modern modeling techniques and/or the information available in edit constraints [EUREDIT (2004a, 2004b), Tempelman (2007); see also Chapter 9]. Since in automatic editing these predictions are used to replace the erroneous values, more accurate predictions will result in wider applicability of automatic methods. Moreover, a selection of different automatic procedures tailored to the characteristics of a survey, such as types of variables, expected systematic errors, and edit rules, can result in powerful automatic editing systems with a much broader application than only the noninfluential errors [cf. Pannekoek and De Waal (2005), EDIMBUS (2007)]. The advantage of increasing the amount of automatic editing is not only that the editing process becomes more efficient. Since automatic procedures are based on formal models and algorithms, automatic editing also enhances the transparency and repeatability of the editing process.

6.3 Micro-selection: The Score Function Approach

6.3.1 THE STRUCTURE OF SCORE FUNCTIONS

The main instrument in the micro-selection process is the score function [cf. Latouche and Berthelot (1992), Lawrence and McDavitt (1994), Lawrence and McKenzie (2000), Farwell and Rain (2000), Hoogland (2002)]. This function assigns to each record a score that measures the expected influence of editing the record on the most important target parameters, records with high scores are the first to be considered for interactive editing.

A score for a record (record or global score) is usually a combination of scores for each of a number of important variables (the local scores). For instance, local scores can be defined that measure the influence of editing an important variable on the estimated total of that variable. The local scores are generally constructed so that they reflect the following two elements that together constitute an influential error: the size and likelihood of a potential error (the "risk" component) and the contribution or influence of that record on the estimated target parameter (the "influence" component). Local scores are then defined as the product of these two components, that is,

$$s_{ij} = F_{ij} \times R_{ij}, \tag{6.1}$$

with F_{ij} the influence component and R_{ij} the risk component for unit i and variable j. The risk component can be measured by comparing the raw value with an approximation to what the value would have been after editing. This value is called the "anticipated value" or "reference value" and is often based on information from previous cycles of the same survey. Large deviations from the anticipated value are taken as an indication that the value may be in error and, if indeed so, that the error is substantial. Small deviations indicate that there is no reason to suspect that the value is in error and, even if it were, the error would be unimportant. The influence component can often be measured as the (relative) contribution of the anticipated value to the estimated total.

A global or unit score is a function that combines the local scores to form a measure for the whole unit. It is a function of the local scores, say

$$S_i = f(s_{i1}, \ldots, s_{iJ}). \tag{6.2}$$

If interactive editing could wait until the data collection was completed, the editing could proceed according to the order of priority implied by the scores until the change in estimates of the principle output parameters would become unimportant or time or resources became exhausted. In micro-selection, however, it is understood that responses are received during a considerable period of time and that starting the time-consuming editing process after the data collection period will lead to an unacceptable delay of the survey process. Therefore, a threshold or cutoff value is determined in advance such that records with scores above the threshold are designated to be not plausible. These records are assigned to the so-called "critical stream" and are edited interactively, whereas the other records, with less important errors, are edited automatically. In this way the decision to edit a record is made without the need to compare scores across records. Formally, this selection is based on the plausibility indicator variable (PI) defined by

$$PI_i = \begin{cases} 1 & \text{(plausible)} & \text{if } S_i \leq C, \\ 0 & \text{(implausible)} & \text{otherwise} \end{cases} \tag{6.3}$$

with C a cutoff or threshold value.

In defining a selective editing strategy we can distinguish three steps that will be described in more detail in the following subsections. These steps can be summarized as follows:

- Defining local scores for parameters of interest such as domain totals and quarterly or yearly changes in these totals using reference values that approximate the true values as well as possible (Section 6.3.2).
- Defining a function that combines the local scores to form a global or record score (Section 6.3.4).
- Determining a threshold value for the global scores (Section 6.3.5).

6.3.2 COMMON WAYS TO CONSTRUCT SCORE FUNCTIONS

Basic Score Functions for Totals. For most business surveys, the principal outputs are totals of a large number of variables such as turnover, employment, and profits. These totals are usually published for a number of domains defined by (a combination of) classifying variables such as type of industry, size class, and geographical region.

A score function for totals should quantify the effect of editing a record on the estimated total. Let x_{ij} denote the value of a variable x_j in record i. The usual estimator of the corresponding total can then be defined as

$$\hat{X}_j = \sum_{i \in D} w_i \hat{x}_{ij} \tag{6.4}$$

with D the data set (sample or census) and i denoting the records or units. The weights w_i correct for unequal inclusion probabilities and/or nonresponse. In the case of a census, nonresponse is of course the only reason to use weights in the estimator. The $\hat{x}_{ij}$ in (6.4) are edited values; that is, they have been subjected to an editing process in which some of the raw values (x_{ij}, say) have been corrected, either by an automated process or by human intervention. For most records, x_{ij} will be (considered) correct and the same as $\hat{x}_{ij}$. The (additive) effect on the total of editing a single record can be quantified as the difference in the total estimated with and without editing record i. The estimated total without editing record i is $\hat{X}_j - w_i(\hat{x}_{ij} - x_{ij}) = \hat{X}_j^{(-i)}$, say, and the difference can be expressed as

(6.5) $$d_i(\hat{X}_j) = \hat{X}_j^{(-i)} - \hat{X}_j = w_i(x_{ij} - \hat{x}_{ij}).$$

The difference $d_i(\hat{X}_j)$ depends on the as yet unknown corrected value $\hat{x}_{ij}$ and can therefore not be calculated. A score function is based on an approximation to $\hat{x}_{ij}$, $\tilde{x}_{ij}$ say, which is referred to as the "anticipated value." The anticipated value serves as a reference for judging the quality of the raw value. Often used sources for anticipated values are:

- Edited data for the same unit from a previous version of the same survey, possibly multiplied by an estimate of the development between the previous and the current time point.
- The value of a similar variable for the same unit from a different source, in particular an administration such as a tax register.
- The mean or median of the target variable in a homogeneous subgroup of similar units for a previous period.

Except for the unknown corrected value, the difference (6.5) depends also on an unknown weight w_i. Because the weights do not only correct for unequal inclusion probabilities but also for nonresponse, they can only be calculated after the data collection is completed and the nonresponse is known. A score function that can be used during the data collection period cannot use these weights and will need an approximation. An obvious solution is to use, as an approximation, "design weights" that only correct for unequal inclusion probabilities and which are defined by the sampling design (the inverse of the inclusion probabilities).

Using these approximations the effect of editing a record i can be quantified by the score function:

(6.6) $$s_{ij} = v_i \left| x_{ij} - \tilde{x}_{ij} \right| = v_i \tilde{x}_{ij} \times \left| x_{ij} - \tilde{x}_{ij} \right| / \tilde{x}_{ij} = F_{ij} \times R_{ij},$$

say, with $\tilde{x}_{ij}$ the anticipated value and v_i the design weight. As (6.6) shows, this score function can be written as the product of an "influence" factor ($F_{ij} = v_i \tilde{x}_{ij}$) and a "risk" factor ($R_{ij} = \left| x_{ij} - \tilde{x}_{ij} \right| / \tilde{x}_{ij}$). The risk factor is a measure for the relative deviation of the raw value from the anticipated value. Large deviations are an indication that the raw value may be in error. If the anticipated value is the true value and the editor is capable of retrieving the true value, it is also the effect of correcting the error in this record. The influence factor is the contribution of the record to the estimated total. Multiplying the risk factor by the influence factor results in a measure for the effect of editing the record on the estimated total. Large values of the score indicate that the record may contain an influential error and that it is worthwhile to spend time and resources on correcting the record. Smaller values of the score indicate that the record does not contain very influential errors and that it can be entrusted to automatic procedures that use approximate solutions for the error detection and correction problems.

For nonnegative variables, which includes most of the variables in economic surveys, a risk factor can also be based on the ratio of the raw value to the anticipated value instead of the absolute difference of these values. To derive this ratio-based risk factor, we start with the risk defined in (6.6) but ignore the absolute value, resulting in

$$(x_{ij} - \tilde{x}_{ij})/\tilde{x}_{ij} = \frac{x_{ij}}{\tilde{x}_{ij}} - 1.$$

In this way the risk is expressed as the ratio of the raw value to the anticipated value, while the added -1 ensures that the risk is zero if both values are equal. This expression is, however, not yet suitable for a risk function because both small and large values of the ratio indicate deviations from the anticipated value. To remedy this problem the following ratio-based risk function can be defined:

$$R_{ij} = \max\left(\frac{\tilde{x}_{ij}}{x_{ij}}, \frac{x_{ij}}{\tilde{x}_{ij}}\right) - 1. \tag{6.7}$$

This definition ensures that upward and downward multiplicative deviations from the anticipated value of the same size will lead to the same scores. Multiplying this risk factor by the influence F_{ij} defined in (6.6) leads to an alternative to the score function (6.6).

Often a scaled version of the score function is used that can be obtained by replacing the influence component F_{ij} by the relative influence $F_{ij}/\sum_i F_{ij}$. Because

$$\sum_i F_{ij} = \sum_i v_i \tilde{x}_{ij} = \tilde{X},$$

the resulting scaled score is the original score divided by an estimate (based on the anticipated values) of the total. This scaling makes the score independent of size and unit of measurement of the target variable. This is an advantage when local scores are combined to form a record or global score (see Section 6.3.4).

Models for Anticipated Values. In general, the anticipated value is a function of auxiliary variables and coefficients:

$$\tilde{x}_{ij} = f(\hat{\mu}_1, \ldots, \hat{\mu}_K, z_{i1}, \ldots, z_{iK}), \tag{6.8}$$

with z_{ik} $(k = 1, \ldots, K)$ the known values of auxiliary variables and $\hat{\mu}_k$ $(k = 1, \ldots, K)$ the estimated coefficients. The auxiliary variables should be free from gross errors; otherwise the corresponding anticipated values can be far from the true values and become useless as reference values. Auxiliary variables and estimates of coefficients can sometimes be obtained from the actual survey but are more often obtained from other sources such as a previous already edited version of the actual survey (usually referred to as $t-1$ data) and/or administrative sources. For business surveys, an administrative source that is commonly available is the business register that serves as the sampling frame for the survey. From this register, typically, the variables "branch of industry" and "size class" (a classification based on the number of employees) can be obtained. An anticipated value based on these auxiliary variables can be an estimate of the mean or median of the target variable in each of the subgroups defined by combinations of "branch of industry" and "size class." This can be expressed as the following linear function:

$$\tilde{x}_{ij} = \hat{\boldsymbol{\mu}}_j^T \mathbf{z}_i \tag{6.9}$$

with $\mathbf{z}_i$ a vector with dummy (0–1) variables identifying the subgroup to which unit i belongs and $\hat{\boldsymbol{\mu}}_j$ a vector with estimated means or medians for each of these subgroups. These estimates can be obtained from data from another survey, usually $t-1$ data, in which case the means or medians are sometimes multiplied by a "development" or "growth" factor to adjust for differences between $t-1$ and t. In the case of means, this group-mean model can be viewed as a standard linear regression function, estimated by least squares, and generalization of this model to more general regression models is obvious. For instance, instead of dummy variables for all combinations of categories of "branch of industry" and "size class" a more parsimonious additive model can be formulated that only includes the main effects of these variables and not their interaction. It is also straightforward to add numerical auxiliary variables to the model.

Numerical auxiliary variables that are strongly correlated with the target variable are also often used to improve the risk factor by first dividing the target variable by the auxiliary variable and then comparing this ratio with an anticipated value for this ratio. For instance, suppose that the number of employees is available as an auxiliary variable, then the ratio could be turnover per employee. The value of turnover can show much variation between units even within the same size class and type of industry, but the value of turnover per employee is typically much less variable across units. Therefore, erroneous values of turnover are better detectable by using this ratio than by using turnover itself. Score functions based on ratios can be obtained by replacing the x_{ij} and $\tilde{x}_{ij}$ in the risk factors in (6.6) and (6.7) by the raw value of the ratio and an anticipated value of the ratio, respectively. Again, possible estimates of the anticipated ratio are the mean and median of the $t-1$ values of this ratio, preferably within homogeneous subgroups. Denoting the vector with estimated means or medians of the target ratio within groups by $\hat{\boldsymbol{\mu}}_j$, the anticipated value of the ratio is

$$\widetilde{\left(\frac{x_{ij}}{y_{ij}}\right)} = \hat{\boldsymbol{\mu}}_j^T \mathbf{z}_i, \tag{6.10}$$

with y_{ij} the numerical auxiliary variable for the target variable x_{ij} and $\mathbf{z}_i$ the vector with dummy variables indicating the group to which unit i belongs. Ratio-based additive and multiplicative risk factors can be defined by substituting the unedited value of the ratio (x_{ij}/y_{ij}) and the corresponding anticipated value (6.10) for $\tilde{x}_{ij}$ and x_{ij} in (6.6) or (6.7), respectively. A special case of the latter option will be discussed below.

Models for the anticipated value that are applied in practice are often not very sophisticated, such as the group mean or median models above, and may not lead to very accurate predictions. However, as Lawrence and McKenzie (2000) argue, for the purpose of significance editing the anticipated values need not be accurate enough to be usable as imputations. It is valid to manually inspect values that are atypical according to a model for the anticipated value with limited predictive accuracy, perhaps just to confirm the correctness of such an atypical value, but it would be inappropriate to use these anticipated values as imputed values in estimators of publication figures.

Scores for Continuing Units. For longitudinal data, obtained by panel surveys or repeated surveys with a sampling fraction of one in some strata (for instance strata corresponding to large size classes), a value of the target variable from a previous period may be available for most units or the most important units. This historical value can be a very effective auxiliary variable that can be used in a ratio-based risk factor. For the multiplicative risk (6.7), this leads to the risk factor proposed in a seminal article by Hidiroglou and Berthelot (1986):

$$(6.11) \qquad R_{ij} = \max\left(\widetilde{\left(\frac{x_{ij,t}}{\hat{x}_{ij,t-1}}\right)} \Big/ \left(\frac{x_{ij,t}}{\hat{x}_{ij,t-1}}\right), \left(\frac{x_{ij,t}}{\hat{x}_{ij,t-1}}\right) \Big/ \widetilde{\left(\frac{x_{ij,t}}{\hat{x}_{ij,t-1}}\right)}\right) - 1,$$

with $x_{ij,t}$ the value of the target variable x_j for unit i in the current survey and $\hat{x}_{ij,t-1}$ the corresponding value for the same unit in a previous, already edited, survey. The ratio is in this case a measure of change or trend and units with changes that are not in line with the anticipated change will be detected by the risk factor.

Hidiroglou and Berthelot proposed their risk factor in an application of selective editing after all data had been collected (see Section 6.4.1) so that the change for all units could be calculated. As an anticipated value of the change, they suggested to use the median of these changes. This proposal does not fit into a micro-selection approach which is applied during the data collection period when a reliable estimate of this median is not yet available. As an alternative, Latouche and Berthelot (1992) propose to use the median of the changes at a previous cycle of the survey (i.e., the changes between $t-2$ and $t-1$), which seems reasonable only if the change between $t-2$ and $t-1$ is similar to the change between $t-1$ and t. This is the case for variables that are gradually and moderately changing over time, such as labor costs per hour. Another way to obtain an anticipated value, especially for short-term statistics, is to estimate a time series model with a seasonal component for a historical series of $x_{ij,t}$ values and to use the prediction from this model for the current value as an anticipated value for $x_{ij,t}$. By dividing this anticipated value for $x_{ij,t}$ by the corresponding edited value for $t-1$, an anticipated value for the change is found that also does not rely on the current data except, of course, for the record for which the score is calculated.

To define a score function, Hidiroglou and Berthelot propose to multiply the risk factor (6.11) by an influence factor given by (the unweighted version of)

$$(6.12) \qquad F_{ij} = \left[\max\left(v_{i,t}x_{ij,t}, w_{i,t-1}\hat{x}_{ij,t-1}\right)\right]^c$$

with $0 \leq c \leq 1$. The parameter c can be used to control the importance of the influence: Higher values give more weight to the influence factor. Latouche and Berthelot (1992) report that in empirical studies at Statistics Canada it was found that 0.5 was a reasonable value for c. The maximum function in (6.12) has the effect that an error in the reported value $x_{ij,t}$ is more likely to lead to an overestimation of the influence than to lead to an underestimation. This is because a too low reported value $x_{ij,t}$ can never result in an influence value smaller than $\hat{x}_{ij,t-1}$, whereas a too high value can increase the influence in principle without limit. A scaled version of a score with influence factor (6.12) can be obtained by dividing $w_{i,t-1}\hat{x}_{ij,t-1}$ and $v_{i,t}x_{ij,t}$ by their respective totals. The total of $w_{i,t-1}\hat{x}_{ij,t-1}$ is simply the population estimate for the previous period, $\hat{X}_{j,t-1} = \sum_{i \in D_{t-1}} w_{i,t-1}\hat{x}_{ij,t-1}$. The current total, however, must be approximated since we cannot assume that all the data are already collected. An obvious approximation is obtained by using the anticipated values resulting in $\tilde{X}_{j,t} = \sum_{i \in D_t} v_{i,t}\tilde{x}_{ij,t}$.

Scaled Scores for Domain Totals. For business statistics, the target parameters are often domain totals. For instance, the total turnover and employment for a large number of branches of industry often crossed with size classes and possibly also with geographic regions. In principle, selective editing could be applied to each of these domains separately. A potentially much more efficient strategy is, however, to treat the domains simultaneously and apply the selection process to the combined data for all domains. This approach requires that in data without gross errors the spread of the scores should be made comparable between domains. Large scores or deviations from the anticipated values are less "suspect" in domains where, in the absence of errors, these deviations already show much variability than in domains with a smaller variability. To compensate for differences in the variability of the scores, scores can be scaled by dividing them by a measure of their dispersion within domains. Similar to the anticipated value, an anticipated measure of dispersion can be obtained from the edited values from a previous period. For instance, the standard deviation of the values of the score function applied to the edited units in a domain. As another example of such an anticipated measure of dispersion, Latouche and Berthelot (1992) propose to use the interquartile range of scores calculated for a previous period.

A Score Function for a Ratio as Target Parameter. In this section we derived score functions that measure the effect of editing a value on the estimate of a total. Although totals are the primary parameters of interest in most business surveys, sometimes ratios of totals are the more important parameters. As an example, Lawrence and McDavitt (1994) report on a survey in which the most important parameters are the average weekly earnings in sectors of the Australian economy. Suppose that the estimate of a ratio can be written as $\hat{Q}_{jk} = \hat{X}_j / \hat{X}_k$

with $\hat{X}_j$ and $\hat{X}_k$ defined as in (6.4). Then the additive effect of editing a single record i on this estimate can be expressed as the difference

$$\begin{aligned} d_i(\hat{Q}_{jk}) &= \hat{Q}_{jk}^{(-i)} - \hat{Q}_{jk} \\ &= \frac{\hat{X}_j^{(-i)}}{\hat{X}_k^{(-i)}} - \frac{\hat{X}_j}{\hat{X}_k} \\ &= \frac{\hat{X}_k\hat{X}_j^{(-i)} - \hat{X}_j\hat{X}_k^{(-i)}}{\hat{X}_k\hat{X}_k^{(-i)}} \\ &= \frac{\hat{X}_j d_i(\hat{X}_k) - \hat{X}_k d_i(\hat{X}_j)}{\hat{X}_k(\hat{X}_k - d_i(\hat{X}_k))}, \end{aligned} \tag{6.13}$$

with $\hat{Q}_{jk}^{(-i)}$ the estimate of the target ratio without editing record i.

The effect of editing record i on the estimated ratio depends on the unknown edited values $\hat{x}_{ij}$ and $\hat{x}_{ik}$ [compare (6.5)] and to transform (6.13) into a score that can be calculated, these unknowns must be replaced by their anticipated values.

6.3.3 OTHER APPROACHES TO CONSTRUCT SCORE FUNCTIONS

The score functions discussed so far measure the impact of editing on estimates and are based on the distance between anticipated values and the raw data. Such score functions are the ones that are now commonly used by national statistical institutes. The construction of score functions has, however, not been limited to this approach. Other approaches have also occasionally been investigated, although these approaches have not (yet) seen such a wide acceptance in practical applications as the more traditional approach. Three alternative approaches are briefly discussed below.

Parametric Models for Data with Errors. One alternative approach is to specify a parametric model for the data that takes possible errors into account. Recent work in this area is based on a model that assumes that the data that are free from errors are from a different distribution than the data with errors. This approach has been followed by Ghosh–Dastidar and Schafer (2006), Di Zio, Guarnera, and Luzi (2008) and Bellisai et al. (2009). These authors assume that the correct data are from a normal distribution with mean μ and variance σ^2 and that the erroneous data are from a normal distribution with the same mean but a variance inflated by a factor $c > 1$. These assumptions lead to the contaminated normal model, which describes the density of the observations on a variable x with the following mixture of normal distributions:

$$f_x = \pi N(\mu, \sigma^2) + (1 - \pi)N(\mu, c\sigma^2)$$

with the mixture probability π equal to the fraction of error free data.

This model is consistent with an additive error mechanism of the form

$$x = x^* + e$$

with x the observed value, x^* the true value, and e an error that may (with probability $(1 - \pi)$) or may not (with probability π) have been realized and $Ex = Ex^* = \mu$, $Ee = 0$, $\text{var}(x^*) = \sigma^2$ and $\text{var}(e) = \sigma^2(c - 1)$. Ghosh-Dastidar and Schafer (2006) use an EM algorithm to estimate the parameters μ and σ for chosen values of π and c. Di Zio, Guarnera, and Luzi (2008) and Bellisai et al. (2009) use a variant of the EM algorithm and estimate the parameters π and c as well as μ and σ.

Using this model, estimates of a number of quantities of interest for selective editing can be derived. One such quantity is the estimated conditional probability, $\hat{\pi}_i$ say, of an observation being free of error, given its observed value, that is, an estimate of $\Pr(x_i = x_i^* | x_i)$. This probability, conditional on the observed data, is called the posterior probability. Ghosh-Dastidar and Schafer suggest to order the observations according to $\hat{\pi}_i$ and to flag the values with $\hat{\pi}_i$ smaller than some appropriate cutoff value as outliers that become candidates for editing. Another quantity of interest, used by Di Zio, Guarnera and Luzi (2008) and Bellisai et al. (2009), is the expected value of the true variable given the observed value $E(x_i^* | x_i)$. An estimate of this "predicted" true value can be used as an anticipated value in a score function, for which Di Zio, Guarnera, and Luzi suggest to use $|x_i - \hat{x}_i^*| / T^*$, with $\hat{x}_i^*$ the estimate of $E(x_i^* | x_i)$ and T^* a robust estimate of the population total based on the "predicted" true values $\hat{x}_i^*$ instead of the observed values.

The model described above is just a basic version of the contaminated normal model. The cited authors actually used more involved versions of this model. Ghosh-Dastidar and Schafer (2006) and Di Zio, Guarnera, and Luzi (2008) used the multivariate version of the model with a vector-valued x variable. Ghosh-Dastidar and Schafer (2006) considered the estimation of the contaminated normal model in the presence of missing values. Di Zio, Guarnera, and Luzi (2008) and Bellisai et al. (2009) applied a log transform to the observations to make their distribution more symmetrical and the normal model more realistic. However, the predicted true values and scores remained in the original scale, leading to a more complicated estimate of $E(x_i^* | x_i)$. Extension of the model to make use of covariates was considered by Bellisai et al. (2009), who let the value of μ vary with a covariate according to a linear regression model.

The Edit-Related Approach. Hedlin (2003) proposes to use the extent to which a record fails edits—that is, how many edits are failed and by how much they are failed—as a criterion for selective editing. Hedlin calls this approach the *edit-related* approach as opposed to the more traditional score function approach, which he refers to as the *estimate-related* approach. The underlying idea of the edit-related approach is that influential errors will lead to the violation of fatal as well as query edits and that the amount of failure can be used as a measure of the importance of interactively editing the record for a range of possible estimates, without estimating the effect of editing on the estimates directly.

To assess how much a record fails the applied edits, one first measures the amount of failure for each edit in some way. For a balance edit, one can, for instance, measure the amount of failure as the absolute difference between the observed total and the sum of its observed components. Query edits are often specified as intervals; for example, turnover divided by the number of employees must be between 0.5 and 2 times the value for a previous period. For such intervals, the amount of failure can, for instance, be measured as the distance to the nearest bound.

In order to combine the scores for each edit into a global score per record, the individual scores must be scaled in some way because the size of the edit failures can be very different. Moreover, the (amount of) failure for the different edits will often be correlated because edits usually show some redundancy: A value that causes the failure of one edit will often also cause the failure of one or more others. To take care of the differences in size and the correlation between the amounts of edit failure, Hedlin (2003) proposes to combine the scores for the edits by the Mahalanobis distance. The edit-related approach has the advantage that it does not focus on a single target variable. It has the drawback that it is obviously dependent on the specified edits. In an empirical study, Hedlin (2003) found that the estimate-related approach performed better than the edit-related approach.

Prediction Model Approach. The traditional use of score functions can be seen as a way to predict the occurrence and size of influential errors in a record, based on a comparison between the raw values and anticipated values. Another approach to the prediction of influential errors is to construct a model that relates the occurrence and size of influential errors in a target variable to other variables that are available in each record (the predictors). This *prediction model approach* was studied at Statistics Netherlands by Van Langen (2002). The approach starts with an already edited "training" data set, containing both the edited and the original values. Using this data set, the influence of editing a value on the total estimate can be calculated for each record.

Based on the influence of the errors, an "error probability," π say, is defined as the variable to be predicted. The error probability is derived from a categorization of the influence of the errors. Van Langen used six categories; the first category corresponds to influence values of zero (no error). The other five categories each contained 20% of the records with errors. The first of these classes, the second class, corresponds to the 20% least influential errors, the third class contains the next 20% influential errors, and so on, until the sixth class, which contains the 20% most inluential errors. The value of π was set to 0, 0.2, 0.4, 0.6, 0.8, and 1 for the records in these six classes, respectively. Next, a model is built to predict π from the predictor variables. A possible choice for such a model is a logistic regression model that has the nice property that the predictions will lie within the required [0,1] interval. The model parameters can be estimated by using the training data. The estimated model is put to use by applying it to a new data set that contains the same predictor variables but is not yet edited. For this data set the model will produce an estimated error probability that can be used to prioritize the records for interactive editing.

Instead of the logistic regression model, other parametric or nonparametric models can also be used for the prediction problem. At Statistics Netherlands, classification and regression trees [see, e.g., Breiman et al. (1984)] have been investigated as nonparametric alternatives to the logistic regression model. Classification trees are used for categorical response variables and result in rules that, based on the values of the predictors, assign each record to one of the categories. This method can be used to predict, for instance, the "error category" as defined by the six categories described above. Regression trees predict, based on the values of the predictors, the value of a numerical response variable and can be used to predict, for instance, the value of π. Beside the "error category" and π, other response variables can be used for the tree-based methods as well. For instance, in the applications at Statistics Netherlands, the magnitude and impact of an error in a single variable was defined as the absolute value of the observed change in the training data, multiplied by the raising weight. And, as a categorical response variable, a dichotomous variable was defined, based on the training data, indicating whether or not a variable is considered to need interactive editing. In addition to these response variables that are targeted at a single variable in a record, response variables can also be defined as global, record level, measures—for instance, a binary variable indicating whether or not a record needs interactive editing or a numerical variable measuring the combined impact of editing a number of important target variables. In empirical studies so far, the prediction model did not outperform the more traditional approach based on simple global score functions.

6.3.4 COMBINING LOCAL SCORES

In order to select a complete record for interactive editing, a score on the record level is needed. This "record score" or "global score" combines the information from the "local scores" that are defined for a number of important target parameters. The global score should reflect the importance of editing the complete record. In order to combine scores, it is important that the scores are measured on comparable scales. It is common, therefore, to scale local scores before combining them into a global score. In the previous subsection we have seen one option for scaling local scores—that is, by dividing by the (approximated) total. Another method is to divide the scores by the standard deviation of the anticipated values [see Lawrence and McKenzie (2000)]. This last approach has the advantage that deviations from anticipated values in variables with large variability will lead to less high scores and are therefore less likely to be designated as suspect values than deviations in variables with less natural variability.

Scaled or standardized local scores have been combined in a number of ways. Often, the global score is defined as the sum of the local scores [cf. Latouche and Berthelot (1992)]. As a result, records with many deviating values will get high scores. This can be an advantage because editing many variables in the same record is relatively less time-consuming than editing the same number of variables in different records, especially if it involves recontacting the respondent. But a consequence of the sum-score is also that records with many, but only moderately deviating, values will have priority for interactive editing over records with only a few strongly deviating values. If it is deemed necessary that strongly deviating values in an otherwise plausible record are treated by specialists, then the sum-score is not the global score of choice.

An alternative for the sum of the local scaled scores, suggested by Lawrence and McKenzie (2000), is to take the maximum of these scores. The advantage of the maximum is that it guarantees that a large value of any one of the contributing local scores will lead to a large global score and hence manual review of the record. The drawback of this strategy is that that it cannot discriminate between records with a single large local score and records with numerous equally large local scores. As a compromise between the sum and max functions, Farwell (2005) proposes the use of the Euclidean metric (the root of the sum of the squared local scores). These three proposals (sum, max, Euclidean metric) are all special cases of the Minkowski metric [cf. Hedlin (2008)] given by

$$S_i^{(\alpha)} = \left(\sum_{j=1}^{J} s_{ij}^{\alpha}\right)^{1/\alpha}, \tag{6.14}$$

with $S_i^{(\alpha)}$ the global score as a function of the parameter α, s_{ij} the jth local score, and J the number of local scores. The parameter α determines the influence of large values of the local scores on the global score, and the influence increases with α. For $\alpha = 1$, (6.14) is the sum of the local scores; for $\alpha = 2$, (6.14) is the Euclidean metric; and for α approaching ∞, (6.14) approaches the maximum of the local scores.

For the extensive and detailed questionnaires that are often used in economic surveys, it may be more important for some variables to be subjected to an interactive editing process than for others. In such cases the local scores in the summation can be multiplied by weights that express their relative importance [cf. Latouche and Berthelot (1992)].

6.3.5 DETERMINING A THRESHOLD VALUE

Record scores are used to split the data into a critical stream of implausible records that will be edited interactively and a noncritical stream of plausible records that will be edited automatically. This selection process is equivalent to determining the value (1, plausible; 0, implausible) of the plausibility indicator PI given by (6.3).

The most prominent method for determining a threshold value C [see (6.3)] is to study, by simulation, the effect of a range of threshold values on the bias in the principal output parameters. Such a simulation study is based on a raw unedited data set and a corresponding fully interactively edited version of the same data set. These data must be comparable with the data to which the threshold values are applied. Often, data from a previous cycle of the same survey are used for this purpose.

The simulation study now proceeds according to the following steps:

- Calculate the global scores according to the chosen methods for the records in the raw version of the data set.
- Simulate that only the first p% of the records is designated for interactive editing. This is done by replacing the values of the p% of the records with the highest scores in the raw data by the values in the edited data. The subset of the p%-edited records is denoted by E_p.
- Calculate the target parameters using both the p%-edited raw data set and the completely edited data set.

These steps are repeated for a range of values of p. The effect of editing p% of the records can be measured by the difference between the estimates of the target parameter based on the p%-edited and the raw values. The absolute value of the relative difference between these estimates is called the absolute pseudo-bias [Latouche and Berthelot (1992)]. For the estimation of the total of variable j, this absolute pseudo-bias is given by

$$AB_j(p) = \frac{1}{\hat{X}_j} \left| \sum_{i \notin E_p} w_i (x_{ij} - \hat{x}_{ij}) \right|. \tag{6.15}$$

As (6.15) shows, the absolute pseudo-bias is determined by the difference in totals of the edited and nonedited values for the records not-selected (for interactive editing). If the editing results in correcting all errors (and only errors), then (6.15) equals the absolute value of the relative bias that remains because not all records have been edited. However, because it is uncertain that editing indeed reproduces the correct values, (6.15) is an approximation to this bias, hence the name "pseudo-bias."

The pseudo-bias at p%-editing can also be interpreted as an estimator of the gain in accuracy that can be attained by also editing the remaining $1 - p$% of the records. By calculating the pseudo-bias for a range of values of p, we can trace the gain in accuracy as a function of p. If sorting the records by their scores has the desired effect, this gain will decline with increasing values of p. At a certain value of p, one can decide that the remaining pseudo-bias is small enough and that it is not worthwhile to pursue interactive editing beyond that point. The record score corresponding to this value of p will be the threshold value.

The pseudo-bias, as described above, is based on a comparison between interactive editing and not editing at all. In most cases the alternative to interactive editing is automatic editing rather than not editing. Assuming that automatic editing does at least not lead to more bias than not editing at all, the value of the pseudo-bias according to (6.15) is an upper bound of the pseudo-bias in situations where automatic editing is applied.

As an alternative to the absolute pseudo-bias, Lawrence and McDavitt (1994) define the relative pseudo-bias given by

$$RB_j(p) = \frac{1}{\text{std.err.}(\hat{X}_j)} \left| \sum_{i \notin E_p} w_i(x_{ij} - \hat{x}_{ij}) \right|. \tag{6.16}$$

In the relative pseudo-bias the bias is compared with the standard error of the estimate, std.err.$(\hat{X}_j)$, rather than the magnitude of the estimate as in the absolute pseudo-bias. This measure can be interpreted as a bias-variance ratio; it compares one source of error, the bias due to not editing $1 - p\%$ of the records, with another source of error, the sampling error. Lawrence and McDavitt (1994) choose a value of p by limiting the bias-variance ratio to 20%.

As Lawrence and McKenzie (2000) argue, the simulation study approach is also a way to check the effectiveness of the selective editing process itself. The simulation study allows us to check if the records with high global scores indeed contain influential errors and, reversely, if records with low scores contain only noninfluential errors. Without a simulation study, score functions can partially be evaluated using data from the current editing process. Using the changes made by the editors, it can be verified whether the records that were selected for interactive editing did contain influential errors. It can, however, not be verified whether the records with low scores that were not edited interactively did not also contain some influential errors that have remained undetected by the score function.

6.4 Selection at the Macro-level

Selection at the macro-level or macro-editing is a process for the selection of records with potentially influential errors that is performed when all or at least a substantial part of the data is collected. The purpose of macro-selection is the same as for micro-selection: the advancement of effective and efficient editing processes by limiting the interactive treatment to those records where this treatment is likely to have a significant effect on estimates of interest. Macro-selection techniques differ from micro-selection procedures in their use of all available data to identify anomalous and suspect values on the micro-level, as opposed to handling each record in isolation which is typical for micro-editing procedures. We discuss two approaches to macro-selection. First, we discuss the aggregate or top-down method that starts with checking preliminary estimates of the publication figures (aggregates) and then drills-down to only those individual records that contribute to a suspicious aggregate. Second, we discuss the distribution method that looks at the distribution of the collected data to identify outlying values that warrant further inspection.

6.4.1 AGGREGATE METHOD

The aggregate method [cf. Granquist (1990, 1995)] starts with applying score functions to aggregates, usually the publication figures. These score functions resemble those applied at the micro level; for instance, a simple macro level score function is

$$S_j = X_j - \tilde{X}_j, \tag{6.17}$$

with X_j the total estimate of variable x_j based on the unedited values and $\tilde{X}_j$ an anticipated value for this estimated total. Possible anticipated values include the value of a previous estimate of the same variable (the $t-1$ value) and estimates based on other sources such as registers or other surveys. In some cases, anticipated values have been based on econometric time series models (Meyer et al., 2008). Just as with scores at the micro-level, it can be more effective for detecting anomalous aggregates to use ratios between aggregates than the aggregates themselves, leading to scores of the form

$$S_j = \frac{X_j}{X_k} - \widetilde{\left(\frac{X_j}{X_k}\right)}. \tag{6.18}$$

Examples of such ratios are the ratio between the total turnover and the total costs for a certain branch of industry or the total of salaries divided by the total number of employees. The discrepancies between observed values of totals (ratios) and their corresponding anticipated values can again be expressed in the form of multiplicative deviations rather than the additive differences shown here.

Once a suspect aggregate has been detected, the search for the record(s) that cause this suspect value can proceed by looking at lower level aggregates, such as size classes within a certain branch of industry. At some point, however, the procedure will drill-down to the record level to identify possible influential errors that will be routed to interactive editors. At that stage the record level scores can in principle be used to identify the microdata with suspect values. There are, however, a few differences when record-level scores are applied in the macro-editing stage.

1. First, as a source for anticipated values, the complete actual data are now available. So, for instance, the median in a homogeneous subgroup of the current data can be used instead of the $t-1$ median or mean. It is important to use measures like medians that are robust against outliers when the current data are used since, contrary to the $t-1$ data, the current data are not yet edited and may still contain influential errors.
2. Second, the actual final weights w_i that will be used in the estimation of publication figures can now be used. It is not necessary to approximate these weights with the design weights v_i as in (6.6).

3. Third, in the macro-editing stage it is not needed to set a cutoff value in advance. The scores provide a prioritation of the records for interactive editing, and this editing can proceed according to this prioritation while tracing the changes in the estimates. The editing can be pursued until these changes become insignificant compared to the desired accuracy of the estimates.

6.4.2 DISTRIBUTION METHOD

The distribution method tries to identify values that do not seem to fit into the observed distribution of a variable. Many methods for detecting such outliers are based on the distance of data points to the center of the distribution of the bulk of the data, without the outliers. To measure this distance, a measure of location must be used that is not (much) influenced by the possible presence of outliers. A common outlier robust measure of location is the median and a measure of "outlyingness" based on the median is $x_{ij} - \text{med}(x_{ij})$, with med() denoting the median. Note that this is similar to using a score function with an anticipated value equal to the median of the data set to be edited. To compare deviations from medians across publication cells, it is common to standardize these deviations by the median of their absolute values, resulting in the measure, for the units in cell c,

$$(6.19) \qquad o_{ij,c} = |x_{ij} - \text{med}_c(x_{ij,c})|/(1.4826 \times \text{MAD}_c(x_{ij,c})),$$

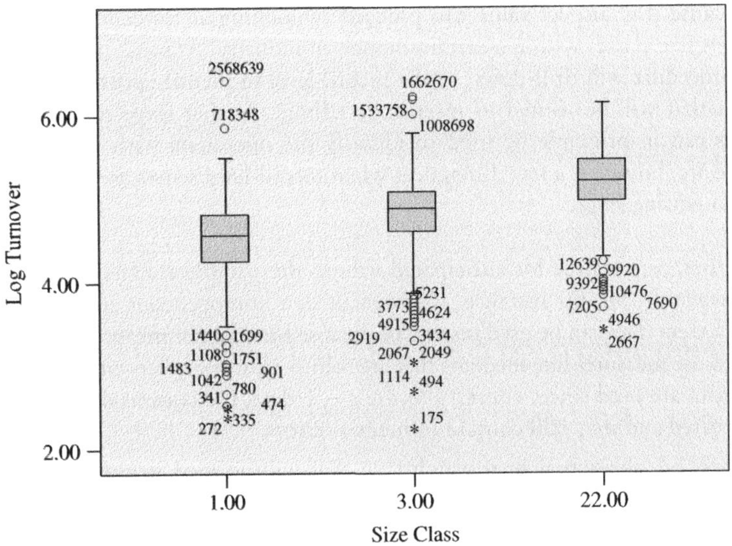

FIGURE 6.2 Boxplots of log(Turnover) of supermarkets in three size classes.

with $\mathrm{med}_c(x_{ij,c})$ the median of the units in cell c and $\mathrm{MAD}_c(x_{ij,c})$ the median absolute deviation for these units given by

$$\mathrm{MAD}_c(x_{ij,c}) = \mathrm{med}_c(|x_{ij} - \mathrm{med}_c(x_{ij,c})|).$$

The factor 1.4826 in (6.19) stems from the fact that for the normal distribution $1.4826 \times \mathrm{MAD}$ is a consistent estimate of the standard deviation (Hoaglin, Mosteller, and Tukey, 1983). Other robust measures of location can also be used to identify outliers—for instance, Winsorized and trimmed means. Nonrobust measures of dispersion such as the variance or standard deviation can also be used for determining outliers in ongoing surveys. It is an indication of the presence of outliers if the current values of these statistics, based on unedited data, are much larger than anticipated values for these statistics—for instance, based on edited $t-1$ data.

Deviations from the median are often displayed graphically by boxplots. Figure 6.2 shows an example of such boxplots. In this figure the values of the logarithm of turnover are plotted in three size classes. A boxplot shows the location (median) and spread of a distribution. To picture the spread, the range of a variable is divided into three parts: The box contains 50% of the observations, and the areas below and above the box indicated by vertical lines are limited by 1.5 times the interquartile range (iqr) beneath the first quartile and 1.5 times the iqr above the third quartile. Values outside the areas covered by the box and vertical lines are considered to be outliers. For these outliers the unit identifiers are shown. To facilitate the editing, software can be used that can display the record in an edit screen by clicking on the record identifier.

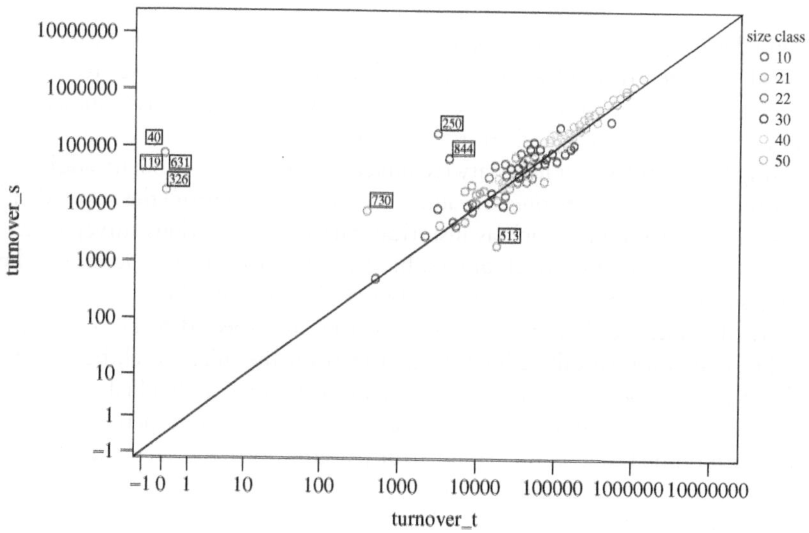

FIGURE 6.3 Scatterplot of current turnover versus turnover in the previous year.

Another basic graphical tool in finding outlying values is the scatterplot illustrated in Figure 6.3. A scatterplot is used to identify outliers with respect to the bivariate distribution of the variables involved, rather than outliers with respect to univariate distributions as is the purpose of the boxplot. Figure 6.3 is a scatterplot in which the current turnover (x axis) of a number of firms is plotted against the turnover in the previous year (y axis). The outliers with uncommon changes in turnover are easily identified in this figure and marked with their unit identifiers to facilitate further processing.

The use of medians, boxplots, and scatterplots to inspect distributions are examples of Exploratory Data Analysis (EDA) techniques [see, e.g., Tukey (1977)]. These techniques emphasize graphical techniques and outlier-resistant descriptive statistics. Applications of EDA techniques and many graphical tools for macro-editing have been described, among others, by Bienias et al. (1997), DesJardins (1997), and Weir, Emery, and Walker (1997).

6.5 Interactive Editing

In this section, some general remarks are made on the manual editing of microdata. Section 6.5.1 introduces the subject. Since nowadays manual editing is nearly always a computer-assisted operation, Section 6.5.2 discusses the use of computers in this context. The term "interactive editing" is commonly used for modern computer-assisted manual editing. Section 6.5.3 sketches potential problems associated with interactive editing. Finally, Section 6.5.4 discusses the design of editing instructions.

6.5.1 MANUAL EDITING

Manual editing is the traditional approach to check and correct microdata at statistical institutes. In an ideal world, manual editing means that survey records are reviewed by experts, who use their extensive knowledge on the survey subject and the survey population to find errors and to impute new values for erroneous and missing fields, and who if necessary recontact the respondent to obtain additional information. To locate erroneous values, they may also compare the respondent's data to reference values, such as historical data from a previous survey or data from external (administrative) sources. In addition, they may have access to other sources of information—for example, through internet searches.

In this ideal setting, manually edited data are expected to be of a higher quality than automatically edited data, at least on the micro-level, because the subject-matter specialist can tailor the editing to each individual record. In practice, however, manual editing is also performed by nonexpert clerks, who have been trained to look for certain error patterns. In that case, manual editing will not necessarily lead to data of a higher quality than automatic editing, even on the micro-level.

At first glance, the natural way of obtaining correct values for fields that were reported erroneously (or not at all) during the original survey may seem to be recontacting the respondent. However, this approach is often considered problematic. Statistical institutes are constantly trying to reduce the response burden and improve the timeliness of their statistics. Recontacts increase the burden on respondents and also tend to slow down the editing process. Moreover, if a respondent was not able to answer a question correctly during the original survey—that is, while filling in a meticulously designed questionnaire or while talking to a professional interviewer—it seems doubtful whether an editor will obtain the correct answer from him. Following Granquist (1997), if recontacts are used, they should focus on working out why the respondent was unable to provide a correct answer in the first place. Such insights into response behavior may then be used to improve the survey.

The nature of the questionnaires tends to be different for business surveys and social surveys. This is reflected in the kinds of operations that are performed during manual editing. Bethlehem (1987) gives the following characterization:

> In most economical surveys the questionnaire is straightforward. The questions are answered one after another without routing directions. Many range checks and consistency checks are carried out. Totals of individual entries have to be made or have to be checked. Often figures of firms are confronted with corresponding figures from previous years. In case of detected errors it is sometimes possible to contact the firm (...). In short, economical surveys can be characterized by simple questionnaires with a lot of accounting and checking.
>
> Questionnaires for social surveys are often very large with complicated routing structures. Error checking mainly concerns the route followed through the questionnaire and some range checks. Only a few consistency checks are carried out. To be able to correct detected errors, contact is necessary with the supplier of the information. Since this is hardly ever possible, generally correction results in imputation of a value "unknown."

Because of its high costs, both in terms of money and time, statistical institutes nowadays use manual editing in combination with selective editing whenever this is possible. In particular for business surveys, selective editing has become common practice. Also, manual editing is nearly always performed with the aid of a computer. Computer-assisted editing is the subject of the next subsection.

6.5.2 COMPUTER-ASSISTED MANUAL EDITING

As decribed in Section 6.2, the advent of the microcomputer in the 1980s enabled computer-assisted interactive editing. The computer runs consistency checks and displays a list of edit violations per record on the screen. Subject-matter specialists perform manual editing directly on the captured data. Whenever a possible correction is typed in, the computer immediately reruns the consistency checks to see whether this correction removes edit violations. Each record is edited separately until it does not violate any edit rules. See Bethlehem (1987) and Pierzchala (1990) for more details.

Generalized survey-processing systems such as *Blaise* [cf. Blaise (2002)] and *CSPro* [cf. CSPro (2008)] can be used for interactive editing. These computer systems provide two data entry modes: "heads down" and "heads up" (Ferguson, 1994). In the former case, no consistency checks are performed during data entry. This means that data can be keyed in by professional typists at great speed. Scanning the paper questionnaires in combination with optical character recognition can also be seen as a form of "heads down" data entry. "Heads up" data entry means that subject-matter specialists key in the data themselves, while editing them at the same time, using the computer system to perform consistency checks. This has the benefit that each record is treated in one go. A drawback of "heads up" data entry is that no file of unedited data is available for a subsequent analysis of the editing process itself. The survey context dictates which data entry mode is to be preferred. For instance, Van de Pol (1995) remarks that "heads down" data entry is a good choice if the quality of incoming data is high—that is, if most records do not contain any errors.

A more recent development is the increasing use of electronic questionnaires, where data arrive at the statistical institute already in digital form. With an electronic questionnaire, data entry is performed during data collection, either directly by the respondent or by an interviewer with a laptop computer. In most cases, electronic data collection is not completely equivalent to "heads down" data entry, because some consistency checks are built into the electronic questionnaire [see, e.g., Van der Loo (2008) and Section 1.4.1 of this book].

EXAMPLE 6.1

The best-known generalized survey-processing system is probably Blaise, developed at Statistics Netherlands from 1986 onward. Figure 6.4 shows two screen shots taken from the Blaise data entry tool in editing mode. The record being edited is the record from Example 3.7, with the same set of edits being used.

The editor can ask Blaise to display a list of all edit violations, as shown in the top image. In addition, all variables involved in currently violated edits are marked, to help the editor find which variable(s) are responsible for the edit violations. The editor makes changes to the record to resolve the edit violations. He can also store remarks to explain the changes he made, and he can read previously made remarks.

The bottom image of Figure 6.4 displays the situation after the editor has made changes to the original record. When the edits are run again, Blaise returns the message that no more errors remain. ■

A good computer system for interactive editing should provide the editor with all relevant auxiliary information for the identification and correction of errors in the data. Apart from a display of consistency checks, this auxiliary information

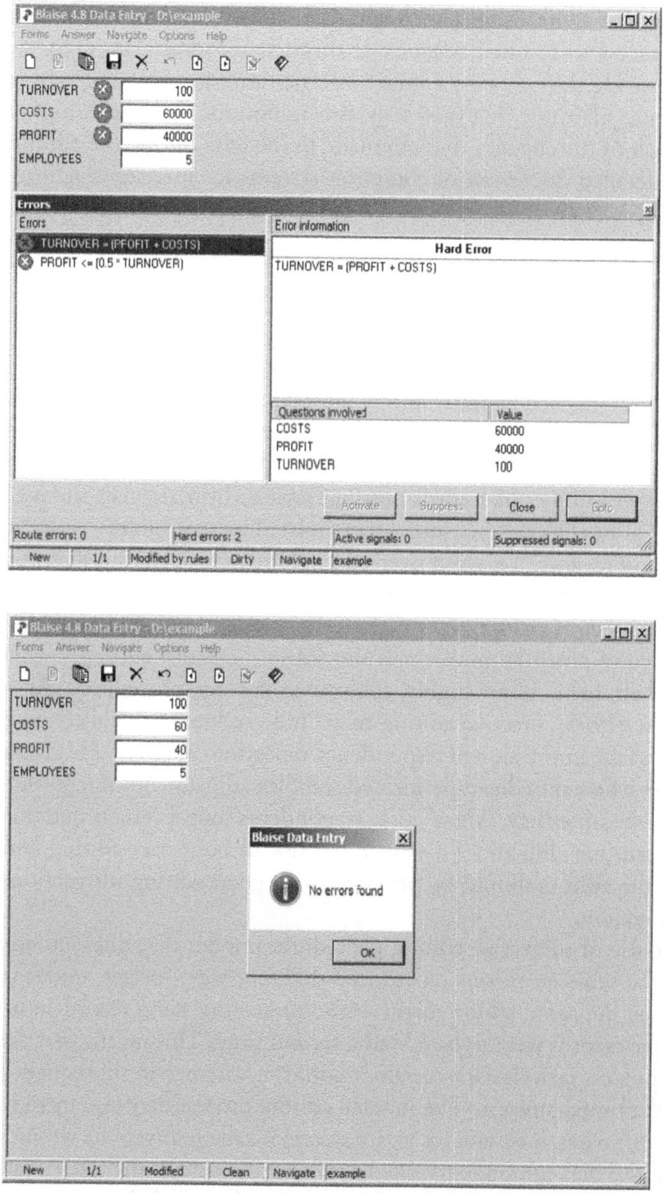

FIGURE 6.4 Screen shots of interactive editing in Blaise.

includes reference data, such as edited data of the same respondent from a previous period, from a different survey, or from an administrative source, or summaries (possibly in graphical form) of data from similar respondents. Comparisons of the observed data with these reference data may be presented to the editor in the form of indicators—for example, score functions. If data collection was done using paper questionnaires, the computer system should provide access to scanned versions of these. If recontacts are used, the system should provide the necessary contact information for each responding unit. For the purpose of future analyses, it is also important that the computer system keeps track of the changes that are made in the data during the editing process. A more detailed discussion of computer systems for interactive editing is given by Pierzchala (1990).

6.5.3 POTENTIAL PROBLEMS

There are several potential problems with interactive editing. The first is the risk of "over-editing." This occurs when "the share of resources and time dedicated to editing is not justified by the resulting improvements in data quality" (Granquist, 1995). The introduction of computer-assisted editing has made it possible to check the data against more edit rules than could be done by hand, resulting in an increased number of records needing review. In particular, the use of query edits—that is, edits that are sometimes violated by true values—can cause many unnecessary reviews. To avoid over-editing, interactive editing should always be limited to the most influential errors; that is, some form of selective editing should be used.

A second potential problem is the risk of "creative editing," when editors invent their own, often highly subjective, editing procedures. As noted by Granquist (1995), creative editing may "hide serious data collection problems and give a false impression of respondents' reporting capacity." He gives examples of surveys where the editors performed complex adjustments of reported financial amounts, because they "knew" that respondents found certain questions on the survey form too difficult. To reduce the risk of both over-editing and creative editing, the editors should be provided with good editing instructions; see the next subsection.

The use of subjective editing procedures implies that different editors may correct the same erroneous record in a different way. In fact, under real-world conditions the same editor might even correct the same record in a different way, if the records were to be edited a second time. This means that the manual editing process introduces a certain amount of variance in subsequent estimates of population parameters. The manual editing process may also introduce a bias into the estimates, if certain errors are systematically resolved the wrong way. The variance and bias components equal zero only if we make the (rather far-fetched) assumption that every editor would correct each record the same way, namely such that the edited record contains the "true" value of each variable. While it is clear that in practice there is thus a nonsampling error due to manual editing, it is difficult to quantify this error.

In any analysis of the quality of the manual editing process, the "true" data have to be estimated by data that have been edited according to some gold standard. Nordbotten (1955) describes an early successful attempt to measure the influence on publication figures of errors that remain after (or indeed are introduced by) manual editing. A random sample of records from the 1953 Industrial Census in Norway was re-edited using every available resource (including recontacts), and the resulting estimates were compared to the corresponding estimates after ordinary editing (without recontacts). No significant deviations were found on the aggregate level. Thus, in this case the less intensive form of manual editing used in practice was sufficient to obtain accurate statistical results. In other words: the experimental "gold standard" editing process would have led to over-editing if it were used in practice.

6.5.4 EDITING INSTRUCTIONS

As we remarked in the previous subsection, it is important to rationalize the interactive editing process by drafting editing instructions. In particular, this helps to reduce the risk of creative editing. Of course, these editing instructions should not attempt to specify exactly how each potential error pattern must be resolved; if this were possible, the entire editing process could be automated and no manual editing would be necessary. Instead, the editing instructions should specify, for the most common error patterns, which explanations should be examined, which external sources should be consulted to verify these explanations, how the data should be adapted if one of the explanations seems plausible, and so on. The actual verification of such potential explanations requires expert knowledge and/or operations that are difficult to automate (such as internet searches), and this is where the merit of interactive editing lies.

If indicators are used, such as ratios of numerical variables, or (local) score functions as discussed in Section 6.3, the editing instructions should explain their interpretation. They should also outline follow-up actions for records with particular (combinations of) indicator scores. Again, ultimately the decision regarding which correction to perform must be left to the individual editor, because it is based on steps that cannot be automated. For example, a large increase or decrease in monthly *turnover* compared to the previous month may be suspicious for enterprises in some economic sectors. In other sectors, a substantial increase or decrease may actually be expected due to natural seasonal variability; for example, in many countries the retail trade sector is known to have a busy period during December. Such subject-matter-specific knowledge should be described in the editing instructions.

If this is relevant, editing instructions should also list the order in which checks should be performed. When editing data from a business survey, for instance, it may be important to start by checking that a responding unit has been given correct codes for *economic activity* and *size class*, since a unit has to be processed in a different way in the event of a classification error. The instructions may also give a hierarchy of variables in terms of their importance; the most important variables should receive the most editing.

6.6 Summary and Conclusions

Statistical editing of survey data is an important part of the processing of survey data: It enhances the general quality of the output, it is indispensible in preventing gross errors in publications, and it is an important learning process that builds up knowledge about possible flaws in the statistical process as a whole, from questionnaire design and data gathering to estimation of publication figures. This knowledge can be used to improve the process in the future—for instance, by improving the wording or layout of the questionnaire, adding new rules for detecting systematic errors, improving imputation methods, and making estimation methods more robust against outliers.

The editing process can be very time- and resource-consuming, and costs and timeliness considerations therefore urge statistical offices to make this editing process as efficient as possible, while maintaining the effectiveness. The cost-effectiveness can be enhanced by applying the time-consuming manual (or interactive) editing only to a limited selection of units that are likely to contain influential errors. In this context, an influential error is defined as an error that has a considerable effect on the publication figures. This approach requires a selection mechanism to divide the records in two streams: a critical stream with records that are suspected to contain influential errors and a noncritical stream with records for which it is likely that interactive editing will not lead to significant effects on publication figures. Only the records in the critical stream are edited interactively. In this chapter we have focused on two selection mechanisms: a micro-selection process that is carried out during the data collection phase and a macro-selection process that is applied after the data collection phase has been completed.

The main instrument in the micro-selection process is the score function. This score function assigns to each record a score that reflects the estimated effects of editing the record on the output parameters of interest. The editing is prioritized according to these scores. Records with low scores, beneath a certain threshold, need not be edited interactively. In this chapter we have seen how to build score functions and how to choose a threshold value. A record or unit score is usually a function of a number of local scores measuring the effect of editing on a particular output parameter, often a total of an important variable, but local scores for ratios of totals have also been defined. The local scores are based on the discrepancy between the observed raw value for a variable and an approximation for that value called the "anticipated value." Alternative ways to measure this discrepancy can be based on differences between raw and anticipated values (additive scores) or ratios between these values (multiplicative scores). Anticipated values can be based on information from different sources; often edited data of the same survey for a previous period can be used, but data obtained from registers or other surveys with similar variables can be used as well.

Macro-selection methods are applied after the data collection phase is finished. We have discussed two approaches to macro-selection. The aggregate method starts with calculating preliminary estimates of publication figures. These estimates of aggregates are checked for implausibilities using macro-level scores. Suspect aggregates can be broken down into more detailed aggregates to further localize the problematic records. As a first selection step the records contributing to a suspect aggregate are isolated. Within this reduced set of records, records with suspect influential values can then be found by sorting them according to a micro-level score function similar to the ones used in micro-selection. The distribution method searches for outlying values in the empirical distribution of the data. Outliers can be found by formulae measuring "outlyingness" or by graphical displays.

Since preliminary estimates of publication figures are available, macro-selection methods have the potential of a more accurate prediction of the records for which interactive editing really makes a difference, but the disadvantage is that the time-consuming follow-up of the selected records starts near the end of the data collection period and can therefore have an undesirable influence on the often important timeliness of the publication figures. This disadvantage does not arise when a large part of the data become available within a short period of time. As mentioned in Section 1.5, this is often the case for administrative data where most of the records are available at once.

REFERENCES

Al-Hamad, A., D. Lewis, and P. L. N. Silva (2008), *Assessing the Performance of the Thousand Pounds Automatic Editing Procedure at the ONS and the Need for an Alternative Approach*. Working Paper No. 21, UN/ECE Work Session on Statistical Data Editing, Vienna.

Bellisai, D., M. Di Zio, U. Guarnera, and O. Luzi (2009), *A selective editing approach based on contamination models: An application to an ISTAT business survey*. Working Paper No. 27, UN/ECE Work Session on Statistical Data Editing, Neuchatel.

Bethlehem, J. G. (1987), *The Data Editing Research Project of the Netherlands Central Bureau of Statistics*. Report 2967-87-M1, Statistics Netherlands, Voorburg.

Bienias, J. L., D. M. Lassman, S. A. Scheleur, and H. Hogan (1997), Improving Outlier Detection in Two Establishment Surveys. In: *Statistical Data Editing*, Volume 2: *Methods and Techniques*. United Nations, Geneva, pp. 76–83.

Blaise (2002), *Blaise for Windows 4.5 Developer's Guide*. Statistics Netherlands, Heerlen.

Breiman, L., J. H. Friedman, R. A. Olsen, and C. J. Stone (1984), *Classification and Regression Trees*. Wadsworth International Group, Belmont.

CSPro (2008), *CSPro User's Guide*, version 4.0. U.S. Census Bureau, Washington, D.C.

DesJardins, D. (1997), *Experiences with Introducing New Graphical Techniques for the Analysis of Census Data*. Working Paper No. 34, UN/ECE Work Session on Statistical Data Editing, Prague.

De Waal, T., R. Renssen, and F. Van de Pol (2000), Graphical Macro-Editing: Possibilities and Pitfalls. In: *Proceedings of the Second International Conference on Establishment Surveys*, Buffalo, pp. 579–588.

Di Zio, M., U. Guarnera, and O. Luzi (2008), *Contamination Models for the Detection of Outliers and Influential Errors in Continuous Multivariate Data*. Working Paper No. 22, UN/ECE Work Session on Statistical Data Editing, Vienna.

EDIMBUS (2007), *Recommended Practices for Editing and Imputation in Cross-Sectional Business Surveys*. Manual prepared by ISTAT, Statistics Netherlands and SFSO.

EUREDIT Project (2004a), *Towards Effective Statistical Editing and Imputation Strategies*. (*Findings of the Euredit Project*, Volume 1) (available at http://www.cs.york.ac.uk/euredit/results/results.html).

EUREDIT Project (2004b), *Methods and Experimental Results from the Euredit Project*, Volume 2 (available at http://www.cs.york.ac.uk/euredit/results/results.html).

Farwell, K. (2005), *Significance Editing for a Variety of Survey Situations*. Paper presented at the 55th session of the International Statistical Institute, Sydney.

Farwell, K., and M. Rain (2000), Some Current Approaches to Editing in the ABS. In: *Proceedings of the Second International Conference on Establishment Surveys*, Buffalo, pp. 529–538.

Ferguson, D. P. (1994), An Introduction to the Data Editing Process. In: *Statistical Data Editing*, Volume 1: *Methods and Techniques*, United Nations, Geneva.

Ghosh-Dastidar, B., and J. L. Schafer (2006), Outlier Detection and Editing Procedures for Continuous Multivariate Data. *Journal of Official Statistics 22*, pp. 487–506.

Granquist, L. (1990), A Review of Some Macro-Editing Methods for Rationalizing the Editing Process. In: *Proceedings of the Statistics Canada Symposium*, pp. 225–234.

Granquist, L. (1995), Improving the Traditional Editing Process. In: *Business Survey Methods*, B.G. Cox, D.A. Binder, B.N. Chinnappa, A. Christianson, M.J. Colledge, and P.S. Kott, eds. John Wiley & Sons, New York, pp. 385–401.

Granquist, L. (1997), The New View on Editing. *International Statistical Review 65*, pp. 381–387.

Granquist, L., and J. Kovar (1997), Editing of Survey Data: How Much Is Enough? In: *Survey Measurement and Process Quality*, L.E. Lyberg, P. Biemer, M. Collins, E.D. De Leeuw, C. Dippo, , N. Schwartz, and D. Trewin, eds. John Wiley & Sons, New York, pp. 415–435.

Hedlin, D. (2003), Score Functions to Reduce Business Survey Editing at the U.K. Office for National Statistics. *Journal of Official Statistics 19*, pp. 177–199.

Hedlin, D. (2008), *Local and Global Score Functions in Selective Editing*. Working Paper No. 31, UN/ECE Work Session on Statistical Data Editing, Vienna.

Hidiroglou, M. A., and J. M. Berthelot (1986), Statistical Editing and Imputation for Periodic Business Surveys. *Survey Methodology 12*, pp. 73–78.

Hoaglin, D. C., F. Mosteller, and J. W. Tukey (1983), *Understanding Robust and Exploratory Data Analysis*. John Wiley & Sons, New York.

Hoogland, J. (2002), *Selective Editing by Means of Plausibility Indicators*. Working Paper No. 33, UN/ECE Work Session on Statistical Data Editing, Helsinki.

Latouche, M., and J. M. Berthelot (1992), Use of a Score Function to prioritise and Limit Recontacts in Editing Business Surveys. *Journal of Official Statistics 8*, pp. 389–400.

Lawrence, D., and C. McDavitt (1994), Significance Editing in the Australian Survey of Average Weekly Earning. *Journal of Official Statistics 10*, pp. 437–447.

Lawrence, D., and R. McKenzie (2000), The General Application of Significance Editing. *Journal of Official Statistics 16*, pp. 243–253.

Meyer, J., J. Shore, P. Weir, and J. Zyren (2008), *The Development of a Macro Editing Approach*. Working Paper No. 30, UN/ECE Work Session on Statistical Data Editing, Vienna.

Nordbotten, S. (1955), Measuring the Error of Editing the Questionnaires in a Census. *Journal of the American Statistical Association 50*, pp. 364–369.

Nordbotten, S. (1963), Automatic Editing of Individual Statistical Observations. *Conference of European Statisticians-Statistical Standards and Studies No. 2*. United Nations, New York.

Pannekoek, J., and T. de Waal (2005), Automatic Edit and Imputation for Business Surveys: The Dutch Contribution to the EUREDIT Project. *Journal of Official Statistics 21*, pp. 257–286.

Pierzchala, M. (1990), A Review of the State of the Art in Automated Data Editing and Imputation. *Journal of Official Statistics 6*, pp. 355–377.

Scholtus, S. (2008), *Algorithms for Correcting Some Obvious Inconsistencies and Rounding Errors in Business Survey Data.* Discussion paper 08015, Statistics Netherlands, The Hague (see also www.cbs.nl).

Scholtus, S. (2009), *Automatic Correction of Simple Typing Errors in Numerical Data with Balance Edits.* Discussion paper 09046, Statistics Netherlands, The Hague (see also www.cbs.nl).

Stuart, W. J. (1966), Computer Editing of Survey Data. Five Years of Experience in BLS Manpower Surveys. *Journal of the American Statistical Association 61*, pp. 375–383.

Tempelman, D. C. G. (2007), *Imputation of Restricted Data.* Ph.D. Thesis, University of Groningen (see also www.cbs.nl).

Tukey, J. W. (1977), *Exploratory Data Analysis.* Addison-Wesley, London.

Van de Pol, F. (1995), *Data Editing of Business Surveys: an Overview.* Report 10718-95-RSM, Statistics Netherlands, Voorburg.

Van der Loo, M. P. J. (2008), *An Analysis of Editing Strategies for Mixed-mode Establishment Surveys.* Discussion paper 08004, Statistics Netherlands, Voorburg (see also www.cbs.nl).

Van Langen, S. (2002), *Selective Editing by Using Logistic Regression* (in Dutch). Report, Statistics Netherlands, Voorburg.

Weir, P., R. Emery, and J. Walker (1997), The Graphical Editing Analysis Query System. In: *Statistical Data Editing*, Volume 2: *Methods and Techniques*, United Nations, Geneva, pp. 96–104.

CHAPTER SEVEN

Imputation

7.1 Introduction

Values are frequently missing from surveys. In many cases a respondent did not answer one or more questions on a survey he was supposed to answer, while he did answer the other questions. This is referred to as *item nonresponse* (or sometimes as *partial nonresponse*). There are various reasons for not answering a question. The respondent may not understand the question, may not know the answer to the question, may forget to answer the question, may refuse to answer the question because he considers the answer to the question as private information, may refuse to answer the question because it takes too much time to answer the complete questionnaire, and so on. Also, in registers missing values frequently occur for items that are supposed to be nonmissing. Finally, values may have been set to missing during the editing phase or data may simply have been lost while processing it at the statistical institute.

Some population units that were asked to reply to a questionnaire may not have responded at all. Similarly, some units belonging to the population the statistical institute would like to report about may be lacking completely from a register. These cases, where entire records of potential respondents to a survey or potential units in a register are missing, are referred to as *unit nonresponse*. Unless stated otherwise, whenever we refer to missing values in this book, we will mean missing values due to item nonresponse rather than unit nonresponse.

In the case of partial response, the researcher has to decide whether a sufficient number of answers have been given to consider the record as response—and hence the missing items as item nonresponse—or whether not enough answers

Handbook of Statistical Data Editing and Imputation, by T. de Waal, J. Pannekoek, and S. Scholtus

have been given and the record has to be considered as unit nonresponse. In the case of unit nonresponse, weighting the response of the survey or the available units in the registers is a valid approach to reduce the effect of nonresponse on population estimates [see, e.g., Bethlehem (2009)].

Another way to deal with missing values is to *impute*—that is, estimate and fill in—a feasible value for a missing value in the data set. This is referred to as *imputation.* Imputation is part of the throughput process—that is, the process that encompasses all editing, imputation, and other actions performed in order to transform the raw data to a statistical data set ready for analysis and tabulation. Imputation is, however, not a necessary step in the throughput process: One may decide to leave some values missing and try to solve the estimation problem later by weighting the survey or by applying analysis techniques. As we will describe later, after imputation certain estimators for population totals give the same results as certain weighting methods.

We make a distinction between imputation and *derivation.* When variables are derived, new variables are created. These variables can be seen as functions of the variables that are already contained in the data set. When imputing missing values, values on already existing variables are created.

During the statistical editing process (see Chapters 1 to 6 of this book), errors are detected and corrected. When a value is considered erroneous and the value itself is considered to play no part in the correction process, replacing this value by a better one is considered to be imputation. In fact, one then first creates a missing value by setting the erroneous value to missing. In some cases, however, the original, erroneous value is considered to play a part in estimating a better value—for instance, for the so-called thousand errors (see Chapter 2). The modification of such values is not called imputation but *correction.*

Some values are correctly missing and should be recognized as such to prevent them from being imputed. For instance, males do not have to—and cannot—answer the question when they gave birth to their first child, and people without a job do not—and cannot—answer where they are employed. Answers such as "don't know," "no opinion," and "unknown" are also valid values if these are answers to questions about the knowledge or opinion of the respondent.

Even when values are unjustifiably missing, one can decide not to impute these missing values. As we already mentioned, instead of solving the problem of missing values in the data set by means of imputation, one can then try to solve this problem in a later estimation phase or analysis phase. In the case of categorical data, one also has the option to introduce a special category for missing data: "unknown." This is a reason why imputation is more frequently applied for numerical than for categorical variables, and hence more frequently for economic statistics than for social statistics.

Important reasons for imputing missing data instead of leaving the corresponding fields empty are to obtain a complete data set and to improve the quality of the data. A complete data set, with complete records, makes it easier to aggregate microdata, construct tables from these microdata, and ensure consistency between the constructed tables. For instance, missing values on a variable *Occupation* may result in a different distribution of a variable *Age* in the table

"*Age* × *Occupation*" than in a table "*Age* × *Sex*," unless the missing values of *Occupation* are coded as a special category "unknown." When in a survey values are missing for a numerical variable *Income*, then one can only estimate the mean *Income* for the subpopulation of persons that responded to the questionnaire and not for the population as a whole. Imputation can overcome this problem, but only when the imputed values are of sufficiently high quality.

When one wants to apply imputation to improve the quality of the data, one has to be clear about what quality aspect of the data one actually wants to improve. Often the main purpose of a statistical institute is to estimate means and totals. In other cases the main purpose is to estimate the distribution of a variable as well as possible—for instance, the distribution of *Income* over various groups in the population. In yet other cases, one wants to have a microdata set that can be used by researchers to perform many different kinds of statistical analyses. Different purposes can lead to different "optimal" imputations. National statistical institutes (NSIs), however, generally prefer (at most) one imputed value per missing value in order to ensure consistency between various tables and other results they publish. In general, the statistical institute that collects and processes the data can determine better estimates for missing values than other organizations, as the statistical institute generally has a lot more background information that can be used as auxiliary data in the estimation process to produce the imputations.

Sometimes the "true" value of a missing value can be determined with certainty from the other characteristics of the unit. In such a case one can apply *deductive imputation.* Deductive imputation is not discussed in this chapter, but will be treated in detail in Chapter 9. For deductive imputation one can use the edit restrictions that are also used in the editing process. If deductive imputation can be used, this method is preferred over any other imputation method. Sometimes this imputation method is also used when the true value cannot be derived with certainty but only with a (very) high probability.

When deductive imputation cannot be applied, there is often still additional information available (auxiliary, or x variables) that enables one to predict missing values on a target y variable quite accurately. If a model that predicts the target variable well can be constructed, one can use *model-based imputation* to improve the quality of the data set or of the estimates of the (population) parameters of interest. The predicted values according to the selected model are the imputations, or estimates of the missing values. Regression models, mainly for numerical variables, are the most often used imputation models. Imputation based on a regression model is referred to as *regression imputation.*

Apart from parametric (regression) models, nonparametric approaches are also often used to obtain imputed values. In particular, hot deck donor methods that copy the value from another record to fill in the missing value are popular alternatives that can be applied to both numerical and categorical variables. The goal of these methods is similar to regression, but they are somewhat easier to apply when several related missing values in one record have to be imputed, and one aims to preserve the relations between the variables. When donor imputation is applied, for each nonrespondent i a donor record d is searched for that is as similar as possible to record i with respect to certain background characteristics that are (considered to be) related to the target variable y. Next, the donor score, y_d, is used to impute the missing value: $\tilde{y}_i = y_d$.

The remainder of this chapter is organized as follows. Section 7.2 describes some general issues concerning the choices to be made about the models and methods to apply when imputing data sets. The next four sections treat different imputation methods. Regression imputation is treated in Section 7.3. The imputation methods examined in Sections 7.4 and 7.5, *ratio imputation* and *mean imputation*, are special cases of regression imputation. When ratio imputation is applied, only one numerical auxiliary variable is used; when mean imputation is applied, no auxiliary data are used, usually because no auxiliary information is available. These methods are discussed separately because of their simplicity and their frequent application in practice. In Section 7.6 several donor methods are described. The different methods are summarized in Section 7.7, where the differences and similarities between methods are highlighted by describing all methods as special cases of a general imputation generating model. Section 7.8 briefly discusses imputation for longitudinal data. In Section 7.9 several approaches are outlined for assessing the variance of an estimate based on imputed data. Finally, Section 7.10 discusses a technique known as *fractional imputation*.

Imputing missing values does not necessarily imply that the data after imputation are internally consistent, in the sense that all edit restrictions are satisfied. One can add edit restrictions as constraints to the imputation process, and thus ensure that only allowed values are imputed and no inconsistencies arise after imputation. Imputation under edit restrictions is examined in Chapter 9. An alternative approach is to first impute the missing values without taking the restrictions into account, and later adjust the imputed values so that all edit restrictions become satisfied. This approach is examined in Chapter 10.

There are many excellent books and articles on imputation. For interested readers we name a few in chronological order: Sande (1982), Kalton (1983), Kalton and Kasprzyk (1986), Rubin (1987), Little (1988), Kovar and Whitridge (1995), Schafer (1997), Allison and Allison (2001), Little and Rubin (2002), Durrant (2005), Longford (2005), Tsiatis (2006), McKnight et al. (2007), Molenberghs and Kenward (2007), and Daniels, Daniels, and Hogan (2008).

7.2 General Issues in Applying Imputation Methods

7.2.1 IMPUTATION MODELS PER SUBPOPULATION

One can construct an imputation model for the entire population, or for subpopulations—for instance, defined by "branch of industry" × "size class" for business statistics, separately. Distinguishing such imputation classes (imputation groups) can be beneficial when there is little variation within these classes with respect to the scores on the target variable y, and the scores between classes do differ strongly. For regression imputation, one can view distinguishing subpopulations as part of the modeling process, since regression analysis can take categorical x variables into account in the imputation model. This can be done by incorporating the categorical variables (and their interaction terms) corresponding to these subpopulations as dummy variables into the regression model. Hot deck donor imputation (see Section 7.6) is meant for categorical x variables—that is, for subpopulations.

7.2.2 WEIGHTING

In most of the imputation methods we will examine in this chapter, one has the option to weight the item respondents, for instance by setting their weights equal to the reciprocals of the sampling probabilities, or to the raising weights that are obtained after correcting the sampling weights for selective unit nonresponse [see, e.g., Särndal and Lundström (2005) and Bethlehem (2009)]. In the case of linear regression imputation, this implies that one uses weighted least squares estimation instead of ordinary least squares estimation to estimate the model parameters, and in the case of random hot deck donor imputation it implies that the potential donors with a higher weight have a higher probability to be selected as donor than do potential donors with a lower weight. Using weights does not have any effect on deductive imputation.

There is no clear-cut advice on whether to use weights or not. From a model-based perspective, each record is measured equally accurately, assuming identically distributed residuals, independent of sampling probabilities or response probabilities. From this perspective, if one believes in the validity of the imputation model, one therefore does not have to use weights, and it is even better not to use weights, because weighting inflates the standard errors. If one includes the variable containing the weights—or the variables that have been used to compute these weights—as explanatory variables into the model, then weighting is unnecessary anyway. While selecting the auxiliary variables for the imputation model, one can take this into consideration.

However, from the point of view of sampling theory the answers of a sample unit are "representative" for some population units that are not included in the sample. Implicitly, it is more or less assumed that these population units would have given the same answers as the sample unit. From this point of view, and assuming selective unit nonresponse, weighting is necessary to obtain unbiased sampling-design based results. The issue of model-based versus design-based inference is extensively discussed in Skinner, Holt, and Smith (1989).

Andridge and Little (2009) show by simulation that when hot deck donor imputation is used for imputing missing values of a numerical variable, with the aim to estimate the population mean of this variable, the best approach is to use the sampling weight as a stratifying variable together with additional auxiliary variables when forming imputation classes.

7.2.3 MASS IMPUTATION

Sometimes one wants to impute values not only for the item nonrespondents, but for all units that do not occur in the (responding part of the) sample. This is referred to as *mass imputation*, even if only one target variable y is to be imputed. After mass imputation it is easy to calculate population totals and means for the target variable y, totals are simply obtained by summing all (observed or imputed) values for y and means by dividing these totals by the number of population units. For weighted hot deck imputation, this corresponds to the use of the so-called post-stratification estimator, and for regression imputation with weighted least squares estimation, this corresponds to the so-called regression estimator; see, for instance, Särndal and Lundström (2005), Bethlehem (2009), and Section 7.3.4 below.

As for imputation of only the item nonrespondents, for mass imputation we again have that weighting becomes less important if more variables used to determine the weight variable are included as auxiliary variables in the imputation model. However, sometimes this is impossible because the variables used to determine the weight variable are only available for the sample units. In that case, weighting is an option to consider.

The experiences with mass imputation as reported in the literature lead to different conclusions. Whereas Kaufman and Scheuren (1997) report that they were disappointed with the performance of mass imputation and that so far it has not delivered on their expectations, Krotki, Black, and Creel (2005) conclude that "mass imputation is becoming one more tool in the survey statistician's toolkit for which there is an ever-increasing demand." Shlomo, De Waal, and Pannekoek (2009) report good evaluation results for mass imputation in a limited evaluation study. Our overall conclusion is that mass imputation is still a relatively unexplored area that offers a lot of opportunities for future research, but alas not yet a solution to all problems with unit nonresponse.

For more on mass imputation we refer to Whitridge, Bureau, and Kovar (1990), Whitridge and Kovar (1990), Kaufman and Scheuren (1997), Kooiman (1998), Fetter (2001), Krotki, Black, and Creel (2005), Shlomo, De Waal, and Pannekoek (2009), and Haslett et al. (2010).

7.2.4 SELECTING AUXILIARY VARIABLES

In this book we do not treat in detail how to select suitable auxiliary variables and interactions. Just like regression analysis, selecting suitable auxiliary variables and interactions for the regression model is part of multivariate analysis on which there is ample literature. The basic idea is to look for auxiliary or x variables that are strongly correlated with the target variable y. It is generally a matter of trial and error in combination with statistical analysis and common sense to select suitable auxiliary variables and interactions for the regression model, but one can also use forward or backward search procedures to automatically add auxiliary variables to the regression model and delete them from the regression model, respectively. There also exist automatic search procedures to construct homogeneous imputation classes for categorical auxiliary variables, such as AID [see Sonquist, Baker and Morgan (1971)], CHAID [see SPSS (1998)], and WAID [see Chambers et al. (2001a, 2001b) and De Waal (2001)]. In general, these automatic search procedures are of a nonparametric nature.

Below we will limit ourselves to giving a few general guidelines for selecting auxiliary variables.

First of all, select those variables as auxiliary variables for the imputation model for which it can be expected that they are also relevant for the item nonrespondents. In general, one will use the item respondents to check whether the auxiliary variables are able to explain the target variable well, since a test of the imputation model for the item nonrespondents is impossible.

Second, do not include too many auxiliary variables in a regression model. The parameter estimates of such a model would then have large standard errors. In order to obtain good imputed values, a model with not too many auxiliary variables is to be preferred.

Third, when donor imputation is applied, it does not matter if many subpopulations are distinguished. Even adding variables that have no explanatory value for the target variable with the sole purpose to obtain a unique donor is no problem. This can, in fact, be seen as an alternative to draw a random donor from an imputation class. In this case, one has to be careful not to draw the same donor too often.

Fourth, when sequentially adding variables to the imputation model, use quality measures such as the increase of R^2, the F-test, AIC, and BIC [see, e.g., Burnham and Anderson (2002) for more on AIC and BIC] to determine the benefits of adding another variable to the model versus not adding this variable. It should be noted that when sequentially adding variables to the imputation model, the order of adding variables to the model is part of the model selection process.

Fifth, often it is important to include design variables—for instance, variables that define sampling strata having differential inclusion probabilities.

7.2.5 OUTLIERS

If outliers on a numerical target variable y occur among the item respondents, one can consider limiting their influence on the imputation process. For instance, one can carry out a robust regression analysis, or give a potential donor with an outlying value on the target variable y—that is, outlying given the values of the auxiliary variables—a smaller probability of being selected as donor. Taking outliers into account during the imputation process leads to smaller standard errors, but may lead to biased results. Therefore, one has to be careful when taking outliers into account during the imputation process and has to carefully consider which estimates one wants to obtain. For instance, robust methods are often appropriate for small subpopulations, because otherwise the standard errors may become too large; for large populations, robust methods are often less appropriate or beneficial. For instance, when someone earns a million euros per year, one can use this person as donor to obtain results for the Netherlands as a whole. However, if this person lives in a poor neighborhood of a small Dutch village, it is obviously not a good idea to use this person as donor in order to obtain results for this neighborhood. To decide on how to deal with outliers, subject-matter knowledge should be utilized in combination with statistical analysis.

7.2.6 FLAGGING IMPUTED DATA

For imputation, it is important that the missing values be clearly indicated in the data set. This can be done by giving missing values a special code, such as −1, 9, or 99, if this does not lead to confusion with possible correct values. It is considered to be bad practice to code missing values by zeros when zeros can be correct answers, or vice versa code zeros by missing values. Both situations sometimes happen for economic surveys. If this is done, no distinction can be made between missing values and true zeros.

When missing values are imputed by a statistical institute in order to obtain a complete data set that is later released to external researchers, it is important to document which values have been imputed and which methods have been applied to do so, including the auxiliary variables included in the imputation model and the used model parameters. This is necessary so the researcher can determine for himself whether he wants to use the values imputed by the statistical institute or whether it is better for his research goals to impute the values that were originally missing himself. Also, to determine correct standard errors, the researcher may need to know which values have been imputed, and which imputation method has been applied.

7.3 Regression Imputation

7.3.1 THE REGRESSION IMPUTATION MODEL

When *regression imputation* is applied, a suitable regression model, based on zero, one, or several auxiliary x variables, is used to predict a value for a missing value y_i of target variable y in record i. An imputation for the missing value is obtained, based on the prediction from the model. The linear regression model is given by

$$y = \alpha + \beta_1 x_1 + \cdots + \beta_p x_p + \varepsilon = \alpha + \mathbf{x}^T \boldsymbol{\beta} + \varepsilon, \tag{7.1}$$

where α, $\beta_1, \ldots, \beta_p$ denote model parameters, and ε denotes a residual. Substituting estimates for the parameters yields a predicted variable:

$$\hat{y} = \hat{\alpha} + \mathbf{x}^T \hat{\boldsymbol{\beta}}. \tag{7.2}$$

This predicted variable $\hat{y}$ is defined for both item respondents as well as item nonrespondents.

For each item nonrespondent on target variable y, either the best prediction can be imputed or a random residual can be added. That is, there are two basic options to determine an imputed value $\tilde{y}_i$ for an item nonrespondent:

1. Without residual:

$$\tilde{y}_i = \hat{y}_i = \hat{\alpha} + \mathbf{x}_i^T \hat{\boldsymbol{\beta}}. \tag{7.3}$$

2. With residual:

$$\tilde{y}_i = \hat{y}_i + \varepsilon_i = \hat{\alpha} + \mathbf{x}_i^T \hat{\boldsymbol{\beta}} + \varepsilon_i. \tag{7.4}$$

If in formula (7.1) no auxiliary variables x are used, this formula becomes $y = \mu + \varepsilon$ with μ the expected value of y, and formula (7.3) reduces to $\tilde{y}_i = \hat{\mu} = \bar{y}$. This is mean imputation (see Section 7.5). We treat this method separately because of its popularity in practice. Imputation without using auxiliary information can only be justified when there are only a few item nonrespondents and the imputations hardly have an effect on the parameters to be estimated.

If no constant term is used and only one numerical auxiliary variable x, formula (7.1) becomes $y = Rx + \varepsilon$, and (7.3) reduces to ratio imputation (see Section 7.4).

Which of the two options (7.3) or (7.4) should be chosen depends on the goal of imputation. For the purpose of estimating means and totals, adding a random residual to the predicted value is unnecessary—and may even lead to biased results, unless the expectation of the residuals equals zero—but if a goal is also to estimate the variation of target variable y, adding a random residual to the imputations is the preferred option.

If one imputes the best possible prediction according to the regression model for all missing values, the imputed data become very smooth; that is, all imputed values fit the regression perfectly well. Apart for estimating totals and means, the imputed data will be quite useless for other kinds of analysis of the microdata or in some cases even of tabular data. A simple example is a national demographic statistic of the population, where for each unknown age of husband or wife one uses the imputation model that the husband is 2 years older than the wife. Such an imputation model may be a good model for the age distribution of husbands and wives, but if external researchers were to examine the microdata later, they might discover an unexpected peak in the distribution of the age difference between men and women.

In general, imputation of the best possible prediction according to the regression model leads to an underestimation of the variation in the scores ("regression toward the mean"). It leads to peaked distributions and tails that are too thin, especially when the target variable y has many missing values and the regression model explains little of the variance of y. The effect is the strongest for mean imputation (see Section 7.5). This forms no problem when one only wants to estimate means and totals, but it does form a problem when one wants to estimate distributions or measures of spread.

If one wants to estimate the distribution as well rather than only totals and means, it is to be advised to add a random residual to the best possible predicted value. For regression imputation the residual ε_i in (7.4) can be determined in two ways:

(a) $\varepsilon_i = \varepsilon_d$ with ε_d the residual of an arbitrary or especially selected donor.

(b) ε_i is a draw from a stochastic distribution—for instance, a normal distribution.

In case **(b)** the expectation of the normal distribution generally equals zero, and the variance is often estimated by the residual error of the regression model (7.1).

In Sections 7.4 and 7.5, on ratio imputation and mean imputation, we discuss only the case without a random residual. Adding a residual to these models is similar to the situation for regression imputation.

Regression imputation can be applied to separate groups or imputation classes—that is, subpopulations. In that case, for each group we estimate separate model parameters or even develop a separate imputation model. Each group is hereby defined in terms of auxiliary variables.

Usually a linear model (7.1) is used, but nonlinear models can, in principle, also be used. A specific kind of nonlinear model is the generalized linear model [cf. McCullagh and Nelder (1989)], which is of the form

$$y = f(\mathbf{x}^T \boldsymbol{\beta}). \tag{7.5}$$

The residual term ε can be explicitly added to the model (7.5) or can implicitly be part of the model.

The auxiliary variables of the regression model can be continuous variables or dummy variables to represent the categories of categorical variables. When only categorical auxiliary variables are included, linear regression analysis is sometimes also called "analysis of variance."

When $\tilde{y}_i$ is imputed by means of formula (7.3), imputations have no effect on the estimation of the population total, if the so-called regression estimator with the same model as the imputation model is used; see Särndal and Lundström (2005). We give a proof of this fact in Section 7.3.4 below.

Regression imputation is mainly applied when y is a numerical variable. When y is a categorical variable, a regression approach can also be used, but then a transformation of the target variable is applied, such as in binary or multinomial logistic regression [see, e.g., Draper and Smith (1998), McCullagh and Nelder (1989), and Example 7.1 in Section 7.3.2]. For a binary y variable with possible scores 0 and 1 the logistic regression model is

$$\ln \frac{pr}{1 - pr} = \alpha + \beta_1 x_1 + \cdots + \beta_p x_p \equiv \alpha + \mathbf{x}^T \boldsymbol{\beta} \tag{7.6}$$

with pr the probability that y assumes the score 1, given the x variables and the posited model. If a y value is missing, one can estimate the parameter $\boldsymbol{\beta}$—for instance, by means of the maximum likelihood approach—and subsequently estimate the probability $\hat{pr}$ that the score 1 is imputed by

$$\hat{pr} = \frac{e^{\hat{\alpha} + \mathbf{x}^T \hat{\boldsymbol{\beta}}}}{1 + e^{\hat{\alpha} + \mathbf{x}^T \hat{\boldsymbol{\beta}}}} = \frac{1}{e^{-(\hat{\alpha} + \mathbf{x}^T \hat{\boldsymbol{\beta}})} + 1}. \tag{7.7}$$

These probabilities can easily be calculated by means of several statistical software packages, such as SPSS and R.

In Section 7.2.2 we pointed out the possibility of a weighted regression analysis, if one wants the regression analysis to reflect that some respondents have a higher raising weight than others. Heterogeneity of the residual terms can be another reason for carrying out a regression analysis by means of weighted least squares estimation, instead of ordinary (unweighted) least squares estimation.

In this book we do not discuss the theory of regression analysis in general. There are many excellent books available that treat regression analysis in detail, [e.g., Draper and Smith (1998)]. With respect to model selection, some general remarks were made in Section 7.2.4.

7.3.2 EXAMPLES OF REGRESSION IMPUTATION

EXAMPLE 7.1 (Dutch Household Statistics)

Each year, Statistics Netherlands receives a version of the so-called Municipal Base Administration. The Municipal Base Administration contains for each address data on the people living at the address, including the family relations. Information on how the households living at the address are exactly composed is, however, lacking. For the annual Household Statistics it is essential to know which persons living at the same address constitute a household according to the definition applied at Statistics Netherlands. From 1999 on, the Municipal Base Administration is used to determine the main variables *Number of households* and *Household composition* from the structure of the family or families living at the address. For more than 90% of the addresses in the Municipal Base Administration the information for these derived variables can be constructed. For the remaining addresses, however, neither the number of households nor the exact household composition can be derived. For these addresses, imputation is used, with separate imputation models for different situations.

Here we discuss the simplest type of addresses with unknown *Household composition*: addresses with two unrelated persons—that is, two persons that are not married or registered as each other's partner and who are not family of each other. For these addresses it is unknown whether the two persons together constitute one household or are both single and each has its own household.

First, deductive imputation (see Section 7.1 and Chapter 9) is applied, by means of a deductive rule: When both persons started to live at the address on the same date according to the Municipal Base Administration, then they are considered to constitute one household. This will lead to a slight overestimation of the true number of households. The remaining addresses are linked to the Labor Force Survey. For 1999 thus 1662 addresses with two persons were obtained. Based on these data an imputation model was built.

By means of information obtained from the interviewers of the Labor Force Survey and the actual data collected by means of the Labor Force Survey, for each of the 1662 addresses it was determined whether the address contained one or two households. Sometimes this was quite complicated because of nonresponse or because there turned out to be a difference between the actual and the registered occupation of the address. The probability of two households turned out to be strongly correlated with the age of both persons—in particular, with the age difference between the two persons—whether or not the persons had the same gender, regardless of the degree of urbanization, and regardless of the number of unmarried people at the address. A logistic regression model (7.6) was developed with these variables as auxiliary variables.

Next, formula (7.7) was used for each address with two unrelated persons in the Municipal Base Administration that did not link to the Labor Force Survey to estimate the probability that the address contains two households. The estimated probability was used to draw either "two households" or "one household."

This is an example of register imputation: All addresses with an unknown score in *Number of households* are imputed. The unknown scores are, moreover, highly selective. Namely, *Number of households* is a derived variable that only for specific groups cannot be determined from the Municipal Base Administration. Only by linking the register to a survey, stochastic information on the number of households for those groups becomes available. ■

EXAMPLE 7.2 (Dutch Public Libraries Survey)

For continuous variables in social surveys linear regression imputation was—and is—often used at Statistics Netherlands. Nonlinear regression imputation was—and is—on the other hand not, or hardly, used at Statistics Netherlands. Exceptions do exist, however. For the Public Libraries Survey several regression imputation models were examined, namely ratio imputation, linear regression imputation, and nonlinear regression imputation [see Israëls and Pannekoek (1999)]. The linear regression model included a quadratic term, which is rather unusual at Statistics Netherlands, resulting in the model

$$y = \beta x + \alpha x^2 + \varepsilon,$$

where as usual y is the target variable, x the auxiliary variable, and α and β the parameters of the regression model. The residual ε is a stochastic error term with expectation zero and variance σ^2.

The following nonlinear regression model was also examined:

$$y = \beta x^{\alpha} + \varepsilon,$$

where y, x, α, β, and ε are defined in the same way as above. For this particular data set the linear imputation models appeared to give better results than the nonlinear model.

This example, as well as several others in this chapter, is taken from De Waal (1999). Although there have been quite some changes with respect to the imputation methods applied at Statistics Netherlands since 1999, the overall picture has not changed much. The main imputation techniques that were applied in 1999 are still applied now. Often the differences between 1999 and now are not so much with respect to the imputation methods themselves, but rather with respect to the auxiliary variables used and the estimation of the model parameters. ■

7.3.3 QUALITY OF REGRESSION IMPUTATION

It is important to measure the quality of imputations. As mentioned before, a fundamental problem is that usually the true values are unknown. In many cases, means before and after imputation differ from each other. This is not necessarily a cause for alarm because the item nonresponse may have been selective. If there is overlap with other surveys, external validations may be carried out to gain an impression of the quality of the imputations. However, often differences in definitions of variables and in populations exist between surveys. This means that the possibilities for such validations are limited in practice.

Since generally the quality of the imputations cannot be tested, the quality indicators for regression imputation described below are only based on the fit of the model on the item respondents.

For linear regression analysis based on least squares estimation, one can use the well-known R^2 measure to measure the fit of the model on the item respondents. This fit measure can be used to compare different imputation models with each other. A prerequisite is that one is able to compare a gain in the R^2 measure to an increase in the number of degrees of freedom. This fit measure can also be applied for donor imputation (see Section 7.6), because donor imputation can be seen as regression on dummy variables. For some nonlinear models the likelihood or a quantity derived from the likelihood, such as AIC and BIC [see, e.g., Burnham and Anderson (2002)] or Nagelkerke's R^2 [see, e.g., Nagelkerke (1991)], can be used as an indicator for the fit. We note that it is theoretically possible that although model A has a better fit than another model B for the item respondents, model A has a worse fit than model B for the item nonrespondents—that is, that the residuals of the model predictions and the true values are on average larger for model A than for B for the item nonrespondents.

Another option to gain an impression of the quality of an imputation model is to carry out a simulation experiment. For such an experiment, one uses a fully observed data set, where the true values of the target variables are known for all elements of the population. Such a data set is obtained by either regarding a previously edited realistic data set as "the target population" or by creating a synthetic data set by drawing from a statistical model. In the simulation study, some of the true values are temporarily deleted and new values are imputed for the deleted values. If the imputed values $\tilde{y}_i$ are close to the original values y_i, the quality of the imputation method is likely to be high. By defining a suitable distance metric, one can select the most appropriate imputation model or model parameters. An example of such a distance metric is the mean absolute deviation between imputed and true values,

$$\frac{1}{I}\sum_{i=1}^{I}|\tilde{y}_i - y_i|$$

with I the number of imputed records. On an aggregate level, one can use as distance metric the mean over the simulation experiments of the mean absolute deviation between the aggregated values with and without imputation,

$$\frac{1}{T}\sum_{t=1}^{T}|\tilde{Y}_t - Y|,$$

where $\tilde{Y}_t$ is the sum of the values of variable y after imputation in the tth experiment, Y is the sum of the true values of variable y, and T is the number of simulation experiments carried out. Chambers (2004) presents a large number of evaluation measures for different aspects of imputation quality. Examples of simulation experiments are described in Section 11.3 and Schulte Nordholt (1998).

At Statistics Canada, a SAS program called the Generalized Simulation System (GENESIS) has been developed for carrying out simulation studies; [cf. Haziza (2003)]. In GENESIS, the user supplies a population file and chooses a sampling design, a missing data mechanism, an imputation technique, and the required number of iterations. The program then runs the requested simulation. Various metrics are computed to assess the quality of the imputations.

Haziza (2006) provides an excellent introduction into the design and use of simulation studies.

7.3.4 CONNECTION BETWEEN IMPUTATION AND WEIGHTING

Suppose that a target population consists of elements numbered $1, \ldots, N$. We are interested in estimating, say, the population total of target variable y:

$$Y = \sum_{i=1}^{N} y_i. \tag{7.8}$$

A sample s is drawn from the target population, and for notational convenience, we assume that this sample consists of the elements numbered $1, \ldots, n$. We use "obs" and "mis" to refer to, respectively, the responding and nonresponding parts of the sample. Again, for convenience, we assume that the elements $1, \ldots, r$ are respondents and that the elements $r+1, \ldots, n$ are nonrespondents.

In this situation, the linear regression model (7.1) can be used in (at least) three different ways to obtain an estimate for (7.8):

I. Weight the respondent data by applying the so-called regression estimator.

II. Impute the nonrespondents in the sample using regression imputation, then weight the sample data by applying the regression estimator.

III. Impute both the nonrespondents and the nonsampled elements using regression imputation.

We shall refer to these strategies as the *weighting approach*, the *combined approach*, and the *mass imputation approach*. Since the same regression model is used in these approaches, one might intuitively expect that they should all yield the same estimate. In this subsection, we show that this is in fact true, under certain conditions. Särndal and Lundström (2005) identify estimates based on weighting and imputation from a slightly different point of view.

We start by introducing some more notation. The vector $\mathbf{y} = (y_1, \ldots, y_N)^T$ contains the values of y in the population, while the smaller vector $\mathbf{y}_{\text{obs}} = (y_1, \ldots, y_r)^T$ contains the y values in the responding part of the sample only. Moreover, a vector of auxiliary information, denoted by $\mathbf{x}_i = (x_{i1}, \ldots, x_{ip})^T$, is available for all $i = 1, \ldots, N$. Let $\mathbf{X} = (x_{ij})$ denote the $N \times p$ matrix with $\mathbf{x}_i^T$ as the ith row, for $i = 1, \ldots, N$. The corresponding submatrices for the sampled elements, the respondents, and the nonrespondents are denoted by, respectively, $\mathbf{X}_s$, $\mathbf{X}_{\text{obs}}$, and $\mathbf{X}_{\text{mis}}$. Note that "mis" refers to the fact that y_i is not observed for these population elements, whereas $\mathbf{x}_i^T$ is completely observed for all elements of the population.

For convenience, we slightly rewrite the linear regression model as

$$\mathbf{y} = \mathbf{X}\boldsymbol{\beta} + \boldsymbol{\varepsilon},$$

with $\boldsymbol{\beta} = (\beta_1, \ldots, \beta_p)^T$ and $\boldsymbol{\varepsilon} = (\varepsilon_1, \ldots, \varepsilon_N)^T$ the vectors of regression coefficients and residuals, respectively. In this notation, the constant term α is taken to be one of the β coefficients, corresponding to a column of ones in $\mathbf{X}$.

Using ordinary least squares (ols), the regression coefficients are found to be

$$\boldsymbol{\beta} = (\mathbf{X}^T\mathbf{X})^{-1}\mathbf{X}^T\mathbf{y}.$$

Initially, only the observed part of the sample can be used to estimate the regression coefficients. Taking the sampling design with inclusion probabilities π_i into account, an asymptotically unbiased estimate of $\boldsymbol{\beta}$ is obtained through weighted least squares (wls):

$$\hat{\boldsymbol{\beta}}_{\text{obs}} = (\mathbf{X}_{\text{obs}}^T \boldsymbol{\Pi}_{\text{obs}}^{-1} \mathbf{X}_{\text{obs}})^{-1} \mathbf{X}_{\text{obs}}^T \boldsymbol{\Pi}_{\text{obs}}^{-1} \mathbf{y}_{\text{obs}}, \qquad (7.9)$$

where $\mathbf{\Pi}_{\text{obs}}^{-1} = \text{diag}(1/\pi_1, \ldots, 1/\pi_r)$ is a diagonal matrix of design weights for the responding elements [cf. Bethlehem and Keller (1987) and Knottnerus (2003, pp. 118–123)]. In the special case where a sample is obtained through simple random sampling (with $\pi_1 = \cdots = \pi_N = n/N$), the matrix of design weights can be left out of this expression.

In the weighting approach, the estimated regression coefficients are used to compute the so-called regression estimator:

$$\hat{Y}_W = \sum_{i=1}^{r} \frac{y_i}{\pi_i} + \left(\sum_{i=1}^{N} \mathbf{x}_i^T - \sum_{i=1}^{r} \frac{\mathbf{x}_i^T}{\pi_i} \right) \hat{\boldsymbol{\beta}}_{\text{obs}} \tag{7.10}$$

[cf. Knottnerus (2003)]. The rationale behind the regression estimator is that the direct estimate for the total of y, based on the design weights $(1/\pi_i)$, is adjusted by adding a correction term, which predicts the error in the direct estimate, using the fact that the true value of $\mathbf{x}_i$ is known for the entire population. Note that the term in brackets is just the difference between the vector of true population totals and the vector of direct estimates for the x variables.

The term "weighting" alludes to the fact that $\hat{Y}_W$ can be written as a weighted sum of observed values:

$$\hat{Y}_W = \sum_{i=1}^{r} w_i y_i,$$

with

$$w_i = \frac{1}{\pi_i} \left[1 + \left(\sum_{k=1}^{N} \mathbf{x}_k^T - \sum_{k=1}^{r} \frac{\mathbf{x}_k^T}{\pi_k} \right) (\mathbf{X}_{\text{obs}}^T \mathbf{\Pi}_{\text{obs}}^{-1} \mathbf{X}_{\text{obs}})^{-1} \mathbf{x}_i \right]$$

[see, e.g., Särndal, Swensson, and Wretman (1992)]. A general treatment of the weighting approach falls outside the scope of this book.

In the combined approach, regression imputation is used to impute the missing data for the nonresponding elements of the sample. For now, we assume that no random residuals are added to the imputations:

$$\tilde{y}_i = \mathbf{x}_i^T \hat{\boldsymbol{\beta}}_{\text{obs}}, \qquad i = r+1, \ldots, n.$$

We then apply the weighting approach, using the observed *and* imputed data to estimate the regression coefficients:

$$\hat{\boldsymbol{\beta}}_s = (\mathbf{X}_s^T \mathbf{\Pi}_s^{-1} \mathbf{X}_s)^{-1} \mathbf{X}_s^T \mathbf{\Pi}_s^{-1} \tilde{\mathbf{y}}_s, \tag{7.11}$$

with $\mathbf{\Pi}_s^{-1} = \text{diag}(1/\pi_1, \ldots, 1/\pi_n)$ a diagonal matrix of design weights for the sampled elements, and $\tilde{\mathbf{y}}_s = (y_1, \ldots, y_r, \tilde{y}_{r+1}, \ldots, \tilde{y}_n)^T$. This yields the following regression estimator for (7.8):

$$\hat{Y}_{IW} = \sum_{i=1}^{r} \frac{y_i}{\pi_i} + \sum_{i=r+1}^{n} \frac{\tilde{y}_i}{\pi_i} + \left(\sum_{i=1}^{N} \mathbf{x}_i^T - \sum_{i=1}^{n} \frac{\mathbf{x}_i^T}{\pi_i} \right) \hat{\boldsymbol{\beta}}_s.$$

Using the fact that $\tilde{y}_i = \mathbf{x}_i^T \hat{\boldsymbol{\beta}}_{\text{obs}}$, this expression can be written as

$$\hat{Y}_{IW} = \sum_{i=1}^{r} \frac{y_i}{\pi_i} + \left(\sum_{i=1}^{N} \mathbf{x}_i^T - \sum_{i=1}^{r} \frac{\mathbf{x}_i^T}{\pi_i} \right) \hat{\boldsymbol{\beta}}_s + \sum_{i=r+1}^{n} \frac{\mathbf{x}_i^T}{\pi_i} (\hat{\boldsymbol{\beta}}_{\text{obs}} - \hat{\boldsymbol{\beta}}_s). \tag{7.12}$$

By comparing (7.10) and (7.12), we observe that the weighting estimate and the combined estimate are identical if it holds that $\hat{\boldsymbol{\beta}}_s = \hat{\boldsymbol{\beta}}_{\text{obs}}$.

Finally, in the mass imputation approach, the regression model is used to impute all unobserved values:

$$\tilde{y}_i = \mathbf{x}_i^T \hat{\boldsymbol{\beta}}_{\text{obs}}, \qquad i = r+1, \ldots, N.$$

No weighting is necessary, and we obtain the following estimate for (7.8):

$$\hat{Y}_I = \sum_{i=1}^{r} y_i + \sum_{i=r+1}^{N} \mathbf{x}_i^T \hat{\boldsymbol{\beta}}_{\text{obs}}. \tag{7.13}$$

In general, this estimate need not be identical to $\hat{Y}_W$ or $\hat{Y}_{IW}$.

Theorem 7.1 below establishes general conditions such that the three estimates $\hat{Y}_W$, $\hat{Y}_{IW}$, and $\hat{Y}_I$ are identical. In preparation of this result, we observe that if the N vector of ones is contained in the column space of $\mathbf{X}$, then the weighting estimate (7.10) can be written concisely as

$$\hat{Y}_W = \sum_{i=1}^{N} \mathbf{x}_i^T \hat{\boldsymbol{\beta}}_{\text{obs}}, \tag{7.14}$$

This follows from the well-known fact that the vector of observed residuals $\hat{\boldsymbol{\varepsilon}}_{\text{obs}} = \mathbf{y}_{\text{obs}} - \mathbf{X}_{\text{obs}} \hat{\boldsymbol{\beta}}_{\text{obs}}$ is orthogonal to each of the columns in $\mathbf{X}_{\text{obs}}$:

$$\mathbf{X}_{\text{obs}}^T \boldsymbol{\Pi}_{\text{obs}}^{-1} \hat{\boldsymbol{\varepsilon}}_{\text{obs}} = \mathbf{0}.$$

Note that because we used wls to obtain $\hat{\boldsymbol{\beta}}_{\text{obs}}$, orthogonality is not defined with respect to the usual Euclidean metric, but with respect to a metric that involves multiplication by $\boldsymbol{\Pi}_{\text{obs}}^{-1}$. Since we have assumed that the column space of $\mathbf{X}$ contains the vector of ones, it follows in particular that

$$\sum_{i=1}^{r} \frac{y_i - \mathbf{x}_i^T \hat{\boldsymbol{\beta}}_{\text{obs}}}{\pi_i} = \sum_{i=1}^{r} \frac{\hat{\varepsilon}_{\text{obs},i}}{\pi_i} = (1, \ldots, 1) \boldsymbol{\Pi}_{\text{obs}}^{-1} \hat{\boldsymbol{\varepsilon}}_{\text{obs}} = 0; \tag{7.15}$$

see also Knottnerus (2003, p. 121). Similarly, if the constant vector of ones is contained in the column space of $\mathbf{X}$, it holds that

$$\hat{Y}_{IW} = \sum_{i=1}^{N} \mathbf{x}_i^T \hat{\boldsymbol{\beta}}_s,$$

By comparing (7.14) and (7.13), we immediately see that

$$\hat{Y}_I - \hat{Y}_W = \sum_{i=1}^{r} y_i - \sum_{i=1}^{r} \mathbf{x}_i^T \hat{\boldsymbol{\beta}}_{\text{obs}} = \sum_{i=1}^{r} \hat{\varepsilon}_{\text{obs},i}. \qquad (7.16)$$

For simple random sampling, this expression equals zero by (7.15), and we conclude that the weighting estimate and the mass imputation estimate are identical under the assumption that the column space of $\mathbf{X}$ contains the vector of ones. This does not follow for general sampling designs, however, because the unweighted sum of the observed residuals need not be equal to zero under wls.

We now present the general result.

THEOREM 7.1

For all sampling designs and all linear regression models, it always holds that $\hat{Y}_W = \hat{Y}_{IW}$. In addition, if both the constant vector of ones and the vector of inclusion probabilities $\boldsymbol{\pi} = (\pi_1, \ldots, \pi_N)^T$ are contained in the column space of the auxiliary matrix $\mathbf{X}$, then it also holds that $\hat{Y}_W = \hat{Y}_{IW} = \hat{Y}_I$.

Proof. To prove the first statement, we already observed that it suffices to show that $\hat{\boldsymbol{\beta}}_s = \hat{\boldsymbol{\beta}}_{\text{obs}}$. Intuitively, it is clear that the parameter estimates of a regression model should not change if we add observations to the regression equations that conform exactly to a previously fitted model. A formal proof proceeds as follows. By definition, the wls estimate $\hat{\boldsymbol{\beta}}_s$ minimizes the quadratic function $(\tilde{\mathbf{y}}_s - \mathbf{X}_s\hat{\boldsymbol{\beta}}_s)^T \boldsymbol{\Pi}_s^{-1} (\tilde{\mathbf{y}}_s - \mathbf{X}_s\hat{\boldsymbol{\beta}}_s)$. This function can be expressed as

$$(\mathbf{y}_{\text{obs}} - \mathbf{X}_{\text{obs}}\hat{\boldsymbol{\beta}}_s)^T \boldsymbol{\Pi}_{\text{obs}}^{-1} (\mathbf{y}_{\text{obs}} - \mathbf{X}_{\text{obs}}\hat{\boldsymbol{\beta}}_s) + (\hat{\boldsymbol{\beta}}_{\text{obs}} - \hat{\boldsymbol{\beta}}_s)^T \mathbf{X}_{\text{mis}}^T \boldsymbol{\Pi}_{\text{mis}}^{-1} \mathbf{X}_{\text{mis}} (\hat{\boldsymbol{\beta}}_{\text{obs}} - \hat{\boldsymbol{\beta}}_s),$$

with $\boldsymbol{\Pi}_{\text{mis}}^{-1} = \text{diag}(1/\pi_{r+1}, \ldots, 1/\pi_n)$. The first term is a quadratic function that is minimized by choosing $\hat{\boldsymbol{\beta}}_s = \hat{\boldsymbol{\beta}}_{\text{obs}}$. The second term is a quadratic function that equals zero for this choice of $\hat{\boldsymbol{\beta}}_s$. Since both terms are nonnegative, we conclude that $\hat{\boldsymbol{\beta}}_s = \hat{\boldsymbol{\beta}}_{\text{obs}}$ is the wls estimate under the combined approach. Thus, the first statement is proved.

To prove the second statement, we use expression (7.16). Under the assumption that the column space of $\mathbf{X}$ contains $\boldsymbol{\pi}$, the orthogonality property implies that

$$\sum_{i=1}^{r} \hat{\varepsilon}_{\text{obs},i} = (\pi_1, \ldots, \pi_r) \boldsymbol{\Pi}_{\text{obs}}^{-1} \hat{\boldsymbol{\varepsilon}}_{\text{obs}} = 0,$$

and hence it follows that $\hat{Y}_I = \hat{Y}_W$. ∎

An obvious way to satisfy the condition that the constant vector of ones is contained in the column space of **X** is by including the constant term in the regression equations. Similarly, the condition that $\boldsymbol{\pi}$ is contained in the column space of **X** can be satisfied by adding the inclusion probabilities to the regression model. For stratified sampling, it suffices to include all auxiliary variables that were used to construct the strata in the regression model. Note that for simple random sampling, the two conditions are equivalent.

EXAMPLE 7.3

Consider a simple random sampling design (with nonresponse) and a regression model that involves only the constant term. From (7.9) the estimated regression coefficient is found to be

$$\hat{\beta}_{\text{obs}} = \frac{1}{r}\sum_{i=1}^{r} y_i,$$

that is, the observed mean of y. Using this model in the weighting approach, the regression estimate is

$$\hat{Y}_W = \frac{N}{r}\sum_{i=1}^{r} y_i.$$

Thus, the design weights N/n are adjusted for nonresponse by a constant factor n/r.

In the combined approach, the nonrespondents are first imputed with the observed mean of y. In this case, the weighting part of the combined approach amounts to using just the design weights N/n, because the regression model does not make use of information from outside the sample. We find

$$\hat{Y}_{IW} = \frac{N}{n}\sum_{i=1}^{r} y_i + \frac{N}{n}\sum_{i=r+1}^{n} \tilde{y}_i = \frac{N}{n}\left(1 + \frac{n-r}{r}\right)\sum_{i=1}^{r} y_i = \frac{N}{r}\sum_{i=1}^{r} y_i.$$

Finally, for the mass imputation approach, each unobserved value of y is imputed with the observed mean of y. We find

$$\hat{Y}_I = \sum_{i=1}^{r} y_i + \sum_{i=r+1}^{N} \tilde{y}_i = \left(1 + \frac{N-r}{r}\right)\sum_{i=1}^{r} y_i = \frac{N}{r}\sum_{i=1}^{r} y_i.$$

Thus, under simple random sampling the regression model that includes only the constant term yields the same estimate for all three approaches. ■

EXAMPLE 7.4

In this example, we consider a general sampling design with an estimate based on *post-stratification*. Post-stratification is widely used in inference based on sample surveys [see, e.g., Bethlehem (2009) or Särndal, Swensson, and Wretman (1992)]. This technique corresponds to a regression model involving, say, L dummy variables: $\mathbf{x}_i = (x_{i1}, \ldots, x_{iL})^T$ with $x_{il} = 1$ if the ith population element falls into the lth post-stratum, and $x_{il} = 0$ otherwise. Assuming that the post-strata are mutually exclusive, we have $\sum_{l=1}^{L} x_{il} = 1$ for all i. This shows that the constant vector of ones is contained in the column space of $\mathbf{X}$, although it is not explicitly included in the model.

From (7.9) the vector of estimated regression coefficients is

$$\hat{\boldsymbol{\beta}}_{\text{obs}} = (\hat{\beta}_{\text{obs},1}, \ldots, \hat{\beta}_{\text{obs},L})^T,$$

with

$$\hat{\beta}_{\text{obs},l} = \frac{\hat{Y}_{\text{obs},l}}{\hat{N}_{\text{obs},l}} = \frac{\sum_{i=1}^{r} x_{il} y_i / \pi_i}{\sum_{i=1}^{r} x_{il} / \pi_i}.$$

Using the weighting approach with post-stratification, we obtain the following estimate of (7.8):

$$\hat{Y}_W = \sum_{l=1}^{L} N_l \frac{\hat{Y}_{\text{obs},l}}{\hat{N}_{\text{obs},l}}.$$

Under the combined approach, the nonresponding elements of the sample, $i = r+1, \ldots, n$, are imputed with $\tilde{y}_i = \sum_{l=1}^{L} x_{il} \hat{Y}_{\text{obs},l} / \hat{N}_{\text{obs},l}$. The regression estimator with post-stratification is then computed based on the imputed sample. Theorem 7.1 guarantees that the vector of estimated regression coefficients $\hat{\boldsymbol{\beta}}_s = (\hat{\beta}_{s,1}, \ldots, \hat{\beta}_{s,L})^T$ is identical to $\hat{\boldsymbol{\beta}}_{\text{obs}}$. This is in fact easy to verify directly:

$$\begin{aligned}
\hat{\beta}_{s,l} &= \frac{\sum_{i=1}^{r} x_{il} y_i / \pi_i + \sum_{i=r+1}^{n} x_{il} \tilde{y}_i / \pi_i}{\sum_{i=1}^{n} x_{il} / \pi_i} \\
&= \frac{\hat{Y}_{\text{obs},l} + (\hat{Y}_{\text{obs},l} / \hat{N}_{\text{obs},l}) \sum_{i=r+1}^{n} x_{il} / \pi_i}{\hat{N}_{\text{obs},l} + \sum_{i=r+1}^{n} x_{il} / \pi_i} \\
&= \frac{\hat{Y}_{\text{obs},l} \left[1 + (1/\hat{N}_{\text{obs},l}) \sum_{i=r+1}^{n} x_{il} / \pi_i\right]}{\hat{N}_{\text{obs},l} \left[1 + (1/\hat{N}_{\text{obs},l}) \sum_{i=r+1}^{n} x_{il} / \pi_i\right]} \\
&= \frac{\hat{Y}_{\text{obs},l}}{\hat{N}_{\text{obs},l}} = \hat{\beta}_{\text{obs},l}.
\end{aligned}$$

It follows that

$$\hat{Y}_{IW} = \sum_{l=1}^{L} N_l \frac{\hat{Y}_{\text{obs},l}}{\hat{N}_{\text{obs},l}},$$

so the weighting approach and the combined approach yield the same post-stratified estimate.

Using the mass imputation approach, each unobserved population element, $i = r+1, \ldots, N$, is imputed with $\tilde{y}_i = \sum_{l=1}^{L} x_{il} \hat{Y}_{\text{obs},l}/\hat{N}_{\text{obs},l}$. This yields the following estimate of (7.8):

$$\hat{Y}_I = \sum_{i=1}^{r} y_i + \sum_{l=1}^{L} (N_l - r_l) \frac{\hat{Y}_{\text{obs},l}}{\hat{N}_{\text{obs},l}},$$

where r_l denotes the number of respondents in the lth post-stratum, with $\sum_{l=1}^{L} r_l = r$.

In general, this estimate need not be equal to the two previous estimates. Suppose, however, that we are dealing with a stratified sample, where the post-strata are smaller than the sample strata, and each post-stratum falls into exactly one sample stratum. In this commonly encountered case, all elements in a post-stratum have the same inclusion probability, which means that the vector of inclusion probabilities is contained in the column space of **X**, and Theorem 7.1 states that the weighting estimate and the mass imputation estimate are identical. We can also verify this directly. Noting that

$$\frac{\hat{Y}_{\text{obs},l}}{\hat{N}_{\text{obs},l}} = \frac{1}{r_l} \sum_{i=1}^{r} x_{il} y_i$$

in this case, we find that

$$\hat{Y}_W = \hat{Y}_{IW} = \sum_{l=1}^{L} \frac{N_l}{r_l} \sum_{i=1}^{r} x_{il} y_i$$

and

$$\hat{Y}_I = \sum_{l=1}^{L} \left(\sum_{i=1}^{r} x_{il} y_i + \frac{N_l - r_l}{r_l} \sum_{i=1}^{r} x_{il} y_i \right) = \sum_{l=1}^{L} \frac{N_l}{r_l} \sum_{i=1}^{r} x_{il} y_i.$$

So in this special case, it holds that $\hat{Y}_W = \hat{Y}_{IW} = \hat{Y}_I$ for the post-stratified estimates.

■

Until now, we have only considered imputation without added residuals. If stochastic imputation is used, with a random residual added to each predicted value, the three approaches no longer yield identical estimates. Assuming that the residuals are drawn from a distribution with mean zero, it still holds that the estimates from the combined approach and the mass imputation approach equal $\hat{Y}_W$ in expectation, under the conditions of Theorem 7.1.

7.4 Ratio Imputation

7.4.1 THE RATIO IMPUTATION MODEL

In the day-to-day practice at statistical institutes, an important special case of regression imputation is *ratio imputation*. When ratio imputation for target variable y is applied, a single auxiliary variable x is used for which the ratio with target variable y is approximately constant. If R denotes the ratio between y and x, then a missing value y_i is imputed by

$$\tilde{y}_i = Rx_i. \tag{7.17}$$

Generally, R is not known and is estimated using the records for which both x and y are known:

$$\hat{R} = \frac{\sum_{k \in \text{obs}} y_k}{\sum_{k \in \text{obs}} x_k} \tag{7.18}$$

where, as before, "obs" denotes the set of observed units. The estimated ratio $\hat{R}$ hence equals the ratio of the means of target variable y and x for the item respondents for variable y. An example is when an unknown turnover (y) is estimated based on the number of employees (x). For R one would then use the mean turnover per employee. To estimate the ratio R, one can decide to weight item respondents with their raising weights.

Substituting (7.18) into (7.17) gives

$$\tilde{y}_i = \hat{R}x_i = \frac{\sum_{k \in \text{obs}} y_k}{\sum_{k \in \text{obs}} x_k} x_i \tag{7.19}$$

The most frequently occurring situation in practice is that x measures the same characteristic as y, but at an earlier moment. In this case, we write the variables y and x as $y^{(t)}$, respectively $y^{(t-1)}$. Formula (7.17) then becomes

$$\tilde{y}_i^{(t)} = Ry_i^{(t-1)} \tag{7.20}$$

where R is the relative increase (or decrease) of variable y from moment $t-1$ to t, and

$$\hat{R} = \frac{\sum_{k \in \text{obs}} y_k^{(t)}}{\sum_{k \in \text{obs}} y_k^{(t-1)}}.$$

One can consider formula (7.17) as a regression equation without an intercept. If a model with an intercept leads to a better fit, or if one wants to add more auxiliary variables to the model, the more general regression imputation method may be more suited. At NSIs, such as Statistics Netherlands, generally no residual is added to (7.17), because for many statistics where ratio imputation is applied, means and totals are the most important products.

Even at NSIs there are a few exceptions, however. In the past, Statistics Netherlands used to publish for a few branches of industry a table with the number of enterprises with a higher turnover compared to the previous year versus the number of enterprises with a lower turnover. If imputation by (7.20) were to be applied and R were estimated to be, say, 1.01, then for all item nonrespondents it would be assumed that their turnover had grown from moment $t-1$ to t, which is highly unlikely. For such tables it is therefore necessary to add a residual to (7.20).

A special case of ratio imputation arises when R is set to 1. This means that the imputed value $\tilde{y}_i$ equals x_i. Variable x is then a "proxy variable" for y. If x is from another data set, this is referred to as "cold deck imputation" (see also Section 7.6). An example is when for a missing value $y_i^{(t)}$ the value from a previous period, $y_i^{(t-1)}$, is imputed. For variables for which the value does not change much over time, this can be a good approach, although one will often prefer to estimate R from the observed data in the present and previous period, rather than just setting R equal to 1.

The ratio $\sum_i y_i / \sum_i x_i$ does not change when ratio imputation is applied. When the so-called ratio estimator is applied to weight a sample to the population level [see, e.g., Särndal, Swensson, and Wretman (1992), Knottnerus (2003), Särndal and Lundström (2005), and Bethlehem (2009)] with x as auxiliary variable for y, then the population estimate is not influenced due to the imputed values.

Just like the more general regression imputation, one can apply ratio imputation (7.19) to separate subpopulations. This is especially beneficial if the ratios between subpopulations differ substantially. In that case, for each subpopulation h a ratio $\hat{R}_h$ is estimated. This is referred to as *group ratio imputation*. Using this imputation method is only sensible when the linear relation between x and y varies strongly, or at least significantly, between subpopulations. The subpopulations should not be too small, because this may lead to large standard deviations for estimators of totals.

For ratio imputation no complex software is required. Formulas (7.19) and (7.20) are easy to calculate once R has been estimated.

7.4.2 EXAMPLE OF RATIO IMPUTATION

EXAMPLE 7.5 (Dutch Structural Business Statistics)

At Statistics Netherlands, for the Structural Business Statistics an automated imputation procedure is used for small and medium-sized enterprises. This imputation procedure is mainly based on ratio imputation. The availability of auxiliary information is examined in a fixed order. This order, from the most preferred auxiliary information to the least preferred auxiliary information, is given by:

1. observation of the same enterprise in the previous year $t-1$ (for all variables);
2. observation of the same enterprise in the Short-Term Statistics of the year t (only if *Turnover* is the target variable y);
3. observations of units in the same stratum (defined by "size of the enterprise" × "branch of industry") in year t.

In other words, if item nonresponse occurs for a certain enterprise, it is first examined whether the enterprise had a valid score on the corresponding variable in the previous year. If so, formula (7.20) is applied with $y^{(t)}$ its value of the corresponding variable in year t, $y^{(t-1)}$ the value in the previous year, and R an estimated trend correction. For *Turnover*, the trend correction equals the growth (or shrinkage) of the total turnover.

However, if $y_i^{(t-1)}$ is unknown, for instance, because the enterprise was not included in the sample of the previous year, the second or the third option is chosen, depending on the target variable. These options are not ratio imputations. In the second option, the total turnover of the enterprise under consideration in year t as observed in the Short-Term Statistics is used to impute the missing observation for *Turnover* in the Structural Business Statistics. Option 3 is an example of group mean imputation (see Section 7.5), with a combination of "size of the enterprise" and "branch of industry" as imputation class. ■

7.5 (Group) Mean Imputation

7.5.1 THE (GROUP) MEAN IMPUTATION MODEL

When *mean imputation* is applied, a missing value is replaced by the mean value of the corresponding variable of the units that have a valid value. That is, the imputed value $\tilde{y}_i$ for an unknown value y_i is given by the observed mean

$$\tilde{y}_i = \bar{y}_{\text{obs}} \equiv \frac{\sum_{k\in\text{obs}} y_k}{r}, \tag{7.21}$$

where y_k is the observed value of the kth item respondent, "obs" is the set of observed units, and r is the number of item respondents for target variable y.

If one wishes, the observed values of target variable y may be given a different weight, for instance because these data are collected with a sampling design with different inclusion weights. Let w_i denote the (raising) weight of item-respondent i. The resulting mean imputation is then given by

$$\tilde{y}_i = \bar{y}^{(w)}_{\text{obs}} \equiv \frac{\sum_{k\in\text{obs}} w_k y_k}{\sum_{k\in\text{obs}} w_k}. \tag{7.22}$$

In general, this is a better, i.e. less biased, estimator for the population mean.

When mean imputation is applied, no auxiliary information is used. The method is only to be recommended when auxiliary information is lacking, or when available auxiliary information is not or hardly related to the target variable y. If the fraction of missing values on a certain variable is very small and the imputations will hardly affect the parameters (e.g., a population total) to be estimated, mean imputation can be a good method. In many cases, however, the approach is a bit too simplistic and leads to data of inferior quality.

As we already mentioned, mean imputation leads to a peaked distribution. The method can therefore only be applied successfully if one only wants to estimate population means and totals. Mean imputation is, however, unsuited for estimation of a distribution of, for instance, income or for estimating a standard deviation.

When no suitable auxiliary variables are available for a categorical target variable y, one can impute the most common value (the mode), or draw from the categories with probabilities proportional to the observed frequencies of these categories. Imputing the mode is generally not to be recommended because this usually leads to biased estimates for population frequencies. Drawing from the categories with probabilities proportional to the observed frequencies of these categories is the same as using random donor imputation (see Section 7.6).

When *group mean imputation* is applied, a missing value is replaced by the mean value of the corresponding variable of the units that have a valid value and in addition belong to the same group as the item nonrespondent. In the case of group mean imputation, one first determines appropriate imputation classes, that is, groups. The imputed value for a missing value in group h is then given by

$$\tilde{y}_{hi} = \bar{y}_{h;\text{obs}} \equiv \frac{\sum_{k\in h\cap\text{obs}} y_k}{r_h}, \tag{7.23}$$

where y_k is the observed value for the kth respondent, and r_h is the number of item respondents for variable y in h.

When group mean imputation is applied the distribution is less peaked than for mean imputation, because the variation between groups is taken into account while imputing the missing data; only the variation within groups is neglected. In other words, whereas mean imputation leads to one large peak, group mean imputation leads to a number of smaller peaks in the distribution of the imputed data. If the ratio of the between variance and the within variance is large, the method can also be used to estimate measures of spread reasonably accurately, assuming the validity of the imputation model.

Group mean imputation leads to the same overall totals and means as the so-called post-stratification estimator, when the strata of the post-stratification estimator are used as imputation classes [see, e.g., Bethlehem (2009) and Example 7.4 above].

In the case of group mean imputation, auxiliary information is used, namely one or more categorical variables to construct the groups. The more homogeneous the groups—also referred to as subpopulations, imputation classes, or imputation strata—are with respect to the variable to be imputed, the better the quality of the imputations. In practice, one can only test the homogeneity of the groups for the item respondents, and one often makes the assumption that the groups are not only homogeneous for the item respondents, but also for the item nonrespondents. Distinguishing different groups generally has more effect for (group) mean imputation than for ratio imputation, because ratios of groups are generally more homogeneous than group means.

As usual, (group) mean imputation can also be applied with a stochastic error term.

As for ratio imputation, a major advantage of (group) mean imputation is that one does not need any special software due to the simplicity of the method. (Group) mean imputation can be applied with almost any statistical software package and with many other software packages.

7.5.2 EXAMPLES OF (GROUP) MEAN IMPUTATION

EXAMPLE 7.5 (Dutch Structural Business Statistics) ***(continued)***

As explained above, in the Dutch Structural Business Statistics, if auxiliary information on an enterprise with incomplete response is lacking, group mean imputation is used. In that case, for a missing value of, say, *Turnover*, the mean turnover in the corresponding imputation class is imputed. Subpopulations are defined by combinations of "size of the enterprise" and "branch of industry." The sampling fraction is too small to distinguish all combinations of "size of enterprise" and "branch of industry." The imputation procedure differs somewhat for large and less large enterprises. Group mean imputation will often be used for, for instance, new enterprises for which obviously no auxiliary information from a previous period is available. ■

EXAMPLE 7.6 (Dutch Automation Survey)

In the Dutch Automation Survey group mean imputation was applied (see De Waal, 1999). In this particular case, the imputation groups were not based on the data themselves, but on the output. That is, the publication cells of the tables that were released defined the imputation groups. In some cases the overall mean instead of the group mean was imputed, for instance when there were only a few observations per publication cell. ■

7.6 Hot Deck Donor Imputation

7.6.1 INTRODUCTION

When *hot deck donor imputation* is applied, for each item nonrespondent i a donor record d in the data set is searched for that has characteristics that are as similar as possible to item nonrespondent i, as far as these characteristics are (considered to be) correlated to the target variable y. From the selected donor the score, y_d, is used to impute the missing value for item nonrespondent i:

$$\tilde{y}_i = y_d. \tag{7.24}$$

The item nonrespondent is referred to as the "recipient." Donor imputation can be applied for both a numerical and a categorical target variable y. If several values are missing in a record, in principle, the same donor is used to impute all these missing values. An exception may be when the data have to satisfy constraints (see Section 9.9).

There are different ways to find a donor. These ways can be subdivided into

1. methods that use imputation classes;
2. methods that search for a donor by minimizing a distance function (nearest-neighbor hot deck).

Examples of the first kind of methods are *random hot deck* and *sequential hot deck* imputation. When random hot deck imputation is applied, imputation classes are formed based on auxiliary variables. From the potential donors in the same imputation class as the recipient—that is, with the same values on the auxiliary variables as the recipient—a donor is randomly selected. When sequential hot deck imputation is applied, one does not explicitly construct imputation classes, but for each item nonrespondent the score on the target variable in the first subsequent record with the same scores on the auxiliary variables is imputed for the missing value.

Random and sequential hot deck donor imputation can be applied when the auxiliary variables are categorical. Numerical variables can be used as auxiliary variables by temporarily categorizing them—that is, by classifying them into a number of categories. For very large data sets on which one wants to apply hot deck imputation, one sometimes applies sequential hot deck simply in order to reduce the computing time and use of computer memory.

When nearest-neighbor donor imputation is applied (see Section 7.6.3), no imputation classes are formed and some differences between the scores on the x

variables of the donor and recipient are allowed. Nearest-neighbor imputation is usually applied when the auxiliary x variables are mainly numerical ones, and information would be lost if these variables were to be temporarily categorized in order to carry out the imputation process. Nearest-neighbor imputation can also take qualitative auxiliary variables into account, as long as a suitable distance function is used. Since a distance function measuring the distance between a potential donor and the recipient is minimized when nearest-neighbor donor imputation is applied, it is essential that the importance of each auxiliary x variable is quantified in the form of an appropriate weight.

A special case of nearest-neighbor imputation is *predictive mean matching*, where the nearest-neighbor donor is determined by means of a predictive value for target variable y, using a suitable regression model (see Section 7.6.3 for more on predictive mean matching).

Besides hot deck imputation, also *cold deck imputation* exists. Here the imputed value is taken from another data set—for instance, the value of the same unit on the same variable at a previous moment. If the imputed value from the other data set is the correct value, we can consider this as deductive imputation (see also Section 7.1 and Chapter 9). If the imputed value is from the same unit at a previous moment, simply using this value is rarely a good idea. In most cases the imputed value will improve if one adds a trend factor to the model. This would lead to ratio imputation (see Sections 7.4 and 7.8).

Donor imputation is also used when several values per record on (strongly) correlated variables are missing. By choosing only one donor for all these missing values, inconsistency between the imputed values is prevented. In such a case, one has to create imputation classes that are homogeneous for several target variables simultaneously. Multivariate donor imputation can be seen as a specific solution for the problem of multivariate imputation (see Chapter 8).

7.6.2 RANDOM AND SEQUENTIAL HOT DECK IMPUTATION

When hot deck imputation is applied, a unit in the same data set is searched for that has the same characteristics as the recipient—for instance, a person of the same gender, in the same age class, living in the same county and working in the same branch of industry. The idea is that if the characteristics of two individuals are the same, the values of the target variable to be imputed will in many cases be similar. When random or sequential hot deck imputation is applied, the donor and the recipient should have exactly the same values on the background characteristics—that is, should be in the same imputation class. If in the above example no donor can be found with the same four characteristics as the item nonrespondent, the imputation class is apparently too small. To impute a value for this item-respondent, at least one of the four characteristics has to be dropped, or imputation classes will have to be combined. If there are several potential donors in the corresponding imputation class, then one donor can be drawn randomly. Instead of drawing a donor randomly, one can also add extra background characteristics, thereby hoping to obtain a single potential donor only.

It is important to avoid that one unit will be the donor for many recipients. Namely, the multiple use of a unit as donor increases the standard errors of means and totals of the target variable y, as outliers might be "magnified." One can circumvent this, for instance, by, in a certain imputation class, allowing the multiple use of a unit as donor only when all units in that imputation class have been used as donor.

In Section 7.6.1 we have already described that when sequential hot deck imputation is applied, for each item nonrespondent the score on the target variable y of the first item respondent in the data set with the same characteristics is imputed. If a number of item nonrespondents from the same imputation class occur in the data set in quick succession, they might possibly obtain their imputed value from the same donor. To prevent this, one can adjust the sequential hot deck donor method by not selecting the next potential donor record with the same background characteristics as the recipient, but the first K potential donor records with the same background characteristics as the recipient and then randomly draw one of these K records as the donor.

Sequential hot deck imputation can be applied after sorting the records in a random order. In this form, the method is sometimes referred to as the *random sequential hot deck imputation method*. Sequential hot deck imputation can also be applied without randomly sorting the records. It then depends on how the data are constructed whether or not the sequential hot deck imputation method will lead to biased means and totals or not. In both cases the values actually imputed depend on the order of the records.

The sequential hot deck and cold deck imputation methods are deterministic imputation methods. After sorting the records of the data set in random order, sequential hot deck imputation becomes a stochastic method. As the name suggests, the random hot deck method is also a stochastic method.

For donor imputation, Kalton (1983) gives a number of methods where the probability of being selected as donor is proportional to the weight. In order to prevent that a unit with a small weight is selected as donor for a recipient with a large weight or vice versa, one often makes sure that the weights of the donor and recipient do not differ much. One way to do this is to use the weight variable, or the variables that are used to calculate the weight variable, as auxiliary variables when selecting the donor.

7.6.3 NEAREST-NEIGHBOR IMPUTATION

To apply nearest-neighbor hot deck imputation, a distance function $D(i,k)$ must be defined that measures the distance between two units i and k, where i is the item nonrespondent and k is an arbitrary item respondent. The distance function $D(i,k)$ can be defined in many different ways. A frequently used general distance function is the so-called Minkowski distance

$$D(i,k) = \left(\sum_j |x_{ij} - x_{kj}|^z \right)^{1/z}, \quad (7.25)$$

where the x variables are numerical, and the sum is taken over all auxiliary variables; x_{ij} (x_{kj}) denotes the value of variable x_j in record i (k). Let the smallest value of $D(i,k)$ be attained for item respondent d [$d = \arg\min_k D(i,k)$], then respondent d is said to be the nearest neighbor of the item nonrespondent i and becomes its donor. For $z = 2$ the Minkowski distance is the Euclidean distance, and for $z = 1$ it is the so-called city-block distance. For larger z, large differences between x_{ij} and x_{kj} are "punished" more heavily.

By letting z tend to infinity, one obtains the so-called *minimax distance* [see e.g., Sande (1982) and Little and Rubin (2002)]. The minimax distance between records i and k is defined by

$$D(i,k) = \max_j \left| x_{ij} - x_{kj} \right|,$$

where the maximum is taken over all auxiliary variables x_j. This distance measure is often applied at NSIs. It was, for instance, the distance chosen for the nearest-neighbor module in the edit and imputation software system GEIS of Statistics Canada (Statistics Canada, 1998).

When the minimax distance is used, a donor record is chosen such that the maximal absolute difference between the values of the auxiliary variables of the donor and the recipient is minimal. This way of selecting a donor ensures that even the value of the most differing auxiliary variable of the donor record is close to the corresponding value of the recipient. The method is therefore robust against the presence of outliers.

An even more general distance function than (7.25) is the weighted distance function given by

$$D_\gamma(i,k) = \left(\sum_j \gamma_j |x_{ij} - x_{kj}|^z \right)^{1/z}. \tag{7.26}$$

The extra factor γ_j is a weight expressing the importance of variable x_j. Since only the relative weight is relevant, we may assume that $\sum_j \gamma_j = 1$. The weight of variable x_j should be related to the importance, for an accurate imputation, of finding a donor with a similar value on this variable. In practice, suitable weights are often easier to determine when the x variables are first normalized so they have variance equal to 1. This prevents an implicit weighting when variables are measured in different units. It is also possible to take the covariances between variables into account when defining $D(i,k)$, but this generally complicates the determination of suitable weights.

A weighted version of the minimax distance function is given by

$$D_\gamma(i,k) = \max_j \gamma_j |x_{ij} - x_{kj}|$$

or, even more generally, by

$$D_\gamma(i,k) = \max_j \gamma_j d(x_{ij}, x_{kj})$$

with $d(.,.)$ some univariate distance measure. Such a distance function is appropriate when one aims to find a donor that does not strongly deviate from the recipient on any of the x variables.

A special case of nearest-neighbor hot deck imputation is the *predictive mean matching* method [see also Little (1988)]. When this imputation method is used, one first carries out a linear regression of the target variable y on several numerical predictor x variables, using the records without item nonresponse. Next, the resulting regression model is used to predict for each record a value for the target variable y by means of formula (7.2). The donor record for item nonrespondent i is then given by that item respondent d for which the predicted value $\hat{y}_d$ is closest to the predicted value $\hat{y}_i$ for the item nonrespondent. Finally, the *observed* value y_d of donor d is imputed; that is, $\tilde{y}_i = y_d$ in accordance with formula (7.24).

That predictive mean matching is a special case of nearest-neighbor imputation can be seen by observing that predictive mean matching minimizes the distance function

$$D(i,k) = |\hat{y}(\mathbf{x}_i) - \hat{y}(\mathbf{x}_k)|, \tag{7.27}$$

with $\mathbf{x}_i$ the vector with predictor variables for the nonresponding unit i and $\mathbf{x}_k$ the vector with the same variables for unit k. The distance (7.27) can be expressed as

$$D(i,k) = |\mathbf{x}_i^T\hat{\boldsymbol{\beta}} - \mathbf{x}_k^T\hat{\boldsymbol{\beta}}| = |\sum_j (x_{ij} - x_{kj})\hat{\beta}_j|, \tag{7.28}$$

where $\hat{\boldsymbol{\beta}}$ is the vector with estimated regression coefficients $\hat{\beta}_j$. Expression (7.28) shows that predictive mean matching is a nearest-neighbor method with a distance function equal to the absolute value of a weighted sum of differences between the predictor variable values. Note that (7.28) implies that it is not necessary to actually calculate the predicted values to apply predictive mean matching.

When nearest-neighbor imputation, including predictive mean matching, is applied, one can also select the K nearest records and draw one of those K records randomly, like we described for sequential hot deck imputation. A slight modification is to draw a potential donor with a small score on the distance function with a higher probability. Taking raising weights into account, as in the weighted random hot deck imputation method, does not have any effect if one limits oneself to selecting one nearest-neighbor. When predictive mean matching is applied, using raising weights is unlikely to have much effect either.

One can combine the random and nearest-neighbor hot deck imputation methods by first constructing imputation classes using one or more characteristics and then applying the nearest-neighbor method within these classes. This is one way to apply nearest-neighbor hot deck imputation if one has both categorical and numerical variables. In this case the categorical auxiliary variables are considered to be more important than the numerical ones. More generally, one can add a distance function for the categorical variables to a distance function such as

(7.26) for numerical variables and use a weighted sum of both distance functions as the combined distance function. The various categorical variables can be given different weights.

In Section 7.1 we made a distinction between "imputation" and the more general term "correction." In the case of imputation, a missing value is replaced by a valid value; correction of an erroneous value by a valid value is only considered to be imputation if the original, erroneous value is considered to play no part in the correction process. Nearest-neighbor hot deck imputation can easily be extended to correction where the original value does play a part. The distance function is then extended by means of adding a term expressing that the new value may not deviate much from the original, erroneous value.

In Section 4.5 we discussed the Nearest-neighbor Imputation Methodology (NIM). This is a nearest-neighbor imputation method that ensures that edit restrictions are satisfied after imputation.

7.6.4 EXAMPLES OF HOT DECK IMPUTATION

EXAMPLE 7.7 (Dutch Housing Demand Survey)

At Statistics Netherlands, donor imputation has in the past frequently been applied to the Housing Demand Survey. For the Dutch Housing Demand Survey the variables were divided into groups, such as "the household," "the current dwelling," "the previous dwelling," "the socioeconomic position of the respondent," and so on. Imputation was carried out per group of variables. Within a group of variables the variables with the lowest fractions of missing values were imputed first. Discrete variables were mainly imputed by means of the random hot deck method. Continuous variables were imputed by means of the random hot deck method or by means of predictive mean matching. Many related variables in different groups were imputed by *record matching*, or by means of *the common donor rule* as this technique is also called [see Schulte Nordholt (1998)]. That is, only one donor record was used to impute all missing values on those related variables per record with item nonresponse. For other variables, variables that were already imputed were used as covariates for imputation of missing values that were not yet imputed. As for record matching, an important reason for this approach was to ensure internal consistency of the resulting imputed record. Schulte Nordholt (1998) gives the example that *Age of the respondent* and *Age of the partner of the respondent* are missing. In such a case the value imputed for the first variable was used as covariate during imputation of the second variable in order to prevent very unlikely age combinations. For more details on the imputation of the Dutch Housing Demand Survey, we refer to Schulte Nordholt (1998) and Schulte Nordholt and Hooft van Huijsduijnen (1997). ■

EXAMPLE 7.8 (Dutch Structure of Earnings Survey)

The Dutch Structure of Earnings Survey was created by matching three data sources: the Survey on Employment and Wages, the registration system of the social security fund, and the Labor Force Survey. A subset of the variables available in the three sources was selected for the Dutch Structure of Earnings Survey. These variables were used in the matching, imputation, and weighting processes. Only exact matchings between the three data sources were used. After matching the three data sources the problem of missing values in the resultant data set arose. For some variables this problem was solved by imputation, for other variables it was solved by weighting. To impute for missing values the sequential hot deck method was applied per imputation class. Related variables were imputed simultaneously using the same imputation model in order to avoid the introduction of inconsistencies within an imputed record. The sequential hot deck method and not the random hot deck method was applied, because the number of records was considered too large to use the latter method efficiently. A random component was introduced in the imputation process by putting the potential donor records in random order before the actual imputation took place. For more information on the imputation of the Dutch Structure of Earnings Survey we refer to Schulte Nordholt (1997, 1998).

Missing data from the Dutch Structure of Earnings Survey have also been imputed by means of a neural network (see Heerschap and Van der Graaf, 1999). In particular, missing values of the variable *Gross annual wage* have been imputed by means of a feed forward multilayer perceptron neural network [see, e.g., Fine (1999) for more on neural networks]. ■

EXAMPLE 7.9 (Dutch Statistics of Mechanization in Agriculture and Horticulture)

Cold deck imputation, in the sense that data of an enterprise available in one data set are used to impute missing data of the same enterprise in another data set, is often used at Statistics Netherlands. It was applied, for instance, for the Statistics of Mechanization in Agriculture and Horticulture, where data from the so-called Agricultural Census were used to impute missing values. On the other hand, hot deck imputation methods are rarely applied at Statistics Netherlands for imputing economic data. ■

7.7 A General Imputation Model

In this section we describe the different imputation methods from the previous sections again, this time as special cases of the same general linear model. In this way, we want to emphasize both the differences and the similarities between these methods. This approach to integrating the different imputation methods is similar to the one described in Kalton and Kasprzyk (1986).

The linear model we consider is an extension of the linear regression model treated in Section 7.3:

$$y = \alpha + \beta_1 x_1 + \cdots + \beta_p x_p + \varepsilon^* = \alpha + \mathbf{x}^T \boldsymbol{\beta} + \varepsilon^*, \tag{7.29}$$

where α, $\beta_1, \ldots, \beta_p$ denote model parameters, and ε^* denotes a residual. The variable y is the target variable for which an imputed value is to be found. The x variables are called the auxiliary or predictor variables and can be continuous variables or binary (0,1) dummy variables representing the categories of a categorical variable. These dummy variables are defined as follows: For a categorical variable Z with K categories, K dummy variables $x_1^{(z)}, \ldots, x_K^{(z)}$ are created, each with value $x_k^{(z)} = 1$ if the element belongs to category k of Z and $x_k^{(z)} = 0$ otherwise. Together, the K dummies for the categories of Z will be denoted by the vector $\mathbf{z}$. Since the categories of categorical auxiliary variables are assumed to be mutually exclusive and exhaustive, one of the dummies will be 1 and the others 0. Apart from categorical and continuous variables, the model can also contain interactions between these two types of variables. These interactions can be included in the model by generating, for a K-category categorical variable, K interaction variables defined as the products $x_k^{(z)}.x_j$, for some continuous variable x_j and $k = 1, \ldots, K$. Together, the K variables for the interaction between Z and x_j will be denoted as the product $\mathbf{z}.x_j$.

In (7.29) a superscript "*" is added to ε^* to indicate that, in contrast to the usual regression model, ε^* is not necessarily a realization of a random variable with some parametric distribution, such as the normal distribution. Instead, ε^* may also be the realization of a nonparametric, empirical distribution defined by the data themselves, and in some cases it may even be calculated in a deterministic manner. By specifying whether an α parameter is used, whether—and how many—predictors are used, and how ε^* is determined, many imputation methods for a single target variable y are covered by this general imputation model.

All special cases considered here (except deductive methods) use estimates of the parameters α and $\boldsymbol{\beta}$, $\hat{\alpha}$ and $\hat{\boldsymbol{\beta}}$, say. In general, these parameters are estimated by wls using the data for which both y and the x variables are observed. The wls estimators of α and β can be written as

$$(\hat{\alpha}, \hat{\boldsymbol{\beta}}^T)^T = \left(\sum_{i \in \text{obs}} \mathbf{x}_i v_i^{-1} \mathbf{x}_i^T \right)^{-1} \sum_{i \in \text{obs}} (\mathbf{x}_i v_i^{-1} y_i), \tag{7.30}$$

where the subscript i denotes the units, "obs" denotes the indices of the units used for estimating the parameters, and v_i denotes an estimate of the variance of ε_i^*. Just as the expected value of y is a function of predictors or auxiliary variables (the mean function), v_i can also be a function of auxiliary variables (the variance function). If v_i is assumed to be a constant, we write $v_i = c$ and (7.30) reduces to the ols estimator. Three other assumptions about v_i will be used.

I. The variance is equal within groups. In this case we have $v_i = \mathbf{z}_i^T \mathbf{c}$, with $\mathbf{c}$ a vector with variance parameters, with length equal to the number of categories (or groups) of the discrete variable Z.

II. The variance is proportional to one of the auxiliary variables, x_j say. We then have $v_i = x_{ij}c$, with c the proportionality constant.

Assumptions I and II can be combined to obtain the third assumption:

III. The variance is proportional to an auxiliary variable within groups. In this case $v_i = (\mathbf{z}_i.x_{ij})^T \mathbf{c}$, with $\mathbf{c}$ a vector with proportionality constants.

Using the general imputation model (7.29), an important distinction that can be made is between deterministic and stochastic imputation. For stochastic imputation methods, the residual ε^* is based on a stochastic process, such as a random draw from a distribution or a random draw from observed residuals. For most deterministic imputation methods, the residual ε^* equals zero. An exception is nearest-neighbor imputation (see below), where the residual is nonzero. The residual for nearest-neighbor imputation is, however, calculated in a deterministic manner. Below we consider imputed values generated by special cases of model (7.29). This leads to a concise description of a number of well-known imputation methods.

Deterministic imputation methods include:

- *Proxy imputation* ($\tilde{y}_i = x_i$). The imputed value is the value of another variable that is observed and, preferably, close to the variable to be imputed. No parameters need to be estimated. This is a form of deductive imputation. Examples are imputing, in an ongoing survey, with a previous value, or imputing the missing value of nationality of a husband with the value of his wife.
- *Deductive imputation using balance edits* ($\tilde{y}_i = \mathbf{x}_i^T \hat{\boldsymbol{\beta}}$). Suppose that some variables in the record have to satisfy a relation of the form

$$x_{i,p} = x_{i,1} + \cdots + x_{i,p-1}$$

  and one of these variables is the missing y_i value while the others are all observed, then the imputed value can be calculated from the observed values in the edit. The parameter estimates are $+1$ or -1 and although they could be estimated by least squares, they can be determined directly without any other data than the record i. See also Section 9.2.2 for generalizations of this kind of imputation.
- *Mean imputation* ($\tilde{y}_i = \hat{\alpha}$; $v_i = c$). For mean imputation, no auxiliary variable x_j is used, and $\hat{\alpha}$ is estimated by ols because $v_i = c$, which means that $\hat{\alpha}$ will be equal to the mean of the units with responses on y, the item respondents.

- *Group mean imputation* ($\tilde{y}_i = \mathbf{z}_i^T \hat{\boldsymbol{\beta}}$; $v_i = c$ or $v_i = \mathbf{z}_i^T \mathbf{c}$). For group mean imputation, the means and variances are constant within each group. For deterministic imputation it makes no difference whether the parameters are estimated by ols ($v_i = c$) or by wls with variance function $v_i = \mathbf{z}_i^T \mathbf{c}$; in both cases is $\tilde{y}_i$ equal to the mean value of the item respondents in the group to which record i belongs. For the stochastic models considered below, the assumptions about the variance structure do make a difference.
- *Ratio imputation* ($\tilde{y}_i = x_{ij}\hat{\beta}_j$; $v_i = x_{ij}c$). For ratio imputation, one auxiliary x variable is used. The parameter $\alpha = 0$ by definition. The wls estimator $\hat{\beta}_j$ is equal to the ratio between the means of y and x_j among the units with both variables observed, as can easily be verified by taking $\mathbf{x}_i$ in (7.30) to be the score x_{ij} on a single variable x_j and substituting $x_{ij}c$ for v_i.
- *Group ratio imputation* ($\tilde{y}_i = (\mathbf{z}.x_{ij})^T \hat{\boldsymbol{\beta}}$; $v_i = (\mathbf{z}.x_{ij})^T \mathbf{c}$). The variances are proportional to x_j but with a different proportionality constant c_k for each group k. The components of the wls estimator $\hat{\boldsymbol{\beta}}$ are the ratios of the within group means of y and x_j. In general, if auxiliary values in both the mean function and the variance function only appear in interaction with the same categorical variable, then the model becomes "uncoupled" between groups and is equivalent to applying the model to each group separately. This is, for instance, also the case for group mean imputation.
- *Regression imputation (without residual)* ($\tilde{y}_i = \hat{\alpha} + \mathbf{x}_i\hat{\boldsymbol{\beta}}_j$; v_i). Mean imputation, ratio imputation, and the grouped variants thereof are all special cases of regression imputation, each corresponding to specific mean and variance functions. The general case covers all other specifications of the mean and variance function as well.
- *Nearest-neighbor imputation* ($\tilde{y}_i = (A(\mathbf{x}_i))^T \hat{\boldsymbol{\beta}} + e_d$). For nearest-neighbor imputation, an approximation $A(\mathbf{x}_i)$ to $\mathbf{x}_i$ is used. This approximation is the $\mathbf{x}$ value of another record (the donor d), with both y and $\mathbf{x}$ observed, which is nearest to $\mathbf{x}_i$ according to some specified distance function. The residual e_d is the realized residual $y_d - \mathbf{x}_d^T \hat{\boldsymbol{\beta}}$ for the donor. Therefore, for any method of estimating $\boldsymbol{\beta}$, the imputed value equals the observed value y_d of the donor.
- *Predictive mean matching* ($\tilde{y}_i = A(\mathbf{x}_i^T \hat{\boldsymbol{\beta}}) + e_d$). Predictive mean matching can be seen as a special case of nearest-neighbor imputation. In this case a donor is sought for such that the predicted value of y, without an error term, is closest to the predicted value for the record i. Imputing with this donor value can also be interpreted as approximating the predicted value for record i by $A(\mathbf{x}_i^T \hat{\boldsymbol{\beta}}) = \mathbf{x}_d^T \hat{\boldsymbol{\beta}}$ and adding the observed residual from the donor.
- *Sequential hot deck imputation* ($\tilde{y}_i = y_d$). For sequential hot deck imputation, the donor is taken to be the first record in the data set after the record under consideration that has a valid value for target variable y and the same values of the categorical predictor variables (see Section 7.6.2). The imputed value simply equals the observed value from the donor. This was true of course for the other donor methods as well.

Stochastic imputation methods include:

- *Regression imputation (with a residual from a parametric distribution)* ($\tilde{y}_i = \hat{\alpha} + \mathbf{x}_i^T\hat{\boldsymbol{\beta}} + \varepsilon_i^*; \varepsilon_i^* \sim (0, v_i)$). For regression imputation with a residual from a parametric distribution, an α parameter and a vector of numerical or categorical x variables with corresponding vector of parameters $\boldsymbol{\beta}$ is used and a residual is added. The residual is drawn from a parametric distribution, usually the normal distribution, with mean zero and variance v_i. By choosing specific mean and variance functions, stochastic versions of the various submodels of the regression imputation model arise.
- *Regression imputation (with an observed residual).* Regression imputation with an observed residual is similar to regression imputation with a residual from a parametric distribution. The only difference is that the residual is now drawn from the set of observed residuals, $\mathbf{e}_{\text{obs}}$, say.
- *Random hot deck imputation* ($\tilde{y}_i = \hat{\alpha} + \varepsilon_i^*; \varepsilon_i^* \sim (\mathbf{e}_{\text{obs}})$). For random hot deck imputation, $\hat{\alpha}$ is estimated by ols and hence equals the overall respondents mean. A residual $y_d - \hat{\alpha}$ is added from a randomly chosen donor. This method can be seen as a variant of stochastic mean imputation with the residuals drawn from the empirical distribution of residuals around the mean.
- *Group random hot deck imputation* ($\tilde{y}_i = \mathbf{z}_i^T\hat{\boldsymbol{\beta}} + \varepsilon_i^*; \varepsilon_i^* \sim (\mathbf{e}_{\text{obs}})$). For group random hot deck imputation, $\hat{\boldsymbol{\beta}}$ consists of the respondent means of the categories of Z (the groups). A residual $y_d - \mathbf{z}_i^T\hat{\boldsymbol{\beta}}$ is added from a randomly selected donor with the same value of $\mathbf{z}$ ($\mathbf{z}_i = \mathbf{z}_d$)—that is, a donor from the same group as the recipient. This method can be seen as a variant of stochastic group mean imputation with the residuals drawn from the empirical distribution of residuals around their group means.

Since the imputed values are randomly drawn for stochastic imputation, they cannot be reproduced in general. In the case of *deterministic imputation* the imputed values can be reproduced, given the selected imputation model. The choice between stochastic and deterministic imputation in many cases depends on whether or not one wants to add a residual to the model. Nearest-neighbor imputation, including predictive mean matching, however, implicitly adds a residual to the imputation model and is deterministic, because the donor selected is deterministically found by means of a distance function.

Another distinction that can be made is between regression imputation (either deterministic or stochastic), including its special cases ratio imputation and (group) mean imputation, and donor imputation techniques. For donor imputation techniques the residuals are somehow drawn from the observed residuals, and auxiliary variables $x_1, \ldots, x_p$ are only used to construct imputation classes or to base a distance function upon. The choice between regression imputation and donor imputation is often not an obvious one. Some considerations in choosing between these two approaches may include the following:

- Because donor values are actually observed (and correct) values, they are always valid values for the target variable. If the target variable is integer-valued or nonnegative, so will be the donor values. Regression models will in general not produce integer predicted values and can result in negative predictions for nonnegative variables.

- Regression and nearest-neighbor imputation can naturally make use of both categorical and continuous auxiliary variables. Random and sequential hot deck methods are applied within groups determined by the combinations of values of categorical variables. Continuous variables can only be used as auxiliary variables if they are first discretisized, which implies a loss of information and also predictive power if the relation between the auxiliary variables and the target variable is linear.
- The amount of auxiliary information is more limited for the methods that use a grouping of the data (i.e., random and sequential hot deck methods) than for regression and nearest-neighbor methods. Because the groups are defined by all possible value combinations of the auxiliary variables, groups with only a few donors or none at all will become increasingly a problem as the number of auxiliary variables (and categories per variable) increases. In regression, the equivalent of the grouping structure is to include dummy variables for all categorical auxiliary variables and all interactions between these variables. This too may lead to problems of empty groups and, in this case, inestimable parameters. However, in the regression approach it is easy, and even customary, to reduce the model to a more parsimonious one by leaving out interactions between variables but keeping the main effects. For the methods based on groups, a variable is either used (with all interactions) or not used at all. Nearest-neighbor methods do not run into technical problems when many auxiliary variables are used; but when the number of auxiliary variables increases, the discrepancy between donor and recipient on each variable will tend to increase because of the need to match as good as possible on more variables simultaneously.
- Donor imputation extends more easily than model-based imputation to multivariate problems where several variables in a record must be imputed, and the relation between these variables should be preserved as well as possible (see also Chapter 8).
- For activities where not everyone contributes, the distribution of the data consists of zeros for part of the population, the nonparticipants, and positive values for the participants. Examples are the amount of money spent on vacation, the number of kilometers driven by car, and the amount of money invested by an enterprise in new machines. Hot deck methods give good results for these types of variables, in the sense that they preserve the distribution. However, if one applies mean imputation, no zero at all will be imputed. Regression imputation has the same problem and, in addition, negative imputed values may occur for such nonnegative variables. If the goal is to estimate population means, this is not a major problem. However, if the goal is to estimate the variation of the target variable or the fraction of participants, one cannot apply these techniques. An option for imputation in such cases, apart from hot deck, is to apply imputation

in two steps. First a logistic regression or a hot deck imputation method is applied to "determine"—that is, impute—whether an item nonrespondent is a participant or not, and then imputed values for the supposed participants are obtained by means of a linear regression imputation model.

7.8 Imputation of Longitudinal Data

Longitudinal data occurs when the same variables of the same units are measured several times, at different moments. For instance, data of panel surveys are longitudinal data, just like data from a population register at different moments. At Statistics Netherlands, a panel survey is used for the Labor Force Survey. An example of a longitudinal register in the Netherlands is the Municipal Base Administration for which each year a new version is constructed. Most registers provide longitudinal information when several versions from different moments are linked.

Longitudinal imputation distinguishes itself from most of the other imputation methods described in this chapter, because for longitudinal imputation, data from the same units are used, in many cases without using information from other units. For each unit, one hence has a time series with one or more missing values that have to be imputed. Longitudinal imputation is, however, closely related to ratio imputation (see Section 7.4).

Missing values in longitudinal data occur in two different ways:

1. Missing values that are spread over time;
2. Missing values due to panel drop-out.

When missing values are due to drop-out, only information from the past of the corresponding units can be used, and extrapolation methods or ratio imputation (see Section 7.4) may be used to predict missing values of numerical data. An often used technique is *last value carried forward* where simply the last observed value from a previous period is imputed. For numerical data, last value carried forward is often applied without a trend correction, but can also be applied with a trend correction (see also Section 7.4 on ratio imputation).

When missing values are spread over time, imputation will often be based on interpolation. However, when the aim is to provide estimates as soon as possible after a new wave of the longitudinal data becomes available, one can again use only data from earlier moments and again extrapolation techniques or ratio imputation need to be applied. Information at later moments can only be used when one has enough time to wait for later waves of the longitudinal data or when one constructs imputations for several moments simultaneously in order to obtain a longitudinal data set that is as good as possible.

For some surveys, one starts with a complete data set, even before any data have been observed for the current period. The imputed values may, for instance, be imputed based on values from a previous period $t-1$. Whenever new, observed data are obtained, these values replace the imputed values, after which the remaining imputed values are updated, using the imputation model with updated model parameters. The statistical process for such an approach differs from the standard approach to imputation of missing values—and has the advantage that one can obtain a population estimate at any moment easily—but from a theoretical point of view there is no major difference between the two approaches.

Social surveys typically involve many categorical data. For categorical data the above-mentioned techniques of ratio imputation, extrapolation, and interpolation do not apply. Instead, hot deck imputation methods can be used where imputation classes involve variables that match in a different wave.

The problems caused by missing values due to panel drop-out can generally also be solved by means of weighting methods. When one wants to estimate a statistic at a certain moment, one can consider the recent drop-out as unit nonresponse and add this to the nonresponse of earlier waves of the panel. Panel drop-out in registers is often justified and may, for instance, be caused by emigration and deaths.

For more on longitudinal imputation we refer to Daniels, Daniels, and Hogan (2008).

An illustration of longitudinal imputation has already been given in Example 7.5. Below we give additional examples.

EXAMPLE 7.10 (Dutch Business Statistics)

Imputation of longitudinal data is often applied at Statistics Netherlands to impute missing data in business statistics that are held on a regular basis (monthly, quarterly, or annually). For instance, missing values in the Structural Business Statistics of the Retail Trade are imputed by values of a previous period, adjusted by a trend factor. The trend factor is usually estimated using other records in the same imputation group.

To impute missing values in annual statistics, often data from corresponding monthly or quarterly statistics are used if available. This approach was used, for instance, for the Petroleum Industry Statistics. ■

EXAMPLE 7.11 (Dutch Construction Industry Survey)

For the Construction Industry Survey, internal data sources and external registers were used to guide the imputation process. The survey consisted of about a hundred mostly numerical variables. In case an enterprise did not respond, the value of either the turnover or the number of employees of that enterprise was determined. This was done by recontacting the enterprise, or by using data that were available from another source. The "structure" of the enterprise—that is, the proportions between the various unknown variables and the observed variable—was estimated. This was done by using data from either the previous year or by using the average structure of the responding enterprises. The missing values were then imputed by preserving that structure. Below we illustrate the procedure used for the Construction Industry Survey by means of a simple example.

Suppose the numbers of employees $N_i^{(t)}$ and $N_i^{(t-1)}$ of enterprise i in this period t and the previous period $t-1$, respectively, are observed. Suppose, furthermore, that the value of variable y_j of enterprise i in the previous year is given by $y_{ij}^{(t-1)}$. The imputed value of variable y_j of enterprise i in this year, $\tilde{y}_{ij}^{(t)}$, is then given by

$$\tilde{y}_{ij}^{(t)} = N_i^{(t)} \times \left(\frac{y_{ij}^{(t-1)}}{N_i^{(t-1)}} \right).$$

When the proportions $y_{ij}^{(t-1)}/N_{ij}^{(t-1)}$ are unknown, the imputed value of variable y_j of enterprise i in this year is calculated as

$$\tilde{y}_{ij}^{(t)} = N_i^{(t)} \times \left(\frac{\bar{y}_j^{(t)}}{\bar{N}^{(t)}} \right),$$

where $\bar{y}_j^{(t)}$ is the average value of variable y_j and $\bar{N}^{(t)}$ the average number of employees. Both averages are taken over all responding enterprises in this year. In practice, the method applied for the Construction Industry was slightly more complicated, but the main idea is covered by the brief description above. ■

EXAMPLE 7.12 (European Community Household Panel Survey)

Schulte Nordholt (1998) describes the imputation strategy that was applied to impute missing values of the European Community Household Panel Survey (ECHP) [see also Schulte Nordholt (1996)]. The Dutch Socio-Economic Panel was part of that survey. The ECHP focused on income and the labor market; other topics include health, education, housing and migration. The survey was carried out at the household level. The advantage of the ECHP in comparison to cross-sectional surveys was that it allowed the study of income mobility patterns. However, the fact that the ECHP was a panel survey complicated the imputation process. Imputing each wave separately from the previous waves might lead to strange income shifts. Instead of this naive imputation strategy a better strategy would be to impute all available waves simultaneously. However, because this imputation strategy would lead to changes in the results of all previous waves after the imputation of a new wave, a simpler strategy was adopted. The chosen strategy was to take imputations of previous waves into account while imputing a new wave, but imputations carried out during previous surveys were not adapted.

To impute the first wave of the ECHP, random hot deck imputation within groups was used. Since income variables are very important in the ECHP, imputation concentrated on these variables. Besides variables such as *Gender* and *Year of birth*, several important labor variables, such as *Employment status* and *Present occupation*, were used as auxiliary variables to impute missing values of income variables. Care was taken to avoid inconsistencies between related variables, such as gross and net income from labor per month. To impute *Gross income from labor per month* and *Net income from labor per month* the method of record matching was used. *Gross income from labor per month* was used as the main variable for imputation. The hot deck method within groups was used to impute missing values of this variable. If the value of the related variable *Net income from labor per month* was also missing in a record, this variable was imputed using the same donor record in order to avoid inconsistencies. ■

7.9 Approaches to Variance Estimation with Imputed Data

7.9.1 THE PROBLEM AND APPROACHES TO DEAL WITH IT

Most imputation methods lead to an underestimation of the variance of the imputed variable if the imputations are treated as no different from "real" observations. As a consequence, standard formulae for the variance and standard error of estimators are biased downwards and confidence intervals become too narrow. The following simple example illustrates the problem.

Suppose that a simple random without replacement sample (SRSWOR) of size n from a population of size N is available to estimate the population mean of y. An unbiased estimator for the population mean is the sample mean $\bar{y}$ with variance estimator given by

$$\hat{V}(\bar{y}) = \left(\frac{1}{n} - \frac{1}{N}\right)\hat{\sigma}^2, \tag{7.31}$$

with $\hat{\sigma}^2$ an estimator for the population variance, σ^2, of y. This estimator is in this case $s^2 = \sum_{i=1}^{n} \left(y_i - \bar{y}\right)^2 / (n-1)$. Now suppose that only n_{obs} of the n units responded and the y value for the nonresponding units is imputed with the mean of the responding units, $\bar{y}_{\text{obs}}$. Without further information the population mean can be estimated with the mean of the imputed data, which equals $\bar{y}_{\text{obs}}$. If the standard variance formula (7.31) is applied to the imputed data, the estimator of the variance of y would become $\sum_{i\in\text{obs}}\left(y_i - \bar{y}_{\text{obs}}\right)^2/(n-1)$, since the residuals $y_i - \bar{y}_{\text{obs}}$ cancel out for the imputed values. This estimator of σ^2 will be downwards biased. A correct estimator for σ^2 can be obtained if we assume that the nonresponse mechanism is uniform—that is, independent and identical response probabilities across sample units—so that the observed nonzero residuals can be seen as a random subsample of the residuals for the full sample. The correct estimator for σ^2 is then

$$s^2_{\text{obs}} = \sum_{i\in\text{obs}}\left(y_i - \bar{y}_{\text{obs}}\right)^2/(n_{\text{obs}}-1). \tag{7.32}$$

It is important to note that, in contrast to deterministic imputation, stochastic imputation can lead to a correct estimator of σ^2 by the standard formula. With the estimate s^2_{obs} the variance estimate (7.31) is approximately equal to the estimate that would have been obtained with full response. However, this is still an unsatisfactory estimate because the loss of information due to nonresponse is expected to result in a decrease in precision compared to the case of full response.

This very simple example already shows a few of the main issues in variance estimation after imputation. First of all, the standard formulae for variances or standard errors of estimates are downwards biased. Furthermore, the estimate of the population variance σ^2 is too low, but this can be compensated for (under certain assumptions) by an adaptation of the formula or by adding random residuals. However, a standard formula with a correct estimate of σ^2 does still underestimate the variance of estimates since it does not take the loss of information due to nonresponse adequately into account.

To compute valid standard errors of estimates, it is thus necessary to use methods that are specifically developed to take the imputations into account. Several approaches to this problem have been proposed, and in this section we distinguish three types of approaches.

The Analytical Approach. In this approach the usual design-based repeated sampling variance formulae are extended to account for nonresponse and subsequent imputation. Such extensions of the standard theory have been considered, among others, by Särndal (1992) and Deville and Särndal (1994). An often taken route to derive analytical variance formulae is the two-phase approach. Two-phase sampling means that a final sample is drawn in two steps: In the first phase a sample is drawn from the population and in the second phase a sample is drawn from the sample realised in the first step. When applied to nonresponse, the first-phase sample corresponds to sampling units (respondents and nonrespondents) and the second-phase sample corresponds to the subsample of responding units. To derive variance estimators an assumption for the selection mechanism in the second-phase (the nonresponse mechanism) is made. Usually, it is assumed that, within classes, the nonresponse mechanism is uniform. An alternative assumption, which can also lead to valid variance estimators, is to assume that the nonresponse mechanism is MAR; it depends on the completely observed auxiliary variables that are used in the imputation model but not on the target variable (cf. Section 1.3). Using this last assumption is also called the model-assisted approach.

The Resampling Approach. Resampling methods for variance estimation based on complex samples without nonresponse have been common practice in survey methodology [see e.g. Wolter (1985)]. Their advantage is that, while analytic variance formulae need to be derived for different kinds of estimators separately, and can become quite complex, resampling methods offer a relatively simple computational procedure for obtaining variance estimates that is general enough to be applicable to many estimation problems. This advantage is particularly relevant for the case of variance estimation with imputed data since the formulae tend to be more complicated than without nonresponse. Therefore resampling methods for variance estimation with imputed data have received considerable attention (Rao and Shao, 1992; Shao and Sitter, 1996). The most commonly studied and applied methods are the jackknife and bootstrap resampling schemes. Besides these methods, some work on balanced repeated replication (BRR) has also appeared (Shao, 2002).

Multiple imputation (MI). In this approach, each missing value is imputed several, say M, times and the variation between the M imputations is used to estimate the increase in variance due to nonresponse and imputation. Multiple imputation was originated by Rubin (1978, 1987). Unlike single imputation (filling in a single value for each missing one), multiple imputation was meant from the outset as a method to provide not only a solution to the missing data problem by imputation but also to reflect the uncertainty inherent in the imputations. The aim of multiple imputation is to provide multiply imputed data sets that should enable researchers to perform different kinds of analyses and obtain inferences with valid standard errors, confidence intervals, and statistical tests, in a simple way. Standard analyses are performed on each of the M data sets, and the results are combined using relatively simple formulae to obtain valid inferences.

The analytical and resampling approaches are similar in that both adhere to the design-based repeated sampling framework for statistical inference. The values of auxiliary variables (x) and target variables (y) are seen as fixed. Randomness is introduced by the random selection of units (according to the specified design) from the population, and inference is with respect to the associated sampling distribution. NSIs apply, almost without exception, design-based inference since it is most in line with their primary task of producing, often simple, descriptive statistics of fixed populations. Multiple imputation, on the other hand, was developed as a Bayesian model-based technique; the variables are viewed as random variables with some distribution specified by a model. The observations are viewed as independent random draws generated by this specified distribution. Nevertheless, Rubin (1987, Chapter 4) explaines that MI inferences are also valid for the design-based repeated sampling framework, provided that the imputations are "proper," which implies, among other things, that the features of the sampling design that influence the repeated sampling variance are accounted for in the imputation model. For complex designs involving stratification, unequal selection probabilities, and clustering, this can lead to rather complex imputation models. An interesting debate about the advantages and disadvantages of the different approaches and their applicability for different purposes appeared in the June 1996 issue of the Journal of the American Statistical Association (Rubin, 1996; Fay, 1996; Rao, 1996).

Below, we give a few details of these different approaches and references to the relevant literature.

7.9.2 THE ANALYTIC APPROACH

The analytic variance estimators that have been derived usually consist of two components (Särndal, 1992), leading to the following decomposition of an analytic variance estimator $\hat{V}_A$:

$$\hat{V}_A = V_{sam} + V_{imp}. \tag{7.33}$$

The first component is the variance of the estimator that would have been obtained if there were no nonresponse and the second component is the additional variance due to nonresponse. More precisely, this second component is the conditional variance due to nonresponse given the realised sample. The target of the analytic approach is to obtain unbiased estimators for each of these components separately and obtain an estimator for the total variance as their sum.

As an illustration, we consider again the estimation of the population mean from an SRSWOR sample under mean imputation and a uniform nonresponse mechanism. The sampling variance of an estimator without nonresponse is in this case given by (7.31) and an unbiased estimate of σ^2 is s_{obs}^2 given by (7.32). Substituting s_{obs}^2 in (7.31) yields an estimator for V_{sam}. To obtain an estimator of the second component, we observe that under the uniform response mechanism, $\bar{y}_{\mathrm{obs}}$ is an estimator of the mean of the full sample and the conditional variance, which essentially treats the sample as the population and treats the respondents as the sample, therefore is $V_{imp} = (1/n_{\mathrm{obs}} - 1/n)\, s_{\mathrm{obs}}^2$. By substituting these expressions for V_{sam} and V_{imp} in (7.33), we obtain

$$\hat{V}_A = \left(\frac{1}{n} - \frac{1}{N}\right) s_{\mathrm{obs}}^2 + \left(\frac{1}{n_{\mathrm{obs}}} - \frac{1}{n}\right) s_{\mathrm{obs}}^2 = \left(\frac{1}{n_{\mathrm{obs}}} - \frac{1}{N}\right) s_{\mathrm{obs}}^2, \tag{7.34}$$

which is readily acceptable because it is the variance of drawing a sample of size n_{obs} from the N population units directly. For more advanced imputation methods and more complicated sampling designs than in this simple example, the derivation of a variance formula can become difficult and the resulting formulae can be quite complex. The case of estimating the population mean when ratio imputation is used is also an example where a simple variance formula can be derived under a uniform nonresponse mechanism (Rao and Sitter, 1995). This estimator is

$$\hat{V}_A = \left(\frac{1}{n} - \frac{1}{N}\right) s^2_{\text{obs}} + \left(\frac{1}{n_{\text{obs}}} - \frac{1}{n}\right) s^2_e, \tag{7.35}$$

where

$$s^2_e = \sum_{i \in \text{obs}} \left(y_i - Rx_i\right)^2 / (n_{\text{obs}} - 1),$$

with x_i the auxiliary variable with observed values for all units, and R the ratio of the means of y and x for the units with y observed. This variance estimator differs from the one for mean imputation only in the second component because V_{sam} estimates the variance if there were no response and is therefore unaffected by the imputation model. The second component is smaller than the one for mean imputation because the residuals of the ratio model are smaller than those of the "mean model." This illustrates that the imputation variance decreases as the predictive accuracy of the imputation model increases; if the residuals become zero, the real data are exactly reproduced and the imputation variance vanishes.

Model-assisted and two-phase approaches for different sampling designs and estimators have been described among many others by Särndal (1992), Deville and Särndal (1994), Rao and Sitter (1995), and Särndal and Lundström (2005). A good overview and many references are provided in Lee, Rancourt, and Särndal (2002).

7.9.3 THE RESAMPLING APPROACH

A jackknife variance estimator for a parameter θ for complete data, assuming that the finite population correction factor can be neglected, has the general form

$$\hat{V}_J(\hat{\theta}) = \frac{n-1}{n} \sum_{j=1}^{n} \left(\hat{\theta}_{(j)} - \hat{\theta}\right)^2, \tag{7.36}$$

with $\hat{\theta}_{(j)}$ the estimator obtained after deleting element j from the sample and $\hat{\theta}$ the estimator obtained from the complete sample.

When this jackknife estimator is applied to imputed data, it has been shown by Rao and Shao (1992) that the variance is underestimated because the variance due to nonresponse and imputation is not taken into account. They also showed that a correct variance estimator can be obtained by adjusting the imputed values. This adjustment procedure is equivalent to applying the imputation procedure again when a respondent is deleted from the sample. For instance, for mean imputation this means that the mean is recalculated each time a respondent is deleted. In general this means that the parameters for parametric imputation models vary between jackknife replicates which captures the estimation variance in these parameters. Similarly, for donor-based methods the donor pool will vary between replicates. This adjusted jackknife method leads to a valid variance estimator under the assumption of uniform nonresponse. It was also shown to be valid under MAR.

The bootstrap resampling method for imputed data has been described, among others, by Shao and Sitter (1996). A bootstrap resampling scheme draws a large number (B, say) of with-replacement samples from the original sample, with size equal to that of the original sample. For each of these B bootstrap samples the parameter of interest is estimated and the variance of the estimator is estimated by the variance of the resulting B bootstrap estimates, as follows:

$$\hat{V}_B(\hat{\theta}) = \frac{1}{B-1}\sum_{b=1}^{B}\left(\hat{\theta}_b - \bar{\hat{\theta}}_b\right)^2 \tag{7.37}$$

with $\hat{\theta}_b$ the estimator for bootstrap sample b and $\bar{\hat{\theta}}_b = \frac{1}{B}\sum_{b=1}^{B}\hat{\theta}_b$.

When imputation has been used, the parameter estimate $\hat{\theta}_b$ is estimated on the imputed bootstrap sample. Just as with the jackknife, it is essential that the imputation process be repeated for each bootstrap sample separately.

Many articles showing the applicability of jackknife and bootstrap methods to different sampling designs and imputation methods have been treated in the literature. These include stratified cluster sampling, (stochastic) regression and ratio imputation, and donor imputation methods. The finite population correction factor, which becomes important if sampling is done within small strata as is often the case in business statistics, can also be incorporated in these resampling methods (see Rao, 1996). An overview of replication methods is given by Shao (2002).

7.9.4 THE MULTIPLE IMPUTATION APPROACH

Multiple imputation (MI) imputes each missing value M times; and to create the M imputations, a Bayesian model is used. A Bayesian model treats the parameters as random variables and assigns to these parameters a prior distribution reflecting the knowledge about the parameters before the data are observed (priors can be chosen that reflect essentially no prior information on the parameters). When the prior information on the parameters is combined with information from the available data, the posterior distribution for the parameters is obtained that contains all available information on the parameters and is the basis for Bayesian inference. The multiple imputations reflect the uncertainty in the parameters by creating imputed values using a two-step procedure. First, parameter values are drawn from their posterior distribution (for example, β, σ_e^2 in the case of regression imputation, with σ_e^2 the residual variance). Then, using these parameter values, a stochastic imputation is performed (for example, the regression prediction with an added random residual). This completes one imputed data set. By drawing M times from the posterior distribution, M imputed data sets can be generated, each imputed by a model with different values for the parameters. For most inferences it appears that a small value for M—five is often mentioned—is sufficient. The hard part in this procedure is to draw from the posterior distribution. Algorithms to perform this step are described in, for example, Schafer (1997).

The multiple imputations are used to obtain point and variance estimators for any parameter θ of interest. This is accomplished by first estimating this parameter and its variance, on each of the imputed data sets, and then combining the M estimates according to the rules defined by Rubin (Rubin, 1987). For scalar θ these rules are as follows.

The MI point estimator of θ, $\hat{\theta}_{MI}$, is the average of the M estimators obtained by applying the estimator $\hat{\theta}$ to each of the M data sets:

$$\hat{\theta}_{MI} = \frac{1}{M}\sum_{m=1}^{M}\hat{\theta}_m, \tag{7.38}$$

with $\hat{\theta}_m$ the point estimator for data set m. The MI variance estimator consists of two components: the within imputation variance and the between imputation variance. The within imputation variance, $\hat{V}_W$, is obtained as the average of the complete data variance estimates for each of the M imputed data sets:

$$V_W = \frac{1}{M}\sum_{m=1}^{M}\hat{V}_m, \tag{7.39}$$

with $\hat{V}_m$ the variance estimate for data set m. The between imputation variance is the variance of the M complete data point estimates:

$$V_B = \frac{1}{M-1}\sum_{m=1}^{M}\left(\hat{\theta}_m - \hat{\theta}_{MI}\right)^2. \tag{7.40}$$

The MI variance estimate of $\hat{\theta}_{MI}$ is obtained as the sum of the within and between imputation variances with a correction factor to account for the finite number of multiple imputations:

$$\hat{V}_{MI} = V_W + \left(1 + \frac{1}{M}\right)V_B. \tag{7.41}$$

The variance due to nonresponse is reflected in the between imputation variance component. In the absence of nonresponse, all $\hat{\theta}_m$ are equal, $V_B = 0$, and the MI variance estimator reduces to the complete data variance estimator.

Although multiple imputation was originally formulated for parametric models, it has been extended to random donor imputation. For random donor imputation a donor is selected from the respondents with valid values on the target variable(s). These respondents are called the *donor pool*. A MI version of random donor imputation, called Approximate Bayesian Bootstrap (ABB), has been devised by Rubin and Schenker (1986). This procedure works as follows. First a bootstrap sample is drawn from the donor pool, the bootstrapped donor pool. Then, for each nonrespondent, a donor is selected, at random and with replacement, from the bootstrapped donor pool and the missing values are imputed. This completes one imputation of the data set. The procedure is repeated M times to create the M multiple imputations. The between imputation variance arises here because the bootstrapped donor pool changes between the M imputations.

Multiple imputation is treated in Rubin (1987), Little and Rubin (2002), and Schafer (1997). Rubin (1996) reviews the many applications of the method in the last 18+ years.

7.10 Fractional Imputation

We conclude this chapter with a short discussion of fractional imputation. We treat this technique directly after our discussion of multiple imputation because fractional imputation is closely related to multiple imputation, in the sense that both multiple and fractional imputation are examples of *repeated imputation.* That is, in both multiple imputation and fractional imputation, multiple values are imputed for a single missing value. Below we first explain the technique of fractional imputation and then point out some differences between multiple and fractional imputation.

Fractional imputation can be applied to numerical data and is generally based on hot deck imputation [see, e.g., Kim and Fuller (2004)]. In fractional imputation, imputations in a recipient record i are determined by two factors: the donor records k that are used and the imputation weights $w^*_{ik} \geq 0$ that are assigned to these donor records. The imputation weights w^*_{ik} satisfy

$$\sum_{k \in R} w^*_{ik} = 1,$$

where R denotes the set of records with an observed value for the variable to be imputed. A weight w^*_{ik} expresses the fraction of the value of donor record k that is used for the missing value of recipient record i. For a recipient record i with a missing value on the variable to be imputed, the missing value is imputed by the weighted mean of the donor values for that record

$$y^*_i = \sum_{k \in R} w^*_{ik} y_k.$$

Given the imputed values for all recipient records, a linear (weighted) estimator $\hat{\theta}$ may be written as

$$\hat{\theta} = \sum_{i \in U} w_i y^*_i = \sum_{i \in U} w_i \left(\sum_{k \in R} w^*_{ik} y_k \right) = \sum_{k \in R} W^*_k y_k,$$

where U denotes the entire data set, w_i the design weights (for instance, determined by the sampling design and correction weights for unit nonresponse), and $W_k^* \equiv \sum_{i \in U} w_i w_{ik}^*$.

A major difference between fractional imputation and multiple imputation is that fractional imputation is defined in a frequentist framework whereas multiple imputation is defined in a Bayesian framework. In fact, fractional imputation may be seen as a form of improper multiple imputation. Another important difference is that the main aim of fractional imputation is to improve the efficiency of the imputed point estimator by reducing the variance arising from the imputation process, whereas the main aim of multiple imputation is to simplify variance estimation of a point estimator. A third difference is that fractional imputation is based on hot deck imputation, at least in its usual, traditional form [see, e.g., Qin, Rao, and Ren (2008) and Kim and Fuller (2008) for recent extensions to regression imputation], whereas multiple imputation is based on drawing values from the posterior predictive distribution of the missing value in a record given the observed values.

For more on fractional imputation we refer to Kalton and Kish (1984), Fay (1996), Kim and Fuller (2004), Durrant (2005), and Durrant and Skinner (2006).

REFERENCES

Andridge, R. R., and R. J. Little (2009), The Use of Sample Weights in Hot Deck Imputation. *Journal of Official Statistics 25*, pp. 21–36.

Allison, P., and P. D. Allison (2001), *Missing Data*. Sage Publications.

Bethlehem, J. G. (2009), *Applied Survey Methods: A Statistical Perspective*. John Wiley & Sons, Hoboken, NJ.

Bethlehem, J. G., and W. J. Keller (1987), Linear Weighting of Sample Survey Data. *Journal of Official Statistics 3*, pp. 141–153.

Burnham, K. P., and D. R. Anderson (2002), *Model Selection and Multi-Model Inference*, second edition. Springer-Verlag, New York.

Chambers, R. (2004), Evaluation Criteria for Statistical Editing and Imputation. In: *Methods and Experimental Results from the EUREDIT Project*, J. R. H. Charlton, ed. (http://www.cs.york.ac.uk/euredit/).

Chambers, R. L., J. Hoogland, S. Laaksonen, D. M. Mesa, J. Pannekoek, P. Piela, P. Tsai, and T. de Waal (2001a), *The AUTIMP-Project: Evaluation of Imputation Software*. Report, Statistics Netherlands, Voorburg.

Chambers, R. L., T. Crespo, S. Laaksonen, P. Piela, P. Tsai, and T. de Waal (2001b), *The AUTIMP-project: Evaluation of WAID*. Report, Statistics Netherlands, Voorburg.

Daniels, J., M. J. Daniels, and J. W. Hogan (2008), *Missing Data in Longitudinal Studies*. Taylor & Francis, Philadelphia.

Deville, J. C., and C.-E. Särndal (1994), Variance Estimation for the Regression Imputed Horvitz-Thompson Estimator. *Journal of Official Statistics 10*, pp. 381–394.

De Waal, T. (1999), *A Brief Overview of Imputation Methods Applied at Statistics Netherlands*. Report, Statistics Netherlands, Voorburg.

De Waal, T. (2001), WAID 4.1: A Computer Program for Imputation of Missing Values. *Research in Official Statistics 4*, pp. 53–70.

Draper, N. R., and H. Smith (1998), *Applied Regression Analysis*. John Wiley & Sons, New York.

Durrant, G. B. (2005), *Imputation Methods for Handling Item-Nonresponse in the Social Sciences: A Methodological Review*. NCRM Methods Review Papers (NCRM/002), University of Southampton.

Durrant, G. B., and C. Skinner (2006), Using Missing Data Methods to Correct for Measurement Error in a Distribution Function. *Survey Methodology 32*, pp. 25–36.

Fay, R. E. (1996), Alternative Paradigms for the Analysis of Imputed Survey Data. *Journal of the American Statistical Association 91*, pp. 490–498.

Fetter, M. (2001), *Mass Imputation of Agricultural Economic Data Missing by Design: A Simulation Study of Two Regression Based Techniques*. Federal Conference on Survey Methodology.

Fine, T. L. (1999), *Feedforward Neural Network Methodology*. Springer-Verlag, New York.

Haslett, S., G. Jones, A. Noble and D. Ballas (2010), *More For Less? Using Statistical Modelling to Combine Existing Data Sources to Produce Sounder, More Detailed, and Less Expensive Official Statistics*. Official Statistics Research Series, 2010-1, Statistics New Zealand.

Haziza, D. (2003), The Generalized Simulation System (GENESIS). *Proceedings of the Section on Survey Research Methods, American Statistical Association.*

Haziza, D. (2006), Simulation Studies in the Presence of Nonresponse and Imputation. *The Imputation Bulletin 6*(1), pp. 7–19.

Heerschap, N., and A. Van der Graaf (1999), *A Test of Imputation by means of a Neural Network on Data of the Structure of Earnings Survey* (in Dutch). Report, Statistics Netherlands, Voorburg.

Israëls, A., and J. Pannekoek (1999), *Imputation of the Survey of Public Libraries* (in Dutch). Report, Statistics Netherlands, Voorburg.

Kalton, G. (1983), *Compensating for Missing Survey Data*. Survey Research Center Institute for Social Research, The University of Michigan.

Kalton, G., and D. Kasprzyk (1986), The Treatment of Missing Survey Data. *Survey Methodology 12*, pp. 1–16.

Kalton, G., and L. Kish (1984), Some Efficient Random Imputation Methods. *Communications in Statistics, Part A, Theory and Methods 13*, pp. 1919–1939.

Kaufman, S., and F. Scheuren (1997), Applying Mass Imputation Using the Schools and Staffing Survey Data. *Proceedings of the American Statistical Association*, pp. 129–134.

Kim, J. K., and W. Fuller (2004), Fractional Hot Deck Imputation. *Biometrica 91*, pp. 559–578.

Kim, J. K., and W. Fuller (2008), Parametric Fractional Imputation for Missing Data Analysis. *Proceedings of the Section on Survey Research Methods, Joint Statistical Meeting*, pp. 158–169.

Knottnerus, P. (2003), *Sample Survey Theory: Some Pythagorean Perspectives*. Springer-Verlag, New York.

Kooiman, P. (1998), *Mass Imputation: Why Not!?* (in Dutch). Research paper 8792-98-RSM, Statistics Netherlands, Voorburg.

Kovar, J., and P. Whitridge (1995), Imputation of Business Survey Data. In: *Business Survey Methods*, B.G. Cox, D.A. Binder, B.N. Chinnappa, A. Christianson, M.J. Colledge, and P.S. Kott, eds. John Wiley & Sons, New York, pp. 403–423.

Krotki, K., S. Black, and D. Creel (2005), Mass Imputation. *ASA Section of Survey Research Methods*, pp. 3266–3269.

Lee, H., E. Rancourt, and C.-E. Särndal (2002), Variance Estimation from Survey Data with Imputed Values. In: *Survey Non-Response*, R.M. Groves, D.A. Dillman, J.L. Eltinge, and R.J.A. Little, eds. John Wiley & Sons, Hoboken, NJ, pp. 315–328.

Little, R. J. A. (1988), Missing-Data Adjustments in Large Surveys. *Journal of Business & Economic Statistics 6*, pp. 287–296.

Little, R. J. A., and D. B. Rubin (2002), *Statistical Analysis with Missing Data*, second edition. John Wiley & Sons, Hoboken, NJ.

Longford, N. T. (2005), *Missing Data and Small-Area Estimation*. Springer-Verlag, New York.

McCullagh, P., and J. A. Nelder (1989), *Generalized Linear Models* (second edition). Chapman & Hall, London.

McKnight, P. E., K. M. McKnight, S. Sidani, and A. J. Figueredo (2007), *Missing Data—A Gentle Introduction*. Guilford Publications.

Molenberghs, G., and M. Kenward (2007), *Missing Data In Clinical Studies*. John Wiley & Sons, Hoboken, NJ.

Nagelkerke, N. J. D. (1991), A Note on a General Definition of the Coefficient of Determination. *Biometrika 78*, pp. 691–692.

Qin, Y., J. N. K. Rao, and Q. Ren (2008), Confidence Intervals for Marginal Parameters under Fractional Linear Regression Imputation for Missing Data. *Journal of Multivariate Analysis 99*, pp. 1232–1259.

Rao, J. N. K. (1996), On Variance Estimation with Imputed Survey Data. *Journal of the American Statistical Association 91*, pp. 499–506.

Rao, J. N. K., and J. Shao (1992), Jackknife Variance Estimation with Survey Data under Hot Deck Imputation. *Biometrika 79*, pp. 811–822.

Rao, J. N. K., and R. R. Sitter (1995), Variance Estimation Under Two-Phase Sampling with Application to Imputation for Missing Data. *Biometrika 82*, pp. 453–460.

Rubin, D. B. (1978), Multiple Imputations in Sample Surveys—A Phenomenological Bayesian Approach to Nonresponse, *Proceedings of the Section on Survey Research Methods, American Statistical Association.*

Rubin, D. B. (1987), *Multiple Imputation for Non-Response in Surveys*. John Wiley & Sons, New York.

Rubin, D. B. (1996), Multiple Imputation after 18+ Years. *Journal of the American Statistical Association 91*, pp. 473–489.

Rubin, D. B., and N. Schenker (1986), Multiple Imputation for Interval Estimation from Simple Random Samples with Ignorable Nonresponse. *Journal of the American Statistical Association 81*, pp. 366–374.

Sande, I. G. (1982), Imputation in Surveys: Coping with Reality. *The American Statistician 36*, pp. 145–152.

Särndal, C.-E. (1992), Methods for Estimating the Precision of Survey Estimates when Imputation Has Been Used. *Survey Methodology 18*, pp. 241–252.

Särndal, C.-E., and S. Lundström (2005), *Estimation in Surveys with Nonresponse*. John Wiley & Sons, New York.

Särndal, C.-E., B. Swensson, and J. H. Wretman (1992), *Model Assisted Survey Sampling*. Springer-Verlag, New York.

Schafer, J. L. (1997), *Analysis of Incomplete Multivariate Data*. Chapman & Hall, London.

Schulte Nordholt, E. (1996), *The Used Techniques for the Imputation of Wave 1 Data of the ECHP*. Research report doc. PAN 66/96, Eurostat, Luxembourg.

Schulte Nordholt, E. (1997), *Imputation in the New Dutch Structure of Earnings Survey (SES)*. Report, Statistics Netherlands, Voorburg.

Schulte Nordholt, E. (1998), Imputation: Methods, Simulation Experiments and Practical Examples. *International Statistical Review 66*, pp. 157–180.

Schulte Nordholt, E., and J. Hooft van Huijsduijnen (1997), The Treatment of Item Nonresponse During the Editing of Survey Results. In: *New Techniques and Technologies for Statistics II. Proceedings of the Second Bonn Seminar*, IOS Press, Amsterdam, pp. 55–61.

Shao, J. (2002), Replication Methods for Variance Estimation in Complex Surveys with Imputed Data. In: *Survey Non-Response*, Groves, R.M., Dillman, D.A., Eltinge, J.L., and Little, R.J.A., eds. John Wiley & Sons, Hoboken, NJ, New York, pp. 303–314.

Shao, J., and R. R. Sitter (1996), Bootstrap for Imputed Survey Data. *Journal of the American Statistical Association 91*, pp. 1278–1288.

Shlomo, N., T. de Waal, and J. Pannekoek (2009), *Mass Imputation for Building a Numerical Statistical Database*. Working Paper No. 31, UN/ECE Work Session on Statistical Data Editing, Neuchâtel.

Skinner, C. J., D. Holt, and T. M. F. Smith (1989), *Analysis of Complex Surveys*. Institute for Social Research, University of Michigan.

Sonquist, J. N., E. L. Baker, and J. A. Morgan (1971), *Searching for Structure*. John Wiley & Sons, New York.

SPSS (1998), *AnswerTree 2.0 User's Guide*. Chicago.

Statistics Canada (1998), *GEIS: Functional Description of the Generalized Edit and Imputation System*. Report, Statistics Canada.

Tsiatis, A. A. (2006), *Semiparametric Theory And Missing Data*. Springer-Verlag, New York.

Whitridge, P., M. Bureau, and J. Kovar (1990), Mass Imputation at Statistics Canada. *Proceedings of the Annual Research Conference*, U.S. Census Bureau, Washington, pp. 666–675.

Whitridge, P., and J. Kovar (1990), Use of Mass Imputation to Estimate for Subsample Variables. *Proceedings of the Business and Economic Statistics Section, American Statistical Association*, pp. 132–137.

Wolter, K. M. (1985), *Introduction to Variance Estimation*. Springer-Verlag, New York.

CHAPTER EIGHT

Multivariate Imputation

8.1 Introduction

The imputation of missing values was treated in Chapter 7 on a variable-by-variable basis. In this approach an imputation is either the predicted conditional mean of the target variable, given (some of the) observed values, or a draw from the conditional distribution of the target variable. When there are multiple variables with missing values for a unit, this single-variable imputation approach can be applied to each of these variables separately. Often, however, a better alternative in such cases is to model the simultaneous distribution of the missing variables and to impute with the conditional multivariate mean of the missing variables or a draw from this distribution. This simultaneous or multivariate approach is the subject of this chapter.

In many applications where there are a number of variables with missing values that require imputation, some variables with missing values are used as predictors in the imputation models for others. In such situations, the single-variable imputation methods have some drawbacks, as will be discussed below, that may be alleviated by the use of multivariate imputation methods.

The structure of a four-variable data set with missing values is illustrated in Table 8.1. The columns of this table represent the four variables, denoted by z, x_1, x_2, and x_3. The rows represent observations with missing values on the same variables, and missing values are indicated by M and observed values by O. The pattern of missing and observed data in each of these rows is called a missing data pattern. In this example, one variable, z, is observed for all units. Variables for which it is certain that they are always observed are variables obtained from the sampling frame and other variables from administrative data that can be linked to the sampling frame. The other three variables in Table 8.1 do have missing

Handbook of Statistical Data Editing and Imputation, by T. de Waal, J. Pannekoek, and S. Scholtus

TABLE 8.1 Missing Data Patterns

z	x_1	x_2	x_3	Pattern
O	M	M	M	A
O	M	M	O	B
O	M	O	M	C
O	O	M	M	D
O	M	O	O	E
O	O	M	O	F
O	O	O	M	G
O	O	O	O	H

values for some units, and the eight different missing data patterns are all possible missing data configurations that exist in this situation.

A single-variable imputation method, such as the regression or logistic regression imputation models discussed in Chapter 7, uses a model for the missing variable in a unit with the observed variables in that unit as predictors. For instance, for imputing x_1 in units with data pattern A, x_1 can be imputed by a model with only z as predictor. In data pattern B, x_1 can be imputed by using z and x_3 as predictors; and in data pattern E, x_1 can be imputed by using all other three variables as predictors. The information to estimate these models becomes more limited as the number of predictors with missing values increases. For instance, the model with only z as predictor can be estimated by using all units with x_1 observed (patterns D, F, G, and H), but the model with z, x_2, and x_3 as predictors can only be estimated using the units with missing data pattern H. This can easily lead to a lack of observations to estimate the model reliably, in which case a simpler model must be chosen that excludes some of the available predictors. The multivariate imputation methods discussed in this chapter also use models that predict the missings given the observed values in a unit, but to estimate these models these methods effectively use the available observed data for all units instead of only the data with all predictors and the target variable observed.

A second important advantage of multivariate imputation methods is that these methods can reproduce correlations between variables much better than single-variable imputation methods. Single-variable imputation methods can, to some extent, preserve the relation between the target variable and the predictor variables since that relation is part of the model, but the relation between a target variable and another variable that is missing and thus not included in the model will not be preserved. Multivariate imputation methods generate imputations from an estimated simultaneous distribution of the variables with missing values that will reflect the correlations between these variables and are therefore intended to produce imputations that preserve these correlations.

A last advantage that should be mentioned is that, in contrast to single-variable imputation methods, appropriate multivariate imputation methods can automatically take some edit constraints into account. For instance, multivariate methods for continuous data based on the multivariate normal distribution respect linear equality constraints such as balance edits as long as the observed data conform to these edits. This particular case and a number of other multivariate models that take edit constraints into account are treated in Chapter 9, which is devoted to imputation under edit constraints. For categorical data, edits that prescribe that certain value combinations cannot occur will be satisfied if the variables involved are imputed simultaneously, and the interaction between these variables is accounted for by the model. For instance, if persons with age category less than 14 years cannot be married, the estimated probability distribution over the combinations of categories of age and marital status will assign zero probability to the combination of values (*age* = less than 14; *marital status* = married) as long as there are no observed data with this combination of values. Consequently, no imputed values will fall into this category combination.

Multivariate imputation is not confined to imputation based on explicit models: Hot-deck approaches extend in an almost trivial way to multivariate situations. With multiple missing values in the unit to be imputed (the recipient), a multivariate hot-deck method simply imputes all missing values from the same donor. This can be seen as a draw from the empirical multivariate distribution of the missing values. Edit rules that involve only imputed variables will be satisfied, assuming that the donor satisfies the edit rules, but edit rules that contain both imputed and observed variables may not be satisfied because the values for these variables come from different units. Donor imputation that respects also the last type of edit rules is discussed in Section 4.5.

This chapter provides an introduction to imputation based on multivariate models for continuous and for categorical data and some background on the estimation of such models in the presence of missing data. Other, more extensive accounts of this methodology are presented in, for example, Little and Rubin (2002) and Schafer (1997). An alternative to finding an appropriate multivariate model is to model the distribution of each variable separately, conditional on the (appropriately imputed) other variables. This so-called sequential imputation approach has advantages for data with variables of different types (categorical, continuous, semi-continuous) and data subject to constraints and is discussed in Chapter 9.

The remainder of this chapter begins by discussing the two multivariate models that are most often applied for imputation purposes, the multivariate normal model for continuous data and the multinomial model for categorical data (Section 8.2). Both models and the imputations derived from them depend on unknown parameters that must be estimated from the incomplete data. The maximum likelihood approach to this estimation problem is the subject of Section 8.3, which includes a discussion of the famous Expectation–Maximization algorithm (EM algorithm) which was presented in its general form by Dempster, Laird, and Rubin (1977) and is widely applied to estimation problems with incomplete data. The final section (8.4) of this chapter discusses an application of a multivariate imputation approach to data from a survey among public libraries.

8.2 Multivariate Imputation Models

8.2.1 IMPUTATION OF MULTIVARIATE CONTINUOUS DATA

Regression imputation, as described in Chapter 7, replaces a missing value of unit i on target variable x_t with a prediction of that value based on the model

$$x_{i,t} = \mathbf{x}_{i,p}^T \boldsymbol{\beta} + \varepsilon_{i,t}, \tag{8.1}$$

with $\mathbf{x}_{i,p}$ the vector with the values of the predictor variables for unit i, $\boldsymbol{\beta}$ the vector with regression parameters, and $\varepsilon_{i,t}$ a random residual with zero expectation. Here we use a notation that is slightly different from the notation used in Chapter 7 (where y was used to represent the target variable) but more convenient for generalization to the multivariate case.

By subtracting the means from both sides of (8.1), this model can also be expressed as

$$x_{i,t} - \mu_t = (\mathbf{x}_{i,p} - \boldsymbol{\mu}_p)^T \boldsymbol{\beta} + \varepsilon_{i,t} \tag{8.2}$$

with μ_t the mean of $x_{i,t}$ and $\boldsymbol{\mu}_p$ the mean vector of the predictor variables. For complete data, the least squares estimator of $\boldsymbol{\beta}$ can now be written as

$$\hat{\boldsymbol{\beta}} = \left[\sum_i (\mathbf{x}_{i,p} - \boldsymbol{\mu}_p)(\mathbf{x}_{i,p} - \boldsymbol{\mu}_p)^T \right]^{-1} \sum_i (\mathbf{x}_{i,p} - \boldsymbol{\mu}_p)(x_{i,t} - \mu_t)$$

$$= \boldsymbol{\Sigma}_{p,p}^{-1} \boldsymbol{\Sigma}_{p,t}, \tag{8.3}$$

with $\boldsymbol{\Sigma}_{p,p}$ the sample covariance matrix of the predictor variables and $\boldsymbol{\Sigma}_{p,t}$ the vector with sample covariances between the predictor variables and the target variable.

Model (8.2) can be extended to the case where there is a vector of target variables and is then written as

$$\mathbf{x}_{i,t} = \boldsymbol{\mu}_t + \mathbf{B}_{t,p}(\mathbf{x}_{i,p} - \boldsymbol{\mu}_p) + \boldsymbol{\varepsilon}_{i,t}, \tag{8.4}$$

with $\mathbf{x}_{i,t}$ the vector of target variable values for unit i, $\boldsymbol{\mu}_t$ the mean vector of the target variables, $\mathbf{B}_{t,p}$ the matrix with regression coefficients with its number of rows equal to the number of target variables and its number of columns equal to the number of predictor variables, and $\boldsymbol{\varepsilon}_{i,t}$ a vector of random disturbances with expectation zero. This notation neglects the fact that the coefficient matrix depends on i in the sense that the predictor variables and variables to be predicted depend on the missing data pattern; but whenever this can lead to confusion, we will make the dependence on the missing data pattern explicit. The relatively compact presentation of this model is possible because it is assumed here that for all target variables the same predictors are used. Models where this is not the case are called Seemingly Unrelated Regression (SUR) models and have a more complicated structure [see, e.g., Amemiya (1985)].

The least squares estimator for $\mathbf{B}_{t,p}$ is built from the same components as in the univariate case (8.3):

$$\hat{\mathbf{B}}_{t,p} = \mathbf{\Sigma}_{t,p}\mathbf{\Sigma}_{p,p}^{-1}, \tag{8.5}$$

but, in this case, $\mathbf{\Sigma}_{t,p}$ is a matrix with the covariances of the predictor variables with the target variables instead of a vector.

To use model (8.4) for imputation purposes, we need the statistics $\boldsymbol{\mu}_t$, $\mathbf{\Sigma}_{p,p}$, and $\mathbf{\Sigma}_{t,p}$, or estimates thereof. As noted before, the variables in the mean vector and covariance matrices will vary between missing data patterns, but in any case these parameters can be obtained as partitions of the mean vector $\boldsymbol{\mu}$, say, and covariance matrix $\mathbf{\Sigma}$, say, of all variables. In the presence of missing data, we could base estimates of these parameters upon the fully observed records, but the number of units without any missing value can be rather small, especially in cases with many variables. And it is especially in these cases that the number of parameters in the covariance matrix is large and the need for a large number of observations to estimate this matrix is greatest. In the next section we describe a more satisfactory approach to estimating the mean vector and covariance matrix. For the moment we assume that we have somehow obtained satisfactory estimates of these parameters. Using these estimates, regression imputations for the missing variables in a record i without a residual can be obtained by the predicted mean

$$\hat{\mathbf{x}}_{i,t} = \hat{\boldsymbol{\mu}}_t + \hat{\mathbf{B}}_{t,p}(\mathbf{x}_{i,p} - \hat{\boldsymbol{\mu}}_p), \tag{8.6}$$

and an imputation with a random residual vector added can be obtained by

$$\hat{\mathbf{x}}_{i,t} = \hat{\boldsymbol{\mu}}_t + \hat{\mathbf{B}}_{t,p}(\mathbf{x}_{i,p} - \hat{\boldsymbol{\mu}}_p) + \mathbf{e}_{i,t}, \tag{8.7}$$

where $\mathbf{e}_{i,t}$ is a random draw from the distribution of the residuals of the regression of $\mathbf{x}_t$ on $\mathbf{x}_p$. If it is assumed that the joint distribution of all variables is multivariate normal, then the conditional distribution of $\mathbf{x}_t$ given $\mathbf{x}_p$ is also multivariate normal with mean vector given by (8.6) and covariance matrix $\mathbf{\Sigma}_{t,t.p}$, say. The residuals are then multivariate normally distributed with mean vector zero and covariance matrix equal to the covariance matrix of the conditional distribution. By properties of the multivariate normal distribution, an estimate of this conditional or residual covariance matrix can be obtained from an estimate of the covariance matrix $\hat{\mathbf{\Sigma}}$ of all variables according to

$$\hat{\mathbf{\Sigma}}_{t,t.p} = \hat{\mathbf{\Sigma}}_{t,t} - \hat{\mathbf{\Sigma}}_{t,p}\hat{\mathbf{\Sigma}}_{p,p}^{-1}\hat{\mathbf{\Sigma}}_{p,t}, \tag{8.8}$$

where the matrices on the right-hand side can all be extracted from $\hat{\mathbf{\Sigma}}$. Adding such residuals should help not only in preserving the variances of the imputed variables but also in preserving the correlations among imputed variables. The estimation of $\hat{\mathbf{\Sigma}}$ in the presence of missing data will be discussed in Section 8.3.2.

8.2.2 IMPUTATION OF MULTIVARIATE CATEGORICAL DATA

Representations of Multivariate Categorical Data. A binary variable has only two possible values and these values can be coded, arbitrarily, as 1 and 0. The observations on a multi-category variable x with K categories can be coded with K dummy variables x_k, each taking values 0 or 1 and $\sum_k x_k = 1$. The probability of an observation in category k of x is the probability that $x_k = 1$ and will be denoted by π_k. When there are multiple categorical variables, the outcomes can be represented by sets of dummy variables representing the categories of each of them. For instance, if we have three variables $x^{(1)}$, $x^{(2)}$, and $x^{(3)}$, with categories denoted by $k = 1, \ldots, K$, $l = 1, \ldots, L$ and $m = 1, \ldots, M$ respectively, K, L, and M dummy variables will be created. An observation in category k of $x^{(1)}$, l of $x^{(2)}$, and m of $x^{(3)}$ will then correspond to the event $x_k^{(1)} = x_l^{(2)} = x_m^{(3)} = 1$. The probability of this event will be denoted by π_{klm}. The probabilities π_{klm} for all k,l,m together represent the probability distribution over all possible joint outcomes of the three variables.

Observations on multiple categorical variables can also be represented in a multi-dimensional cross-classification. Notation for the fully general case is somewhat cumbersome, and we therefore use a three-dimensional example here to convey the idea. Generalization to more dimensions (variables) will then be straightforward. For three variables with numbers of categories K, L, and M, the possible outcomes can be arranged in a three-dimensional cross-classification with K rows, L columns and M layers. Each of the $K \times L \times M$ cells in this three-dimensional array corresponds to one of the possible joint outcomes of the three variables, and each unit can be classified into one and only one of the cells. If we denote the cell corresponding to $x^{(1)} = k$, $x^{(2)} = l$, and $x^{(3)} = m$ by klm, the probability that a unit falls in a cell klm is the cell probability π_{klm}. The marginal bivariate probabilities corresponding to the category combinations of $x^{(1)}$ and $x^{(2)}$ can be obtained as $\pi_{kl} = \sum_m \pi_{klm}$ for all k and l, and the other bivariate marginal probabilities can similarly be obtained by summation over the categories of the third variable. The univariate marginal probabilities are denoted by π_k, π_l, and π_m and can be expressed as sums over two variables of the trivariate probabilities or sums over one variable of the bivariate probabilities.

For a sample of n units the number of units falling into each cell of the cross-classification can be counted. The cross-classification with counts in the cells is called a contingency table. When the cell probabilities are equal for each unit and the scores on the variables are independent between units, the counts in the cross-classification are multinomially distributed. For the three-variable example the multinomial distribution gives the probability of observing the realized vector of cell counts $\mathbf{n} = (n_{111}, n_{112}, \ldots, n_{KLM})$. For general multi-way tables it is more convenient to represent the cell counts (and probabilities) by a single index as $\mathbf{n} = (n_1, \ldots, n_j, \ldots, n_C)$, with C the number of cells. In this notation, the multinomial distribution can be written as

$$P(\mathbf{n}|\boldsymbol{\pi}) = \frac{n!}{\prod_{j=1}^{C} n_j!} \prod_{j=1}^{C} \pi_j^{n_j}, \tag{8.9}$$

with $\boldsymbol{\pi} = (\pi_1, \ldots, \pi_C)$ the vector with cell probabilities. An important property of the multinomial distribution, which is used by the imputation methods based on this distribution, is that the conditional distribution over a subset of the cells is also a multinomial distribution. Suppose that it is given that n_s units fall into a subset S of the cells and that $\boldsymbol{\pi}_s$ is the part of the vector $\boldsymbol{\pi}$ corresponding to these cells, then the distribution of the counts in the C_s cells belonging to S is multinomial with parameter $\boldsymbol{\pi}_s^* = \boldsymbol{\pi}_s / \sum_{j \in S} \pi_j$, with the subscript s denoting the part of the parameter vector corresponding to the cells in S.

Representations of Missing Data on Categorical Variables. A unit for which some of the variables are missing cannot be classified in one of the category combinations or cells defined by all variables considered, but it can be partially classified according to the observed variables. For instance, if the third variable is missing in the three variable example, the unit can be classified in the bivariate $K \times L$ marginal table, but not in the full three-dimensional table. More generally, all units with the same observed variables can be classified into a contingency table corresponding to those variables. Such tables are different for each missing data pattern and are all marginal tables of the complete data contingency table. For instance, for the three-variable case, units with one variable missing can be classified in one of the bivariate marginal tables denoted as $x^{(1)} \times x^{(2)}$, $x^{(1)} \times x^{(3)}$, and $x^{(2)} \times x^{(3)}$, and units with two variables missing can be classified in one of the univariate marginal tables corresponding to $x^{(1)}$, $x^{(2)}$, and $x^{(3)}$. The data that are summarized in such a set of contingency tables are referred to as a contingency table with supplemental margins (Little and Rubin, 2002).

A simple example of such data is given in Table 8.2. The data are fictitious but similar to data from a survey on movement behavior. The example concerns data on two variables: Possession of a car (*Car-owner*) and Possession of a driving license (*License*), obtained from 90 individuals of 18 years and older. For 60 of these 90 persons, responses on both variables are obtained, for 20 persons only the response on *Car-owner* is obtained, and 10 persons only responded on the variable *License*. For this example it is assumed that the bivariate cell probabilities given in the last panel of Table 8.2 are known. In general, this is of course not the case, but estimates for these probabilities can be obtained from the incomplete data. This estimation problem is discussed in Section 8.3.

Imputation for Multinomial Variables. Suppose that we want to impute the missing value of *Car Owner* for the 5 respondents with score Yes on *License*. For this problem we consider the conditional distribution of the missing variable, given the observed one. From the cell probabilities we see that the conditional distribution over the categories of *Car Owner*, given *License* = Yes, is $\{0.588/0.743, 0.155/0.743\} = \{0.79, 0.21\}$. More formally, if we index the rows of the table with k and the columns with l, this conditional distribution can be written as $\pi_{k|l=1} = \pi_{k1} / \sum_k \pi_{k1}$. Using the single index notation for the four cells ($j = 1, \ldots, 4$) corresponding to $k, l = (1,1)(1,2)(2,1)(2,2)$, the conditional distribution for the units with *License* = Yes is the distribution over the two cells corresponding to $j = 1$ and $j = 3$. The expected values for the

TABLE 8.2 Two-by-Two Table with Supplemental Margins

Fully Observed Data

	License	
Car Owner	1. Yes	2. No
1. Yes	40	0
2. No	10	10

Partially Observed Data

Car Owner		*License*	
1. Yes	10	1. Yes	5
2. No	10	2. No	5

Cell Probabilities

	License	
Car Owner	1. Yes	2. No
1. Yes	0.588	0
2. No	0.155	0.257

dummy variables corresponding to the four cells, given that *License* = Yes, are equal to the conditional probabilities 0.79 and 0.21 for x_1 and x_3, respectively, and zero for the other cells.

These conditional expected values can be used as imputations, just as conditional expected values are used as imputations for continuous variables. But contrary to the conditional expectations for continuous variables, the conditional expectations for dummy variables are obviously different from real observations; they are not equal to zero or one and there are multiple dummies with nonzero values. However, for the purpose of estimating the cell totals in a contingency table, this need not be a problem. An alternative is, just as with continuous variables, to draw an imputed value from the conditional distribution. In this case, this entails to draw from the multinomial distribution over the two cells with nonzero conditional probability and with $n = 1$. This will result in a count of 1 in one of these two cells, along with a value of 1 for the corresponding dummy variable.

In the general case, we consider, for each cell in a supplemental margin, the conditional distribution over the cells that contribute to the total in that marginal cell. Imputation of the missing information can then be done by setting the dummies corresponding to the contributing cells equal to their expected values or to draw from the conditional multinomial distribution over these cells.

The imputation methods described above will result in a completed data set at the unit level. In this respect, these approaches are similar to the approaches for continuous data. For categorical data, however, there is the alternative of imputing at the aggregate level. In the aggregated approach, no values for dummy variables are imputed, but a completed full-dimensional contingency table is created in which all units are classified. This is accomplished by distributing each cell total in each supplemental margin over the cells in the full table that contribute to that cell. In the example above, the five units with score Yes on *License* can be distributed over the cells with $j = 1$ and $j = 3$ by using a draw from a multinomial distribution with $n = 5$ and the same probabilities as before.

8.3 Maximum Likelihood Estimation in the Presence of Missing Data

The EM algorithm is a general iterative computational procedure designed to obtain maximum likelihood estimates of parameters from incomplete data. In this section we give a short introduction of this algorithm and its application to two standard imputation approaches for continuous and discrete variables. Since the algorithm is primarily designed as a way to maximize the likelihood function in the presence of missing data, we first outline some basic concepts of likelihood theory. More on likelihood theory can be found in many standard texts on mathematical statistics, including, for example, Cox and Hinkley (1974), Wilks (1962), and Stuart and Ord (1991).

8.3.1 MAXIMUM LIKELIHOOD ESTIMATION WITH COMPLETE DATA

Maximum likelihood estimation assumes that a statistical model has been specified that describes the stochastic process that generated the data. In the multivariate case the data consist of n observations $\mathbf{x}_i (i = 1, \ldots, n)$ on J variables each. And, as a possible statistical model for such data, it may be assumed that these data can be described as n independent realizations of a random vector $\mathbf{x}$ that is multivariate normally distributed with mean vector $\boldsymbol{\mu}$ and covariance matrix $\boldsymbol{\Sigma}$. More generally, a statistical model specifies the distribution or joint probability density function of the n observations, up to an unknown vector of parameters $\boldsymbol{\theta}$. If it is assumed that the n observations are independently and identically distributed (iid) with density function $f(\mathbf{x}|\boldsymbol{\theta})$, then the joint density is given by

$$f(\mathbf{X}|\boldsymbol{\theta}) = \prod_{i=1}^{n} f(\mathbf{x}_i|\boldsymbol{\theta}). \tag{8.10}$$

The likelihood function is similar to $f(\mathbf{X}|\boldsymbol{\theta})$ but with the role of $\mathbf{X}$ and $\boldsymbol{\theta}$ reversed; while $f(\mathbf{X}|\boldsymbol{\theta})$ describes the probability density of varying values of $\mathbf{X}$

for a given value of $\boldsymbol{\theta}$, the likelihood function describes the probability density of one value of $\mathbf{X}$, the actually observed one, for varying values of $\boldsymbol{\theta}$. To express this, the likelihood function is defined as $l(\boldsymbol{\theta}|\mathbf{X}) = f(\mathbf{X}|\boldsymbol{\theta})$. The maximum likelihood estimator $\hat{\boldsymbol{\theta}}$ of $\boldsymbol{\theta}$ is defined as the value of $\boldsymbol{\theta}$ that maximizes the likelihood function over the parameter space Ω of $\boldsymbol{\theta}$. Maximizing the likelihood function is equivalent to maximizing the logarithm of the likelihood function, which is often more convenient, and so we can write

$$\hat{\boldsymbol{\theta}} = \arg\max_{\boldsymbol{\theta}\in\Omega} L(\boldsymbol{\theta}|\mathbf{X}) = \arg\max_{\boldsymbol{\theta}\in\Omega} \sum_{i=1}^{n} L(\boldsymbol{\theta}|\mathbf{x}_i), \tag{8.11}$$

with $L(\boldsymbol{\theta}|\mathbf{x}_i) = \ln l(\boldsymbol{\theta}|\mathbf{x}_i)$.

Generally, this maximization is carried out by setting the first-order derivatives with respect to the parameters to zero. The resulting equations are called the likelihood equations. If the likelihood equations cannot be solved in closed form, an iterative procedure is used for this purpose. Typically, Newton–Raphson or Fisher scoring algorithms are applied for maximizing loglikelihood functions.

8.3.2 MAXIMUM LIKELIHOOD ESTIMATION WITH INCOMPLETE DATA

When the data set $\mathbf{X}$ is incomplete, the missing values can be specified by a missing value indicator matrix $\mathbf{M}$ with the same dimensions as $\mathbf{X}$ and with elements $m_{ij} = 1$ if x_{ij} is observed and $m_{ij} = 0$ if x_{ij} is missing. The value of a row of $\mathbf{M}$, $\mathbf{m}_i$ say, identifies the pattern of observed and missing values for all variables and is called a response pattern. An observation $\mathbf{x}_i$ can, after a permutation of its elements, be partitioned as $\mathbf{x}_i = (\mathbf{x}_{i,\text{obs}}^T, \mathbf{x}_{i,\text{mis}}^T)^T$ with $\mathbf{x}_{i,\text{obs}}$ the vector with observed values and $\mathbf{x}_{i,\text{mis}}$ the vector with missing values.

In the presence of missing data, the data that are observed are $\mathbf{x}_{i,\text{obs}}$ and $\mathbf{m}_i$. The process that generated the missing data is called the missing data mechanism. Generally, it is assumed that the observed $\mathbf{m}_i$ are realizations of a random variable with a distribution referred to as the distribution of the missing data mechanism [see Rubin (1987) and Little and Rubin (2002)]. To proceed with likelihood inference, we now have to specify a statistical model that specifies the joint distribution of the observed data up to some unknown parameters. To build such a model, we first write down the joint density of $\mathbf{x}_i$ and $\mathbf{m}_i$ and factorize this density by using standard rules for conditional probabilities in the following way:

$$f(\mathbf{x}_{i,\text{obs}}, \mathbf{x}_{i,\text{mis}}, \mathbf{m}_i|\boldsymbol{\theta}, \boldsymbol{\phi}) = f(\mathbf{x}_{i,\text{obs}}, \mathbf{x}_{i,\text{mis}}|\boldsymbol{\theta}) f(\mathbf{m}_i|\mathbf{x}_{i,\text{obs}}, \mathbf{x}_{i,\text{mis}}, \boldsymbol{\phi}). \tag{8.12}$$

The first factor on the right-hand side specifies the density of $\mathbf{x}_i$ in the absence of missing data, and the second factor specifies the distribution of the missing data indicator. The distribution of the missing data depends on an unknown parameter vector $\boldsymbol{\phi}$ and, in general, also on both the observed and missing x values. The density of the observations $(\mathbf{x}_{i,\text{obs}}, \mathbf{m}_i)$ can now be obtained by integrating (8.12) over the distribution of the missing values, thus arriving at

$$(8.13) \quad f(\mathbf{x}_{i,\text{obs}}, \mathbf{m}_i|\boldsymbol{\theta}, \boldsymbol{\phi}) = \int f(\mathbf{x}_{i,\text{obs}}, \mathbf{x}_{i,\text{mis}}|\boldsymbol{\theta}) f(\mathbf{m}_i|\mathbf{x}_{i,\text{obs}}, \mathbf{x}_{i,\text{mis}}, \boldsymbol{\phi}) d\mathbf{x}_{i,\text{mis}}.$$

This expression still depends on the unknown missing values, and likelihood inference cannot proceed without further assumptions. Typically, assumptions are made about the dependence of the missing data mechanism on the x values. With respect to this dependence, the following three cases are distinguished (see also Section 1.3):

1. The missing data mechanism does not depend on the x values at all; this mechanism is called Missing Completely at Random (MCAR).
2. The missing data mechanism does depend on the observed x values but not on the missing x values; this is called Missing at Random (MAR).
3. The missing data mechanism depends on both the missing and the observed x values; this is called Not Missing at Random (NMAR).

Clearly, assumption 1 is a stronger one than 2, which is in turn stronger than 3. The missing data mechanisms corresponding to these assumptions are extensively discussed by Little and Rubin (2002) and Rubin (1976), who also coined the MCAR, MAR, NMAR terminology. The most important and widely used assumption is MAR, since it results in practical methods for inference without having to assume too much. The MAR assumption allows us to re-express (8.13) as

$$\begin{aligned}(8.14) \quad f(\mathbf{x}_{i,\text{obs}}, \mathbf{m}_i|\boldsymbol{\theta}, \boldsymbol{\phi}) &= \int f(\mathbf{x}_{i,\text{obs}}, \mathbf{x}_{i,\text{mis}}|\boldsymbol{\theta}) f(\mathbf{m}_i|\mathbf{x}_{i,\text{obs}}, \boldsymbol{\phi}) d\mathbf{x}_{i,\text{mis}} \\ &= f(\mathbf{m}_i|\mathbf{x}_{i,\text{obs}}, \boldsymbol{\phi}) \int f(\mathbf{x}_{i,\text{obs}}, \mathbf{x}_{i,\text{mis}}|\boldsymbol{\theta}) d\mathbf{x}_{i,\text{mis}} \\ &= f(\mathbf{m}_i|\mathbf{x}_{i,\text{obs}}, \boldsymbol{\phi}) f(\mathbf{x}_{i,\text{obs}}|\boldsymbol{\theta}).\end{aligned}$$

The joint distribution of the observations $(\mathbf{x}_{i,\text{obs}}, \mathbf{m}_i)$ now factorizes in two factors, the first one depending on $\boldsymbol{\phi}$ and the second one depending on $\boldsymbol{\theta}$. If these parameters are distinct in the sense that they are not functionally related, the missing data mechanism is said to be ignorable, and Rubin (1976) showed that inference on $\boldsymbol{\theta}$ can be based on the second factor only—that is, via the observed data likelihood or loglikelihood

$$L(\boldsymbol{\theta}|\mathbf{X}_{\text{obs}}) = \sum_i \ln f(\mathbf{x}_{i,\text{obs}}|\boldsymbol{\theta}).$$

The importance of this result is that, without the need of specifying the missing data mechanism, the observed data likelihood can provide valid inference on $\boldsymbol{\theta}$ not only if the missing mechanism is purely random (MCAR), but also if this mechanism depends on the x values as long as this dependence is limited to the observed x values (MAR).

8.3.3 THE EM ALGORITHM

Even after the simplification caused by assuming MAR and despite its concise appearance, the observed data likelihood $L(\boldsymbol{\theta}|\mathbf{X}_{\text{obs}})$ is still a complicated function, in general. This is due to the fact that the $\mathbf{x}_{i,\text{obs}}$ are vectors of varying length containing observations on different sets of observed variables x_j which makes the observed data loglikelihood much more complex than the corresponding complete data loglikelihood. Consequently, the derivatives of the likelihood or loglikelihood needed to maximize this function are complicated functions as well. The Expectation–Maximization (EM) algorithm is an iterative algorithm that circumvents these difficulties by filling in expected values for the functions of the missing values that appear in the likelihood in one step (the expectation step or E-step) and maximizing the in this way completed likelihood in the other step (the maximization step or M-step) and iterating these two steps.

To describe the EM algorithm, we first express the density of a data vector as

$$
\begin{aligned}
f(\mathbf{x}_i|\boldsymbol{\theta}) &= f(\mathbf{x}_{i,\text{obs}}, \mathbf{x}_{i,\text{mis}}|\boldsymbol{\theta}) \\
&= f(\mathbf{x}_{i,\text{obs}}|\boldsymbol{\theta}) f(\mathbf{x}_{i,\text{mis}}|\mathbf{x}_{i,\text{obs}}, \boldsymbol{\theta}).
\end{aligned} \tag{8.15}
$$

The contribution of unit i to the loglikelihood can then be written as

$$
L(\boldsymbol{\theta}|\mathbf{x}_i) = \ln f(\mathbf{x}_{i,\text{obs}}|\boldsymbol{\theta}) + \ln f(\mathbf{x}_{i,\text{mis}}|\mathbf{x}_{i,\text{obs}}, \boldsymbol{\theta}), \tag{8.16}
$$

and, by defining $\ln f(\mathbf{X}_{\text{mis}}|\mathbf{X}_{\text{obs}}, \boldsymbol{\theta}) = \sum_i \ln f(\mathbf{x}_{i,\text{mis}}|\mathbf{x}_{i,\text{obs}}, \boldsymbol{\theta})$ the loglikelihood, can be written as

$$
L(\boldsymbol{\theta}|\mathbf{X}) = L(\boldsymbol{\theta}|\mathbf{X}_{\text{obs}}) + \ln f(\mathbf{X}_{\text{mis}}|\mathbf{X}_{\text{obs}}, \boldsymbol{\theta}). \tag{8.17}
$$

This complete data loglikelihood cannot be evaluated since $L(\boldsymbol{\theta}|\mathbf{X})$ and $\ln f(\mathbf{X}_{\text{mis}}|\mathbf{X}_{\text{obs}}, \boldsymbol{\theta})$ depend on the unobserved data. However, the expectation over the missing data of this loglikelihood is a function that can be maximized, and this maximization is used by the EM algorithm as a device for maximizing the observed data loglikelihood. By taking expectations of the terms that involve the missing data, we obtain

$$
\begin{aligned}
&\int f(\mathbf{X}_{\text{mis}}|\mathbf{X}_{\text{obs}}, \boldsymbol{\theta}) L(\boldsymbol{\theta}|\mathbf{X}) d\mathbf{X}_{\text{mis}} \\
&\qquad = L(\boldsymbol{\theta}|\mathbf{X}_{\text{obs}}) + \int f(\mathbf{X}_{\text{mis}}|\mathbf{X}_{\text{obs}}, \boldsymbol{\theta}) \ln f(\mathbf{X}_{\text{mis}}|\mathbf{X}_{\text{obs}}, \boldsymbol{\theta}) d\mathbf{X}_{\text{mis}},
\end{aligned}
$$

say,

$$
Q(\boldsymbol{\theta}) = L(\boldsymbol{\theta}|\mathbf{X}_{\text{obs}}) + H(\boldsymbol{\theta}). \tag{8.18}
$$

The expectations in (8.18) are taken with respect to the density of the missing data given $\mathbf{X}_{\text{obs}}$ and $\boldsymbol{\theta}$. In the EM iterations $\boldsymbol{\theta}^{(t)}$, the current estimate of $\boldsymbol{\theta}$ is taken as the value of this given $\boldsymbol{\theta}$ and, following Little and Rubin (2002), this can be expressed by writing

$$Q(\boldsymbol{\theta}|\boldsymbol{\theta}^{(t)}) = L(\boldsymbol{\theta}|\mathbf{X}_{\text{obs}}) + H(\boldsymbol{\theta}|\boldsymbol{\theta}^{(t)}) \qquad \text{(E-step).} \tag{8.19}$$

The calculation of the expectation of the complete data loglikelihood $L(\boldsymbol{\theta}|\mathbf{X})$ is the E-step of the EM algorithm. The next step, the M-step, is the maximization of this expected loglikelihood, that is,

$$\boldsymbol{\theta}^{(t+1)} = \arg\max_{\boldsymbol{\theta}} Q(\boldsymbol{\theta}|\boldsymbol{\theta}^{(t)}) \qquad \text{(M-step)} \tag{8.20}$$

A key result concerning the convergence of the EM algorithm is that a sequence of E- and M-steps has the property that $L(\boldsymbol{\theta}^{(t+1)}|\mathbf{X}_{\text{obs}}) \geq L(\boldsymbol{\theta}^{(t)}|\mathbf{X}_{\text{obs}})$; that is, the observed data loglikelihood increases in each iteration. This result was proved by Dempster, Laird, and Rubin (1977). To see this, we write the difference between the observed data loglikelihood at two consecutive iterations as

$$\begin{aligned} L(\boldsymbol{\theta}^{(t+1)}|\mathbf{X}_{\text{obs}}) - L(\boldsymbol{\theta}^{(t)}|\mathbf{X}_{\text{obs}}) &= Q(\boldsymbol{\theta}^{(t+1)}|\boldsymbol{\theta}^{(t)}) - Q(\boldsymbol{\theta}^{(t)}|\boldsymbol{\theta}^{(t)}) \\ &\quad + H(\boldsymbol{\theta}^{(t)}|\boldsymbol{\theta}^{(t)}) - H(\boldsymbol{\theta}^{(t+1)}|\boldsymbol{\theta}^{(t)}). \end{aligned} \tag{8.21}$$

The first difference on the right-hand side of (8.21) is nonnegative because of the maximization step. That the second difference is also nonnegative can be shown by using the following form of Jensen's inequality: For a random variable y with probability density $p(y)$ and a convex function φ, it holds that

$$\varphi\left(\int p(y)g(y)\,dy\right) \leq \int p(y)\varphi(g(y))dy,$$

with the inequality sign reversed for concave φ. To apply this inequality, we first expand the second difference in (8.21), by using (8.18), as

$$\begin{aligned} &H(\boldsymbol{\theta}^{(t)}|\boldsymbol{\theta}^{(t)}) - H(\boldsymbol{\theta}^{(t+1)}|\boldsymbol{\theta}^{(t)}) \\ &\quad = \int -f(\mathbf{X}_{\text{mis}}|\mathbf{X}_{\text{obs}}, \boldsymbol{\theta}^{(t)}) \ln \frac{f(\mathbf{X}_{\text{mis}}|\mathbf{X}_{\text{obs}}, \boldsymbol{\theta}^{(t+1)})}{f(\mathbf{X}_{\text{mis}}|\mathbf{X}_{\text{obs}}, \boldsymbol{\theta}^{(t)})} d\mathbf{X}_{\text{mis}}, \end{aligned}$$

which is of the form $\int p(y)\varphi(g(y))dy$ with $\varphi = -\ln$ and therefore larger than or equal to

$$-\ln \int f(\mathbf{X}_{\text{mis}}|\mathbf{X}_{\text{obs}}, \boldsymbol{\theta}^{(t+1)})d\mathbf{X}_{\text{mis}} = -\ln 1 = 0.$$

Thus we have shown that both differences on the right-hand side of (8.21) are larger than or equal to zero. The equality holds only if $Q(\boldsymbol{\theta}^{(t+1)}|\boldsymbol{\theta}^{(t)}) = Q(\boldsymbol{\theta}^{(t)}|\boldsymbol{\theta}^{(t)})$, which means that the maximization step cannot further increase the observed data likelihood.

The EM Algorithm for Exponential Families. Many statistical models are based on probability distributions or densities that are members of the regular exponential family. This family includes a wide variety of distributions, including the normal, Bernoulli, binomial, negative binomial, exponential, gamma, and multinomial distributions [see, e.g., McCullagh and Nelder (1989), Cox and Hinkley (1974), and Andersen (1980)]. The exponential family is of interest for theoretical statistics because any properties derived for this family directly carry over to the many members of this distribution that are used in practical applications. One such property that is relevant for the EM algorithm is that the loglikelihood can be written in a way that makes it particularly easy to take the expectation required for the E-step [cf. Little and Rubin (2002) and Schafer (1997)]. For n iid observations $\mathbf{x}_i$, this exponential family loglikelihood is

$$L(\boldsymbol{\theta}|\mathbf{X}) = \eta(\boldsymbol{\theta})^T T(\mathbf{X}) + ng(\boldsymbol{\theta}) + c, \tag{8.22}$$

with $\eta()$ a function that transforms the parameter vector $\boldsymbol{\theta}$ to what is called the canonical parameter. This is just a re-parameterization that does not affect the number of parameters. The function $T(\ )$ extracts from the data $T(\mathbf{X})$ the sufficient statistics for estimating $\eta(\boldsymbol{\theta})$ or, equivalently, $\boldsymbol{\theta}$ and results in a vector with the same number of elements as $\boldsymbol{\theta}$. The constant c does not contain the parameters and can be ignored for maximizing the loglikelihood. The loglikelihood without this constant is called the kernel of the loglikelihood function.

Since the kernel of the complete data loglikelihood (8.22) depends on the data only via the sufficient statistics and is a linear function of these statistics, the expectation over the missing data of the loglikelihood kernel is obtained by replacing the sufficient statistics by their expectations. The E-step can thus be performed by replacing the components $T_k(\mathbf{X})$ of $T(\mathbf{X})$, with $k = 1, \ldots, K$ and K the number of parameters, in the complete data loglikelihood (8.22) by their expectations $\mathrm{E}_{\mathbf{X}_{\text{mis}}} T_k(X)$. For regular exponential family distributions, the likelihood equations take a particular simple form; they equate the sufficient statistics to their expected values under the model $f(\mathbf{X}|\boldsymbol{\theta})$. The maximum likelihood estimator is then obtained as the solution for $\boldsymbol{\theta}$ of these likelihood equations—that is, as the solution to

$$\mathrm{E}(T(\mathbf{X})) = T(\mathbf{X}), \tag{8.23}$$

where E denotes the expectation over $\mathbf{X}$ under the model. In the M-step of the EM algorithm the maximization is applied to the loglikelihood with the sufficient statistics replaced by their expected values. This results in solving the equations

$$\mathrm{E}(T(\mathbf{X})) = \mathrm{E}_{\mathbf{X}_{\text{mis}}} T(\mathbf{X}). \tag{8.24}$$

In summary, to apply the EM algorithm to an exponential family loglikelihood, we first identify the sufficient statistics for estimating the parameters in the complete data case. Then, the following two steps are iterated:

E-step: Evaluate the expectation over the missing data of the sufficient statistics.

M-step: Solve the complete data likelihood equations with the sufficient statistics replaced by their expected values.

EM for Multivariate Normal Data. We will now apply the EM algorithm to estimating the mean vector and covariance matrix of a multivariate normal distribution. The likelihood of the multivariate normal density, or more precisely the kernel of the likelihood since we ignore constants, can be written as

$$l(\boldsymbol{\theta}|\mathbf{X}) = |\boldsymbol{\Sigma}|^{-n/2} \exp\left(-\frac{1}{2}\sum_{i=1}^{n} (\mathbf{x}_i - \boldsymbol{\mu})^T \boldsymbol{\Sigma}^{-1} (\mathbf{x}_i - \boldsymbol{\mu})\right),$$

and for the loglikelihood we obtain [see, e.g., Schafer (1997, Chapter 5)]

$$L(\boldsymbol{\theta}|\mathbf{X}) = -\frac{n}{2}\ln|\boldsymbol{\Sigma}| + \sum_{i=1}^{n} \boldsymbol{\mu}^T\boldsymbol{\Sigma}^{-1}\mathbf{x}_i - \frac{1}{2}\sum_{i=1}^{n} \mathbf{x}_i^T\boldsymbol{\Sigma}^{-1}\mathbf{x}_i - \frac{n}{2}\boldsymbol{\mu}^T\boldsymbol{\Sigma}^{-1}\boldsymbol{\mu}$$

$$(8.25) \qquad = \boldsymbol{\mu}^T\boldsymbol{\Sigma}^{-1}\sum_{i=1}^{n} \mathbf{x}_i - \frac{1}{2}\sum_{i=1}^{n} \mathrm{tr}(\boldsymbol{\Sigma}^{-1}\mathbf{x}_i\mathbf{x}_i^T) - \frac{n}{2}\left(\ln|\boldsymbol{\Sigma}| - \boldsymbol{\mu}^T\boldsymbol{\Sigma}^{-1}\boldsymbol{\mu}\right),$$

where we have used that the scalar $\mathbf{x}_i^T\boldsymbol{\Sigma}^{-1}\mathbf{x}_i$ equals its trace and that tr(**AB**) is tr(**BA**) if **A** and **B** are matrices with dimensions such that both products are defined. The last expression is of the form (8.22) with sufficient statistics

$$(8.26) \qquad \mathbf{T}_1 = \sum_{i=1}^{n} \mathbf{x}_i, \qquad \mathbf{T}_2 = \sum_{i=1}^{n} \mathbf{x}_i\mathbf{x}_i^T$$

and corresponding expected values

$$(8.27) \qquad \mathrm{E}\mathbf{T}_1 = n\boldsymbol{\mu}, \qquad \mathrm{E}\mathbf{T}_2 = n\left(\boldsymbol{\Sigma} + \boldsymbol{\mu}\boldsymbol{\mu}^T\right).$$

The maximum likelihood estimators for $\boldsymbol{\mu}$ and $\boldsymbol{\Sigma}$ are obtained by equating the sufficient statistics to their expectations, with the well known result that $\boldsymbol{\mu}$ and $\boldsymbol{\Sigma}$ are estimated by the sample mean vector and sample covariance matrix.

To apply the E-step of the EM-algorithm we need to take the expectation over the missing data of the sufficient statistics. Since the sufficient statistics are sums of x values ($\mathbf{T}_1$) or sums of products of x values ($\mathbf{T}_2$), calculation of the expected value of the sufficient statistics requires the calculation of the expectation of missing x-values and the expectation of products of x values involving missing values. For the conditional expectation of the missing values in $\mathbf{x}_i$, $\mathbf{x}_{i,\mathrm{mis}}$, given

$\mathbf{x}_{i,\text{obs}}$ and the current estimates of $\boldsymbol{\mu}$ and $\boldsymbol{\Sigma}$ we have from (8.5) and (8.6)

$$\begin{aligned} E(\mathbf{x}_{i,\text{mis}}|\mathbf{x}_{i,\text{obs}}, \hat{\boldsymbol{\mu}}, \hat{\boldsymbol{\Sigma}}) &= \hat{\mathbf{x}}_{i,\text{mis}} \\ &= \hat{\boldsymbol{\mu}}_{\text{mis}} + \hat{\mathbf{B}}_{\text{mis,obs}}(\mathbf{x}_{i,\text{obs}} - \hat{\boldsymbol{\mu}}_{\text{obs}}) \\ &= \hat{\boldsymbol{\mu}}_{\text{mis}} + \hat{\boldsymbol{\Sigma}}_{\text{mis,obs}}\hat{\boldsymbol{\Sigma}}^{-1}_{\text{obs,obs}}(\mathbf{x}_{i,\text{obs}} - \hat{\boldsymbol{\mu}}_{\text{obs}}), \end{aligned} \tag{8.28}$$

which are the regression predictions for the missing variables in record i with the variables observed in that record as predictors. The conditional expectation over the missing data of $\mathbf{T}_1$, $\mathbf{T}_1^*$ say, can now be written as

$$\mathbf{T}_1^* = E_{\mathbf{X}_{\text{mis}}}\mathbf{T}_1 = \sum_{i=1}^{n} \begin{pmatrix} \hat{\mathbf{x}}_{i,\text{mis}} \\ \mathbf{x}_{i,\text{obs}} \end{pmatrix} = \sum_{i=1}^{n} \mathbf{x}_i^*. \tag{8.29}$$

To evaluate the conditional expectation of products of x variables, we first consider the decomposition

$$x_{i,j} = \hat{x}_{i,j} + \hat{e}_{i,j},$$

with $\hat{e}_{i,j}$ the residual $x_{i,j} - \hat{x}_{i,j}$. A product of two x values can then be expressed as

$$x_{i,j}x_{i,k} = \hat{x}_{i,j}\hat{x}_{i,k} + \hat{x}_{i,j}\hat{e}_{i,k} + \hat{x}_{i,k}\hat{e}_{i,j} + \hat{e}_{i,j}\hat{e}_{i,k}.$$

Since the residuals have expectation zero, terms with one residual vanish when taking the expectation. The expectation of the product of the two residuals equals the residual covariance. However, this residual or conditional covariance is zero if one or both of the values $x_{i,j}$ and $x_{i,k}$ are observed because the conditioning on the observed values implies that $\hat{x}_{i,j} = x_{i,j}$ and $\hat{e}_{i,j} = 0$ if $x_{i,j}$ is observed. Thus, the conditional expectation over the missing data of $\mathbf{T}_2$ and $\mathbf{T}_2^*$, say, can be expressed as

$$\mathbf{T}_2^* = E_{\mathbf{X}_{\text{mis}}}\mathbf{T}_2 = \sum_{i=1}^{n} \mathbf{x}_i^* \left(\mathbf{x}_i^*\right)^T + \mathbf{V}_i, \tag{8.30}$$

with $\mathbf{V}_i$ the matrix with elements $(\mathbf{V}_i)_{jk}$ that are equal to the corresponding elements of the residual covariance matrix given by (8.8) if both $x_{i,j}$ and $x_{i,k}$ are missing, and zero otherwise.

The EM algorithm now proceeds as follows:

E-step: Given the current estimates $\boldsymbol{\mu}^{(t)}$ and $\boldsymbol{\Sigma}^{(t)}$ of $\boldsymbol{\mu}$ and $\boldsymbol{\Sigma}$, calculate the conditional expectations of the sufficient statistics $\mathbf{T}_1^{*(t)}$ and $\mathbf{T}_2^{*(t)}$.

M-step: Using these in this way completed sufficient statistics, update the parameters by the complete data likelihood equations:

$$\boldsymbol{\mu}^{(t+1)} = \frac{1}{n}\mathbf{T}_1^{*(t)}, \tag{8.31}$$

(8.32)
$$\boldsymbol{\Sigma}^{(t+1)} = \frac{1}{n}\mathbf{T}_2^{*(t)} - \boldsymbol{\mu}^{(t+1)}\left(\boldsymbol{\mu}^{(t+1)}\right)^T.$$

EM for Multinomial Data. The EM algorithm for multinomial data is much simpler than that for multivariate normal data and follows closely the aggregated imputation of categorical data discussed in Section 8.2.2. From the multinomial distribution (8.9), we see that the kernel of the multinomial loglikelihood can be written as

(8.33)
$$L(\boldsymbol{\pi}\,|\,\mathbf{n}) = \sum_{j=1}^{C} n_j \ln \pi_j.$$

The multinomial is a member of the exponential family, the sufficient statistics are the cell counts n_j, and the multinomial loglikelihood is clearly linear in the sufficient statistics. Thus, the maximum likelihood estimators for the cell probabilities are obtained by equating the cell counts to their expectations and are hence given by

(8.34)
$$\hat{\pi}_j = \frac{n_j}{n}.$$

Now, consider a set of cells S that sum up to a cell in a supplemental margin as described in Section 8.2.2. Let the number of counts of the fully observed units in these cells be denoted by $n_{j,\mathrm{obs}}(j \in S)$. Furthermore, let the count in the supplemental marginal cell be $n_{s,\mathrm{mis}}$; then the expected value of the cell count n_j, given $n_{j,\mathrm{obs}}$ and $n_{s,\mathrm{mis}}$, is given by

(8.35)
$$\mathrm{E}\left(n_j|\hat{\pi}_j, n_{j,\mathrm{obs}}, n_{s,\mathrm{mis}}\right) = n_{j,\mathrm{obs}} + n_{s,\mathrm{mis}} \times \hat{\pi}_j^S \qquad \text{for } j \in S,$$

with the $\hat{\pi}_j^S$ the estimated conditional probabilities $\hat{\pi}_j / \sum_{j\in S}\hat{\pi}_j$. Thus, we distribute the total count in the cell of the supplemental margin over the cells contributing to it according to the current estimate of the conditional distribution over these cells. This process is repeated for each cell in a supplemental margin; thus all observations, with missing values or not, are classified in the full-dimensional contingency table. This completes the E-step.

The M-step then simply calculates new estimates of the parameters π_j using (8.34) for the completed data obtained in the E-step.

The result of the EM algorithm is a completed contingency table. In this sense the missing values are imputed at an aggregate level: The missing cell counts have been imputed, but the missing values in the underlying records are not (see also Section 8.2.2).

This algorithm is illustrated using the example involving the variables *Car owner* and *License* introduced in Section 8.2.2. Table 8.3 first shows the counts for the fully observed records. In addition to the 60 fully observed units, there

TABLE 8.3 EM for *Car Owner* and *License* Data

Fully Observed Data

	License	
Car Owner	1. Yes	2. No
1. Yes	40	0
2. No	10	10

E-Step for *Car Owner* Missing

	Conditional Probabilities		Imputed Data	
	1. Yes	2. No	1. Yes	2. No
1. Yes	0.8	0	48	0
2. No	0.2	1	12	20

E-Step for *License* Missing

	Conditional Probabilities		Imputed Data	
	1. Yes	2. No	1. Yes	2. No
1. Yes	1	0	53	0
2. No	0.5	0.5	14.5	22.5

Estimated Cell Counts after 5 Iterations

	License	
Car Owner	1. Yes	2. No
1. Yes	52.91	0
2. No	13.97	23.12

were 20 units with *License* missing, 10 for each category of *Car Owner*, and 10 units with *Car owner* missing, 5 for each category of *License*. The counts for the fully observed units are used to obtain initial estimates of the cell probabilities and conditional probabilities over the categories of missing variables. These estimates are then used to perform the E-step. For instance, as shown in the second panel of Table 8.3, for the 10 units with *License* = Yes but *Car Owner* missing, the expected counts for the categories of *Car Owner* are obtained by multiplying the conditional probabilities by 10. These expected counts are then added to the counts in the table with fully observed units to obtain the first column of the "imputed table." After the same is done for the 10 units with *License* = No but *Car Owner* missing, the 20 units with *Car Owner* missing are imputed. Then, the conditional probabilities for the categories of *License*, given the category of *Car Owner*, are calculated and the 10 units with *License* missing are imputed.

All missing values have now been imputed and the total of the (fractional) counts equals the total number of 90 units. Based on the imputed table, new cell probabilities and conditional probabilities are estimated (M-step).

In this case the algorithm converges (in two decimals) after five iterations. After imputation, the marginal distributions of *License* and *Car Owner* have become more uniform, which reflects that the additional margins for both *License* and *Car Owner* are, in this illustration, uniform.

8.4 Example: The Public Libraries

This example concerns data from a survey held among public libraries in the Netherlands in 1998. Among others, the three most important variables in this survey are:

Collection (Coll): The total number of books and other items in possession of the library.

Personnel (Pers): The total number of worked hours by all personnel on a monthly basis.

Turnover (Turn): The gross amount of all payments received.

To demonstrate the effect of missing data and different imputation methods on estimates, we took 586 records with complete data and created incomplete data by setting some values to "missing." This will allow us to compare the estimates based on incomplete data with estimates based on the known complete data. The missing data patterns for the created incomplete data are shown in Table 8.4 in the same format as Table 8.1 in the introduction of this chapter. The total amount of missing data and also the distribution of missing data over the different missing data patterns is realistic. Most records are complete; the "missing data pattern" D, with all variables observed, contains 68% of the records. We did assume that the number of hours worked was available from administrative data (or could be approximated accurately from such an external source) and therefore did not create missing values for this variable.

A standard imputation method for such data could be ratio imputation with *Pers* as an auxiliary variable. This is a very often applied imputation model

TABLE 8.4 Missing Data Patterns in Public Library Data

Pers	Turn	Coll	Pattern	n
O	M	M	A	22
O	M	O	B	86
O	O	M	C	82
O	O	O	D	396

for business statistics. As explained in Chapter 7, ratio imputation estimates a missing value in a target variable by the auxiliary variable multiplied by the ratio of the means of the target variable and the auxiliary variable, where the means are estimated from the units with both these variables observed. Ratio imputation can also be seen as a form of regression imputation where there is no constant term and one predictor (the auxiliary variable), and the variance of the residuals is assumed to be proportional to the value of the auxiliary variable. Besides this standard approach, two alternatives are considered. The first alternative is to use regression imputation with the observed variables as predictors. This means that for records in missing data pattern A we use *Pers* as a predictor for both *Turn* and *Coll*, for missing data pattern B we use *Pers* and *Coll* as predictors for *Turn*, and for missing data pattern C we use *Pers* and *Turn* as predictors for *Coll*. These regression models all included an intercept and were each estimated from the subset of the data with the predictors and target variable for that model observed. Both the ratio and the regression approaches were applied without adding a residual. The second alternative is to apply the EM algorithm for multivariate normal data as discussed in Section 8.3.3. Contrary to the other methods, this algorithm directly produces estimates of the means and covariance matrix from the incomplete data, without actually imputing the missing values. If so desired, regression imputations can be obtained by using the EM-estimated means and covariance matrix as discussed in Section 8.2. If, in this case, residuals are added, the variances and covariances of the imputed data should be similar to the direct estimates from the EM algorithm.

The three methods were applied to the 586 records described above and, since in this experiment the true values were known, we can compare estimated means, standard deviations, and correlations with the values obtained from the complete data. Since the most important parameters for this survey are totals or, equivalently, means, we first compare the estimated means with the complete data means. These results are in Table 8.5.

The results show that the EM-estimated means are very accurate and approximate the means of the complete data better than both other methods. Note that the ratio method underestimates both means while the regression method overestimates them. The estimated standard deviations for the three methods are displayed in Table 8.6. Here it appears that the EM-estimated standard deviations are better than the estimates obtained by the standard ratio method, but the regression method is best for one of the variables (*Coll*). Also note that even without adding a residual, the standard deviations for the ratio and regression methods are too large and adding a residual will make this even worse.

TABLE 8.5 **Means for Turnover and Collection**

	Complete	EM	Ratio	Reg
Turnover	921.3	921.1	915.4	929.1
Collection	58511	58445	58251	58763

TABLE 8.6 Standard Deviations for Turnover and Collection

	Complete	EM	Ratio	Reg
Turnover	1294.0	1291.8	1308.0	1324.0
Collection	67322.1	68373.2	74061.7	67552.5

TABLE 8.7 Complete Data and Estimated Correlations

	Complete			EM		
	Coll	Turn	Pers	Coll	Turn	Pers
Coll	1			1		
Turn	0.959	1		0.953	1	
Pers	0.963	0.982	1	0.964	0.980	1
	Ratio			Regression		
	Coll	Turn	Pers	Coll	Turn	Pers
Coll	1			1		
Turn	0.952	1		0.932	1	
Pers	0.964	0.984	1	0.961	0.950	1

Table 8.7 shows the correlations obtained from the complete data, estimated by the EM algorithm and obtained from imputed data by the ratio and regression methods. The correlations for the complete data are very well preserved by the EM and ratio methods. The correlations for the regression method are somewhat underestimated.

The high correlations between these variables also explain why all methods perform reasonably well. But still, the EM results are overall clearly better than those for the other methods. This good performance can be expected when the data are multivariate normal, as assumed for the EM algorithm, but in this case the variables are rather skewed and the good performance of EM demonstrates some robustness against violation of this assumption.

REFERENCES

Amemiya, T. (1985), *Advanced Econometrics*. Basil Blackwell, Oxford.

Andersen, E. B. (1980), *Discrete Statistical Models with Social Science Applications*. North-Holland, Amsterdam.

Cox, D. R., and D. V. Hinkley (1974), *Theoretical Statistics*. Chapman & Hall, London.

Dempster, A. P., N. M. Laird, and D. B. Rubin (1977), Maximum Likelihood from Incomplete Data via the EM Algorithm (with discussion). *Journal of the Royal Statistical Society, Series B (Methodological) 39*, pp. 1–38.

Little, R. J. A., and D. B. Rubin (2002), *Statistical Analysis with Missing Data*, second edition. John Wiley & Sons, Hoboken, NJ.

McCullagh, P., and J. A. Nelder (1989), *Generalized Linear Models*, second edition. Chapman & Hall, London.

Rubin, D. B. (1976), Inference and Missing Data. *Biometrika 63*, pp. 581–592.

Rubin, D. B. (1987), *Multiple Imputation for Nonresponse in Surveys*. John Wiley & Sons, New York.

Schafer, J. L. (1997), *Analysis of Incomplete Multivariate Data*. Chapman & Hall, London.

Stuart, A., and J. K. Ord (1991), *Kendall's Advanced Theory of Statistics*, Vol. 2, fifth edition. Oxford University Press, New York.

Wilks, S. S. (1962), *Mathematical Statistics*. John Wiley & Sons, New York.

CHAPTER NINE

Imputation Under Edit Constraints

9.1 Introduction

In Chapters 7 and 8 we have discussed the problem of finding good imputations for missing values. At National Statistical Institutes (NSIs) and some other statistical organizations the imputation problem is further complicated owing to the existence of constraints in the form of edit restrictions, or edits for short, that have to be satisfied by the data. While imputing a record, we aim to take these edits into account, and thus ensure that the final, imputed record satisfies all edits. The imputation problem we consider in this chapter is given by: Impute the missing data in the data set under consideration in such a way that the statistical distribution of the data is preserved as well as possible, subject to the condition that all edits are satisfied by the imputed data.

For academic statisticians the wish of NSIs to let the data satisfy specified edits may be difficult to understand. Statistically speaking, there is indeed hardly a reason to let a data set satisfy edits. However, as Pannekoek and De Waal (2005) explain, NSIs have the responsibility to supply data for many different, both academic and nonacademic, users in society. For the majority of these users, inconsistent data are incomprehensible. They may reject the data as being an invalid source or make adjustments themselves. This hampers the unifying role of NSIs in providing data that are undisputed by different parties such as policy makers in government, opposition, trade unions, employer organizations, and so on. As mentioned by Särndal and Lundström (2005, p. 176): "Whatever the imputation method used, the completed data should be subjected to the usual checks for internal consistency. All imputed values should undergo the editing checks normally carried out for the survey."

Handbook of Statistical Data Editing and Imputation, by T. de Waal, J. Pannekoek, and S. Scholtus

Simple sequential imputation of the missing data, where edits involving fields that have to be imputed subsequently are not taken into account while imputing a field, may lead to inconsistencies. Consider, for example, a record where the values of two variables, x and y, are missing. Suppose these variables have to satisfy three edits saying that $x \geq 50$, $y \leq 100$, and $y \geq x$. Now, if x is imputed first without taking the edits involving y into account, one might impute the value 150 for x. The resulting set of edits for y—that is, $y \leq 100$ and $y \geq 150$—cannot be satisfied. Conversely, if y is imputed first without taking the edits involving x into account, one might impute the value 40 for y. The resulting set of edits for x—that is, $x \geq 50$ and $40 \geq x$—cannot be satisfied.

Despite the fact that much research on imputation techniques has been carried out, imputation under edits is still a rather neglected area. As far as we are aware, apart from some research at NSIs [see, e.g., Tempelman (2007)], hardly any research on general approaches to imputation under edit restrictions has been carried out. An exception is imputation based on a truncated multivariate normal model [see, e.g., Geweke (1991) and Tempelman (2007)]. For numerical data, imputation based on a truncated multivariate normal model can take the edit restrictions we consider in this chapter into account (see Section 9.5). Using this model has two drawbacks, however. First of all, the truncated multivariate model is computationally very demanding and complex to implement in a software program. Second, it is obviously only suited for data that (approximately) follow a truncated multivariate normal distribution, not for data that follow other distributions. Another exception is sequential regression imputation [see Section 9.7, Van Buuren and Oudshoorn (1999, 2000), Raghunathan et al. (2001), and Tempelman (2007)]. For categorical data, Winkler (2003) develops an imputation methodology that satisfies edits. This methodology generalizes hot deck imputation. Finally, some software packages developed by NSIs, such as GEIS (Kovar and Whitridge, 1990), SPEER (Winkler and Draper, 1997) and SLICE (De Waal, 2001), also ensure that edits are satisfied after imputation. However, these packages only apply relatively simple imputation models, whereas the approaches described in this chapter allow more complicated imputation models.

There are two general approaches to imputation under edit restrictions. One approach is to first impute the missing data without taking the edits into account, and then adapt the imputed values in a separate step, in such a way that the adapted values satisfy all edits. This approach will be the subject of Chapter 10. In the current chapter, we consider methods that follow the second approach, which is to take the edits into account during the imputation step itself. The second approach has the advantage that it removes the need to adapt the imputed values later on. A potential disadvantage of the second approach is the complexity of the resulting methods. Also, asking that the imputations automatically satisfy a set of edit restrictions can seriously reduce the range of available imputation models.

Note that in Section 4.5 we already described the Nearest-neighbor Imputation Methodology (NIM). This methodology also ensures that after imputation, edits are satisfied. The method has been described in Section 4.5, because, unlike the methods discussed in this chapter, it can also be used to detect and correct errors. In fact, in practice the NIM is mainly used to detect and correct errors rather than to impute missing data.

The remainder of the chapter is organized as follows. Methods for deductive imputation—that is, deriving the correct value of a missing item from the edits with certainty—are discussed in Section 9.2. We then consider two relatively simple methods, in Sections 9.3 and 9.4, that can only handle one balance edit for nonnegative variables. The first of these is an extension of the hot deck imputation method from Section 7.6, and the second method assumes that the data follow a Dirichlet distribution. The remaining sections of this chapter are devoted to methods that can handle a more general set of edits. In Section 9.5 we examine the use of the truncated multivariate normal distribution for imputation of missing numerical data subject to linear edits. This approach is a relatively traditional approach for this problem [see, e.g., Geweke (1991) and Tempelman (2007)]. Section 9.6 describes a relatively simple approach for imputation of missing numerical data, based on the Fourier–Motzkin elimination technique that we also used in several algorithms for automatic editing of random errors (see Chapters 3 to 5). Sequential imputation methods are a well-known class of imputation methods [see, e.g., Van Buuren and Oudshoorn (1999, 2000), Raghunathan et al. (2001), and Rubin (2003)]. Sequential imputation methods can be relatively easily extended to ensure that they satisfy edits. In Section 9.7 we describe such a sequential imputation method that indeed satisfies edits [see also Tempelman (2007)]. In Section 9.8 we further extend the sequential imputation method so that besides satisfying edits also known totals are preserved. Finally, in Section 9.9 we develop analogous imputation methods for categorical data, namely hot deck imputation methods that ensure that edits are satisfied and known totals are preserved.

Because imputation under edit restrictions is a rather new area of research, the material in this chapter is a bit more experimental and offers more room for further improvement than the material in other chapters of this book. The reader is encouraged to elaborate the ideas expressed in this chapter and to develop new ideas for imputation methods under edit restrictions.

9.2 Deductive Imputation

All imputation methods discussed so far impute predicted values for missing data items, based on some explicitly or implicitly assumed model. The same holds for the advanced imputation methods that we shall discuss in the remainder of this chapter. However, in some cases it is possible to deduce imputed values directly from the observed data items, using deterministic rules without model parameters that need to be estimated. Such methods for deductive imputation will be examined in this section.

Deductive imputations are in a way comparable to deductive corrections, which we discussed in Chapter 2. Both methods result in adjusted data that are correct with certainty, as long as the underlying assumptions are met. An important difference is that deductive corrections can be performed at the very beginning of the data editing process, when the data may still contain all kinds of errors. In fact, the object of using methods for deductive correction is to resolve some of these errors. In contrast to this, a crucial assumption for performing deductive imputations is that the data contain only missing values, but no errors. Thus, we assume that all erroneous values have been identified and replaced by missing values; the error localization methods of Chapters 3 to 5 may have been used to achieve this. Once we have obtained an error-free data set with missing values, performing deductive imputations is a natural first step. After this step, model-based or donor imputation methods may be used to impute predictions for the remaining missing values. Note that these predictions are strengthened by the fact that more nonmissing values are available after deductive imputation.

9.2.1 RULE-BASED IMPUTATION

A type of deductive imputation that is often applied in practice is based on if–then rules that are specified by subject-matter specialists. Such rules express that under the condition stated in the if part of the rule, the correct value for a certain variable is uniquely determined, and this is stated in the then part of the rule.

For instance, given that a person is less than 16 years old, it is known (from the Dutch civil code) that this person cannot be married. Thus, we have the following imputation rule:

$$\text{IF } \left(\textit{Marital Status} = \text{“Unknown” AND } \textit{Age} < 16\right),$$
$$\text{THEN } \textit{Marital Status} = \text{“Unmarried.”}$$

Another example is this rule, taken from a business survey:

$$\text{IF } \left(\textit{Total Labor Costs} = \text{“Unknown” AND } \textit{Number Of Employees} = 0\right)$$
$$\text{THEN } \textit{Total Labor Costs} = 0.$$

In order to specify if–then rules, subject-matter knowledge is required. Such knowledge can, at least in part, be represented in the form of edits. In the remainder of this section, we examine methods that can be used to extract deductive imputations from the edits automatically. We first describe two methods that can be applied to numerical data, in Sections 9.2.2 and 9.2.3. Section 9.2.4 is dedicated to a deductive imputation method for categorical data.

9.2.2 USING BALANCE EDITS

Numerical data often have to satisfy many balance edits. These edits state, for instance, that a total value must be equal to the sum of its parts. Clearly, if exactly one variable involved in a balance edit has a missing value, the edit can be used to compute the correct value of this variable from the observed values of the other variables. This is an important example of a situation where a missing value is uniquely determined by the observed values and thus may be imputed deductively. In practice, the situation is more complex, because variables are typically involved in several interrelated balance edits, and hence it is not immediately clear which variables with missing values are uniquely determined by the observed values. Using matrix algebra, these variables may still be identified, as we explain below. This description is based on Pannekoek (2006).

Suppose that the data consist of records with p numerical variables $x_1, \ldots, x_p$ that should conform to r balance edits. Writing a record as a column vector $\mathbf{x} = (x_1, \ldots, x_p)^T$, these balance edits can be represented by a system of linear equations:

$$\mathbf{Ax} = \mathbf{b}, \tag{9.1}$$

with $\mathbf{A}$ the $r \times p$ matrix of coefficients and $\mathbf{b} = (b_1, \ldots, b_r)^T$ the vector of constant terms that appear in the edits. We refer to Example 2.7 (and Example 9.1 below) for an illustration of such a linear system.

For a given record we can, by permuting the elements of $\mathbf{x}$, partition this vector as $\mathbf{x} = (\mathbf{x}_{\text{mis}}^T, \mathbf{x}_{\text{obs}}^T)^T$, with $\mathbf{x}_{\text{mis}}$ the vector of missing values and $\mathbf{x}_{\text{obs}}$ the vector of observed values. By partitioning $\mathbf{A}$ in the same way, system (9.1) can be written as

$$(\mathbf{A}_{\text{mis}}\ \mathbf{A}_{\text{obs}}) \begin{pmatrix} \mathbf{x}_{\text{mis}} \\ \mathbf{x}_{\text{obs}} \end{pmatrix} = \mathbf{b},$$

and hence

$$\mathbf{A}_{\text{mis}}\mathbf{x}_{\text{mis}} = \mathbf{b} - \mathbf{A}_{\text{obs}}\mathbf{x}_{\text{obs}} \equiv \mathbf{b}^*. \tag{9.2}$$

Note that $\mathbf{b}^* = \mathbf{b} - \mathbf{A}_{\text{obs}}\mathbf{x}_{\text{obs}}$ can be computed from the observed values in the record. This leaves a system of linear equations in the variables with missing values $\mathbf{x}_{\text{mis}}$.

In general, a linear system such as (9.2) has either no solution, exactly one solution, or an infinite number of solutions. In the first case, the system is called inconsistent, and it can be shown [see, e.g., Theorem 7.2.1 of Harville (1997)] that $\text{rank}(\mathbf{A}_{\text{mis}}) \neq \text{rank}(\mathbf{A}_{\text{mis}}, \mathbf{b}^*)$. Assuming that the edits do not contradict each other, this case can only occur if the observed values contain errors. As mentioned before, we assume here that all erroneous values have been identified and replaced by missing values. Hence (9.2) cannot be inconsistent.

The second case, where (9.2) has exactly one solution, only occurs if the matrix $\mathbf{A}_{\text{mis}}$ is nonsingular. The unique solution is then given by $\hat{\mathbf{x}}_{\text{mis}} = \mathbf{A}_{\text{mis}}^{-1}\mathbf{b}^*$. In this situation, all variables in $\mathbf{x}_{\text{mis}}$ can be imputed deductively from $\hat{\mathbf{x}}_{\text{mis}}$, since for each variable there is only one value that satisfies the system of balance edits. This ideal situation is rarely encountered in practice.

The third case, where $\mathbf{A}_{\text{mis}}$ is singular and (9.2) has an infinite number of solutions, is more common. In this situation, the only variables that can be imputed deductively are those that have the same value in every possible solution. The solutions to (9.2) can be written as

$$\hat{\mathbf{x}}_{\text{mis}} = \mathbf{A}_{\text{mis}}^{-}\mathbf{b}^* + (\mathbf{A}_{\text{mis}}^{-}\mathbf{A}_{\text{mis}} - \mathbf{I})\mathbf{z} \equiv \mathbf{A}_{\text{mis}}^{-}\mathbf{b}^* + \mathbf{Cz}, \tag{9.3}$$

where $\mathbf{A}_{\text{mis}}^{-}$ is a generalized inverse of $\mathbf{A}_{\text{mis}}$, $\mathbf{I}$ is the identity matrix, and $\mathbf{z}$ is an arbitrary vector of the same length as $\mathbf{x}_{\text{mis}}$ [see, e.g., Rao (1973), page 25]. By definition, any matrix $\mathbf{A}_{\text{mis}}^{-}$ that satisfies

$$\mathbf{A}_{\text{mis}}\mathbf{A}_{\text{mis}}^{-}\mathbf{A}_{\text{mis}} = \mathbf{A}_{\text{mis}} \tag{9.4}$$

is called a generalized inverse of $\mathbf{A}_{\text{mis}}$ [see, e.g., Chapter 9 of Harville (1997)]. All solutions to (9.2) can be found by varying the choice of $\mathbf{z}$ in (9.3). When $\mathbf{A}_{\text{mis}}$ is nonsingular, the generalized inverse is just the regular inverse, so the second term in (9.3) vanishes and (9.2) has a unique solution.

It is easy to see that if the jth row of $\mathbf{C}$ contains only zeros, then the jth element of $\hat{\mathbf{x}}_{\text{mis}}$ equals the jth element of $\mathbf{A}_{\text{mis}}^{-}\mathbf{b}^*$ in every solution generated by (9.3). The value of the jth missing variable is then uniquely determined and can be imputed deductively by the jth element of $\mathbf{A}_{\text{mis}}^{-}\mathbf{b}^*$. Conversely, it is also clear that elements of $\hat{\mathbf{x}}_{\text{mis}}$ corresponding with nonzero rows of $\mathbf{C}$ are not uniquely determined by (9.3) and cannot be imputed deductively. Thus, in order to apply deductive imputation, we just have to identify the rows of $\mathbf{C}$ that contain only zeros.

EXAMPLE 9.1

We use the same set of edits as in Examples 2.6 and 2.7, where records of 11 variables $x_1, \ldots, x_{11}$ should conform to five balance edits:

$$\begin{aligned} x_1 + x_2 &= x_3, \\ x_2 &= x_4, \\ x_5 + x_6 + x_7 &= x_8, \\ x_3 + x_8 &= x_9, \\ x_9 - x_{10} &= x_{11}. \end{aligned}$$

Written in the form (9.1), this yields $\mathbf{Ax} = \mathbf{0}$, with

$$\mathbf{A} = \begin{pmatrix} 1 & 1 & -1 & 0 & 0 & 0 & 0 & 0 & 0 & 0 & 0 \\ 0 & 1 & 0 & -1 & 0 & 0 & 0 & 0 & 0 & 0 & 0 \\ 0 & 0 & 0 & 0 & 1 & 1 & 1 & -1 & 0 & 0 & 0 \\ 0 & 0 & 1 & 0 & 0 & 0 & 0 & 1 & -1 & 0 & 0 \\ 0 & 0 & 0 & 0 & 0 & 0 & 0 & 0 & 1 & -1 & -1 \end{pmatrix}.$$

Now suppose that we are given the following record for imputation:

x_1	x_2	x_3	x_4	x_5	x_6	x_7	x_8	x_9	x_{10}	x_{11}
145	—	155	—	—	—	—	86	—	217	—

where "—" denotes a missing value. In this instance $\mathbf{x}_{\text{obs}} = (x_1, x_3, x_8, x_{10})^T$ and $\mathbf{x}_{\text{mis}} = (x_2, x_4, x_5, x_6, x_7, x_9, x_{11})^T$. From (9.2), we compute

$$\mathbf{b}^* = -\mathbf{A}_{\text{obs}}\mathbf{x}_{\text{obs}} = -\begin{pmatrix} 1 & -1 & 0 & 0 \\ 0 & 0 & 0 & 0 \\ 0 & 0 & -1 & 0 \\ 0 & 1 & 1 & 0 \\ 0 & 0 & 0 & -1 \end{pmatrix} \begin{pmatrix} 145 \\ 155 \\ 86 \\ 217 \end{pmatrix} = \begin{pmatrix} 10 \\ 0 \\ 86 \\ -241 \\ 217 \end{pmatrix}.$$

Thus, we obtain the system $\mathbf{A}_{\text{mis}}\mathbf{x}_{\text{mis}} = \mathbf{b}^*$ given by

$$\begin{pmatrix} 1 & 0 & 0 & 0 & 0 & 0 & 0 \\ 1 & -1 & 0 & 0 & 0 & 0 & 0 \\ 0 & 0 & 1 & 1 & 1 & 0 & 0 \\ 0 & 0 & 0 & 0 & 0 & -1 & 0 \\ 0 & 0 & 0 & 0 & 0 & 1 & -1 \end{pmatrix} \begin{pmatrix} x_2 \\ x_4 \\ x_5 \\ x_6 \\ x_7 \\ x_9 \\ x_{11} \end{pmatrix} = \begin{pmatrix} 10 \\ 0 \\ 86 \\ -241 \\ 217 \end{pmatrix}.$$

The reader may verify that

$$\mathbf{A}_{\text{mis}}^- = \begin{pmatrix} 1 & 0 & 0 & 0 & 0 \\ 1 & -1 & 0 & 0 & 0 \\ 0 & 0 & 1 & 0 & 0 \\ 0 & 0 & 0 & 0 & 0 \\ 0 & 0 & 0 & 0 & 0 \\ 0 & 0 & 0 & -1 & 0 \\ 0 & 0 & 0 & -1 & -1 \end{pmatrix}$$

satisfies (9.4) and hence is a generalized inverse of $\mathbf{A}_{\text{mis}}$. A simple computation reveals that

$$\mathbf{C} = \mathbf{A}_{\text{mis}}^-\mathbf{A}_{\text{mis}} - \mathbf{I} = \begin{pmatrix} 0 & 0 & 0 & 0 & 0 & 0 & 0 \\ 0 & 0 & 0 & 0 & 0 & 0 & 0 \\ 0 & 0 & 0 & 1 & 1 & 0 & 0 \\ 0 & 0 & 0 & -1 & 0 & 0 & 0 \\ 0 & 0 & 0 & 0 & -1 & 0 & 0 \\ 0 & 0 & 0 & 0 & 0 & 0 & 0 \\ 0 & 0 & 0 & 0 & 0 & 0 & 0 \end{pmatrix}.$$

The first, second, sixth, and seventh elements of $\mathbf{x}_{\text{mis}}$ correspond with zero rows of $\mathbf{C}$ and hence are determined uniquely by the observed values. This means that we can impute the values of x_2, x_4, x_9, and x_{11} deductively. To find these values, we compute

$$\mathbf{A}_{\text{mis}}^{-}\mathbf{b}^* = \begin{pmatrix} 1 & 0 & 0 & 0 & 0 \\ 1 & -1 & 0 & 0 & 0 \\ 0 & 0 & 1 & 0 & 0 \\ 0 & 0 & 0 & 0 & 0 \\ 0 & 0 & 0 & 0 & 0 \\ 0 & 0 & 0 & -1 & 0 \\ 0 & 0 & 0 & -1 & -1 \end{pmatrix} \begin{pmatrix} 10 \\ 0 \\ 86 \\ -241 \\ 217 \end{pmatrix} = \begin{pmatrix} 10 \\ 10 \\ 86 \\ 0 \\ 0 \\ 241 \\ 24 \end{pmatrix}.$$

Thus, $\hat{x}_2 = 10$, $\hat{x}_4 = 10$, $\hat{x}_9 = 241$, and $\hat{x}_{11} = 24$. After imputing these values, we are left with the following record:

x_1	$\hat{x}_2$	x_3	$\hat{x}_4$	x_5	x_6	x_7	x_8	$\hat{x}_9$	x_{10}	$\hat{x}_{11}$
145	10	155	10	—	—	—	86	241	217	24

The remaining missing values could not be imputed deductively and have to be imputed by another method. ■

9.2.3 USING NONNEGATIVITY CONSTRAINTS

Another possibility for deriving deductive imputations from a system of balance edits arises if some of the numerical variables are constrained to be nonnegative. For instance, suppose that five data items should sum up to a total, that the total value is observed along with two of the item values, and that moreover these two item values sum up to the reported total. If the three missing data items are constrained to be nonnegative, we know that each of these values is bounded from above by their sum, which in this case equals zero. Since the missing values are also bounded from below by zero, they are in fact determined uniquely, and we can impute zeros for the missing data items.

To describe this method more generally, we consider again (9.2). Suppose that, say, the lth element of $\mathbf{b}^*$ equals zero. That is, it should hold that $\mathbf{a}_{\text{mis},l}^T\mathbf{x}_{\text{mis}} = 0$, where $\mathbf{a}_{\text{mis},l}^T = (a_{\text{mis},l1}, \ldots, a_{\text{mis},lm})$ denotes the lth row of $\mathbf{A}_{\text{mis}}$. Suppose moreover that the following two properties hold:

1. Each $a_{\text{mis},lj} \neq 0$ has the same sign; that is, the nonzero elements of $\mathbf{a}_{\text{mis},l}^T$ are either all negative or all positive.
2. Each $a_{\text{mis},lj} \neq 0$ corresponds with a variable $x_{\text{mis},j}$ that is constrained to be nonnegative.

Then it is easy to see that we can impute $\hat{x}_{\text{mis},j} = 0$ for all j such that $a_{\text{mis},lj} \neq 0$.

EXAMPLE 9.2

We consider the same set of edits as in Example 9.1, but we start with the following record:

x_1	x_2	x_3	x_4	x_5	x_6	x_7	x_8	x_9	x_{10}	x_{11}
145	—	155	—	86	—	—	86	—	217	—

That is, this time the value of x_5 is also observed. We also assume that all variables with the exception of x_{11} should be nonnegative.

The reader may verify that by performing the same steps as in Example 9.1, we arrive at the system $\mathbf{A}_{\text{mis}}\mathbf{x}_{\text{mis}} = \mathbf{b}^*$ given by

$$\begin{pmatrix} 1 & 0 & 0 & 0 & 0 & 0 \\ 1 & -1 & 0 & 0 & 0 & 0 \\ 0 & 0 & 1 & 1 & 0 & 0 \\ 0 & 0 & 0 & 0 & -1 & 0 \\ 0 & 0 & 0 & 0 & 1 & -1 \end{pmatrix} \begin{pmatrix} x_2 \\ x_4 \\ x_6 \\ x_7 \\ x_9 \\ x_{11} \end{pmatrix} = \begin{pmatrix} 10 \\ 0 \\ 0 \\ -241 \\ 217 \end{pmatrix}.$$

Note that the third element of $\mathbf{b}^*$ equals zero, while the nonzero elements of the third row of $\mathbf{A}_{\text{mis}}$ have the same sign. Since we also assumed that the corresponding variables x_6 and x_7 are nonnegative, by the deductive method described above we can impute zeros for these variables. The second element of $\mathbf{b}^*$ also equals zero, but we cannot derive a deductive imputation from this, because the corresponding row of $\mathbf{A}_{\text{mis}}$ contains both a positive and a negative coefficient.

Since we have imputed some of the missing values, we can now form a new partition of $\mathbf{x}$ into $\mathbf{x}_{\text{obs}}$ and $\mathbf{x}_{\text{mis}}$. The reader may verify that by performing similar steps as in Example 9.1, we eventually obtain the following imputed record:

x_1	$\hat{x}_2$	x_3	$\hat{x}_4$	x_5	$\hat{x}_6$	$\hat{x}_7$	x_8	$\hat{x}_9$	x_{10}	$\hat{x}_{11}$
145	10	155	10	86	0	0	86	241	217	24

Thus, in this example all missing values could be imputed deductively. ■

Two remarks have to be made with respect to the two deductive methods we just described. First, it may appear from the examples that for small systems of equations, the deductive imputations could in fact easily be determined ad hoc, without resort to formal matrix algebra. In practice, we encounter much larger systems—for instance, 30 balance edits involving about 100 variables, and then it becomes necessary to formalize the procedure as we did above.

Second, it may be useful to apply the two methods repeatedly. Namely, when both methods have been applied one time, the second method may allow additional deductive imputations that can be identified using the first method. Next, the new application of the first method may allow additional deductive imputations that can be identified using the second method. Thus, we should iterate both methods in turn until no new deductive imputations are found.

9.2.4 DEDUCTIVE IMPUTATION FOR CATEGORICAL DATA

We shall now describe a deductive method for imputing categorical data. This method makes use of the elimination procedure of Fellegi and Holt (1976), which we already considered in the context of error localization in Chapter 4. As in Chapter 4, the categorical variables are denoted by $v_1, \ldots, v_m$. Each variable v_j has a domain of allowed values D_j. The edits we consider are written in the so-called *normal form*,

$$F_1^k \times \cdots \times F_m^k,$$

with $F_j^k \subseteq D_j$. This means that a record fails edit k if and only if $v_j \in F_j^k$ for all $j = 1, \ldots, m$. If $F_j^k = D_j$, then v_j is said to be not involved in edit k, because in that case this edit is satisfied or failed regardless of the value of v_j.

We recall from Chapter 4 that in the procedure of Fellegi and Holt (1976), a variable v_g is eliminated from the original set of edits Ω_0 by considering all minimal index sets of edits S such that

$$\bigcup_{k \in S} F_g^k = D_g$$

and

$$\bigcap_{k \in S} F_j^k \neq \emptyset \qquad \text{for } j = 1, \ldots, g-1, g+1, \ldots, m.$$

The implied edit

$$\bigcap_{k \in S} F_1^k \times \cdots \times \bigcap_{k \in S} F_{g-1}^k \times D_g \times \bigcap_{k \in S} F_{g+1}^k \times \cdots \times \bigcap_{k \in S} F_m^k \tag{9.5}$$

does not involve v_g. If we replace the original set of edits Ω_0 with the new set Ω_1, which consists of the original edits that do not involve v_g and all implied edits of the form (9.5), then we have obtained a set of edits that does not involve v_g. It follows from Theorem 4.3 (as a special case) that the edits in Ω_1 can be satisfied by $v_1, \ldots, v_{g-1}, v_{g+1}, \ldots, v_m$, if and only if the edits in Ω_0 can be satisfied by $v_1, \ldots, v_m$. A repeated application of this result yields the following theorem.

THEOREM 9.1

Let Ω_{m-1} be the set of edits obtained by successively eliminating $v_1, \ldots, v_{m-1}$ from Ω_0. If there exists a value u_m for v_m that satisfies the edits in Ω_{m-1}, then there also exist values $u_1, \ldots, u_{m-1}$ for $v_1, \ldots, v_{m-1}$, such that $u_1, \ldots, u_m$ satisfy the edits in Ω_0. Conversely, every choice of values $u_1, \ldots, u_m$ for $v_1, \ldots, v_m$ that satisfy the edits in Ω_0 has the property that u_m satisfies the edits in Ω_{m-1}.

Proof. The proof is very similar to that of Theorem 4.4. Basically, we repeatedly apply Theorem 4.3, but this time we do not eliminate the final variable. ■

Now consider a record $(v_1, \ldots, v_m)$ with missing values, along with a set of edits Ω. We assume that the missing values can be imputed consistently, i.e. such that all edits in Ω become satisfied. This is guaranteed if the error localization methods from Chapter 4 have been used to delete all original erroneous values. We suggest the following algorithm for deductive imputation:

1. Let M be the index set of variables with missing values, and define $T := \emptyset$. Fill in the original values of the nonmissing variables into Ω; this yields a reduced set of edits Ω_0 that involves only the missing variables.
2. If $M \backslash T \neq \emptyset$: Choose any $g \in M \backslash T$. Otherwise: stop, because all missing variables have been treated.
3. Eliminate all variables in $M \backslash \{g\}$ from Ω_0, using the procedure outlined above. Denote the resulting set of univariate edits for v_g by Ω^*.
4. If there exist several values in D_g that satisfy all edits in Ω^*: Define $T := T \cup \{g\}$ and return to step 2.
5. If there exists exactly one value in D_g that satisfies all edits in Ω^*: Impute this value for v_g in the record, update Ω_0 by filling in this value for v_g, define $M := M \backslash \{g\}$, and return to step 2.

In step 5 a deductive imputation is performed.

The correctness of this algorithm follows from Theorem 9.1. Since we have assumed that the record can be imputed consistently, there exist values for v_j, $j \in M$, such that all edits in Ω_0 are satisfied. According to Theorem 9.1, every such choice of values for v_j, $j \in M$, has the property that the value of v_g satisfies all edits in Ω^*. In step 5 of the algorithm, there happens to be exactly one value in D_g that satisfies all edits in Ω^*. Therefore, we conclude that v_g has the same value in every consistently imputed version of the record. In other words: v_g may be imputed deductively.

EXAMPLE 9.3

To illustrate the algorithm we use an example taken from Kartika (2001). In this example, records consist of four categorical variables with the following domains: $D_1 = \{1, 2, 3, 4\}$, $D_2 = D_3 = \{1, 2, 3\}$, and $D_4 = \{1, 2\}$. There are four edits:

$$D_1 \times \{3\} \times \{1, 2\} \times \{1\}, \tag{9.6}$$

$$D_1 \times \{2, 3\} \times D_3 \times \{2\}, \tag{9.7}$$

$$\{1, 2, 4\} \times \{1, 3\} \times \{2, 3\} \times D_4, \tag{9.8}$$

$$\{3\} \times D_2 \times \{2, 3\} \times \{1\}. \tag{9.9}$$

There exist $4 \times 3 \times 3 \times 2 = 72$ different records. Out of these 72 records, there are 20 records that satisfy edits (9.6) to (9.9), namely:

$$\begin{array}{llll} (1,1,1,1), & (2,1,1,1), & (3,1,1,1), & (4,1,1,1), \\ (1,1,1,2), & (2,1,1,2), & (3,1,1,2), & (4,1,1,2), \\ (1,2,1,1), & (2,2,1,1), & (3,1,2,2), & (4,2,1,1), \\ (1,2,2,1), & (2,2,2,1), & (3,1,3,2), & (4,2,2,1), \\ (1,2,3,1), & (2,2,3,1), & (3,2,1,1), & (4,2,3,1). \end{array}$$

Consider the record (3,2,—,—), where "—" represents a missing value. Filling in the observed values of v_1 and v_2 into (9.6) to (9.9), the reduced edits for v_3 and v_4 are

$$D_3 \times \{2\}, \tag{9.10}$$

$$\{2, 3\} \times \{1\}. \tag{9.11}$$

We select the first missing variable, v_3. To determine whether v_3 can be imputed deductively, we eliminate all other missing variables from (9.10) and (9.11). In this case we only have to eliminate v_4, which yields one implied edit for v_3:

$$\{2, 3\}. \tag{9.12}$$

The only value from D_3 that satisfies (9.12) is $v_3 = 1$. This means that v_3 may be imputed deductively. We obtain the revised record $(3, 2, 1, -)$ and a further reduced set of edits for v_4:

$$\{2\}.$$

Since v_4 is now the only missing variable, we do not have to eliminate any variables to determine whether v_4 can be imputed deductively. We immediately observe that $v_4 = 1$ is the only value that satisfies the reduced set of edits. Hence, in this case the algorithm returns the fully imputed record (3,2,1,1). Indeed, it can be seen in the list of feasible records given above that this is the only way to impute the original record $(3, 2, -, -)$ in a consistent manner.

Now consider the record $(-, -, -, 2)$. The missing variables v_1, v_2, and v_3 have to satisfy the following reduced set of edits:

$$D_1 \times \{2, 3\} \times D_3, \tag{9.13}$$

$$\{1, 2, 4\} \times \{1, 3\} \times \{2, 3\}. \tag{9.14}$$

To determine whether the first missing variable, v_1, can be imputed deductively, we have to eliminate v_2 and v_3 from (9.13) and (9.14). This yields an empty set of edits. Since any value from D_1 satisfies the empty set of edits, we cannot impute v_1 deductively.

We select the next missing variable, v_2. By eliminating v_1 and v_3 from (9.13) and (9.14), we obtain the following edit for v_2:

$$\{2, 3\}. \tag{9.15}$$

This means that v_2 may be imputed deductively, because $v_2 = 1$ is the only value from D_2 that satisfies (9.15). We obtain the revised record $(-, 1, -, 2)$. Filling in $v_2 = 1$ in (9.13) and (9.14) yields the following reduced set of edits for v_1 and v_3:

$$\{1, 2, 4\} \times \{2, 3\}.$$

Finally, we select the missing variable v_3. It turns out that this variable cannot be imputed deductively. Therefore, the algorithm returns the partially imputed record $(-, 1, -, 2)$. Indeed, it can be seen in the list of feasible records that every consistent record with $v_4 = 2$ satisfies $v_2 = 1$. The values of v_1 and v_3 are not uniquely determined, so these variables have to be imputed by a different method. ■

9.3 The Ratio Hot Deck Method

In this section and the next, we consider the situation that a set of nonnegative numerical variables has to satisfy exactly one balance edit. More precisely, it should hold that

$$a_1 x_1 + \cdots + a_p x_p = a_{p+1} x_{p+1}, \tag{9.16}$$

$$x_j \geq 0, \qquad j = 1, \ldots, p + 1, \tag{9.17}$$

where $a_1, \ldots, a_{p+1}$ are positive coefficients. In vector form, we may write the balance edit as $\mathbf{a}^T\mathbf{x} = 0$, with $\mathbf{a} = (a_1, \ldots, a_p, -a_{p+1})^T$ and $\mathbf{x} = (x_1, \ldots, x_{p+1})^T$. Moreover, we assume that variable x_{p+1} is either always observed or imputed previously, whereas some variables from $x_1, \ldots, x_p$ may be missing. A typical example of this situation occurs for items $x_1, \ldots, x_p$ that should sum up to a total x_{p+1}, in which case $a_j = 1$ for each $j = 1, \ldots, p+1$. If the total value is reported or already imputed, but some of the items are missing, then we need a method that imputes the missing items in such a way that the balance edit is satisfied.

We shall describe two such methods, one nonparametric and one parametric. The first method is a simple extension of the hot deck imputation method from Section 7.6. Basically, instead of directly imputing values from a donor record, we impute the ratio of each missing item to the total missing amount. This ensures that the imputed record satisfies the balance edit.

Suppose that a record $\mathbf{x}_i = (\mathbf{x}_{i,\text{mis}}^T, \mathbf{x}_{i,\text{obs}}^T, x_{i,p+1})^T$ is to be imputed, where $\mathbf{x}_{i,\text{mis}}$ and $\mathbf{x}_{i,\text{obs}}$ contain the missing and observed items from $x_1, \ldots, x_p$, respectively. Similarly, we partition[1] $\mathbf{a} = (\mathbf{a}_{\text{mis}}^T, \mathbf{a}_{\text{obs}}^T, -a_{p+1})^T$. Compute

$$r_i = a_{p+1}x_{i,p+1} - \mathbf{a}_{\text{obs}}^T\mathbf{x}_{i,\text{obs}}.$$

Note that edit (9.16) will hold, if the imputed values for the missing items in the ith record satisfy $\mathbf{a}_{\text{mis}}^T\mathbf{x}_{i,\text{mis}} = r_i$. One could say that r_i represents the amount to distribute among the missing items.

Assuming that the record does not contain any errors, we have $r_i \geq 0$. If $r_i = 0$, then the deductive method of Section 9.2.3 may be used to impute zeros for all missing items. Therefore, we assume that $r_i > 0$.

Similarly to the original hot deck imputation method, a donor is found for record i among the completely observed records that satisfy (9.16) and (9.17). For instance, we could find the nearest neighbor according to some distance function. Denote the donor record by $\mathbf{x}_d$ and partition this vector just as we did $\mathbf{x}_i$. The ratio hot deck method imputes values

$$\hat{x}_{ij} = \frac{r_i x_{dj}}{\mathbf{a}_{\text{mis}}^T\mathbf{x}_{d,\text{mis}}}, \qquad j \in \text{mis}, \tag{9.18}$$

or, more formally,

$$\hat{\mathbf{x}}_{i,\text{mis}} = \frac{r_i}{\mathbf{a}_{\text{mis}}^T\mathbf{x}_{d,\text{mis}}}\mathbf{x}_{d,\text{mis}}.$$

It is easy to see that these imputations satisfy $\mathbf{a}_{\text{mis}}^T\hat{\mathbf{x}}_{i,\text{mis}} = r_i$. Therefore, a record satisfies balance edit (9.16) when it is imputed by the ratio hot deck method. Since the donor record satisfies (9.17), the same holds for the imputed record.

[1] Note that the partition into $\mathbf{a}_{\text{mis}}$ and $\mathbf{a}_{\text{obs}}$ depends on the pattern of missingness in $\mathbf{x}_i$, and hence on i. To keep the notation simple, we do not make this dependence explicit.

EXAMPLE 9.4

Consider a balance edit $3x_1 + 2x_2 + 8x_3 = 5x_4$, where all variables are nonnegative, and suppose that we are given the following record to impute by the ratio hot deck method:

x_{i1}	x_{i2}	x_{i3}	x_{i4}
6	—	—	12

where "–" denotes a missing value. The amount to distribute among the two missing variables is $r_i = 5 \times 12 - 3 \times 6 = 42$. Suppose moreover that we want to use the following donor record:

x_{d1}	x_{d2}	x_{d3}	x_{d4}
12	3	1	10

Using (9.18), we obtain the following imputations:

$$\hat{x}_{i2} = 42 \times \frac{3}{14} = 9,$$

$$\hat{x}_{i3} = 42 \times \frac{1}{14} = 3.$$

(The denominator is obtained from $2 \times 3 + 8 \times 1 = 14$.) It is easily verified that the imputed record

x_{i1}	$\hat{x}_{i2}$	$\hat{x}_{i3}$	x_{i4}
6	9	3	12

satisfies the balance edit. ■

It might happen that all items in $\mathbf{x}_{d,\text{mis}}$ are zero, which means that we cannot apply (9.18). Such a donor record does not contain information on how to distribute r_i among the missing items. Hence, we have to choose a different donor in this situation—for instance, the next nearest neighbor.

For an application of the ratio hot deck method in practice, see Section 11.3 of this book, Pannekoek and Van Veller (2004), or Pannekoek and De Waal (2005).

9.4 Imputing from a Dirichlet Distribution

As an alternative to the ratio hot deck method, we consider a parametric imputation method that also takes the edits (9.16) and (9.17) into account. This method, which uses the Dirichlet distribution as a model for the data, is due to Tempelman (2007).

Like before, we assume that the value of x_{p+1} is observed or already imputed. Note that if $x_{p+1} = 0$, zeros may be imputed for all missing items, using the deductive imputation method of Section 9.2.3. Therefore, we assume without essential loss of generality that $x_{p+1} > 0$. Define

$$y_j = \frac{a_j x_j}{a_{p+1} x_{p+1}}, \qquad j = 1, \ldots, p. \tag{9.19}$$

Under this transformation, edits (9.16) and (9.17) become

$$y_1 + y_2 + \cdots + y_p = 1, \tag{9.20}$$

$$y_j \geq 0, \qquad j = 1, \ldots, p. \tag{9.21}$$

Thus the edits restrict the vector $\mathbf{y} = (y_1, \ldots, y_p)^T$ to the p-dimensional simplex.

Recall that the gamma function is given by $\Gamma(z) = \int_0^\infty v^{z-1} e^{-v} dv$. The $(p-1)$-dimensional Dirichlet distribution is defined on the p-dimensional simplex by the following probability density function:

$$f(\mathbf{y}|\boldsymbol{\alpha}) = \frac{\Gamma\left(\sum_{j=1}^p \alpha_j\right)}{\prod_{j=1}^p \Gamma(\alpha_j)} \prod_{j=1}^p y_j^{\alpha_j - 1}, \tag{9.22}$$

$$\mathbf{y} = (y_1, \ldots, y_p)^T, \qquad y_j \geq 0, \quad j = 1, \ldots, p, \quad \sum_{j=1}^p y_j = 1,$$

$$\boldsymbol{\alpha} = (\alpha_1, \ldots, \alpha_p)^T, \qquad \alpha_j > 0, \quad j = 1, \ldots, p.$$

This distribution is $(p-1)$-dimensional because we lose one dimension by forcing the elements of the p vector $\mathbf{y}$ to sum up to one. Its first-order moments are

$$\mathrm{E}(y_j) = \frac{\alpha_j}{\sum_{j=1}^p \alpha_j}, \qquad j = 1, \ldots, p. \tag{9.23}$$

In the one-dimensional case, the Dirichlet distribution simplifies to the perhaps more familiar beta distribution given by

$$f(y|\alpha, \beta) = \frac{\Gamma(\alpha + \beta)}{\Gamma(\alpha)\Gamma(\beta)} y^{\alpha-1}(1-y)^{\beta-1}, \qquad 0 \leq y \leq 1, \quad \alpha > 0, \quad \beta > 0.$$

Note that we have incorporated the sum restriction into the density function here, to obtain the standard representation of the beta distribution.

Using the Dirichlet distribution to find imputations that satisfy (9.20) and (9.21) is attractive for two reasons. In the first place, this distribution is very flexible in the kinds of data patterns it can model. The Dirichlet distribution accommodates many shapes for various choices of $\boldsymbol{\alpha}$ in (9.22); see Tempelman

(2007) for some examples. Second, and perhaps more importantly, the Dirichlet distribution turns out to have the following property, which is crucial for our present application.

THEOREM 9.2

Suppose that $\mathbf{y}$ follows a $(p-1)$-dimensional Dirichlet distribution with parameter vector $\boldsymbol{\alpha}$. If we make partitions $\mathbf{y} = (\mathbf{y}_1^T, \mathbf{y}_2^T)^T$ and $\boldsymbol{\alpha} = (\boldsymbol{\alpha}_1^T, \boldsymbol{\alpha}_2^T)^T$, where $\mathbf{y}_1$ and $\boldsymbol{\alpha}_1$ contain p_1 elements, then the conditional distribution of $\mathbf{y}_1^* \equiv (1 - \mathbf{1}^T\mathbf{y}_2)^{-1}\mathbf{y}_1$, given $\mathbf{y}_2$, is a $(p_1 - 1)$-dimensional Dirichlet distribution with parameter vector $\boldsymbol{\alpha}_1$. Here, $\mathbf{1}$ denotes a vector of ones of length $p - p_1$.

Thus, if we take a subvector of a vector following a Dirichlet distribution, then, conditional on the discarded elements, the subvector also follows a Dirichlet distribution, after we rescale the elements to sum up to one. For a proof of Theorem 9.2, see, for example, Wilks (1962).

In the context of missing data, this theorem has the following consequence. Suppose that a set of records $\mathbf{y}_i$, $i = 1, \ldots, n$, is distributed according to a $(p-1)$-dimensional Dirichlet distribution, a fact that we summarize by writing $\mathbf{y}_i \sim \mathrm{Dir}_{p-1}(\boldsymbol{\alpha})$, $i = 1, \ldots, n$. For each record, we partition the data vector into a missing part and an observed part $\mathbf{y}_i = (\mathbf{y}_{i,\mathrm{mis}}^T, \mathbf{y}_{i,\mathrm{obs}}^T)^T$, and denote the number of missing elements by m_i. Then by Theorem 9.2

$$\mathbf{y}_{i,\mathrm{mis}}^* | \mathbf{y}_{i,\mathrm{obs}}, \boldsymbol{\alpha} \sim \mathrm{Dir}_{m_i - 1}(\boldsymbol{\alpha}_{\mathrm{mis}}), \tag{9.24}$$

where $\mathbf{y}_{i,\mathrm{mis}}^* = (1 - \mathbf{1}^T\mathbf{y}_{i,\mathrm{obs}})^{-1}\mathbf{y}_{i,\mathrm{mis}}$, $\mathbf{1}$ is a vector of $p - m_i$ ones, and $\boldsymbol{\alpha}_{\mathrm{mis}}$ contains the parameters associated with $\mathbf{y}_{i,\mathrm{mis}}$.

This property can be exploited in an application of the EM algorithm to obtain imputations for $\mathbf{y}_{i,\mathrm{mis}}$, given the observed data. Recall that the EM algorithm was introduced in Section 8.3.3 as a device for obtaining maximum likelihood estimates in the presence of missing data. For the Dirichlet distribution, it is easy to derive the loglikelihood function from (9.22):

$$L(\boldsymbol{\alpha}|\mathbf{y}_1, \ldots, \mathbf{y}_n) = n \ln \Gamma(\sum_{j=1}^{p} \alpha_j) - n \sum_{j=1}^{p} \ln \Gamma(\alpha_j) + \sum_{i=1}^{n}\sum_{j=1}^{p} (\alpha_j - 1) \ln y_{ij},$$

for a data set consisting of records $\mathbf{y}_i = (y_{i1}, \ldots, y_{ip})^T$, $i = 1, \ldots, n$. To find the maximum likelihood estimate of $\boldsymbol{\alpha}$, the vector $\mathbf{s}(\boldsymbol{\alpha})$ of first-order partial derivatives of $L(\boldsymbol{\alpha}|\mathbf{y}_1, \ldots, \mathbf{y}_n)$ must be set equal to zero. We need to solve

$$s_j(\boldsymbol{\alpha}) = \frac{\partial L(\boldsymbol{\alpha}|\mathbf{y}_1, \ldots, \mathbf{y}_n)}{\partial \alpha_j} = n\Psi(\sum_{l=1}^{p} \alpha_l) - n\Psi(\alpha_j) + \sum_{i=1}^{n} \ln y_{ij} = 0, \tag{9.25}$$

for $j = 1, \ldots, p$, with $\Psi(z) \equiv d \ln \Gamma(z)/dz = \Gamma'(z)/\Gamma(z)$, the so-called digamma function. These equations cannot be solved in closed form, so we need to use an iterative method. Tempelman (2007) suggests using the Newton–Raphson method, which yields a sequence of estimates $\boldsymbol{\alpha}^{(0)}, \boldsymbol{\alpha}^{(1)}, \ldots$ given by

$$\boldsymbol{\alpha}^{(t+1)} = \boldsymbol{\alpha}^{(t)} + \mathbf{I}^{-1}(\boldsymbol{\alpha})|_{\boldsymbol{\alpha}=\boldsymbol{\alpha}^{(t)}} \mathbf{s}(\boldsymbol{\alpha})|_{\boldsymbol{\alpha}=\boldsymbol{\alpha}^{(t)}}, \qquad t = 0, 1, \ldots,$$

with $\mathbf{I}(\boldsymbol{\alpha})$ the observed information matrix, which is just the negative of the Hessian matrix of second-order partial derivatives of $L(\boldsymbol{\alpha}|\mathbf{y}_1, \ldots, \mathbf{y}_n)$. Note that in this case the second-order derivatives $\partial^2 L(\boldsymbol{\alpha}|\mathbf{y}_1, \ldots, \mathbf{y}_n)/\partial\alpha_j\partial\alpha_l = \partial s_j(\boldsymbol{\alpha})/\partial\alpha_l$ do not depend on the values of y_{ij}.

When there are missing data, we cannot evaluate expression (9.25) directly. The EM algorithm replaces each missing term $\ln y_{ij}$ in (9.25) by its expected value, conditional on the observed elements of $\mathbf{y}_i$. Tempelman (2007) shows that for $\mathbf{y} \sim \mathrm{Dir}_{p-1}(\boldsymbol{\alpha})$, it holds that

$$\text{(9.26)} \qquad \mathrm{E}(\ln y_j|\boldsymbol{\alpha}) = \Psi(\alpha_j) - \Psi(\sum_{l=1}^{p} \alpha_l), \qquad j = 1, \ldots, p.$$

Now suppose that we are in an expectation step of the EM algorithm and that the current vector of parameter estimates is $\boldsymbol{\alpha}^{(t)}$. Then, by (9.24) and (9.26) we know that

$$\mathrm{E}(\ln y_{ij}^*|\mathbf{y}_{i,\mathrm{obs}}, \boldsymbol{\alpha}^{(t)}) = \Psi(\alpha_j^{(t)}) - \Psi(\sum_{l \in \mathrm{mis}} \alpha_l^{(t)}), \qquad j \in \mathrm{mis},$$

with $y_{ij}^* = (1 - \mathbf{1}^T\mathbf{y}_{i,\mathrm{obs}})^{-1} y_{ij}$ and $\mathbf{1}$ a vector of $p - m_i$ ones. It follows that

$$\mathrm{E}(\ln y_{ij}|\mathbf{y}_{i,\mathrm{obs}}, \boldsymbol{\alpha}^{(t)}) = \ln(1 - \mathbf{1}^T\mathbf{y}_{i,\mathrm{obs}}) + \Psi(\alpha_j^{(t)}) - \Psi(\sum_{l \in \mathrm{mis}} \alpha_l^{(t)}), \qquad j \in \mathrm{mis}.$$

The EM algorithm now consists of the following steps. We start with initial parameter estimates $\boldsymbol{\alpha}^{(0)}$. In the first expectation step, the missing terms $\ln y_{ij}$ in (9.25) are replaced by $\mathrm{E}(\ln y_{ij}|\mathbf{y}_{i,\mathrm{obs}}, \boldsymbol{\alpha}^{(0)})$, making it possible to evaluate $\mathbf{s}(\boldsymbol{\alpha})|_{\boldsymbol{\alpha}=\boldsymbol{\alpha}^{(0)}}$. In the first maximization step, we perform one step of the Newton–Raphson method to obtain new parameter estimates $\boldsymbol{\alpha}^{(1)}$. Subsequently, the expectation and maximization steps are alternated until the algorithm "converges"—for instance, when the parameter estimates do not show any substantial change between iterations anymore.

After convergence of the EM algorithm, the Dirichlet distribution with the final parameter estimates is used to impute the missing data items. We can choose either to impute the expected values (9.23) or to draw from the conditional distribution (9.24). The first choice yields a deterministic imputation method, whereas the second choice yields a stochastic imputation method. We observe that, for both methods, the imputed values automatically satisfy (9.20) and (9.21), because these edits are incorporated into the Dirichlet distribution. Finally, we transform the imputed records back to the original data space by applying the inverse transformation of (9.19).

For a more detailed derivation and a discussion of the EM algorithm in this context, see Tempelman (2007). Tempelman also reports on an application of the Dirichlet imputation method to real-world data.

EXAMPLE 9.4 (*continued*)

Consider again the balance edit $3x_1 + 2x_2 + 8x_3 = 5x_4$, where all variables are nonnegative, and the record

x_{i1}	x_{i2}	x_{i3}	x_{i4}
6	—	—	12

where "–" denotes a missing value. This time we want to obtain imputations for the missing values by drawing from a Dirichlet distribution.

We begin by transforming the data according to (9.19). This yields $\mathbf{y}_i = (y_{i1}, y_{i2}, y_{i3})^T = (3/10, -, -)^T$. Suppose that an application of the EM algorithm to the incomplete data set to which this record belongs reveals that the transformed data follow a Dirichlet distribution with parameter vector $\boldsymbol{\alpha} = (2, 3, 4)^T$. According to (9.24),

$$(y_{i2}^*, y_{i3}^*)^T = \frac{10}{7}(y_{i2}, y_{i3})^T \sim \text{Dir}_1(3, 4).$$

That is, after rescaling, the two missing values in $\mathbf{y}_i$ follow a one-dimensional Dirichlet distribution (i.e., a beta distribution) with parameter vector $(3, 4)^T$.

Suppose that we want to apply the deterministic version of the Dirichlet imputation method. We compute the expected values of y_{i2}^* and y_{i3}^* from (9.23): $\hat{y}_{i2}^* = 3/7$ and $\hat{y}_{i3}^* = 4/7$. Dividing out the rescaling factor, we find $\hat{\mathbf{y}}_i = (y_{i1}, \hat{y}_{i2}, \hat{y}_{i3})^T = (3/10, 3/10, 4/10)^T$. Finally, we transform the imputed values back:

$$\hat{x}_{i2} = \frac{3}{10} \times \frac{60}{2} = 9,$$

$$\hat{x}_{i3} = \frac{4}{10} \times \frac{60}{8} = 3.$$

Thus we obtain the following imputed record:

x_{i1}	$\hat{x}_{i2}$	$\hat{x}_{i3}$	x_{i4}
6	9	3	12

This happens to be the same imputed record as the one found using the ratio hot deck method. We already noted that it satisfies the balance edit. ■

9.5 Imputing from a Singular Normal Distribution

In this section, we consider the situation where a record $\mathbf{x} = (x_1, \ldots, x_p)^T$ of numerical variables has to satisfy r linear balance edits as well as q linear inequality edits. These edits can be written as

$$\mathbf{Ax} = \mathbf{b}, \tag{9.27}$$

$$\mathbf{Bx} \geq \mathbf{c}, \tag{9.28}$$

with $\mathbf{A}$ an $r \times p$ matrix of coefficients, $\mathbf{b}$ an r vector of constants, $\mathbf{B}$ a $q \times p$ matrix of coefficients, and $\mathbf{c}$ a q vector of constants. We assume that $\mathbf{A}$ has full row rank—that is, that there are no redundant balance edits.

If $r = 1$ and if all inequality edits are nonnegativity edits, the ratio hot deck method of Section 9.3 and the Dirichlet method of Section 9.4 provide valid imputations. When $r > 1$ or when other inequality edits are specified, these methods cannot be applied in general. If the structure of the balance edits allows this, we could in principle try to use these methods in a hierarchical fashion, considering one balance edit at a time: First impute an overall total value, then impute the subtotals that contribute to the overall total, then the items contributing to each subtotal, and so on. We may run into problems, however, for instance when the previously imputed value for a subtotal is less than the sum of the observed values for some of the contributing items. Also, inequality edits other than nonnegativity edits still have to be taken into account somehow. Clearly, this approach can easily become very messy.

Tempelman (2007) establishes an imputation method based on assuming a multivariate normal distribution for the data, such that the imputed values automatically satisfy edits (9.27) and (9.28).[2] Ultimately, the underlying idea is the same as for the Dirichlet imputation method: We draw the imputations from a distribution with the edits incorporated into it. We shall first consider the case that only balance edits are specified, in Section 9.5.1. The data are then modeled by a singular normal distribution—that is, a normal distribution with a singular covariance matrix. In Section 9.5.2 we adapt the method to also take

[2] Actually, Tempelman (2007) considers $\mathbf{Ax} = \mathbf{0}$ instead of (9.27), but the method is also applicable in our more general situation.

inequality edits into account. In this case, the data are modeled by a truncated singular normal distribution.

9.5.1 USING THE SINGULAR NORMAL DISTRIBUTION

The multivariate singular normal distribution is an extension of the ordinary multivariate normal distribution, when the covariance matrix $\mathbf{\Sigma}$ does not have full rank. The singular normal distribution seems more appropriate for modeling data subject to balance edits than the ordinary normal distribution, because the edit structure introduces linear dependence in the covariance matrix. The fact that the covariance matrix of a distribution that incorporates the edits (9.27) is singular follows from

$$\mathbf{\Sigma A}^T = \mathrm{E}\left[(\mathbf{x}-\boldsymbol{\mu})(\mathbf{x}-\boldsymbol{\mu})^T\right]\mathbf{A}^T = \mathrm{E}\left[(\mathbf{x}-\boldsymbol{\mu})(\mathbf{Ax}-\mathbf{A}\boldsymbol{\mu})^T\right] = \mathbf{O},$$

with $\mathbf{O}$ a matrix of zeros. Here, we use that $\mathbf{Ax} = \mathbf{A}\boldsymbol{\mu} = \mathbf{b}$. Since the covariance matrix cannot be inverted the ordinary way, the density of the p-dimensional singular normal distribution cannot be defined on the whole of $\mathbb{R}^p$, but only on a subspace defined by the balance edits.

Suppose that we have a data set of records $\mathbf{x}_i$, $i = 1, \ldots, n$, that follow a p-dimensional singular normal distribution with mean $\boldsymbol{\mu}$ and covariance matrix $\mathbf{\Sigma}$, a fact which we abbreviate by $\mathbf{x}_i \sim N_p(\boldsymbol{\mu}, \mathbf{\Sigma})$, $i = 1, \ldots, n$. Here, $\boldsymbol{\mu}$ is a vector such that $\mathbf{A}\boldsymbol{\mu} = \mathbf{b}$—that is, $\boldsymbol{\mu}$ satisfies (9.27)—and $\mathbf{\Sigma}$ is a positive semidefinite, symmetric matrix with the column span of $\mathbf{A}^T$ as its null space, which we denote by L:

$$L = \left\{\mathbf{z} \in \mathbb{R}^p : \mathbf{\Sigma z} = \mathbf{0}\right\} = \left\{\mathbf{z} \in \mathbb{R}^p : \mathbf{z} = \mathbf{A}^T\mathbf{v} \text{ for some } \mathbf{v} \in \mathbb{R}^r\right\}. \tag{9.29}$$

We assume for now that $\mathbf{x}_i$, $i = 1, \ldots, n$, all satisfy the system of balance edits given by (9.27). Obviously, this assumption needs to be relaxed when we consider missing values in the data set; we shall therefore assume later that only the completely observed records satisfy (9.27). Moreover, we assume throughout that system (9.27) describes all linear equalities that should hold for the records in our data set. The latter is merely a technical restriction needed below; in practice, one would probably want the system of balance edits to capture all structural linear equalities anyway.

Since the $r \times p$ matrix $\mathbf{A}$ has full row rank, we find that L is an r-dimensional vector space. This implies that rank $[\mathbf{\Sigma}] = p - r$. Hence each added (nonredundant) balance edit lowers rank $[\mathbf{\Sigma}]$ by one.

The orthogonal complement of L, denoted by $L^\perp$, is defined as the set of all vectors in $\mathbb{R}^p$ that are orthogonal to L:

$$L^\perp = \left\{\mathbf{z} \in \mathbb{R}^p : \mathbf{z}^T\mathbf{w} = 0 \text{ for each } \mathbf{w} \in L\right\}.$$

A standard result in linear algebra states that $L^\perp$ is a $(p - r)$-dimensional vector space. Khatri (1968) and Tempelman (2007) show that the probability density function of $\mathbf{x} \sim N_p(\boldsymbol{\mu}, \mathbf{\Sigma})$ is defined on $\boldsymbol{\mu} + L^\perp$, which is an affine subspace of

$\mathbb{R}^p$, by the following expression:

$$\varphi(\mathbf{x}|\boldsymbol{\mu}, \boldsymbol{\Sigma}) = (2\pi)^{-(p-r)/2}|\boldsymbol{\Sigma}|_{p-r}^{-1/2} \exp\left[-\frac{1}{2}(\mathbf{x}-\boldsymbol{\mu})^T\boldsymbol{\Sigma}^+(\mathbf{x}-\boldsymbol{\mu})\right], \tag{9.30}$$

with $|\boldsymbol{\Sigma}|_{p-r}$ the product of the nonzero eigenvalues of $\boldsymbol{\Sigma}$, of which there are $p-r$. This is called a pseudo-determinant. Also, $\boldsymbol{\Sigma}^+$ denotes the Moore–Penrose inverse of $\boldsymbol{\Sigma}$. This is the unique generalized inverse of $\boldsymbol{\Sigma}$ that has the following properties:

1. $\boldsymbol{\Sigma}\boldsymbol{\Sigma}^+\boldsymbol{\Sigma} = \boldsymbol{\Sigma}$.
2. $\boldsymbol{\Sigma}^+\boldsymbol{\Sigma}\boldsymbol{\Sigma}^+ = \boldsymbol{\Sigma}^+$.
3. $(\boldsymbol{\Sigma}\boldsymbol{\Sigma}^+)^T = \boldsymbol{\Sigma}\boldsymbol{\Sigma}^+$.
4. $(\boldsymbol{\Sigma}^+\boldsymbol{\Sigma})^T = \boldsymbol{\Sigma}^+\boldsymbol{\Sigma}$.

For a proof of the uniqueness of $\boldsymbol{\Sigma}^+$, see, for example, Harville (1997). Note that $\varphi(\mathbf{x}|\boldsymbol{\mu}, \boldsymbol{\Sigma})$ becomes the density function of the ordinary, nonsingular normal distribution if $r = 0$, and in that case $L^\perp = \mathbb{R}^p$.

As L is just the column span of $\mathbf{A}^T$, it follows that

$$\mathbf{z} \in L^\perp \Leftrightarrow \mathbf{z}^T\mathbf{A}^T = \mathbf{0}^T \Leftrightarrow \mathbf{A}\mathbf{z} = \mathbf{0}.$$

Since $\mathbf{A}\boldsymbol{\mu} = \mathbf{b}$, we conclude that $\mathbf{A}\mathbf{x} = \mathbf{b}$ for each $\mathbf{x} \in \boldsymbol{\mu} + L^\perp$. Thus, every vector drawn from $N_p(\boldsymbol{\mu}, \boldsymbol{\Sigma})$ automatically satisfies the balance edits (9.27). Conversely, it is easy to see that if $\mathbf{x}$ satisfies $\mathbf{A}\mathbf{x} = \mathbf{b}$, then $(\mathbf{x} - \boldsymbol{\mu}) \in L^\perp$. This means that any vector in $\mathbb{R}^p$ that satisfies (9.27) lies in the affine subspace $\boldsymbol{\mu} + L^\perp$ and can therefore be drawn from $N_p(\boldsymbol{\mu}, \boldsymbol{\Sigma})$, at least theoretically.

EXAMPLE 9.5

Consider a set of vectors $\mathbf{x}_i = (x_{i1}, x_{i2}, x_{i3})^T$, $i = 1, \ldots, n$, that should conform to one balance edit:

$$x_1 + x_2 = x_3. \tag{9.31}$$

Suppose that $\mathbf{x}_i \sim N_3(\boldsymbol{\mu}, \boldsymbol{\Sigma})$, with

$$\boldsymbol{\mu} = \begin{pmatrix} 2 \\ 3 \\ 5 \end{pmatrix}, \qquad \boldsymbol{\Sigma} = \begin{pmatrix} 3 & 1 & 4 \\ 1 & 3 & 4 \\ 4 & 4 & 8 \end{pmatrix}.$$

Note that $\boldsymbol{\mu}$ satisfies edit (9.31).

It is easily verified that the null space of $\boldsymbol{\Sigma}$ is

$$L = \left\{\mathbf{z} \in \mathbb{R}^3 \ : \ \mathbf{z} = s(1, 1, -1)^T, \ s \in \mathbb{R}\right\}.$$

The orthogonal complement $L^\perp$ consists of all vectors that are orthogonal to $(1, 1, -1)^T$, that is,

$$L^\perp = \left\{\mathbf{z} \in \mathbb{R}^3 \ : \ z_1 + z_2 = z_3\right\}.$$

The affine subspace $\boldsymbol{\mu} + L^\perp$ of $\mathbb{R}^3$ is now defined as

$$\boldsymbol{\mu} + L^\perp = \left\{\mathbf{x} \in \mathbb{R}^3 \ : \ \mathbf{x} = \begin{pmatrix} 2 + z_1 \\ 3 + z_2 \\ 5 + z_3 \end{pmatrix}, z_1 + z_2 = z_3 \right\},$$

and each $\mathbf{x} \in \boldsymbol{\mu} + L^\perp$ clearly satisfies (9.31). Since $N_3(\boldsymbol{\mu}, \boldsymbol{\Sigma})$ is defined on $\boldsymbol{\mu} + L^\perp$, it follows that every $\mathbf{x} \sim N_3(\boldsymbol{\mu}, \boldsymbol{\Sigma})$ satisfies (9.31). ■

For the multivariate (singular or nonsingular) normal distribution, we have the following analog of Theorem 9.2 and (9.24).

THEOREM 9.3

Suppose that $\mathbf{x}_i \sim N_p(\boldsymbol{\mu}, \boldsymbol{\Sigma})$ is partitioned into an observed part and a missing part, $\mathbf{x}_i = (\mathbf{x}_{i,\text{mis}}^T, \mathbf{x}_{i,\text{obs}}^T)^T$, and that $\boldsymbol{\mu}$ and $\boldsymbol{\Sigma}$ are partitioned accordingly:

$$\boldsymbol{\mu} = \begin{pmatrix} \boldsymbol{\mu}_{\text{mis}} \\ \boldsymbol{\mu}_{\text{obs}} \end{pmatrix}, \qquad \boldsymbol{\Sigma} = \begin{pmatrix} \boldsymbol{\Sigma}_{\text{mis,mis}} & \boldsymbol{\Sigma}_{\text{mis,obs}} \\ \boldsymbol{\Sigma}_{\text{obs,mis}} & \boldsymbol{\Sigma}_{\text{obs,obs}} \end{pmatrix}.$$

If $\mathbf{x}_{i,\text{mis}}$ contains m_i elements, then it holds that

$$\mathbf{x}_{i,\text{mis}} | \mathbf{x}_{i,\text{obs}}, \boldsymbol{\mu}, \boldsymbol{\Sigma} \sim N_{m_i}(\boldsymbol{\mu}_{i,\text{mis.obs}}, \boldsymbol{\Sigma}_{\text{mis,mis.obs}}),$$

with

$$\boldsymbol{\mu}_{i,\text{mis.obs}} = \boldsymbol{\mu}_{\text{mis}} + \boldsymbol{\Sigma}_{\text{mis,obs}} \boldsymbol{\Sigma}_{\text{obs,obs}}^{-} (\mathbf{x}_{i,\text{obs}} - \boldsymbol{\mu}_{\text{obs}}),$$

$$\boldsymbol{\Sigma}_{\text{mis,mis.obs}} = \boldsymbol{\Sigma}_{\text{mis,mis}} - \boldsymbol{\Sigma}_{\text{mis,obs}} \boldsymbol{\Sigma}_{\text{obs,obs}}^{-} \boldsymbol{\Sigma}_{\text{obs,mis}}.$$

Here, $\boldsymbol{\Sigma}_{\text{obs,obs}}^{-}$ denotes a generalized inverse of $\boldsymbol{\Sigma}_{\text{obs,obs}}$.

For a proof of Theorem 9.3, see, for example, Anderson (1984). We can use this result to impute the missing values in our data set, once we have obtained estimates of $\boldsymbol{\mu}$ and $\boldsymbol{\Sigma}$. Khatri (1968) and Tempelman (2007) show that the maximum likelihood estimates of $\boldsymbol{\mu}$ and $\boldsymbol{\Sigma}$ for data from a singular normal distribution are identical to the corresponding estimates for data from a nonsingular normal distribution (see also Section 8.3.3):

$$\hat{\boldsymbol{\mu}} = \frac{1}{n}\sum_{i=1}^{n} \mathbf{x}_i, \tag{9.32}$$

$$\hat{\boldsymbol{\Sigma}} = \frac{1}{n}\sum_{i=1}^{n} (\mathbf{x}_i - \hat{\boldsymbol{\mu}})(\mathbf{x}_i - \hat{\boldsymbol{\mu}})^T. \tag{9.33}$$

These estimates concur with the balance edits in the following sense: $\mathbf{A}\hat{\boldsymbol{\mu}} = \mathbf{b}$ and the null space of $\hat{\boldsymbol{\Sigma}}$ is L from (9.29). Note that both properties are important, because otherwise the density functions of the singular normal distribution with the estimated parameters and the true distribution would be defined on different affine subspaces of $\mathbb{R}^p$.

The proof of the first identity is more or less trivial:

$$\mathbf{A}\hat{\boldsymbol{\mu}} = \frac{1}{n}\sum_{i=1}^{n} \mathbf{A}\mathbf{x}_i = \frac{1}{n}\sum_{i=1}^{n} \mathbf{b} = \mathbf{b}.$$

To prove the second identity, we begin by noting that, since $\mathbf{A}\mathbf{x}_i = \mathbf{A}\hat{\boldsymbol{\mu}} = \mathbf{b}$,

$$\hat{\boldsymbol{\Sigma}}\mathbf{A}^T = \frac{1}{n}\sum_{i=1}^{n} (\mathbf{x}_i - \hat{\boldsymbol{\mu}})(\mathbf{x}_i - \hat{\boldsymbol{\mu}})^T\mathbf{A}^T = \frac{1}{n}\sum_{i=1}^{n} (\mathbf{x}_i - \hat{\boldsymbol{\mu}})(\mathbf{A}\mathbf{x}_i - \mathbf{A}\hat{\boldsymbol{\mu}})^T = \mathbf{O}.$$

This shows that L, being the column span of $\mathbf{A}^T$, is contained in the null space of $\hat{\boldsymbol{\Sigma}}$. To prove the reverse inclusion, suppose that $\hat{\boldsymbol{\Sigma}}\mathbf{z} = \mathbf{0}$ for some vector $\mathbf{z} \notin L$. Then

$$\frac{1}{n}\sum_{i=1}^{n} (\mathbf{x}_i - \hat{\boldsymbol{\mu}})(\mathbf{x}_i - \hat{\boldsymbol{\mu}})^T\mathbf{z} = \mathbf{0}.$$

Introduce $e_i = (\mathbf{x}_i - \hat{\boldsymbol{\mu}})^T\mathbf{z} = \mathbf{z}^T(\mathbf{x}_i - \hat{\boldsymbol{\mu}})$. Then, by premultiplying the previous display with $\mathbf{z}^T$, we find that $\frac{1}{n}\sum_{i=1}^{n} e_i^2 = 0$. This implies that $e_i = 0$ for $i = 1, \ldots, n$, that is,

$$\mathbf{z}^T\mathbf{x}_i = \mathbf{z}^T\hat{\boldsymbol{\mu}} \equiv b, \qquad i = 1, \ldots, n,$$

for some constant b. In other words: $\mathbf{z}^T\mathbf{x}_i = b$ describes a linear equality that holds for all records in our data set. By the assumption made earlier that (9.27) describes all linear equalities that should hold for the records in our data set, $\mathbf{z}$ must be contained in the column span of $\mathbf{A}^T$. This completes the proof that the null space of $\hat{\boldsymbol{\Sigma}}$ is identical to L.

Replacing $\boldsymbol{\mu}$ and $\boldsymbol{\Sigma}$ by their maximum likelihood estimates $\hat{\boldsymbol{\mu}}$ and $\hat{\boldsymbol{\Sigma}}$, we can use Theorem 9.3 to derive imputations. Conditional on the observed values in record $\mathbf{x}_i$, we can either impute the expected values $\hat{\boldsymbol{\mu}}_{i,\text{mis.obs}}$ for $\mathbf{x}_{i,\text{mis}}$, which yields a deterministic imputation method, or we can draw from

$$N_{m_i}(\hat{\boldsymbol{\mu}}_{i,\text{mis.obs}}, \hat{\boldsymbol{\Sigma}}_{\text{mis,mis.obs}}),$$

which yields a stochastic imputation method. For both methods, the imputed records satisfy (9.27), as we shall see below.

Expressions (9.32) and (9.33) are based on a completely observed data set, so they cannot be used for the purpose of imputation. We are therefore left with the problem of finding maximum likelihood estimates of $\boldsymbol{\mu}$ and $\boldsymbol{\Sigma}$ in the presence of missing data. Just as with the Dirichlet method, the EM algorithm can be used for this.

In this case, an iteration of the EM algorithm works as follows [see also Schafer (1997), Tempelman (2007), and Section 8.3.3 of this book]. Suppose that the current estimates of $\boldsymbol{\mu}$ and $\boldsymbol{\Sigma}$ are $\boldsymbol{\mu}^{(t)}$ and $\boldsymbol{\Sigma}^{(t)}$. Define

$$\mathbf{x}_i^{*(t)} = \begin{pmatrix} \boldsymbol{\mu}_{i,\text{mis.obs}}^{(t)} \\ \mathbf{x}_{i,\text{obs}} \end{pmatrix}.$$

In $\mathbf{x}_i^{*(t)}$ the missing values $\mathbf{x}_{i,\text{mis}}$ are replaced by their conditional expectation $\boldsymbol{\mu}_{i,\text{mis.obs}}^{(t)}$ from Theorem 9.3, based on $\mathbf{x}_{i,\text{obs}}$, $\boldsymbol{\mu}^{(t)}$ and $\boldsymbol{\Sigma}^{(t)}$. Also, define a $p \times p$ matrix $\mathbf{V}_i^{(t)}$ as follows: The (j,l)th element of $\mathbf{V}_i^{(t)}$ equals the corresponding element of $\boldsymbol{\Sigma}_{\text{mis,mis.obs}}^{(t)}$ if the jth and lth elements of $\mathbf{x}_i$ are both missing, and zero otherwise. Then the estimates of $\boldsymbol{\mu}$ and $\boldsymbol{\Sigma}$ are updated as follows:

$$\boldsymbol{\mu}^{(t+1)} = \frac{1}{n}\sum_{i=1}^{n} \mathbf{x}_i^{*(t)}, \tag{9.34}$$

$$\boldsymbol{\Sigma}^{(t+1)} = \frac{1}{n}\sum_{i=1}^{n} \left[(\mathbf{x}_i^{*(t)} - \boldsymbol{\mu}^{(t+1)})(\mathbf{x}_i^{*(t)} - \boldsymbol{\mu}^{(t+1)})^T + \mathbf{V}_i^{(t)}\right]. \tag{9.35}$$

Starting from initial values $\boldsymbol{\mu}^{(0)}$ and $\boldsymbol{\Sigma}^{(0)}$, the EM algorithm now consists of iterating (9.34) and (9.35) until convergence.

THEOREM 9.4

If the initial parameter estimates $\boldsymbol{\mu}^{(0)}$ and $\boldsymbol{\Sigma}^{(0)}$ concur with system (9.27), in the sense that $\mathbf{A}\boldsymbol{\mu}^{(0)} = \mathbf{b}$ and $\boldsymbol{\Sigma}^{(0)}\mathbf{A}^T = \mathbf{O}$, then the same holds for the final parameter estimates from the EM algorithm.

Proof. The proof works by induction on the number of iterations. Suppose that $\boldsymbol{\mu}^{(t)}$ and $\boldsymbol{\Sigma}^{(t)}$ concur with the balance edits; then we must show that the same holds for $\boldsymbol{\mu}^{(t+1)}$ and $\boldsymbol{\Sigma}^{(t+1)}$.

For $\mathbf{A}\boldsymbol{\mu}^{(t+1)} = \mathbf{b}$ to hold, it suffices by (9.34) that $\mathbf{A}\mathbf{x}_i^{*(t)} = \mathbf{b}$ for $i = 1, \ldots, n$, or equivalently

$$\mathbf{A}_{\text{mis}}\boldsymbol{\mu}^{(t)}_{i,\text{mis.obs}} = \mathbf{b} - \mathbf{A}_{\text{obs}}\mathbf{x}_{i,\text{obs}},$$

where $\mathbf{A}$ is partitioned into $\mathbf{A}_{\text{mis}}$ and $\mathbf{A}_{\text{obs}}$. By definition,

$$\mathbf{A}_{\text{mis}}\boldsymbol{\mu}^{(t)}_{i,\text{mis.obs}} = \mathbf{A}_{\text{mis}}\boldsymbol{\mu}^{(t)}_{\text{mis}} + \mathbf{A}_{\text{mis}}\boldsymbol{\Sigma}^{(t)}_{\text{mis,obs}}\boldsymbol{\Sigma}^{(t),-}_{\text{obs,obs}}(\mathbf{x}_{i,\text{obs}} - \boldsymbol{\mu}^{(t)}_{\text{obs}}),$$

in obvious notation. Since $\mathbf{A}\boldsymbol{\mu}^{(t)} = \mathbf{b}$, we have $\mathbf{A}_{\text{mis}}\boldsymbol{\mu}^{(t)}_{\text{mis}} = \mathbf{b} - \mathbf{A}_{\text{obs}}\boldsymbol{\mu}^{(t)}_{\text{obs}}$. Moreover, it follows from $\mathbf{A}\boldsymbol{\Sigma}^{(t)} = \mathbf{O}$ that $\mathbf{A}_{\text{mis}}\boldsymbol{\Sigma}^{(t)}_{\text{mis,obs}} = -\mathbf{A}_{\text{obs}}\boldsymbol{\Sigma}^{(t)}_{\text{obs,obs}}$. Hence

$$\mathbf{A}_{\text{mis}}\boldsymbol{\mu}^{(t)}_{i,\text{mis.obs}} = \mathbf{b} - \mathbf{A}_{\text{obs}}\boldsymbol{\mu}^{(t)}_{\text{obs}} - \mathbf{A}_{\text{obs}}\boldsymbol{\Sigma}^{(t)}_{\text{obs,obs}}\boldsymbol{\Sigma}^{(t),-}_{\text{obs,obs}}(\mathbf{x}_{i,\text{obs}} - \boldsymbol{\mu}^{(t)}_{\text{obs}}).$$

In the case that $\boldsymbol{\Sigma}^{(t)}_{\text{obs,obs}}$ is nonsingular, $\boldsymbol{\Sigma}^{(t),-}_{\text{obs,obs}}$ is just the ordinary inverse, and it follows immediately that $\mathbf{A}_{\text{mis}}\boldsymbol{\mu}^{(t)}_{i,\text{mis.obs}} = \mathbf{b} - \mathbf{A}_{\text{obs}}\mathbf{x}_{i,\text{obs}}$. Tempelman (2007) shows that the statement is also true if $\boldsymbol{\Sigma}^{(t)}_{\text{obs,obs}}$ is singular, by examining the spectral decomposition. We omit this part of the proof here.

To show that $\boldsymbol{\Sigma}^{(t+1)}\mathbf{A}^T = \mathbf{O}$, or equivalently $\mathbf{A}\boldsymbol{\Sigma}^{(t+1)} = \mathbf{O}$, we begin by observing that

$$\mathbf{A}\mathbf{V}_i^{(t)} = (\mathbf{A}_{\text{mis}}, \mathbf{A}_{\text{obs}}) \begin{pmatrix} \boldsymbol{\Sigma}^{(t)}_{\text{mis,mis.obs}} & \mathbf{O} \\ \mathbf{O} & \mathbf{O} \end{pmatrix} = \left(\mathbf{A}_{\text{mis}}\boldsymbol{\Sigma}^{(t)}_{\text{mis,mis.obs}}, \mathbf{O}\right).$$

From the definition of $\boldsymbol{\Sigma}^{(t)}_{\text{mis,mis.obs}}$, we see that

$$\mathbf{A}_{\text{mis}}\boldsymbol{\Sigma}^{(t)}_{\text{mis,mis.obs}} = \mathbf{A}_{\text{mis}}\boldsymbol{\Sigma}^{(t)}_{\text{mis,mis}} - \mathbf{A}_{\text{mis}}\boldsymbol{\Sigma}^{(t)}_{\text{mis,obs}}\boldsymbol{\Sigma}^{(t),-}_{\text{obs,obs}}\boldsymbol{\Sigma}^{(t)}_{\text{obs,mis}}.$$

As noted before, it follows from $\mathbf{A}\boldsymbol{\Sigma}^{(t)} = \mathbf{O}$ that $\mathbf{A}_{\text{mis}}\boldsymbol{\Sigma}^{(t)}_{\text{mis,obs}} = -\mathbf{A}_{\text{obs}}\boldsymbol{\Sigma}^{(t)}_{\text{obs,obs}}$. By the same token, it also holds that $\mathbf{A}_{\text{mis}}\boldsymbol{\Sigma}^{(t)}_{\text{mis,mis}} = -\mathbf{A}_{\text{obs}}\boldsymbol{\Sigma}^{(t)}_{\text{obs,mis}}$. Hence

$$\mathbf{A}_{\text{mis}}\boldsymbol{\Sigma}^{(t)}_{\text{mis,mis.obs}} = -\mathbf{A}_{\text{obs}}\boldsymbol{\Sigma}^{(t)}_{\text{obs,mis}} + \mathbf{A}_{\text{obs}}\boldsymbol{\Sigma}^{(t)}_{\text{obs,obs}}\boldsymbol{\Sigma}^{(t),-}_{\text{obs,obs}}\boldsymbol{\Sigma}^{(t)}_{\text{obs,mis}},$$

from which it follows that $\mathbf{A}_{\text{mis}}\boldsymbol{\Sigma}^{(t)}_{\text{mis,mis.obs}} = \mathbf{O}$ if $\boldsymbol{\Sigma}^{(t)}_{\text{obs,obs}}$ is nonsingular. For a proof of the same statement in the case that $\boldsymbol{\Sigma}^{(t)}_{\text{obs,obs}}$ is singular, we refer again to Tempelman (2007). We conclude that $\mathbf{A}\mathbf{V}_i^{(t)} = \mathbf{O}$ for each $i = 1, \ldots, n$.

Furthermore, we already established that

$$\mathbf{A}\mathbf{x}_i^{*(t)} = \mathbf{A}_{\text{mis}}\boldsymbol{\mu}^{(t)}_{i,\text{mis.obs}} + \mathbf{A}_{\text{obs}}\mathbf{x}_{i,\text{obs}} = \mathbf{b} = \mathbf{A}\boldsymbol{\mu}^{(t+1)},$$

and so we find that

$$(\mathbf{Ax}_i^{*(t)} - \mathbf{A}\boldsymbol{\mu}^{(t+1)})(\mathbf{x}_i^{*(t)} - \boldsymbol{\mu}^{(t+1)})^T = \mathbf{O},$$

for each $i = 1, \ldots, n$. It now follows from (9.35) that $\mathbf{A}\boldsymbol{\Sigma}^{(t+1)} = \mathbf{O}$. ■

COROLLARY 9.5

Provided that $\boldsymbol{\mu}^{(0)}$ and $\boldsymbol{\Sigma}^{(0)}$ are chosen as in Theorem 9.4, both the deterministic and the stochastic imputation methods based on the final parameter estimates of the EM algorithm yield imputations that satisfy the balance edits (9.27).

Proof. Denote the final parameter estimates by $\tilde{\boldsymbol{\mu}}$ and $\tilde{\boldsymbol{\Sigma}}$. The deterministic imputation method imputes $\hat{\mathbf{x}}_{i,\text{mis}} = \tilde{\boldsymbol{\mu}}_{i,\text{mis.obs}}$, and we already showed in the proof of Theorem 9.4 that these imputations satisfy $\mathbf{A}_{\text{mis}}\hat{\mathbf{x}}_{i,\text{mis}} + \mathbf{A}_{\text{obs}}\mathbf{x}_{i,\text{obs}} = \mathbf{b}$.

The stochastic imputation method draws $\hat{\mathbf{x}}_{i,\text{mis}}$ from the conditional distribution $N_{m_i}(\tilde{\boldsymbol{\mu}}_{i,\text{mis.obs}}, \tilde{\boldsymbol{\Sigma}}_{\text{mis,mis.obs}})$. Consequently, $\hat{\mathbf{x}}_{i,\text{mis}} - \tilde{\boldsymbol{\mu}}_{i,\text{mis.obs}}$ always lies in the orthogonal complement of the null space of $\tilde{\boldsymbol{\Sigma}}_{\text{mis,mis.obs}}$. Since it was shown in the proof of Theorem 9.4 that $\mathbf{A}_{\text{mis}}\tilde{\boldsymbol{\Sigma}}_{\text{mis,mis.obs}} = \mathbf{O}$ (or, equivalently, $\tilde{\boldsymbol{\Sigma}}_{\text{mis,mis.obs}}\mathbf{A}_{\text{mis}}^T = \mathbf{O}$), it follows that the column span of $\mathbf{A}_{\text{mis}}^T$ is contained in the null space of $\tilde{\boldsymbol{\Sigma}}_{\text{mis,mis.obs}}$. In particular, for all $\mathbf{z}$ in the orthogonal complement of this null space, it holds that $\mathbf{z}^T\mathbf{A}_{\text{mis}}^T = \mathbf{0}^T$, that is, $\mathbf{A}_{\text{mis}}\mathbf{z} = \mathbf{0}$. Hence

$$\mathbf{A}_{\text{mis}}\hat{\mathbf{x}}_{i,\text{mis}} = \mathbf{A}_{\text{mis}}\tilde{\boldsymbol{\mu}}_{i,\text{mis.obs}} = \mathbf{b} - \mathbf{A}_{\text{obs}}\mathbf{x}_{i,\text{obs}},$$

for any $\hat{\mathbf{x}}_{i,\text{mis}} \sim N_{m_i}(\tilde{\boldsymbol{\mu}}_{i,\text{mis.obs}}, \tilde{\boldsymbol{\Sigma}}_{\text{mis,mis.obs}})$. ■

Theorem 9.4 and Corollary 9.5 show that provided that the initial parameter values $\boldsymbol{\mu}^{(0)}$ and $\boldsymbol{\Sigma}^{(0)}$ concur with the balance edits, using the EM algorithm as described above results in imputations that satisfy the balance edits (9.27). We already showed that for the complete case maximum likelihood estimates $\hat{\boldsymbol{\mu}}_{CC}$ and $\hat{\boldsymbol{\Sigma}}_{CC}$, it holds that $\mathbf{A}\hat{\boldsymbol{\mu}}_{CC} = \mathbf{b}$ and the null space of $\hat{\boldsymbol{\Sigma}}_{CC}$ is L. Therefore, the complete case estimates may be used for $\boldsymbol{\mu}^{(0)}$ and $\boldsymbol{\Sigma}^{(0)}$—in any case as far as the balance edits (9.27) are concerned.

EXAMPLE 9.5 (*continued*)

Suppose that a particular record $\mathbf{x}_i = (x_{i1}, x_{i2}, x_{i3})^T$ has both x_{i1} and x_{i2} missing, with x_{i3} observed, i.e. $\mathbf{x}_{i,\text{mis}} = (x_{i1}, x_{i2})^T$ and $x_{i,\text{obs}} = x_{i3}$. The appropriate partitions of $\boldsymbol{\mu}$ and $\boldsymbol{\Sigma}$ are:

$$\boldsymbol{\mu}_{\text{mis}} = \begin{pmatrix} 2 \\ 3 \end{pmatrix}, \qquad \mu_{\text{obs}} = 5,$$

$$\boldsymbol{\Sigma}_{\text{mis,mis}} = \begin{pmatrix} 3 & 1 \\ 1 & 3 \end{pmatrix}, \qquad \boldsymbol{\Sigma}_{\text{mis,obs}} = \boldsymbol{\Sigma}^T_{\text{obs,mis}} = \begin{pmatrix} 4 \\ 4 \end{pmatrix}, \qquad \Sigma_{\text{obs,obs}} = 8.$$

The parameters of the conditional distribution of $\mathbf{x}_{i,\text{mis}}$, given $x_{i,\text{obs}}$, are computed from Theorem 9.3 as follows:

$$\boldsymbol{\mu}_{i,\text{mis.obs}} = \begin{pmatrix} 2 \\ 3 \end{pmatrix} + \begin{pmatrix} 4 \\ 4 \end{pmatrix} \frac{1}{8}(x_{i3} - 5) = \begin{pmatrix} \frac{1}{2}x_{i3} - \frac{1}{2} \\ \frac{1}{2}x_{i3} + \frac{1}{2} \end{pmatrix},$$

$$\boldsymbol{\Sigma}_{\text{mis,mis.obs}} = \begin{pmatrix} 3 & 1 \\ 1 & 3 \end{pmatrix} - \begin{pmatrix} 4 \\ 4 \end{pmatrix} \frac{1}{8}(4, 4) = \begin{pmatrix} 1 & -1 \\ -1 & 1 \end{pmatrix}.$$

Thus,

$$\begin{pmatrix} x_{i1} \\ x_{i2} \end{pmatrix} \Bigg| x_{i3}, \boldsymbol{\mu}, \boldsymbol{\Sigma} \sim N_2 \left(\begin{pmatrix} \frac{1}{2}x_{i3} - \frac{1}{2} \\ \frac{1}{2}x_{i3} + \frac{1}{2} \end{pmatrix}, \begin{pmatrix} 1 & -1 \\ -1 & 1 \end{pmatrix} \right).$$

Note that the expected values of the conditional distribution satisfy

$$\text{E}(\hat{x}_{i1} + \hat{x}_{i2} | x_{i3}) = x_{i3}.$$

This shows that the deterministic imputation method based on the singular normal distribution yields records that satisfy balance edit (9.31) in this example. The same holds for the stochastic imputation method. This can be seen by noting that the null space of $\boldsymbol{\Sigma}_{\text{mis,mis.obs}}$ is spanned by the vector $(1, 1)^T$. From the definition of the singular normal distribution, this means that the vector

$$\begin{pmatrix} \hat{x}_{i1} - \text{E}(\hat{x}_{i1} | x_{i3}) \\ \hat{x}_{i2} - \text{E}(\hat{x}_{i2} | x_{i3}) \end{pmatrix}$$

is always orthogonal to $(1, 1)^T$. Thus

$$\hat{x}_{i1} - \text{E}(\hat{x}_{i1} | x_{i3}) + \hat{x}_{i2} - \text{E}(\hat{x}_{i2} | x_{i3}) = 0,$$

which shows that $\hat{x}_{i1} + \hat{x}_{i2} = x_{i3}$ for any imputed vector drawn from the conditional distribution.

As a second example, suppose that another record has $x_{i,\text{mis}} = x_{i1}$ and $\mathbf{x}_{i,\text{obs}} = (x_{i2}, x_{i3})^T$. In this case, we partition $\boldsymbol{\mu}$ and $\boldsymbol{\Sigma}$ as follows:

$$\mu_{\text{mis}} = 2, \ \boldsymbol{\mu}_{\text{obs}} = \begin{pmatrix} 3 \\ 5 \end{pmatrix},$$

$$\Sigma_{\text{mis,mis}} = 3, \ \boldsymbol{\Sigma}^T_{\text{mis,obs}} = \boldsymbol{\Sigma}_{\text{obs,mis}} = \begin{pmatrix} 1 \\ 4 \end{pmatrix}, \ \boldsymbol{\Sigma}_{\text{obs,obs}} = \begin{pmatrix} 3 & 4 \\ 4 & 8 \end{pmatrix}.$$

Note that

$$\boldsymbol{\Sigma}^{-1}_{\text{obs,obs}} = \frac{1}{8}\begin{pmatrix} 8 & -4 \\ -4 & 3 \end{pmatrix}.$$

This time, we obtain the following parameters of the conditional distribution from Theorem 9.3:

$$\mu_{i,\text{mis.obs}} = 2 + (1, \ 4)\frac{1}{8}\begin{pmatrix} 8 & -4 \\ -4 & 3 \end{pmatrix}\begin{pmatrix} x_{i2} - 3 \\ x_{i3} - 5 \end{pmatrix} = x_{i3} - x_{i2},$$

$$\Sigma_{\text{mis,mis.obs}} = 3 - (1, \ 4)\frac{1}{8}\begin{pmatrix} 8 & -4 \\ -4 & 3 \end{pmatrix}\begin{pmatrix} 1 \\ 4 \end{pmatrix} = 0.$$

Thus,

$$x_{i1} \left| \begin{pmatrix} x_{i2} \\ x_{i3} \end{pmatrix}, \boldsymbol{\mu}, \boldsymbol{\Sigma} \sim N_1(x_{i3} - x_{i2}, 0). \right.$$

Hence, in this case the imputed value $\hat{x}_{i1} = x_{i3} - x_{i2}$ is the same for the deterministic and the stochastic imputation methods, because there is only one value of x_{i1} that satisfies the balance edit. We could therefore also have applied the deductive imputation technique of Section 9.2.2. In fact, one could say that the imputation method using the singular normal distribution degenerates into a deductive imputation method in this example. ■

9.5.2 USING THE TRUNCATED SINGULAR NORMAL DISTRIBUTION

We now consider data that have to satisfy both balance edits (9.27) and inequality edits (9.28). Following Tempelman (2007), we show how consistent records may be obtained by imputing from a so-called truncated singular normal distribution.

In general, truncation of a distribution to a subset of its original domain works as follows. Suppose that $f(\mathbf{x}|\boldsymbol{\theta})$ is the probability density of a nontruncated distribution on $\mathbb{R}^p$, and let $G \subset \mathbb{R}^p$. Then the probability density function of the associated truncated distribution on G is defined as

$$f(\mathbf{x}|\boldsymbol{\theta}; G) = \begin{cases} \dfrac{f(\mathbf{x}|\boldsymbol{\theta})}{\int_G f(\mathbf{x}|\boldsymbol{\theta})\, d\mathbf{x}} & \text{if } \mathbf{x} \in G, \\ 0 & \text{if } \mathbf{x} \notin G. \end{cases} \tag{9.36}$$

Note that the denominator in the first case corresponds to $\mathrm{P}(\mathbf{x} \in G)$, i.e., the probability that $\mathbf{x} \in G$. Dividing out this integral normalizes the truncated distribution, so that $f(\mathbf{x}|\boldsymbol{\theta}; G)$ integrates to one.

For our present application, we define

$$G = \left\{\mathbf{x} \in \mathbb{R}^p \ : \ \mathbf{B}\mathbf{x} \geq \mathbf{c}\right\},$$

which is the subset of vectors in $\mathbb{R}^p$ that satisfy all inequality edits in (9.28). We then truncate the singular normal distribution $N_p(\boldsymbol{\mu}, \boldsymbol{\Sigma})$ of Section 9.5.1 to G to obtain a distribution of vectors that satisfy both (9.27) and (9.28). Note that the density function of the truncated singular normal distribution is only defined and nonzero on

$$H = (\boldsymbol{\mu} + L^{\perp}) \cap G = \left\{\mathbf{x} \in \mathbb{R}^p \ : \ \mathbf{A}\mathbf{x} = \mathbf{b},\ \mathbf{B}\mathbf{x} \geq \mathbf{c}\right\},$$

which is the subset of vectors in $\mathbb{R}^p$ that satisfy both (9.27) and (9.28).

Although the truncated distribution $N_p(\boldsymbol{\mu}, \boldsymbol{\Sigma}; H)$ is characterized by the same parameters $\boldsymbol{\mu}$ and $\boldsymbol{\Sigma}$ as the nontruncated normal distribution, these parameters in general do not correspond to the mean and covariance of the truncated distribution. This is easily seen from an example. Consider a univariate normal distribution $N(\mu, \sigma^2)$, with $\mu > 0$, truncated to the positive real line. Since negative realizations are discarded, the sample mean will be positively biased for μ. In addition, the sample standard deviation underestimates σ, because some of the natural variation is lost due to truncation.

Because $\boldsymbol{\mu}$ and $\boldsymbol{\Sigma}$ do not correspond to the mean and covariance of the truncated normal distribution, obtaining maximum likelihood estimates of these parameters is less straightforward for truncated normal data than for nontruncated data. Based on a data set of records $\mathbf{x}_i \sim N_p(\boldsymbol{\mu}, \boldsymbol{\Sigma}; H)$, $i = 1, \ldots, n$, the loglikelihood of the truncated singular normal distribution is [cf. (9.30) and (9.36)]

$$\begin{aligned} L(\boldsymbol{\mu}, \boldsymbol{\Sigma}|\mathbf{x}_1, \ldots, \mathbf{x}_n; H) = {} & -\frac{n(p-r)}{2} \ln 2\pi - \frac{n}{2} \ln |\boldsymbol{\Sigma}|_{p-r} \\ & -\frac{1}{2} \sum_{i=1}^{n} (\mathbf{x}_i - \boldsymbol{\mu})^T \boldsymbol{\Sigma}^{+} (\mathbf{x}_i - \boldsymbol{\mu}) - n \ln \Phi(H), \end{aligned}$$

with

$$\Phi(H) = \int_H \varphi(\mathbf{x}|\boldsymbol{\mu}, \boldsymbol{\Sigma})\, d\mathbf{x}$$

The difficulty of maximizing $L(\boldsymbol{\mu}, \boldsymbol{\Sigma}|\mathbf{x}_1, \ldots, \mathbf{x}_n; H)$ lies in the evaluation of $\Phi(H)$, which is a multi-dimensional integral that cannot be expressed in closed form.

The singular matrix $\boldsymbol{\Sigma}$ can be decomposed into $\boldsymbol{\Sigma} = \mathbf{C}\boldsymbol{\Lambda}\mathbf{C}^T$, with $\mathbf{C}$ the orthogonal matrix of eigenvectors and $\boldsymbol{\Lambda}$ the diagonal matrix of eigenvalues of $\boldsymbol{\Sigma}$. We write

$$\boldsymbol{\Lambda} = \begin{pmatrix} \boldsymbol{\Lambda}_1 & \mathbf{O} \\ \mathbf{O} & \mathbf{O} \end{pmatrix},$$

with $\boldsymbol{\Lambda}_1$ the diagonal matrix of $p - r$ nonzero eigenvalues of $\boldsymbol{\Sigma}$. Partitioning $\mathbf{C} = (\mathbf{C}_1, \mathbf{C}_2)$ accordingly, it follows that $\boldsymbol{\Sigma} = \mathbf{C}_1\boldsymbol{\Lambda}_1\mathbf{C}_1^T$. In addition, it can be shown that $\boldsymbol{\Sigma}^+ = \mathbf{C}_1\boldsymbol{\Lambda}_1^{-1}\mathbf{C}_1^T$. Hence

$$\begin{aligned} L(\boldsymbol{\mu}, \boldsymbol{\Sigma}|\mathbf{x}_1, \ldots, \mathbf{x}_n; H) = {} & -\frac{n(p-r)}{2}\ln 2\pi - \frac{n}{2}\ln|\boldsymbol{\Lambda}_1| \\ & -\frac{1}{2}\sum_{i=1}^{n}(\mathbf{x}_i - \boldsymbol{\mu})^T\mathbf{C}_1\boldsymbol{\Lambda}_1^{-1}\mathbf{C}_1^T(\mathbf{x}_i - \boldsymbol{\mu}) - n\ln\Phi(H). \end{aligned}$$

Written in this form, the loglikelihood is a function of three parameters: $\boldsymbol{\mu}$, $\boldsymbol{\Lambda}_1^{-1}$ and $\mathbf{C}_1$. The maximum is found by setting the first-order partial derivatives with respect to these parameters equal to zero.

Tempelman (2007) gives the following expressions for the partial derivatives:

$$\frac{\partial L(\boldsymbol{\mu}, \boldsymbol{\Lambda}_1^{-1}, \mathbf{C}_1|\mathbf{x}_1, \ldots, \mathbf{x}_n; H)}{\partial \boldsymbol{\mu}} = \mathbf{C}_1\boldsymbol{\Lambda}_1^{-1}\mathbf{C}_1^T\left[\sum_{i=1}^{n}\mathbf{x}_i - n\mathrm{E}(\mathbf{x} \mid \mathbf{x} \in H, \boldsymbol{\mu}, \boldsymbol{\Sigma})\right],$$

$$\begin{aligned} \frac{\partial L(\boldsymbol{\mu}, \boldsymbol{\Lambda}_1^{-1}, \mathbf{C}_1|\mathbf{x}_1, \ldots, \mathbf{x}_n; H)}{\partial \boldsymbol{\Lambda}_1^{-1}} = {} & -\frac{1}{2}\sum_{i=1}^{n}\mathbf{C}_1^T(\mathbf{x}_i - \boldsymbol{\mu})(\mathbf{x}_i - \boldsymbol{\mu})^T\mathbf{C}_1 \\ & + \frac{n}{2}\mathrm{E}[\mathbf{C}_1^T(\mathbf{x} - \boldsymbol{\mu})(\mathbf{x} - \boldsymbol{\mu})^T\mathbf{C}_1 \mid \mathbf{x} \in H, \boldsymbol{\mu}, \boldsymbol{\Sigma}], \end{aligned}$$

$$\begin{aligned} \frac{\partial L(\boldsymbol{\mu}, \boldsymbol{\Lambda}_1^{-1}, \mathbf{C}_1|\mathbf{x}_1, \ldots, \mathbf{x}_n; H)}{\partial \mathbf{C}_1} = {} & -\sum_{i=1}^{n}(\mathbf{x}_i - \boldsymbol{\mu})(\mathbf{x}_i - \boldsymbol{\mu})^T\mathbf{C}_1\boldsymbol{\Lambda}_1^{-1} \\ & + n\mathrm{E}[(\mathbf{x} - \boldsymbol{\mu})(\mathbf{x} - \boldsymbol{\mu})^T\mathbf{C}_1\boldsymbol{\Lambda}_1^{-1} \mid \mathbf{x} \in H, \boldsymbol{\mu}, \boldsymbol{\Sigma}]. \end{aligned}$$

Note that the subset H does not depend on the parameters.

In order to evaluate the truncated expectations $\mathrm{E}(\mathbf{x} \mid \mathbf{x} \in H, \boldsymbol{\mu}, \boldsymbol{\Sigma})$ and $\mathrm{E}[(\mathbf{x} - \boldsymbol{\mu})(\mathbf{x} - \boldsymbol{\mu})^T \mid \mathbf{x} \in H, \boldsymbol{\mu}, \boldsymbol{\Sigma}]$, Tempelman (2007) suggests using Monte Carlo integration. This means that we obtain a large number, say S, of random draws $\mathbf{v}_s$, $s = 1, \ldots, S$, from the $N_p(\boldsymbol{\mu}, \boldsymbol{\Sigma}; H)$ distribution, and compute

$$\hat{\mathrm{E}}(\mathbf{x} \mid \mathbf{x} \in H, \boldsymbol{\mu}, \boldsymbol{\Sigma}) = \frac{1}{S}\sum_{s=1}^{S}\mathbf{v}_s,$$

$$\hat{\mathrm{E}}[(\mathbf{x}-\boldsymbol{\mu})(\mathbf{x}-\boldsymbol{\mu})^T \mid \mathbf{x} \in H, \boldsymbol{\mu}, \boldsymbol{\Sigma}] = \frac{1}{S}\sum_{s=1}^{S}(\mathbf{v}_s-\boldsymbol{\mu})(\mathbf{v}_s-\boldsymbol{\mu})^T.$$

Thus, the expectation of each expression is estimated by its sample mean in a random sample drawn from the truncated singular normal distribution.

Tempelman (2007) mentions two ways to obtain draws from the truncated singular normal distribution. The first method, called Acceptance/Rejection sampling, simply consists of drawing from the nontruncated distribution and discarding each vector that does not belong to H. The main advantage of this method is its simplicity. A potential drawback is that the method may be very inefficient, especially if H is relatively small or if $\boldsymbol{\mu}$ is located near the boundary of H.

The second method is an application of Gibbs sampling, which exploits the fact that it is relatively easy to draw from a univariate truncated normal distribution. For each vector $\mathbf{v}_s$, we start with initial values $\mathbf{v}_s^{(0)} = (v_{s1}^{(0)}, \ldots, v_{sp}^{(0)})^T$, and repeatedly perform the following steps:

$$\text{draw } v_{s1}^{(t+1)} \sim N_1(\mu_{s,1.(2,3,\ldots,p)}, \Sigma_{11.(2,3,\ldots,p)}; H(v_{s2}^{(t)}, v_{s3}^{(t)}, \ldots, v_{sp}^{(t)})),$$

$$\text{draw } v_{s2}^{(t+1)} \sim N_1(\mu_{s,2.(1,3,\ldots,p)}, \Sigma_{22.(1,3,\ldots,p)}; H(v_{s1}^{(t+1)}, v_{s3}^{(t)}, \ldots, v_{sp}^{(t)})),$$

$$\vdots$$

$$\text{draw } v_{sp}^{(t+1)} \sim N_1(\mu_{s,p.(1,2,\ldots,p-1)}, \Sigma_{pp.(1,2,\ldots,p-1)}; H(v_{s1}^{(t+1)}, v_{s2}^{(t+1)} \ldots, v_{s,p-1}^{(t+1)})).$$

Thus, for each element of the random vector we draw from a conditional distribution, while taking the current values of the other elements into account. Here $\mu_{s,j.(1,\ldots,j-1,j+1,\ldots,p)}$ and $\Sigma_{jj.(1,\ldots,j-1,j+1,\ldots,p)}$ are defined analogous to $\boldsymbol{\mu}_{i,\text{mis.obs}}$ and $\boldsymbol{\Sigma}_{\text{mis,mis.obs}}$ in Theorem 9.3, and

$$H(v_{s1}^{(t+1)}, \ldots, v_{s,j-1}^{(t+1)}, v_{s,j+1}^{(t)}, \ldots, v_{sp}^{(t)})$$

is the region in $\mathbb{R}$ found by setting all elements of $\mathbf{v}_s$ except v_{sj} to their current value in the definition of H. This iterative process is repeated, say, t_G times. We then set $\mathbf{v}_s = \mathbf{v}_s^{(t_G)}$. Note that to obtain S draws from a p-dimensional truncated singular normal distribution using the Gibbs sampler, we need to perform pt_GS univariate draws in total.

For more details on Acceptance/Rejection sampling and Gibbs sampling in this context, see Tempelman (2007).

If the data set contains missing values, then the other terms in the partial derivatives of $L(\boldsymbol{\mu}, \boldsymbol{\Lambda}_1^{-1}, \mathbf{C}_1 | \mathbf{x}_1, \ldots, \mathbf{x}_n; H)$ also cannot be evaluated exactly. The EM algorithm provides a way to find approximate maximum likelihood estimates of the parameters in this case. Here, we only give an outline of this application of the EM algorithm, and refer to Tempelman (2007) for a more detailed description.

Suppose that the current parameter estimates are $\boldsymbol{\mu}^{(t)}$, $\boldsymbol{\Lambda}_1^{-1(t)}$, and $\mathbf{C}_1^{(t)}$. Basically, the E-step of the EM algorithm replaces the missing terms in the partial derivatives by their expected values, conditional on the observed data and the current parameter estimates. In the M-step, we solve the system of equations found by setting the estimated partial derivatives equal to zero. Hence, we obtain the following equations in $\boldsymbol{\mu}$, $\boldsymbol{\Lambda}_1^{-1}$ and $\mathbf{C}_1$:

$$\begin{aligned}
\mathbf{0} &= \mathbf{C}_1\boldsymbol{\Lambda}_1^{-1}\mathbf{C}_1^T \sum_{i=1}^{n} \mathrm{E}(\mathbf{x}_i \mid \mathbf{x}_{i,\text{obs}}, \boldsymbol{\mu}^{(t)}, \boldsymbol{\Lambda}_1^{-1(t)}, \mathbf{C}_1^{(t)}) \\
&\quad - n\mathbf{C}_1\boldsymbol{\Lambda}_1^{-1}\mathbf{C}_1^T \mathrm{E}(\mathbf{x} \mid \mathbf{x} \in H, \boldsymbol{\mu}^{(t)}, \boldsymbol{\Lambda}_1^{-1(t)}, \mathbf{C}_1^{(t)}), \\
\mathbf{O} &= -\frac{1}{2}\sum_{i=1}^{n} \mathbf{C}_1^T \mathrm{E}[(\mathbf{x}_i - \boldsymbol{\mu})(\mathbf{x}_i - \boldsymbol{\mu})^T \mid \mathbf{x}_{i,\text{obs}}, \boldsymbol{\mu}^{(t)}, \boldsymbol{\Lambda}_1^{-1(t)}, \mathbf{C}_1^{(t)}]\mathbf{C}_1 \\
&\quad + \frac{n}{2}\mathbf{C}_1^T \mathrm{E}[(\mathbf{x} - \boldsymbol{\mu})(\mathbf{x} - \boldsymbol{\mu})^T \mid \mathbf{x} \in H, \boldsymbol{\mu}^{(t)}, \boldsymbol{\Lambda}_1^{-1(t)}, \mathbf{C}_1^{(t)}]\mathbf{C}_1, \\
\mathbf{O} &= -\sum_{i=1}^{n} \mathrm{E}[(\mathbf{x}_i - \boldsymbol{\mu})(\mathbf{x}_i - \boldsymbol{\mu})^T \mid \mathbf{x}_{i,\text{obs}}, \boldsymbol{\mu}^{(t)}, \boldsymbol{\Lambda}_1^{-1(t)}, \mathbf{C}_1^{(t)}]\mathbf{C}_1\boldsymbol{\Lambda}_1^{-1} \\
&\quad + n\mathrm{E}[(\mathbf{x} - \boldsymbol{\mu})(\mathbf{x} - \boldsymbol{\mu})^T \mid \mathbf{x} \in H, \boldsymbol{\mu}^{(t)}, \boldsymbol{\Lambda}_1^{-1(t)}, \mathbf{C}_1^{(t)}]\mathbf{C}_1\boldsymbol{\Lambda}_1^{-1}.
\end{aligned}$$

Apart from the truncated expectations that we already considered above, these equations also require that we evaluate

$$\mathrm{E}(\mathbf{x}_{i,\text{mis}} \mid \mathbf{x}_{i,\text{obs}}, \boldsymbol{\mu}^{(t)}, \boldsymbol{\Lambda}_1^{-1(t)}, \mathbf{C}_1^{(t)}) \tag{9.37}$$

and

$$\mathrm{E}(\mathbf{x}_{i,\text{mis}}\mathbf{x}_{i,\text{mis}}^T \mid \mathbf{x}_{i,\text{obs}}, \boldsymbol{\mu}^{(t)}, \boldsymbol{\Lambda}_1^{-1(t)}, \mathbf{C}_1^{(t)}). \tag{9.38}$$

Note that $\mathrm{E}(\mathbf{x}_{i,\text{obs}} \mid \mathbf{x}_{i,\text{obs}}, \boldsymbol{\mu}^{(t)}, \boldsymbol{\Lambda}_1^{-1(t)}, \mathbf{C}_1^{(t)}) = \mathbf{x}_{i,\text{obs}}$.

Elaborating on Theorem 9.3, it is seen that conditional on $\mathbf{x}_{i,\text{obs}}$, $\mathbf{x}_{i,\text{mis}}$ follows a truncated singular normal distribution, with the truncation region given by

$$\begin{aligned}
H(\mathbf{x}_{i,\text{obs}}) = \{&\mathbf{x}_{i,\text{mis}} \in \mathbb{R}^{m_i} : \mathbf{A}_{\text{mis}}\mathbf{x}_{i,\text{mis}} = \mathbf{b} - \mathbf{A}_{\text{obs}}\mathbf{x}_{i,\text{obs}}, \\
&\mathbf{B}_{\text{mis}}\mathbf{x}_{i,\text{mis}} \geq \mathbf{c} - \mathbf{B}_{\text{obs}}\mathbf{x}_{i,\text{obs}}\},
\end{aligned}$$

where $\mathbf{A}$ and $\mathbf{B}$ are partitioned analogously to $\mathbf{x}_i$, and m_i denotes the number of missing items in $\mathbf{x}_i$. This means that we can again use either Acceptance/Rejection sampling or Gibbs sampling to obtain estimates of (9.37) and (9.38).

After all the multiple integrals have been evaluated numerically, we are left with a system of equations in $\boldsymbol{\mu}$, $\boldsymbol{\Lambda}_1^{-1}$ and $\mathbf{C}_1$. The updated parameter estimates $\boldsymbol{\mu}^{(t+1)}$, $\boldsymbol{\Lambda}_1^{-1(t+1)}$, and $\mathbf{C}_1^{(t+1)}$ are found by solving this system. In this fashion, the

E- and M-steps of the EM algorithm are iterated until the parameter estimates have converged.

As before, the parameter estimates from the EM algorithm can be used for both deterministic and stochastic imputation. For the deterministic imputation method, we impute the conditional expectation of $\mathbf{x}_{i,\text{mis}}$, given $\mathbf{x}_{i,\text{obs}}$ and the final parameter estimates $\tilde{\boldsymbol{\mu}}$, $\tilde{\boldsymbol{\Lambda}}_1^{-1}$, and $\tilde{\mathbf{C}}_1$. This requires another estimation of the form (9.37), which can once again be done using Acceptance/Rejection sampling or Gibbs sampling, as described above. For the stochastic imputation method, we draw from the conditional distribution of $\mathbf{x}_{i,\text{mis}}$, given $\mathbf{x}_{i,\text{obs}}$ and the final parameter estimates $\tilde{\boldsymbol{\mu}}$, $\tilde{\boldsymbol{\Lambda}}_1^{-1}$, and $\tilde{\mathbf{C}}_1$. This can be done by another application of either Acceptance/Rejection sampling or Gibbs sampling, with $S = 1$ this time. It follows from Corollary 9.5 and the definition of $H(\mathbf{x}_{i,\text{obs}})$ that both the deterministic and the stochastic imputation methods yield consistent records with respect to (9.27) and (9.28).

EXAMPLE 9.5 *(continued)*

Consider again $\mathbf{x}_i \sim N_3(\boldsymbol{\mu}, \boldsymbol{\Sigma})$ with

$$\boldsymbol{\mu} = \begin{pmatrix} 2 \\ 3 \\ 5 \end{pmatrix}, \qquad \boldsymbol{\Sigma} = \begin{pmatrix} 3 & 1 & 4 \\ 1 & 3 & 4 \\ 4 & 4 & 8 \end{pmatrix},$$

and suppose that, in addition to balance edit (9.31), the data also have to satisfy four inequality edits:

$$x_1 \geq 0, \qquad x_2 \geq 0, \qquad x_3 \geq 0, \qquad x_2 \geq x_1. \tag{9.39}$$

To take the inequality edits into account as well, we assume that

$$\mathbf{x}_i \sim N_3(\boldsymbol{\mu}, \boldsymbol{\Sigma}; H),$$

with the truncation region defined by

$$H = \left\{ \mathbf{x} \in \mathbb{R}^3 \; : \; \mathbf{A}\mathbf{x} = \mathbf{0}, \; \mathbf{B}\mathbf{x} \geq \mathbf{0} \right\},$$

with $\mathbf{A} = (1, 1, -1)$ and

$$\mathbf{B} = \begin{pmatrix} 1 & 0 & 0 \\ 0 & 1 & 0 \\ 0 & 0 & 1 \\ -1 & 1 & 0 \end{pmatrix}.$$

Note that $\boldsymbol{\mu} \in H$.

We found previously that, for a record $\mathbf{x}_i$ with x_{i1} and x_{i2} missing but x_{i3} observed, the conditional distribution of $(x_{i1}, x_{i2})^T$ is a bivariate singular normal distribution with parameters

$$\boldsymbol{\mu}_{i,\text{mis.obs}} = \begin{pmatrix} \frac{1}{2}x_{i3} - \frac{1}{2} \\ \frac{1}{2}x_{i3} + \frac{1}{2} \end{pmatrix}, \qquad \boldsymbol{\Sigma}_{\text{mis,mis.obs}} = \begin{pmatrix} 1 & -1 \\ -1 & 1 \end{pmatrix}.$$

It follows from our previous discussion of this example that any vector drawn from this conditional distribution is of the form

$$\begin{pmatrix} \hat{x}_{i1} \\ \hat{x}_{i2} \end{pmatrix} = \begin{pmatrix} \frac{1}{2}x_{i3} - \frac{1}{2} + z \\ \frac{1}{2}x_{i3} + \frac{1}{2} - z \end{pmatrix},$$

with $z \in \mathbb{R}$. Depending on the choice of z, this vector may or may not satisfy the inequality edits (9.39). For instance, the edit $x_2 \geq x_1$ is violated if $z > \frac{1}{2}$.

To make sure that the imputations $\hat{x}_{i1}$ and $\hat{x}_{i2}$ also satisfy the inequality edits, we truncate the conditional distribution to the following region:

$$H(x_{i3}) = \left\{ \begin{pmatrix} x_{i1} \\ x_{i2} \end{pmatrix} \in \mathbb{R}^2 \; : \; x_{i1} + x_{i2} = x_{i3}, \; x_{i1} \geq 0, \; x_{i2} \geq 0, \; x_{i2} \geq x_{i1} \right\}.$$

Basically, this means that we only allow values of z that do not cause violations of inequality edits. In this case, we must have

$$\frac{1}{2} - \frac{1}{2}x_{i3} \leq z \leq \frac{1}{2},$$

as is easily verified. ■

From the sketchy description given in this section, it should be clear that obtaining imputations from a truncated singular normal distribution is a complex and computationally intensive operation. Tempelman (2007) mentions that the method is difficult to apply in practice. In particular, the EM algorithm may not converge if the data do not fit the model assumptions. Note that transforming the data to obtain a closer resemblance to a normal distribution is not an option here, because a nonlinear transformation does not respect the linear structure of the edits given by (9.27) and (9.28). (We shall return to this point in Section 9.7.)

In the next three sections, we examine different approaches to obtain consistent imputations for numerical data with respect to (9.27) and (9.28), which may be more useful in practice.

9.6 An Imputation Approach Based on Fourier–Motzkin Elimination

In this section we develop a general approach for imputation of missing numerical data that ensures that edits are satisfied, while at the same time allowing one to specify a statistical imputation model. This approach allows one to separate the imputation model from how the edits are handled. For this approach a broad class of imputation models can be applied.

We refer to this imputation approach as the FM approach since a fundamental role in this approach is played by Fourier–Motzkin elimination. For a description of Fourier–Motzkin elimination, see Section 3.4.3. The edits we consider in this section are (9.27) and (9.28). These are linear edits, just like in Chapter 3.

To illustrate the approach, we will assume in this section that the data are approximately multivariately normally distributed. In fact, in our calculations we will treat the unknown distribution of the data as being a multivariate normal distribution exactly. For data that have to satisfy edits defined by linear inequalities, this is surely incorrect, because at best the data could follow a truncated normal distribution but never a regular normal distribution. Our simplification makes it relatively easy to determine marginal and conditional distributions, which are needed for the imputation approach examined in this section. We only use the (approximate) multivariate normal model to illustrate how our general approach can actually be applied in practice. We have selected the (approximate) multivariate normal model for computational convenience. We certainly do not want to suggest that this model is the most appropriate one for data sets encountered in practice. Another computationally convenient choice would have been to use hot deck imputation instead of the (approximate) multivariate normal model.

In order to estimate the parameters of the multivariate normal distribution, the EM algorithm can be used, with the observed means and covariance matrix of the complete cases as starting values. See Section 8.3.3 of this book and Schafer (1997) for a description of the EM algorithm.

This section is based on Coutinho, De Waal, and Remmerswaal (2007).

9.6.1 THE FM APPROACH

The FM approach consists of the following steps:

0. Assume a statistical imputation model for the data, and—if necessary for the model—estimate the model parameters.

We order the variables to be imputed from the variable with the most missing values to the variable with the least missing values. For each record to be imputed, we apply Steps 1 to 5 below. We repeat this process until all records have been imputed.

1. Fill in the values of the nonmissing data into the edits. This leads to a set of edits $E(0)$ involving only the variables to be imputed.
2. Use Fourier–Motzkin elimination (see Section 3.4.3) to eliminate the variables to be imputed from these edits in the fixed order for the record under consideration until only one variable remains. The set of edits after the jth variable to be imputed has been eliminated is denoted by $E(j)$. The final set of edits defines a feasible interval for the remaining variable. Set k equal to the number of variables to be imputed for the record under consideration.
3. Draw a value for the kth variable to be imputed.
4. If the drawn value lies inside the feasible interval $E(k-1)$, accept it and go to Step 5. If it lies outside the feasible interval, reject it and return to Step 3.
5. If $k = 1$, all variables have been imputed and we stop. Otherwise, we fill in the drawn value for the selected variable, say the kth eliminated variable, into the edits in $E(k-2)$. This defines a feasible interval for the $(k-1)$th eliminated variable. We update k by $k := k - 1$, and go to Step 3.

Note that the theory of Fourier–Motzkin elimination (see Section 3.4.3) implies that if the record to be imputed can be imputed consistently, the feasible interval determined in Step 2 or 5 is never empty.

In Step 0, one can assume either an implicitly defined statistical imputation model—for instance, when one wants to apply hot deck imputation—or an explicitly defined imputation model, such as the multivariate normal model like we do in the illustration below. In both cases we suggest to draw a value for the variable to be imputed from the conditional distribution of the selected variable given all known values—either observed or already imputed ones.

If the feasible interval determined in Step 2 has width 0, there is only one feasible value for the variable under consideration. In this case it is not necessary to draw a value in Step 3. Instead we immediately impute the only feasible value. In some other cases the width of the feasible interval determined in Step 2 may be rather small. In those cases many values may need to be drawn before a value inside the feasible interval is drawn. We therefore set a limit, N_d, on the number of times that a value for a particular variable may be drawn. If this limit is reached, and no value inside the feasible interval has been drawn, the last value drawn is set to the nearest value of the feasible interval. By means of N_d, one can indirectly control the number of imputed records on the boundary of the feasible region defined by the edits. If N_d is set to a low value, relatively many imputed records will be on this boundary; if N_d is set to a high value, relatively few imputed records will be on the boundary.

The variables are imputed in reverse order of elimination. Since we have ordered the variables to be imputed from the variable with the most missing values to the variable with the least missing values before applying Steps 1 to 5 of the above algorithm, the variables are imputed in order of increasing number of missing values. That is, the variable with the least missing values is imputed first and the variable with the most missing values last.

It is well known that in the worst case, Fourier–Motzkin elimination can be computationally very expensive. However, the imputation problems arising in practice at statistical offices only have a limited number of variables and edits. The largest problems we are aware of have a few hundreds of variables and slightly more than 100 edits. For realistic problems of this limited size, Fourier–Motzkin elimination is generally sufficiently fast. In fact, it has been shown for the related—but computationally much more demanding—error localization problem of the same size in terms of variables and edits that in practical cases arising at statistical offices the computational performance of Fourier–Motzkin elimination is generally acceptable [see De Waal and Coutinho (2005), De Waal (2005), and especially Chapter 3 of the present book]. In an evaluation study of the FM approach, described by Coutinho, De Waal, and Remmerswaal (2007), Fourier–Motzkin elimination again performed sufficiently fast. In fact, the bulk of the computing time in the FM approach was spent on drawing values from the multivariate normal distribution rather than on Fourier–Motzkin elimination.

The main reason for developing the FM approach are the promising results obtained by sequential imputation methods. Sequential imputation methods are a well-known class of imputation methods [see, e.g., Section 9.7, Van Buuren and Oudshoorn (1999, 2000), Raghunathan et al. (2001), and Rubin (2003)]. These imputation methods sequentially impute the variables and allow a separate imputation model to be specified for each variable. By imputing all variables containing missing data in turn and iteratively repeating this process several times, the statistical distribution of the imputed data generally converges to an unspecified multivariate distribution. The main strength of sequential imputation is its flexibility: Rather than using one multivariate imputation model for all variables simultaneously, which is generally computationally demanding and complex to handle, one can specify a different imputation model for each variable. Sequential imputation methods can be extended to ensure that they satisfy edits. In principle, the FM approach can be implemented as a sequential imputation approach that allows such an extension, although in our illustration we assume a multivariate normal rather than separate imputation models for the variables to be imputed [see also Sections 9.7 and 9.8 and Tempelman (2007), for other extensions to ensure that edits are satisfied].

9.6.2 ILLUSTRATION OF THE FM APPROACH

In this subsection we illustrate the FM approach by means of an example, taken from Coutinho, De Waal, and Remmerswaal (2007). In this example, we assume that we are given a data set with some missing values, that there are four variables, T, P, C, and N, and that the edits are given by

$$T = P + C, \tag{9.40}$$

$$T \geq 0, \tag{9.41}$$

$$P \leq 0.5T, \tag{9.42}$$

$$(9.43)\qquad -0.1T \leq P,$$

$$(9.44)\qquad T \leq 550N.$$

As mentioned before, to illustrate our approach we assume that the data are multivariately normally distributed. We assume that the model parameters, means $\boldsymbol{\mu}$, and covariance matrix $\boldsymbol{\Sigma}$, estimated in Step 0 of our approach are given by

$$\boldsymbol{\mu} = (1000, 200, 500, 4)^T$$

and

$$\boldsymbol{\Sigma} = \begin{pmatrix} 13500 & 3000 & 10500 & 60 \\ 3000 & 2500 & 500 & 10 \\ 10500 & 500 & 10000 & 50 \\ 60 & 10 & 50 & 1 \end{pmatrix}.$$

Here the first column/row corresponds to T, the second column/row to P, the third column/row to C, and the fourth column/row to N.

Now, suppose that for a certain record in our data set we have $N = 5$ and that the values for T, P, and C are missing. We first fill in the observed value for N into the edits (9.40) to (9.44) (Step 1 of our approach). We obtain (9.40) to (9.43) and

$$(9.45)\qquad T \leq 2750.$$

Next, we sequentially eliminate the variables for which the values are missing from the edits. We start by eliminating P from (9.40) to (9.43) and (9.45). This leads to the edits (9.41), (9.45), and

$$(9.46)\qquad T - C \leq 0.5T \text{ (equivalently: } 0.5T \leq C),$$

$$(9.47)\qquad -0.1T \leq T - C \text{ (equivalently: } C \leq 1.1T).$$

Edits (9.41), (9.45), (9.46), and (9.47) have to be satisfied by C and T.

We next eliminate variable C, and obtain (9.41), (9.45), and

$$(9.48)\qquad 0.5T \leq 1.1T.$$

Edit (9.48) is equivalent to (9.41). The edits that have to be satisfied by T are hence given by (9.41) and (9.45). The feasible interval for T is therefore given by [0,2750]. We have now completed Step 2 of our approach.

To impute T, we determine the distribution of T, conditional on the value for variable N. The distribution of T turns out to be $N(1060,9900)$, the normal distribution with mean 1060 and variance 9900. We draw values from this distribution until we draw a value inside the feasible interval (Steps 3 and 4 of the approach). Suppose we draw the value 1200.

We fill in the imputed value for T into the edits for C and T—that is, edits (9.41), (9.45), (9.46), and (9.47) (Step 5 of the approach). We obtain

$$1200 \geq 0,$$

$$1200 \leq 2750,$$

$$600 \leq C,$$

$$C \leq 1320.$$

The feasible interval for C is hence given by [600,1320]. We determine the distribution of C, conditional on the values for variables N and T. This distribution turns out to be $N(656.06, 1818.18)$. We draw values from this distribution until we draw a value inside the feasible interval (Steps 3 and 4 of the approach). Suppose we draw the value 700.

We fill in the imputed values for C and T into the edits that have to be satisfied by C, T, and P—that is, edits (9.40) to (9.43) and (9.45) (Step 5 of the approach). We obtain

$$1200 = P + 700,$$

$$1200 \geq 0,$$

$$P \leq 600,$$

$$-120 \leq P,$$

$$1200 \leq 2750.$$

There is only one feasible value for P, namely 500. The imputed record we obtain is given by $T = 1200$, $P = 500$, $C = 700$, and $N = 5$.

9.7 A Sequential Regression Approach

In Section 9.5 the (truncated) singular normal distribution was considered as a joint model for a data set with numerical variables that should satisfy linear balance edits and inequality edits. Before that we considered, in Section 9.4, the Dirichlet distribution as a joint model for a data set with nonnegative numerical variables involved in exactly one balance edit. This approach—that is, fitting a joint model to the data that incorporates the edit structure and therefore provides imputations that automatically satisfy the edits—is attractive from a theoretical point of view, but the underlying assumptions may be too strong for practical purposes. In particular, we rarely encounter real-life numerical data that can be said to follow (even approximately) a multivariate normal distribution or any other standard multivariate distribution from the literature. For this reason, statisticians often apply nonlinear transformations to a data set, such as taking the logarithm, the cube root, or a Box–Cox transformation, in order to obtain a closer resemblance to a standard distribution. As we already noted at the end of Section 9.5, this is not possible here, because the linear edit structure would be lost under such transformations.

An alternative to trying to find an appropriate joint model for the data is to consider instead the univariate distribution of each variable separately, conditional on the other variables. This leads to a method called sequential regression. Elaborating on work by Raghunathan et al. (2001) and Van Buuren, Boshuizen, and Knook (1999), Tempelman (2007) describes an imputation method based on sequential regression that takes linear edits of the form (9.27) and (9.28) into account. For convenience, we shall first consider only inequality edits.

For the sequential regression imputation method, we define vectors $\mathbf{x}_j = (x_{1j}, \ldots, x_{nj})^T$, $j = 1, \ldots, p$, which contain the variable scores across the records in a data set. These vectors are partitioned into a missing part and an observed part: $\mathbf{x}_j = (\mathbf{x}_{j,\text{mis}}^T, \mathbf{x}_{j,\text{obs}}^T)^T$. Note that this notation differs from the previous sections of this chapter, because we now consider missingness within a variable rather than within a record. Instead of trying to specify an explicit joint model by a multivariate density function $f(\mathbf{x}_1, \ldots, \mathbf{x}_p|\boldsymbol{\theta})$, we specify a univariate density for each variable $\mathbf{x}_j$, conditional on the values of the other variables: $f_j(\mathbf{x}_j|\mathbf{x}_1, \ldots, \mathbf{x}_{j-1}, \mathbf{x}_{j+1}, \ldots, \mathbf{x}_p, \boldsymbol{\theta}_j)$, for $j = 1, \ldots, p$. The univariate model is chosen such that x_{ij}, the entries of $\mathbf{x}_j$, satisfy the appropriate inequality restrictions. This can be achieved through truncation [cf. Tempelman (2007) and Example 9.6 below].

The sequential regression imputation method of Tempelman (2007) now proceeds as follows. Suppose that we have a current imputed version of the data set, given by $\mathbf{x}_1^{(t)}, \ldots, \mathbf{x}_p^{(t)}$, where the imputations satisfy (9.28). Then one iteration of the algorithm constructs a new imputed data set $\mathbf{x}_1^{(t+1)}, \ldots, \mathbf{x}_p^{(t+1)}$ by performing the following steps:

$$
\begin{aligned}
&\text{estimate } \boldsymbol{\theta}_1^{(t+1)} \text{ by regressing } \mathbf{x}_1^{(t)} \text{ on } \mathbf{x}_2^{(t)}, \ldots, \mathbf{x}_p^{(t)};\\
&\text{draw } \mathbf{x}_{1,\text{mis}}^{(t+1)} \text{ from } f_1(\mathbf{x}_1|\mathbf{x}_2^{(t)}, \ldots, \mathbf{x}_p^{(t)}, \boldsymbol{\theta}_1^{(t+1)});\\
&\vdots\\
&\text{estimate } \boldsymbol{\theta}_j^{(t+1)} \text{ by regressing } \mathbf{x}_j^{(t)} \text{ on } \mathbf{x}_1^{(t+1)}, \ldots, \mathbf{x}_{j-1}^{(t+1)}, \mathbf{x}_{j+1}^{(t)}, \ldots, \mathbf{x}_p^{(t)};\\
&\text{draw } \mathbf{x}_{j,\text{mis}}^{(t+1)} \text{ from } f_j(\mathbf{x}_j|\mathbf{x}_1^{(t+1)}, \ldots, \mathbf{x}_{j-1}^{(t+1)}, \mathbf{x}_{j+1}^{(t)}, \ldots, \mathbf{x}_p^{(t)}, \boldsymbol{\theta}_j^{(t+1)});\\
&\vdots\\
&\text{estimate } \boldsymbol{\theta}_p^{(t+1)} \text{ by regressing } \mathbf{x}_p^{(t)} \text{ on } \mathbf{x}_1^{(t+1)}, \ldots, \mathbf{x}_{p-1}^{(t+1)};\\
&\text{draw } \mathbf{x}_{p,\text{mis}}^{(t+1)} \text{ from } f_p(\mathbf{x}_p|\mathbf{x}_1^{(t+1)}, \ldots, \mathbf{x}_{p-1}^{(t+1)}, \boldsymbol{\theta}_p^{(t+1)}).
\end{aligned}
$$

This algorithm is iterated until the parameter estimates have stabilized.

Some remarks are in order. First of all, if the algorithm converges to a stable solution, it converges to a joint model for the data that follows implicitly from the assumed conditional models. However, such an implied joint model might not exist, and hence the algorithm is not guaranteed to converge. Conditional distributions without an implied joint distribution are called incompatible [see, e.g., Arnold and Press (1989)]. According to Tempelman (2007), the theoretical consequences of incompatibility are as of yet unclear, but it does not appear to pose a problem in practice.

Second, note that the algorithm requires an initial imputed data set, where the imputations satisfy the edit restrictions. Hence the sequential regression imputation method can only be used in combination with another, preferably less involved method for consistent imputation. The main purpose of using sequential regression is to improve upon the initial imputations, by constructing an imputed data set with better statistical properties. For instance, we could use the truncated singular normal distribution to obtain initial imputations, even though the data clearly do not follow a multivariate normal distribution. By choosing appropriate univariate models, we may then apply the sequential regression method to replace the initial imputations by better imputations while maintaining consistency with respect to the edit restrictions.

Third, the main reason why the sequential regression method may return imputations with better statistical properties is that it offers flexibility in the specification of the univariate conditional distributions. For each variable, the best model $f_j(\mathbf{x}_j|\mathbf{x}_1, \ldots, \mathbf{x}_{j-1}, \mathbf{x}_{j+1}, \ldots, \mathbf{x}_p, \boldsymbol{\theta}_j)$ can be selected based on particular features of that variable. We can, for instance, use linear regression for one variable and use logistic regression for another. Moreover, specifying different types of conditional distributions should not lead to additional difficulties in applying the algorithm, because the underlying joint distribution (if it exists) is never explicitly needed.

As an example, many numerical variables encountered in practice are semi-continuous; that is, the underlying distribution is continuous with a spike at a particular value. Often, the spike is located at zero: A substantial part of the responding units has a zero score on the variable, and the scores of the other responding units follow a continuous distribution. This occurs, for instance, with certain questions on expenses in business surveys, which do not necessarily apply to each business. A semi-continuous distribution cannot be modeled well by a continuous distribution, because there the probability of drawing any particular value is by definition zero. Therefore, it is better to model a semi-continuous variable by a two-stage procedure, where the probability of drawing the spiked value is modeled in the first stage, and the continuous distribution of the other values is modeled in the second stage. Sequential regression offers the flexibility to use this two-stage procedure for semi-continuous variables. See, for example, Tempelman (2007) or Drechsler (2009) for more details and other examples.

Another advantage of the sequential regression method is the fact that for univariate modeling a large class of nonlinear transformations can be applied to the data. We remarked previously that such transformations cannot be used for a joint model of the data, because if T is a nonlinear transformation and

$$b_{k1}x_1 + b_{k2}x_2 + \cdots + b_{kp}x_p \geq c_k$$

is a linear inequality edit, then it does not follow that

$$b'_{k1}T(x_1) + b'_{k2}T(x_2) + \cdots + b'_{kp}T(x_p) \geq c'_k$$

for some transformed coefficients. For the conditional univariate distribution of x_j given $x_1, \ldots, x_{j-1}, x_{j+1}, \ldots, x_p$, however, the edit can be written as

$$x_j \geq C(x_1, \ldots, x_{j-1}, x_{j+1}, \ldots, x_p),$$

for some expression C which is constant with respect to the conditional distribution. (We assume for convenience that $b_{kj} > 0$.) Under a nonlinear, monotone increasing transformation T, we find

$$T(x_j) \geq T[C(x_1, \ldots, x_{j-1}, x_{j+1}, \ldots, x_p)],$$

which is a linear edit in the transformed data $T(x_j)$, conditional on the values of $x_1, \ldots, x_{j-1}, x_{j+1}, \ldots, x_p$. (If T is monotone decreasing, we obtain the same edit but with "$\leq$" instead of "$\geq$".) As a consequence, we can improve the quality of the univariate modeling by applying appropriate transformations (e.g., Box–Cox transformations) to the data [see, e.g. Tempelman (2007)].

EXAMPLE 9.6

Suppose that we want to use the linear regression model for the conditional distribution of $\mathbf{x}_j$, but with a modification to take the inequality edits $\mathbf{Bx} \geq \mathbf{c}$ into account. Let $\mathbf{x}_{i,-j}$ denote the ith record without x_{ij}. Also, partition $\mathbf{B}$ into $\mathbf{B}_j$, its jth column, and $\mathbf{B}_{-j}$, its other columns. Then the value of x_{ij} has to satisfy the set of restrictions given by $\mathbf{B}_j x_{ij} \geq \mathbf{c} - \mathbf{B}_{-j}\mathbf{x}_{i,-j}$. It is easily verified that this is equivalent to

$$l_{ij} \leq x_{ij} \leq u_{ij}, \tag{9.49}$$

with

$$l_{ij} = \max_{k\,:\,b_{kj} > 0} \frac{1}{b_{kj}} \left(c_k - \sum_{l \neq j} b_{kl} x_{il} \right),$$

$$u_{ij} = \min_{k\,:\,b_{kj} < 0} \frac{1}{b_{kj}} \left(c_k - \sum_{l \neq j} b_{kl} x_{il} \right),$$

where b_{kj} and c_k denote elements of $\mathbf{B}$ and $\mathbf{c}$.

The ordinary linear regression model of $\mathbf{x}_j$ with $\mathbf{x}_1, \ldots, \mathbf{x}_{j-1}, \mathbf{x}_{j+1}, \ldots, \mathbf{x}_p$ as predictors is

$$\mathbf{x}_j = \beta_0 \mathbf{1} + \sum_{l \neq j} \beta_l \mathbf{x}_l + \boldsymbol{\varepsilon}_j, \qquad \boldsymbol{\varepsilon}_j \sim N(\mathbf{0}, \sigma_j^2 \mathbf{I}_n),$$

where β_0, $\boldsymbol{\beta}_{-j} = (\beta_1, \ldots, \beta_{j-1}, \beta_{j+1}, \ldots, \beta_p)^T$ and σ_j^2 are model parameters, $\mathbf{1}$ is an n-vector of ones, and $\mathbf{I}_n$ is the $n \times n$-dimensional identity matrix. Under this model, the variable x_{ij} follows an $N(\beta_0 + \boldsymbol{\beta}_{-j}^T \mathbf{x}_{i,-j}, \sigma_j^2)$ distribution.

If we want to take the edits into account, then x_{ij} needs to follow a truncated normal distribution $N(\beta_0 + \boldsymbol{\beta}_{-j}^T \mathbf{x}_{i,-j}, \sigma_j^2; G)$, with G defined by (9.49). The density of x_{ij} is given by

$$f(x_{ij} | \beta_0 + \boldsymbol{\beta}_{-j}^T \mathbf{x}_{i,-j}, \sigma_j^2; G) = \begin{cases} \dfrac{\varphi(x_{ij} | \beta_0 + \boldsymbol{\beta}_{-j}^T \mathbf{x}_{i,-j}, \sigma_j^2)}{\int_{l_{ij}}^{u_{ij}} \varphi(x | \beta_0 + \boldsymbol{\beta}_{-j}^T \mathbf{x}_{i,-j}, \sigma_j^2) dx} & \text{if } l_{ij} \leq x_{ij} \leq u_{ij}, \\ 0 & \text{otherwise} \end{cases}$$

[cf. (9.36).]. Here, φ denotes the density of the univariate normal distribution.

During an iteration of the sequential regression algorithm, we have to estimate the parameters β_0, $\boldsymbol{\beta}_{-j}$ and σ_j^2 of $f(x_{ij} | \beta_0 + \boldsymbol{\beta}_{-j}^T \mathbf{x}_{i,-j}, \sigma_j^2; G)$, by regressing the current imputed vector $\mathbf{x}_j^{(t)}$ on the current imputed predictors

$$\mathbf{x}_1^{(t+1)}, \ldots, \mathbf{x}_{j-1}^{(t+1)}, \mathbf{x}_{j+1}^{(t)}, \ldots, \mathbf{x}_p^{(t)}.$$

Due to truncation, estimating these parameters is not as straightforward as for ordinary linear regression. In particular, there is no closed form solution to the maximum likelihood estimation problem. Tempelman (2007) describes how maximum likelihood estimates of β_0, $\boldsymbol{\beta}_{-j}$, and σ_j^2 can be approximated by an iterative algorithm (see also Section 9.5). ■

Until now, we assumed that no balance edits were specified for our data set. Balance edits are a complicating factor when using the sequential regression method. Namely: If a variable occurs in a balance edit, then its univariate distribution, conditional on the values of the other variables, automatically becomes degenerate, because there is only one value that will satisfy the balance edit. This means that the sequential regression algorithm remains stuck on the initially imputed values for variables that occur in a balance edit. Therefore, the sequential regression method needs to be modified in order to handle balance edits as well as inequality edits.

Tempelman (2007) suggests to eliminate some variables from the data set by means of Fourier–Motzkin elimination (see Section 3.4.3) before running the algorithm, in such a way that the reduced data set does not contain any singularities. When the sequential regression algorithm has finished, the imputations for the removed variables can be derived from the other imputations by using the balance edits. A drawback of this approach is that we lose some information from the original observed data, which is detrimental to the quality of the resulting imputations. To spread the loss of information across all variables, Tempelman advocates randomly choosing a different set of variables to remove for each record.

9.8 Calibrated Imputation of Numerical Data Under Linear Edit Restrictions

9.8.1 INTRODUCTION

In the previous sections we examined the case where numerical imputed data have to satisfy edit restrictions. An additional problem is that numerical data sometimes have to sum up to known population totals. In this section we aim to take these edits and known totals into account while imputing a record.

The problem of imputing missing data in records having to satisfy edits such that at the same time known totals are satisfied can arise in the context of a survey amongst a subpopulation of enterprises. Often large enterprises—that is, enterprises with a number of employees exceeding a certain threshold value—are integrally observed. Some of those enterprises may, however, not provide answers to all questions, and some may even not answer any question at all. Totals corresponding to this subpopulation of enterprises may be known from other sources—for example, from available register data—or may already have been estimated. Because data of enterprises usually have to satisfy edits, imputation of such a data set then naturally leads to the problem we consider in the present section.

The remainder of this section is organized as follows. Section 9.8.2 illustrates the problem and describes some technical preparations that are required for our algorithms. Sections 9.8.3 and 9.8.4 develop two imputation algorithms for our problem. The edits we consider in this section are again linear edits (9.27) and (9.28).

This section is based on Pannekoek, Shlomo, and De Waal (2008).

9.8.2 ILLUSTRATION OF THE PROBLEM AND TECHNICAL PREPARATIONS

As in the previous section, the imputation approach in this section is sequential regression imputation. This is an iterative procedure that uses regression imputation to impute, in each iteration, all variables with missing values one by one using all other variables as predictors, with missing values in the predictors being replaced by their current imputed values. At the start of this procedure,

TABLE 9.1 Illustration of a Data Set with Edits and Known Totals

x_{11}	x_{12}	x_{13}
x_{21}	x_{22}	x_{23}
$\vdots$	$\vdots$	$\vdots$
x_{n1}	x_{n2}	x_{n3}
X_1	X_2	X_3

no "current imputed values" are available and an initialization round is applied in which imputations are derived from models that use only the variables with observed values in a record as predictors for the missing value of the target variable in that record. In this section we propose two sequential regression imputation methods that can deal with both edit constraints as well as consistency with known population totals.

To illustrate how to deal with edit restrictions and (population) totals, we consider a case where we have n records with only three variables (Table 9.1). These columns contain missing values that require imputation. Suppose that the data have to satisfy the edit restrictions

$$x_{i1} + x_{i2} = x_{i3}, \tag{9.50}$$

$$x_{i1} \geq x_{i2}, \tag{9.51}$$

$$x_{i3} \geq 3x_{i2}, \tag{9.52}$$

$$x_{ij} \geq 0 \qquad (j = 1, 2, 3) \tag{9.53}$$

and that in addition the following (population) total restrictions have to be satisfied:

$$\sum_{i=1}^{n} x_{ij} = X_j \qquad (j = 1, 2, 3), \tag{9.54}$$

where we assume that the population totals are given and consistent with each other; that is, the totals X_j $(j = 1, 2, 3)$ satisfy the edits (9.50) to (9.53).

As we already mentioned, in this section we use a sequential imputation method; that is, we impute the variables with missing data one by one. Suppose we want to impute variable x_j. In order to impute a missing field x_{ij} in record i, we first fill in the observed and previously imputed values for the other variables in record i into the edits. This leads to a reduced set of edits involving only the variables to be imputed. For instance, if in the above example (9.50) to (9.53) the observed value of variable x_1 in record i equals 10 and the values of variables x_2 and x_3 are missing, then the reduced set of edits is given by

$$10 + x_{i2} = x_{i3}, \tag{9.55}$$

(9.56) $$10 \geq x_{i2},$$

(9.57) $$x_{i3} \geq 3x_{i2},$$

(9.58) $$x_{ij} \geq 0 \qquad (j = 2, 3).$$

Once the reduced set of edits has been determined for a record i, we eliminate all equations from this reduced set of edits. That is, we sequentially select an equation and one of the variables x involved in this equation. We then express x in terms of the other variables in the selected equation and substitute this expression for x into the other edits in which x is involved. For instance, in the above example (9.55) to (9.58), we can eliminate x_{i3} by substituting the expression $x_{i3} = 10 + x_{i2}$ into the other edits (9.56) to (9.58). In this way we obtain a set of edits involving only inequality restrictions for the remaining variables. Later, once we have obtained imputation values for the variables involved in the set of inequalities, we can find values for the variables we have eliminated by back-substitution. For instance, in our example where we have eliminated x_{i3} from the edits (9.56) to (9.58), once we have obtained an imputation value for x_{i2} we can obtain a consistent value for x_{i3}—that is, a value satisfying all edits—by filling in the imputation value for x_{i2} into (9.55).

Next, we eliminate any remaining fields except x_{ij} itself from the set of edits by means of Fourier–Motzkin elimination [see, e.g., Section 3.4.3 of the present book and also De Waal and Coutinho (2005)]. The restrictions for x_{ij} can then be expressed as interval constraints:

(9.59) $$l_{ij} \leq x_{ij} \leq u_{ij}.$$

The problem for variable x_j now is to fill in the missing values with imputations, such that the sum constraint (9.54) and the interval constraints (9.59) are satisfied.

Below we present two different approaches to solving this problem. These approaches are based on standard regression imputation techniques, but with (slight) adjustments to the imputed values such that they satisfy the constraints (9.54) and (9.59).

9.8.3 ADJUSTED PREDICTED MEAN IMPUTATION

The idea of this algorithm is to obtain predicted mean imputations that satisfy the sum constraint and then adjust these imputations such that they also satisfy the interval constraints. To illustrate this idea, we use a simple regression model with one predictor, but generalization to multiple regression models is straightforward.

Standard Regression Imputation. Suppose that we want to impute a target column $\mathbf{x}_t$ using as a predictor a column $\mathbf{x}_a$, where the subscript "a" refers to auxiliary variables. The standard regression imputation approach is based on the model:

$$\mathbf{x}_t = \beta_0 \mathbf{1} + \beta \mathbf{x}_a + \boldsymbol{\varepsilon},$$

where $\mathbf{1}$ is the vector of appropriate length with ones in every entry—that is, $\mathbf{1} = (1, \ldots, 1)^T$—and $\boldsymbol{\varepsilon}$ is a vector with random residuals.

We assume that the predictor is either completely observed or already imputed, so there are no missing values in the predictor anymore. There are of course missing values in $\mathbf{x}_t$ and to estimate the model we can only use the records for which both $\mathbf{x}_t$ and $\mathbf{x}_a$ are observed. The data matrix for estimation consists of the columns $\mathbf{x}_{t,\text{obs}}, \mathbf{x}_{a,\text{obs}}$, where "obs" denotes the records with $\mathbf{x}_t$ observed (and "mis" will denote the opposite). With the ordinary least squares (ols) estimators of the parameters, $\hat{\beta}_0$ and $\hat{\beta}$, we obtain predictions for the missing values in $\mathbf{x}_t$ using

$$\hat{\mathbf{x}}_{t,\text{mis}} = \hat{\beta}_0 \mathbf{1} + \hat{\beta} \mathbf{x}_{a,\text{mis}},$$

where $\mathbf{x}_{a,\text{mis}}$ contains the $\mathbf{x}_a$ values for the records with $\mathbf{x}_t$ missing and $\hat{\mathbf{x}}_{t,\text{mis}}$ are the predictions for the missing $\mathbf{x}_t$ values in those records. The imputed column $\tilde{\mathbf{x}}_t$ consists of the observed values and the predicted values filled in for the missing values: $\tilde{\mathbf{x}}_t = (\mathbf{x}_{t,\text{obs}}^T, \hat{\mathbf{x}}_{t,\text{mis}}^T)^T$.

These imputed values will not satisfy the sum constraint, but a slightly modified regression approach can ensure that they do and will be described next.

Extension to Satisfy the Sum Constraint. This approach adds to the observed data the known totals of the missing data for the target variable as well as the predictor. These totals are

$$X_{t,\text{mis}} = X_t - \sum_i x_{t,\text{obs},i}$$

and

$$X_{a,\text{mis}} = X_a - \sum_i x_{a,\text{obs},i},$$

respectively, where the summation is over the records with observed values for the target variable. The total $X_{t,\text{mis}}$ is added to the column $\mathbf{x}_{t,\text{obs}}$ and the total $X_{a,\text{mis}}$ is added to the column $\mathbf{x}_{a,\text{obs}}$. Furthermore, the regression model is extended with a separate constant term for the record with the totals of the missing data. The model for these observed data can then be written as

$$\begin{aligned} \mathbf{x}_{t,\text{obs}} &= \beta_0 \mathbf{1} + \beta \mathbf{x}_{a,\text{obs}} + \boldsymbol{\varepsilon}, \\ X_{t,\text{mis}} &= \beta_1 n_{\text{mis}} + \beta X_{a,\text{mis}} \end{aligned} \tag{9.60}$$

with n_{mis} the number of records with missing values for target variable $\mathbf{x}_t$. Note that the second equation in (9.60) can be seen as an aggregated version of an underlying system of n_{mis} regression equations for the records with missing $\mathbf{x}_t$ values, where the sum of the residuals equals zero.

We apply ols to estimate the model parameters that will be used to predict and impute the missing values in $\mathbf{x}_t$, that is,

$$\hat{\mathbf{x}}_{t,\text{mis}} = \hat{\beta}_1 \mathbf{1} + \hat{\beta} \mathbf{x}_{a,\text{mis}}, \tag{9.61}$$

and so the sum of the predicted values over the records with missing values for target variable $\mathbf{x}_t$ will equal

$$\hat{X}_{t,\text{mis}} = \sum_i \hat{x}_{t,\text{mis},i} = \hat{\beta}_1 n_{\text{mis}} + \hat{\beta} X_{a,\text{mis}}.$$

In order to demonstrate the property of this model that the imputed values will sum up to the known total, we re-express model (9.60) for the observed data with the known totals added as

$$\begin{pmatrix} \mathbf{x}_{t,\text{obs}} \\ X_{t,\text{mis}} \end{pmatrix} = \begin{pmatrix} \mathbf{1} & \mathbf{0} & \mathbf{x}_{a,\text{obs}} \\ 0 & n_{\text{mis}} & X_{a,\text{mis}} \end{pmatrix} \begin{pmatrix} \beta_0 \\ \beta_1 \\ \beta \end{pmatrix} + \begin{pmatrix} \boldsymbol{\varepsilon} \\ 0 \end{pmatrix}$$

or

$$\begin{pmatrix} \mathbf{x}_{t,\text{obs}} \\ X_{t,\text{mis}} \end{pmatrix} = \mathbf{Z}\boldsymbol{\beta} + \begin{pmatrix} \boldsymbol{\varepsilon} \\ 0 \end{pmatrix}.$$

If this model is estimated by ols estimation, the residuals are orthogonal to each of the columns of the model matrix **Z**. Thus, for the second column we obtain $n_{\text{mis}}(X_{t,\text{mis}} - \hat{X}_{t,\text{mis}}) = 0$ and hence $\hat{X}_{t,\text{mis}} = \sum_i \hat{x}_{t,\text{mis},i} = X_{t,\text{mis}}$, which implies that the sum of the imputed values equals the known value of this total.

Adjustment to Satisfy the Sum Constraint and the Interval Constraints. Since the interval constraints have not been considered in obtaining the predicted values, it can be expected that a number of these predictions are not within their admissible intervals. One way to remedy this situation is to calculate adjusted predicted values defined by

$$\hat{\mathbf{x}}^{\text{adj}}_{t,\text{mis}} = \hat{\mathbf{x}}_{t,\text{mis}} + \mathbf{a}_t \tag{9.62}$$

such that the adjusted predictions satisfy both the sum constraint (which is equivalent to $\sum_i a_{t,i} = 0$) and the interval constraints and the adjustments are as small as possible. One way to find such a value for $\mathbf{a}_t$ is to solve the quadratic programming problem

$$\text{minimize } \mathbf{a}_t^T \mathbf{a}_t, \quad \text{subject to} \quad \mathbf{1}^T \mathbf{a}_t = 0 \quad \text{and} \quad \mathbf{l}_t \leq \hat{\mathbf{x}}_{t,\text{mis}} + \mathbf{a}_t \leq \mathbf{u}_t,$$

or we can minimize the sum of the absolute values of the $a_{t,i}$ instead and solve the resulting linear programming problem.

A simple alternative approach to determine the $a_{t,i}$ is described in Pannekoek, Shlomo, and De Waal (2008).

9.8.4 REGRESSION IMPUTATION WITH RANDOM RESIDUALS ADDED

It is well known that, in general, predictive mean imputations show less variability than the true values that they are replacing. In order to better preserve the variance of the true data, random residuals can be added to the predicted means. The adjusted predictive mean imputations considered in the previous subsection will also be hampered by this drawback because these adjustments are intended to be as close as possible to the predicted means and not to reflect the variance of the original data.

In order to better preserve the variance of the true data, we start with the predicted values $\hat{\mathbf{x}}_{t,\text{mis}}$ obtained from (9.61) that already satisfy the sum constraint, and our purpose is to add random residuals to these predicted means such that the distribution of the data is better preserved and in addition both the interval and sum constraints are satisfied. These residuals serve the same purpose (satisfying the constraints) as the adjustments $a_{t,i}$; but in contrast to the $a_{t,i}$, they are not as small as possible, because they are intended to also reflect the true variability around the predicted means.

A simple way to obtain residuals is to draw each of the n_{mis} residuals by Acceptance/Rejection (AR) sampling [see, e.g., Robert and Casella (1999) for more on AR sampling] from a normal distribution with mean zero and variance equal to the residual variance of the regression model. This means that we repeatedly draw from this normal distribution until a residual is drawn that satisfies the interval constraint.

The residuals obtained by AR sampling may not sum to zero so that the imputed values do not satisfy the sum constraint. We may then adjust these residuals to sum to zero by applying a "shift" operation. This shift operation is defined as follows. First of all, we divide the n_{mis} units in three sets, L_t, U_t, O_t, with numbers of elements $n_{\text{mis},L}$, $n_{\text{mis},U}$, $n_{\text{mis},O}$, according to whether the current adjusted value $\hat{\mathbf{x}}^{\text{adj}}_{t,\text{mis}}$ is on the lower boundary, upper boundary, or neither boundary. Let the current sum of the $a_{t,i}$ be S_t, then zero sum adjustments can be obtained as

$$(9.63)\qquad a^{(1)}_{t,i} = a_{t,i} - S_t/(n_{\text{mis},U} + n_{\text{mis},O}) \qquad \text{for all } i \in U_t \cup O_t \text{ if } S_t > 0$$

or

$$(9.64)\qquad a^{(1)}_{t,i} = a_{t,i} + S_t/(n_{\text{mis},L} + n_{\text{mis},O}) \qquad \text{for all } i \in L_t \cup O_t \text{ if } S_t < 0.$$

We add or subtract a constant to the $a_{t,i}$ to make them sum to zero, thereby taking care not to subtract anything from $a_{t,i}$ that already set the $\hat{\mathbf{x}}^{\text{adj}}_{t,\text{mis}}$ on their lower boundary and not to add anything to $a_{t,i}$ that already set the $\hat{\mathbf{x}}^{\text{adj}}_{t,\text{mis}}$ on their upper boundary. After this step it may be that some of the $a^{(1)}_{t,i}$ cause their corresponding $\hat{\mathbf{x}}^{\text{adj}}_{t,\text{mis}}$ to cross their interval boundaries. In that case each prediction outside its admissible interval will be moved to the closest boundary value by an appropriate adjustment, which is the smallest possible adjustment to satisfy the interval constraints, that is,

$$a_{t,i}^{(2)} = l_{t,i} - \hat{x}_{t,\text{mis},i} \text{ if } \hat{x}_{t,\text{mis},i} < l_{t,i}, \tag{9.65}$$

$$a_{t,i}^{(2)} = u_{t,i} - \hat{x}_{t,\text{mis},i} \text{ if } \hat{x}_{t,\text{mis},i} > u_{t,i}, \tag{9.66}$$

$$a_{t,i}^{(2)} = 0 \text{ if } l_{t,i} \leq \hat{x}_{t,\text{mis},i} \leq u_{t,i}. \tag{9.67}$$

Both step (9.63) and (9.64) and step (9.65) to (9.67) are repeated until both the interval constraint and the sum constraint are satisfied.

9.9 Calibrated Hot Deck Imputation Subject to Edit Restrictions

9.9.1 INTRODUCTION

In this section we basically consider the same problem as in Section 9.8, with the major difference that in Section 9.8 we focused on numerical data whereas in the present section we focus on categorical data. That is, in the present section we aim to let imputed categorical data satisfy edits while at the same time preserving certain known totals, which in the case of categorical data take the form of known population frequencies.

A population frequency of a category may, for instance, be known from an available related register. Alternatively, previously *estimated* frequencies may be known. In the Dutch Social Statistical Database, estimated frequencies are fixed and later used to calibrate estimates of other quantities. In fact, this strategy of fixing frequencies and later using these fixed frequencies to calibrate other quantities to be estimated forms the basis of the so-called repeated weighting method: a weighting method designed to obtain unified estimates when combining data from different sources [see Houbiers (2004) and Knottnerus and Van Duin (2006)]. In our case, we do not use a weighting approach but instead aim to take these edits and known frequencies into account while imputing a record.

The remainder of this section is organized as follows. Section 9.9.2 introduces (the notation of) the edits and frequencies we consider in this section. Sections 9.9.3 to 9.9.7 describe the imputation algorithms we have developed for our problem.

This section is based on Coutinho, De Waal, and Shlomo (2010). A similar approach based on nearest-neighbor hot deck imputation that (approximately) preserves population frequencies, but that does not necessarily satisfy edits, is described by Zhang and Nordbotten (2008) and in some more detail by Zhang (2009).

9.9.2 EDITS AND FREQUENCIES FOR CATEGORICAL DATA

Edits for Categorical Data. As in Chapter 4, we denote the categorical variables by $v_1, \ldots, v_m$. Furthermore, we denote the domain—that is, the set of all allowed values of variable v_j by D_j. In the case of categorical data, an edit k is usually written in so-called *normal form*—that is, as a Cartesian product of sets F_j^k $(j = 1, \ldots, m)$:

$$F_1^k \times \cdots \times F_m^k,$$

meaning that if for a record with values $(v_1, \ldots, v_m)$ we have $v_j \in F_j^k$ for all $j = 1, \ldots, m$, then the record fails edit k, otherwise the record satisfies edit k (see also Section 4.3).

For instance, suppose we have three variables: *Marital status*, *Age*, and *Relation to head of household*. The possible values of *Marital status* are "Married," "Unmarried," "Divorced," and "Widowed"; the possible values of *Age* are "< 16" and "≥ 16"; and the possible values of *Relation to head of household* "Spouse," "Child," and "Other." Suppose we have two edits. The first edit states that someone who is less than 16 years old cannot be married, and the second edit states that someone who is not married cannot be the spouse of the head of household. In normal form the first edit can be written as

$$\{\text{Married}\} \times \{<16\} \times \{\text{Spouse, Child, Other}\} \tag{9.68}$$

and the second one can be written as

$$\{\text{Unmarried, Divorced, Widowed}\} \times \{<16, \geq 16\} \times \{\text{Spouse}\}. \tag{9.69}$$

Frequencies for Categorical Data. When a frequency for categorical data is known—for instance, because it has already been estimated—this simply means that one knows how many units in a certain subpopulation should have a specific value for a certain variable. For instance, one may know how many people in a subpopulation have a certain age and how many people in a subpopulation are married, even though some values of the variable *Age* and the variable *Marital status* are missing in an observed, but incomplete, data set. In this section we assume that for several categories such frequencies are known, and our aim is to obtain a fully imputed data set that preserves these frequencies. In this section we will also refer to known frequencies as totals.

9.9.3 THE BASIC IDEA OF THE IMPUTATION METHODS

The imputation methods we apply in this section are all based on a hot deck donor approach. When hot deck donor imputation is used, for each record containing missing values, the so-called recipient record, one uses the values of one or more other records, the so-called donor record(s), to impute these missing values (see also Section 7.6).

Usually, hot deck donor imputation is applied multivariately—that is, several missing values in a record are imputed simultaneously—using the same donor record. The Nearest-neighbor Imputation Methodology (NIM; see Section 4.5) is a multivariate hot deck approach that ensures that edits are satisfied after imputation. Our goal—to satisfy edits and simultaneously preserve totals—is more general and complex than the problem that the NIM aims to solve. In order to attain our goal, we want a bit more freedom than the NIM and other multivariate hot deck approaches allow. We, therefore, do not necessarily use a single donor record per recipient, but allow the use of multiple donor records per recipient. In principle, we allow a different donor record for each missing value in a recipient. Nevertheless, our methods do try to limit the number of donor records per recipient, and—if possible—we try to use a single donor record for all missing values in a recipient.

Even if we were able to find single donor records for all records requiring imputation, this would solve only part of our problem as the totals would only be preserved in very rare cases. We therefore apply sequential univariate hot deck donor imputation, where for each missing value in a record requiring imputation in principle a different donor record may be selected. The variables with missing values are imputed sequentially. For each variable, the records for which the value of the variable under consideration is missing are imputed one by one. Once all records for the variable under consideration have been imputed, the next variable with missing values is considered. The univariate hot deck imputation methods we apply are described in Sections 9.9.4 and 9.9.5. These univariate hot deck imputation methods are used to order the possible values for a certain missing field. Whether a value is actually used to impute the missing field depends on the order of the possible values, on whether the edits can be satisfied, and on whether the totals can be preserved.

While imputing a missing value, care is taken to ensure that the record can satisfy all edits. Only values of donor records that can result in a consistent record are eligible to be used. In Section 9.9.6. we explain how we determine whether a value is eligible to be used for imputation. For each record we make a list of eligible values for imputation.

Before an eligible value is actually used to impute a value, we first check whether the corresponding total can be preserved. If this total cannot be preserved, the value is rejected and a new value is drawn. This process goes on until we find an eligible value such that the corresponding total can be preserved.

In the next subsections we develop two univariate hot deck donor imputation methods: a nearest-neighbor approach and a random hot deck approach.

9.9.4 NEAREST-NEIGHBOR HOT DECK IMPUTATION

Suppose we want to impute a certain variable v_* in a record i_0. In the nearest-neighbor approach, we calculate for each other record i for which the value of v_* is not missing a distance given by

$$D(i_0, i) = \sum_j w_j(v_{i_0 j}, v_{ij}),$$

where the sum is taken over all variables apart from v_*, v_{i_0j} denotes the original value of the jth variable in record i_0, v_{ij} the original value of the jth variable in record i, and $0 \leq w_j(v_{i_0j}, v_{ij}) \leq 1$ a weight expressing how serious one considers a difference between v_{i_0j} and v_{ij} to be. The weight $w_j(v_{i_0j}, v_{ij})$ equals zero, if $v_{i_0j} = v_{ij}$. The weight $w_j(v_{i_0j}, v_{ij})$ is large if one considers the difference between v_{i_0j} and v_{ij} to be important, and it is small if one considers the difference to be unimportant. The original value of the jth variable in record i_0, v_{i_0j}, or the original value of the jth variable in record i, v_{ij}, may be missing. If v_{i_0j} or v_{ij} is missing, we set $w_j(v_{i_0j}, v_{ij})$ equal to 1.

To impute a missing value for v_* in i_0, we first select the potential donor value from the record with the smallest distance. If that value is not allowed, we try the second smallest value, and so on, until we find a donor value that is allowed. If none of the potential donor records has an allowed value, we try all values not occurring in the donor records in a random order. In this way we can construct an ordered list of potential donor values.

Note that the potential donor records for i_0 are ordered in the same way for different variables with missing values. So, if possible, multivariate hot deck imputation, using several values from the first potential donor record in this list, will be used. Only if a value of the first potential donor record cannot be used, because this would lead to failed edits or nonpreserved totals, a value from another potential donor record is used.

9.9.5 RANDOM HOT DECK IMPUTATION

When random hot deck imputation is applied, a random donor record is selected, often within certain subgroups defined by auxiliary data. In our case we implement random hot deck by sorting the potential donor records for a recipient i_0 in a random order. The same random order is then fixed for all variables with missing values in i_0; that is, for all variables with missing values in i_0 we use the same order. We thus obtain a list of potential donor records for recipient i_0.

To impute the missing value of a variable v_* in record i_0, we select the corresponding value of the first potential donor record on the list of potential donor records for recipient i_0. If that value is not allowed, we try the corresponding value of the second potential donor record on the list of potential donor records for recipient i_0, and so on, until we find a donor value that is allowed. If none of the potential donor records has an allowed value, we try all values not occurring in the donor records in a random order. As for nearest-neighbor imputation, we can thus construct an ordered list of potential donor values.

Since we sort the potential donor records in the same way for each variable, our random hot deck approach will use—if possible—multivariate donor imputation, just like in the case of nearest-neighbor imputation. That is, if possible, our random hot deck approach will use several values from the first potential donor record on the sorted list of potential donor records.

9.9.6 SATISFYING EDIT RESTRICTIONS

In order to ensure that the set of edits can be satisfied, we derive so-called implied edits (see also Section 4.3). These implied edits are necessary to ensure that whenever we impute the current variable to be imputed, the variables remaining to be imputed can indeed be imputed consistently. Example 9.7 below illustrates the use of implied edits in the context of imputing missing categorical data.

EXAMPLE 9.7

We assume that we have a data set with the three variables *Marital status*, *Age*, and *Relation to head of household* and their categories as defined in Section 9.9.2. We also assume that these variables have to satisfy edits (9.68) and (9.69). Now suppose that both *Marital status* and *Age* in a certain record are missing and that the value of *Relation to head of household* equals "Spouse." Suppose that we first impute *Age* and subsequently *Marital status*. In this case we cannot simply ignore the edits involving the variable to be imputed later, *Marital status*, while imputing *Age*. If we were to ignore the edits involving *Marital status*—that is, both edits—we could impute the value "< 16" for *Age*. In that case there would be no value for *Marital status* such that all edits are satisfied.

The edits (9.68) and (9.69) have an implied edit

(9.70)
$$\{\text{Married, Unmarried, Divorced, Widowed}\} \times \{< 16\} \times \{\text{Spouse}\},$$

which expresses that someone who is less than 16 years of age cannot be the spouse of the head of household. If we take this implied edit into account while imputing the missing value for *Age*, we find that we cannot impute the value "< 16" and that only "≥ 16" is allowed. When "≥ 16" is imputed for *Age*, then *Marital status* can be imputed in a consistent manner. ■

To determine the set of edits for the remaining variables to be imputed while imputing the current variable, we use the method proposed by Fellegi and Holt (1976) to eliminate a variable.

To eliminate a variable v_r we start by determining all index sets S such that

$$\bigcup_{k \in S} F_r^k = D_r \tag{9.71}$$

and

$$\bigcap_{k \in S} F_j^k \neq \emptyset \qquad \text{for all } j \neq r. \tag{9.72}$$

From these index sets we select the *minimal* ones—that is, the index sets S that obey (9.71) and (9.72), but none of their proper subsets obey (9.71). Given such a minimal index set S, we construct the implied edit

$$\bigcap_{k\in S} F_1^k \times \ldots \times \bigcap_{k\in S} F_{r-1}^k \times D_r \times \bigcap_{k\in S} F_{r+1}^k \times \ldots \times \bigcap_{k\in S} F_m^k.$$

For example, if we eliminate variable *Marital status* from the edits (9.68) and (9.69), we obtain the implied edit (9.70).

By adding the implied edits resulting from all minimal sets S to the set of edits and then removing all edits involving the eliminated variable, one obtains a set of edits for the remaining variables. It can be shown [see Chapter 4 of the present book, Fellegi and Holt (1976), and De Waal and Quere (2003)] that if and only if this set of edits for the remaining variables can be satisfied, a value for the eliminated variable exists such that the original set of edits can be satisfied. We call this the lifting property: A property—namely that the corresponding set of edits can be satisfied—for a certain number of variables is "lifted" to a higher number of variables (see also Section 4.3).

For records where multiple values are missing, we now order these variables in some order that we describe later in Section 9.9.7. Next, we eliminate the variables according to this order. Let us assume that, say, the values of variables v_1 to v_t are missing. We first substitute the values of the remaining variables into the original set of edits. This gives a set of edits E_0 that have to be satisfied by variables v_1 to v_t. We then eliminate variable v_1 from E_0 and obtain a set of edits E_1 that have to be satisfied by variables v_2 to v_t. Next, we eliminate variable v_2 from E_1 and obtain a set of edits E_2 that have to be satisfied by variables v_3 to v_t. We continue this process until we eliminate v_{t-1} from E_{t-2}, and obtain a set of edits E_{t-1} for variable v_t. For a single variable, edits simply define a set of allowed values for that variable. So, for variable v_t we now know which values are eligible for imputation. By a repeated application of the lifting property, it can be shown that the original set of edits can be satisfied if and only if v_t satisfies E_{t-1} (see Chapter 4).

Once we have determined the edit sets E_k $(k = 0, \ldots, t-1)$, we impute the variables in reverse order. That is, we impute v_t by drawing values by means of one of our hot deck imputation methods (see Sections 9.9.4 and 9.9.5) until we have selected an eligible value that can also preserve the total for this variable (see Section 9.9.7). We fill in this value for v_t into the edits in E_{t-2}. This gives us a set of eligible values for variable v_{t-1}. We continue this procedure until we have imputed all variables. What is important here is that whenever we want to impute a certain variable, we know the set of eligible values for that variable. We will use this property to preserve totals (see Section 9.9.7). Example 9.8 illustrates the procedure.

EXAMPLE 9.8

Suppose that in Example 9.7 we order the variables as follows: *Marital status* and then *Age*. We substitute the value of *Relation to head of household*—that is, "Spouse"—into the edits and obtain the edits

(9.73) $\{\text{Married}\} \times \{< 16\}$

and

(9.74) $\{\text{Unmarried, Divorced, Widowed}\} \times \{< 16, \geq 16\}$

for *Marital status* and *Age*. In this very simple case we now only have to eliminate one variable, *Marital status*, and obtain the edit

(9.75) $\{< 16\}$,

which has to be satisfied by *Age*. Edit (9.75) defines the set of eligible values for *Age*: In this case only the value "≥ 16" is allowed. If we impute "≥ 16" for the missing value of *Age*, we can be sure that a value for *Marital status* exists such that all edits are satisfied. Imputing the value "≥ 16" for *Age* and substituting this value into edits (9.73) and (9.74), gives the edit

(9.76) $\{\text{Unmarried, Divorced, Widowed}\}$

for *Marital status*. The set of allowed values for *Marital status* hence consists of only one value: "Married." ■

Implied edits are often used to automatically identify erroneous fields in a data set [see Chapter 4, and also Fellegi and Holt (1976)]. It is well known that in that case the number of implied edits may be immense. In our case the number of implied edits is much less high, however. In order to identify erroneous fields automatically, one basically has to generate implied edits for every possible subset of the variables (see Section 4.3). In our case, one only has to consider a limited number of possible subsets, because the variables are eliminated in a fixed order. For instance, if there are five variables, one has to consider 32 subsets (ranging from eliminating no variables to eliminating all 5 variables) for automatic error localization, and only six subsets (ranging from eliminating no variable, eliminating variable 1, eliminating variables 1 and 2, and so on, until eliminating variables 1, 2, 3, 4, and 5) for our problem.

9.9.7 PRESERVING TOTALS

In the previous subsection we have explained that whenever we want to impute a certain variable in a record, we know the set of eligible values for this variable. We now explain how to check whether an eligible value can preserve the known total. Suppose the variable to be imputed has k_0 categories C_1 to C_{k_0}. We can then summarize the situation for the variable under consideration in a table such as Table 9.2.

TABLE 9.2 Illustration of the Sets of Eligible Values

	Cat. C_1	Cat. C_2	...	Cat. C_{k_0}
Record 1	*	0	...	*
Record 2	1	0	...	0
Record 3	0	*	...	*
...	...	...	...	...
Record n	*	*	...	0
Frequency	T_1	T_2		T_{k_0}

In Table 9.2 a 0 denotes that the category is not eligible, a "*" denotes that the category is eligible, and a 1 denotes that this value occurs in the corresponding record (i.e., the value of the variable under consideration is not missing for that record). The T_l ($l = 1, \ldots, k_0$) denote the known totals.

Now, we impute the variable under consideration record by record. We draw a value from the set of eligible values for the variable to be imputed for record 1, using one of our hot deck imputation approaches. After a category C_x has been drawn, we perform the following checks:

1. Do we not have too many records with the selected category C_x? If so, we reject the selected category C_x and select a new one. If not, we perform the second check.
2. Will it be possible to preserve the totals for this variable if we accept the selected category C_x? If so, we accept this value and go to the next record to be imputed. If not, we reject the selected category C_x and select a new one, which is again subjected to the same checks.

Checking whether we will not have too many records with the selected category C_x is trivial: We simply check whether the number of records so far with the value C_x exceeds the total T_x or not.

Checking whether the totals for the variable under consideration can be satisfied if we accept the selected category C_x is a well-known problem from combinatorial mathematics. It is called the *Harem Problem*. In the Harem Problem, several men (the categories in our case) have to choose a specified number (the T_l in our case) of wives (the records in our case) into their harem. The men all specify which women they would like to have in their harem and which women they do not want in their harem (the *'s and the 0's in our case). The 1's in our case correspond to women these men already have in their harem.

Conditions, and a constructive algorithm, for solving this problem are given by Anderson (1989). The underlying idea of the algorithm is to assign records to categories in a simple manner until one gets stuck. Once that happens, a specific algorithm is applied with the aim to assign one more record to the categories by reshuffling the assignments of records to categories. This algorithm is repeatedly applied until either all records are assigned to categories or until one again gets stuck.

In the first case we have constructed a solution to this instance of the Harem Problem, and we have shown that it is possible to preserve the totals if we accept the selected category C_x. In the second case a solution to this instance of the Harem Problem is not possible.

Note that if, for a certain variable to be imputed, the first record with a missing value has a solution to the Harem Problem, then by construction of our algorithm all records to be imputed for that variable can also be imputed.

Obviously, for the first record it is generally easier to find solutions to the Harem Problem than for later records. That is, for later records it is generally harder to find suitable imputation values. To avoid that for the same records, it is hard to find suitable imputation values for different variables, we first randomize the records each time before we start imputing a new variable.

Below we illustrate the Harem Problem and our approach to the imputation problem by means of a simple example.

EXAMPLE 9.9

Suppose that for a certain variable to be imputed we have the Harem Problem summarized in Table 9.3.

TABLE 9.3 **An Example of the Harem Problem**

	Cat. C_1	Cat. C_2	Cat. C_3
Record 1	0	*	*
Record 2	*	*	*
Record 3	0	0	*
Record 4	*	*	*
Record 5	*	0	*
Frequency	3	1	1

Now if we select category C_3 for the first record, the Harem Problem for the remaining records turns out be infeasible. This is easy to see: To categories C_1 and C_2 in total, all four records have to be assigned in some way. However, record 3 cannot be assigned to either of these categories. This means that category C_3 is rejected for record 1, and we have to impute category C_2 for this record. The Harem Problem for the remaining records is then feasible. In fact, in this case there is only one solution: Assign category C_2 to record 1, category C_3 to record 3, and category C_1 to records 2, 4, and 5. ■

As noted above, with respect to satisfying edits and totals, there may be a problem only for the first record to be imputed for each new variable. If we cannot find a suitable imputation value for that record, we would have to backtrack. That is, we would have to return to a previously imputed variable and impute one or more missing values for that variable in another way. This would lead to an extremely complicated process, because it is hard to specify beforehand which missing values would have to be imputed in another way and how they would have to be imputed. We would more or less have to explore all possibilities, which is obviously very time-consuming.

In an attempt to avoid this situation, we use a simple rule of thumb. We order the variables according to the average number of missing values per category. We first impute the variable for which this value is the lowest, and we end by imputing the variable for which this value is the highest. The higher the average number of values per category to be imputed, the more "freedom" one has while imputing. If a certain value cannot be imputed in a record, it is likely that due to this "freedom" there will be another record where this value is permitted.

If the number of edits is high or when the edits are rather restricting, one can consider *deactivating* edits in order to avoid having to backtrack to previously imputed variables. Edits can be deactivated by as soon as possible filling in values for missing values such that edits cannot be violated anymore by values to be imputed later. Deactivating edits can be implemented by taking the first record on the sorted list of potential donor records that is allowed according to the edits and the Harem problem and that also leads to (a sufficiently high number of) deactivated edits, instead of simply taking the first record on the list that is allowed according to the current edits and Harem Problem.

REFERENCES

Anderson, T. W. (1984), *An Introduction to Multivariate Statistical Analysis*, second edition. John Wiley & Sons, New York.

Anderson, I. (1989), *A First Course in Combinatorial Mathematics*, second edition. Oxford University Press, Oxford.

Arnold, B. C., and S. J. Press (1989), Compatible Conditional Distributions. *Journal of the American Statistical Association 84*, pp. 152–156.

Coutinho, W., T. de Waal, and M. Remmerswaal (2007), *Imputation of Numerical Data under Linear Edit Restrictions*. Discussion paper 07012, Statistics Netherlands (see also www.cbs.nl).

Coutinho, W., T. de Waal, and N. Shlomo (2010), *Calibrated Hot Deck Imputation Subject to Edit Restrictions*. Discussion paper 201016, Statistics Netherlands (see also www.cbs.nl).

De Waal, T. (2001), SLICE: Generalised Software for Statistical Data Editing. In: *Proceedings in Computational Statistics*, J. G. Bethlehem and P. G. M. Van der Heijden, eds. Physica-Verlag, New York, pp. 277–282.

De Waal, T. (2005), Automatic Error Localisation for Categorical, Continuous and Integer Data. *Statistics and Operations Research Transactions 29*, pp. 57–99.

De Waal, T., and W. Coutinho (2005), Automatic Editing for Business Surveys: An Assessment of Selected Algorithms. *International Statistical Review 73*, pp. 73–102.

De Waal, T., and R. Quere (2003), A Fast and Simple Algorithm for Automatic Editing of Mixed Data. *Journal of Official Statistics 19*, pp. 383–402.

Drechsler, J. (2009), *Far from Normal—Multiple Imputation of Missing Values in a German Establishment Survey*. Working Paper No. 21, UN/ECE Work Session on Statistical Data Editing, Neuchâtel.

Fellegi, I. P., and D. Holt (1976), A Systematic Approach to Automatic Edit and Imputation. *Journal of the American Statistical Association 71*, pp. 17–35.

Geweke, J. (1991), *Efficient Simulation from the Multivariate Normal and Student-t Distributions Subject to Linear Constraints and the Evaluation of Constraint Probabilities*. Report, University of Minnesota.

Harville, D. A. (1997), *Matrix Algebra from a Statistician's Perspective*. Springer-Verlag, New York.

Houbiers, M. (2004), Towards a Social Statistical Database and Unified Estimates at Statistics Netherlands. *Journal of Official Statistics 20*, pp. 55–75.

Kartika, W. (2001), *Consistent Imputation of Categorical and Numerical Data*. Report, Statistics Netherlands, Voorburg.

Khatri, C. G. (1968), Some Results for the Singular Normal Multivariate Regression Models. *Sankhyā Series A 30*, pp. 267–280.

Knottnerus, P., and C. Van Duin (2006), Variances in Repeated Weighting with an Application to the Dutch Labour Force Survey. *Journal of Official Statistics 22*, pp. 565–584.

Kovar, J., and P. Whitridge (1990), Generalized Edit and Imputation System; Overview and Applications. *Revista Brasileira de Estadistica 51*, pp. 85–100.

Pannekoek, J. (2006), *Regression Imputation with Linear Equality Constraints on the Variables*. Working Paper No. 28, UN/ECE Work Session on Statistical Data Editing, Bonn.

Pannekoek, J., and T. de Waal (2005), Automatic Edit and Imputation for Business Surveys: the Dutch Contribution to the EUREDIT Project. *Journal of Official Statistics 21*, pp. 257–286.

Pannekoek, J., N. Shlomo, and T. de Waal (2008), *Calibrated Imputation of Numerical Data under Linear Edit Restrictions*. Working Paper No. 23, UN/ECE Work Session on Statistical Data Editing, Vienna.

Pannekoek, J. and M. G. P. Van Veller (2004), Regression and Hot-Deck Imputation Strategies for Continuous and Semi-Continuous Variables. In: *Methods and Experimental Results from the EUREDIT Project*, J. R. H. Charlton, eds. (http://www.cs.york.ac.uk/euredit/).

Raghunathan, T. E., J. M. Lepkowski, J. Van Hoewyk, and P. Solenberger (2001), A Multivariate Technique for Multiply Imputing Missing Values Using a Sequence of Regression Models. *Survey Methodology 27*, pp. 85–95.

Rao, C. R. (1973), *Linear Statistical Inference and its Applications*, second edition. John Wiley & Sons, New York.

Robert, C. P., and G. Casella (1999), *Monte Carlo Statistical Methods*. Springer-Verlag, New York.

Rubin, D. B. (2003), Nested Multiple Imputation of NMES via Partially Incompatible MCMC. *Statistica Neerlandica 57*, pp. 3–18.

Särndal, C.-E., and S. Lundström (2005), *Estimation in Surveys with Nonresponse*. John Wiley & Sons, Chichester.

Schafer, J. L. (1997), *Analysis of Incomplete Multivariate Data*. Chapman & Hall, London.

Tempelman, D. C. G. (2007), *Imputation of Restricted Data*. Ph.D. Thesis, University of Groningen (see also www.cbs.nl).

Van Buuren, S., H. C. Boshuizen, and D. L. Knook (1999), Multiple Imputation of Missing Blood Pressure Covariates in Survival Analysis. *Statistics in Medicine 18*, pp. 681–694.

Van Buuren, S. and C. G. M. Oudshoorn (1999), *Flexible Multivariate Imputation by MICE*. Report TNO/PG 99.054, TNO Preventie en Gezondheid, Leiden.

Van Buuren, S. and C. G. M. Oudshoorn (2000), *Multivariate Imputation by Chained Equations: MICE V1.0 User's Manual*. Report PG/VGZ/00.038, TNO Preventie en Gezondheid, Leiden.

Wilks, S. S. (1962), *Mathematical Statistics*. John Wiley & Sons, New York.

Winkler, W. E. (2003), *A Contingency-Table Model for Imputing Data Satisfying Analytic Constraints*. Working Paper No. 26, UN/ECE Work Session on Statistical Data Editing, Madrid.

Winkler, W. E., and L. A. Draper (1997), The SPEER Edit System. In: *Statistical Data Editing (Volume 2); Methods and Techniques*, United Nations, Geneva.

Zhang, L.-C. (2009), *A Triple-Goal Imputation Method for Statistical Registers*. Working Paper No. 28, UN/ECE Work Session on Statistical Data Editing, Neuchâtel.

Zhang, L.-C., and S. Nordbotten (2008), *Prediction and Imputation in ISEE: Tools for More Efficient Use of Combined Data Sources*. Working Paper No. 11, UN/ECE Work Session on Statistical Data Editing, Vienna.

CHAPTER TEN

Adjustment of Imputed Data

10.1 Introduction

Constraints on the data are a form of external information, already available before the survey or administrative data set is observed, with which the observed data should comply. In the process of data editing, this external information is referred to as the edit rules and is used in a number of different ways to improve data quality. In Chapters 2 to 6 the edit rules have been used to detect and localize errors, in Chapter 2 the edit rules have also been instrumental in finding the solution to an error, and in Chapter 9 we have seen ways to restrict the set of possible imputations to those that are admissible in the sense that they satisfy certain edit rules. In this chapter, the edit rules will be used in a two-step approach to consistent imputation. In this approach, the imputations are first generated, using any imputation method that preserves the relevant statistical properties as well as possible. During this step the edits are not necessarily taken into account and, as a result, some edit rules may be violated after imputation. In the second step, the adjustment step, the imputed values are modified as little as possible, such that all edit rules become satisfied and a consistent record results. The nonimputed, original values are not modified. By keeping the modified imputed values as close as possible to the original imputed values, the statistical properties of the imputations are preserved as well as possible. Modifying the imputed values in this sense will be called the *adjustment* problem. Throughout this chapter we assume that the imputed values can indeed be modified in such a way

Handbook of Statistical Data Editing and Imputation, by T. de Waal, J. Pannekoek, and S. Scholtus

that a consistent record results. This is, for instance, the case if the (generalized) Fellegi–Holt paradigm [see Fellegi and Holt (1976) and Chapters 3 to 5] has been adopted in order to find a solution to the error localization problem.

Adjustment of data to satisfy edit constraints is not only applied as a last step in the imputation process. It sometimes happens that data that have been edited, imputed, and made consistent are modified again at a later stage. One reason can be that the data quality can be further improved by using new information or as a result of specific analyses, such as confrontation with other sources during the compilation of the national accounts. This modification can result in violations of edit rules that may call for a secondary application of adjustment methods. Another reason, not related to data quality, for modifying the edited data is to protect the privacy of respondents when micro-data files are released to outside researchers for further analysis. In order to prevent the identification of individual respondents in the data set and the disclosure of sensitive information, disclosure protection techniques are applied. These techniques include, among many others, data perturbation methods that modify (slightly) the values in the data set, for instance by adding noise. Again, these modifications may lead to edit failures, and adjustment procedures can be applied to restore consistency. The problem of preserving consistency while applying measures to protect the confidentiality is discussed in Shlomo and De Waal (2008).

In this chapter, solutions to the adjustment problem are discussed, primarily motivated by the problem of adjustment of imputed values as a step in the edit and imputation process. However, these algorithms can be applied to many other related "adjustment" problems, and some examples of such related problems are given. In Section 10.2 of this chapter, solutions to the adjustment problem for numerical variables are treated. In Section 10.3 the adjustment problem is addressed for a mix of categorical and numerical data.

10.2 Adjustment of Numerical Variables

Apart from edit rules, additional information can also be available in the form of data from other sources than the current data set. The values of some key variables for the responding units may be available from other sources, such as a more recent but less detailed short-term survey or an administrative data set. The problem then becomes to adjust the values in the current data set such that they are consistent with the newly obtained external values. Algorithms for the adjustment of numerical variables can often not only be used to adjust imputed data to satisfy edit rules but also to adjust to external information in the form of data and the combination of both these forms of additional information. In Section 10.2.1, examples will be given that illustrate the kind of adjustment problems that the algorithms can be applied to. In the next two subsections, algorithms are described that are based on two different criteria for measuring the amount of adjustment applied to the original data. In Section 10.2.2 a least squares criterion is used, leading to additive adjustments, and in Section 10.2.3 the Kullback–Leibler discriminating information is used which leads to multiplicative adjustments. In

TABLE 10.1 **An Imputed Business Record**

Variable	Description	Value
x_1	Profit	200
x_2	Number of employees	20
x_3	Turnover main activity	1000
x_4	Turnover other activities	***30***
x_5	Total turnover	***1030***
x_6	Labor costs	***500***
x_7	Costs of purchases	200
x_8	Total costs	***700***

Section 10.2.4 the algorithms are applied to the examples of problems introduced in the beginning of this section to illustrate some properties of the different algorithms.

10.2.1 EXAMPLES OF PROBLEMS

EXAMPLE 10.1 **Adjustment of Imputations in a Business Record**

As an example of an adjustment problem for numerical variables, we consider the data displayed in Table 10.1. The data shown are part of a record from a business survey. The four values printed in italics have been imputed, and the other four values are original observed values.

Suppose that for this survey it is known that the following relations, or edit rules, must hold:

$$e_1: x_1 - x_5 + x_8 = 0 \quad (\textit{Profit} = \textit{Total turnover} - \textit{Total costs}),$$

$$e_2: x_5 - x_3 - x_4 = 0 \quad (\textit{Total turnover} = \textit{Turnover main} + \textit{Turnover other}),$$

$$e_3: x_8 - x_6 - x_7 = 0 \quad (\textit{Total costs} = \textit{Labor costs} + \textit{Costs of purchases}).$$

In the example, one edit rule is violated, $e_1 = -130$, and we want to adjust the imputed values such that all edit rules become satisfied. Since only the imputed values are to be changed, the restrictions on the complete data vector are first translated to restrictions on the imputed values. The restrictions for the eight variables can be expressed as

$$\mathbf{Ax} = \mathbf{0},$$

with $\mathbf{A}$ a 3×8 matrix with ones, minus ones, and zeros and $\mathbf{x}$ the data vector displayed in Table 10.1. If we partition the vector $\mathbf{x}$, after a permutation of elements, as $\mathbf{x} = (\mathbf{x}_m^T, \mathbf{x}_o^T)^T$, with $\mathbf{x}_m^T$ the vector containing the imputed values and $\mathbf{x}_o^T$ the vector with observed values and partition the matrix $\mathbf{A}$ accordingly, the restrictions can be written as

$$\mathbf{A}\mathbf{x} = \mathbf{A}_m\mathbf{x}_m + \mathbf{A}_o\mathbf{x}_o = \mathbf{0},$$

and so

$$\mathbf{A}_m\mathbf{x}_m = -\mathbf{A}_o\mathbf{x}_o = \mathbf{b},$$

say. Thus, the edit rules are satisfied by adjusted imputed values $\check{\mathbf{x}}_m$, say, for which $\mathbf{A}_m\check{\mathbf{x}}_m = \mathbf{b}$. The adjustment problem now becomes

$$\text{Minimize } D(\check{\mathbf{x}}_m, \mathbf{x}_m)$$
$$\text{Subject to } \mathbf{A}_m\check{\mathbf{x}}_m = \mathbf{b},$$

with D a function measuring the distance or discrepancy between the adjusted and original imputed values. Here, only edits formulated as equality constraints are used, but in most applications inequality constraints are present as well. The most obvious one being that most variables should be nonnegative. ■

EXAMPLE 10.2 Micro-level Adjustment to New Data

Suppose that at some point in time, new information is obtained for the same firm that provided the data for the record in Table 10.1, and, more specifically, let that information consist of more recent values for *Number of employees*, *Total turnover*, and *Total costs*. These values could have been obtained from an administrative source, such as the tax register, which is more up to date but less detailed than the yearly survey. The problem is now: How do we combine these two data sources to create a new record that is both up to date and detailed? One possible solution to this problem is to cast the construction of an up-to-date and complete record as an imputation problem with edit constraints. The newly observed values for *Number of employees*, *Total turnover*, and *Total costs* are the observed values and all other values are considered to be imputed ones. The imputed values will then be adjusted to satisfy the edit rules. The edit rules are in this case not used to identify possible errors; they are used as a model, describing the relation between the three new values and the other values, with the purpose of bringing the other values up to date. Note that this means that edit rules should be specified for every variable that needs to be updated. ■

EXAMPLE 10.3 Macro-level Adjustment to New Data

Adjustment to new data is predominantly applied at macro-level. A particularly well documented problem in this area is the adjustment of input–output tables in macroeconomics. Table 10.2 is a very much simplified example of an input–output table.

TABLE 10.2 An Input–Output Table

	Input to Economic Activity			
Economic Activity	Manufacturing	Transportation	Consumption	Total Output
Manufacturing	200	100	300	600
Transportation	100	0	50	150
Labor	300	100	0	400
Total input	600	200	350	1150

An entry (i, j) of this table is the monetary value of the output of economic activity i that is used by economic activity (or industry) j, during some period t. Information on the row and column totals may be routinely gathered by production surveys. Data on the entries are harder to come by and may be based on a combination of different surveys and expert judgment. When new information on the margins becomes available, a fast approximation to the entries can be obtained by adjusting a previous $(t-1)$ table to conform to the new row and column totals. This adjustment, in this context referred to as *benchmarking*, can be cast as a minimal adjustment problem:

$$\text{Minimize } D(\mathbf{x}_{t-1}, \mathbf{x}_t)$$
$$\text{Subject to } \mathbf{A}\mathbf{x}_t = \mathbf{b},$$

with $\mathbf{x}_{t-1}$ and $\mathbf{x}_t$ vectors containing the entries of, respectively, the previous and updated tables, $\mathbf{A}$ a matrix with ones and zeros that generates the row and column totals of a table when the vectorized table is premultiplied by it, and $\mathbf{b}$ the vector with row and column totals for period t. ■

10.2.2 LEAST SQUARES ADJUSTMENTS

In this subsection we consider the least squares criterion to find an adjusted $\mathbf{x}$ vector that is closest to the original unadjusted data. We start with two ways to obtain the solution of the problem for the case with equality constraints only. The first way is a direct solution by an explicit formula that takes all constraints into account simultaneously, and the second way is an iterative procedure in which the original data vector is successively adjusted to each of the constraints separately. This last procedure is then extended to problems with inequality constraints.

Equality Constraints. The least squares solution, $\check{\mathbf{x}}$ say, to the equality constraint adjustment problem is the solution of the following minimization problem:

$$\min_{\mathbf{x}} \tfrac{1}{2}(\mathbf{x}-\mathbf{x}^0)^T(\mathbf{x}-\mathbf{x}^0) \quad \text{subject to } \mathbf{A}\mathbf{x}=\mathbf{b} \tag{10.1}$$

with $\mathbf{x}^0$ the original data. It is assumed here that the constraints are formulated such that there are no redundant constraints and the row rank of $\mathbf{A}$ equals K, the number of rows. This assumption is not necessary and can be circumvented by using a generalized inverse instead of the regular inverse in the formulae below. We shall also see that in the iterative approach where no matrix inversion takes place, the problem of redundant constraints does not occur.

In order to find the constrained minimum we first form the Lagrangian for this problem, which can be expressed as

$$L(\mathbf{x},\boldsymbol{\lambda}) = \tfrac{1}{2}(\mathbf{x}-\mathbf{x}^0)^T(\mathbf{x}-\mathbf{x}^0) + \boldsymbol{\lambda}^T(\mathbf{A}\mathbf{x}-\mathbf{b}), \tag{10.2}$$

with $\boldsymbol{\lambda}$ a vector of *Lagrange multipliers* or *dual variables*. The constrained minimum can then be found as a stationary point of the Lagrangian, by equating the partial derivatives of $L(\mathbf{x},\boldsymbol{\lambda})$ with respect to $\mathbf{x}$ and $\boldsymbol{\lambda}$ to zero, thus by solving the equations

$$L'_{\mathbf{x}}(\mathbf{x},\boldsymbol{\lambda}) = \mathbf{x}-\mathbf{x}^0+\mathbf{A}^T\boldsymbol{\lambda} = \mathbf{0}, \tag{10.3}$$

$$L'_{\boldsymbol{\lambda}}(\mathbf{x},\boldsymbol{\lambda}) = \mathbf{A}\mathbf{x}-\mathbf{b} = \mathbf{0}, \tag{10.4}$$

with $L'_{\mathbf{x}}(\mathbf{x},\boldsymbol{\lambda})$ and $L'_{\boldsymbol{\lambda}}(\mathbf{x},\boldsymbol{\lambda})$ denoting the partial derivatives of $L(\mathbf{x},\boldsymbol{\lambda})$ with respect to $\mathbf{x}$ and $\boldsymbol{\lambda}$, respectively.

Solving (10.3) for $\mathbf{x}$ and substituting the resulting $\check{\mathbf{x}}$ in (10.4) yields

$$\boldsymbol{\lambda} = (\mathbf{A}\mathbf{A}^T)^{-1}\mathbf{A}(\mathbf{x}^0-\check{\mathbf{x}}) = (\mathbf{A}\mathbf{A}^T)^{-1}(\mathbf{A}\mathbf{x}^0-\mathbf{b})$$

and hence

$$\check{\mathbf{x}} = \mathbf{x}^0 - \mathbf{A}^T(\mathbf{A}\mathbf{A}^T)^{-1}(\mathbf{A}\mathbf{x}^0-\mathbf{b}) \tag{10.5}$$

An alternative approach to solving (10.1) is to use an iterative procedure in which the constraints are used one at a time. Such iterative procedures have been advocated for several reasons: They avoid the matrix inversion in (10.5) which can become unstable when the number of constraints is large, they can easily be adapted to handle large and sparse constraint matrices, and, the main reason that we follow this approach here, they are relatively easy to extend to inequality constraints and other objective functions than least squares.

An iterative procedure for the least squares problem, often attributed to Kaczmarz (1937), starts by minimizing (10.1) subject to one of the constraints. The resulting approximate solution is then updated such that a next constraint is satisfied and the difference with the previous approximation is minimized. When all constraints are visited, the first iteration is completed and the next iteration starts that will again sequentially adjust the current approximation to satisfy each of the constraints. The minimization carried out in each step solves the problem

$$\min_{\mathbf{x}} \tfrac{1}{2}(\mathbf{x} - \mathbf{x}^{t,k-1})^T(\mathbf{x} - \mathbf{x}^{t,k-1})$$

$$\text{subject to } \mathbf{a}_k^T \mathbf{x} = b_k, \tag{10.6}$$

with t denoting the iterations and k the constraints.

The solution of this problem, $\mathbf{x}^{t,k}$, is by definition the projection of $\mathbf{x}^{t,k-1}$ on the hyperplane defined by $\mathbf{a}_k^T\mathbf{x} = b_k$. The method is therefore also referred to as a cyclic or successive projection algorithm. Calculations similar to those leading to (10.5) show that this projection is given by

$$\lambda_k^t = \frac{b_k - \mathbf{a}_k^T \mathbf{x}^{t,k-1}}{\mathbf{a}_k^T \mathbf{a}_k}, \tag{10.7}$$

$$\mathbf{x}^{t,k} = \mathbf{x}^{t,k-1} + \lambda_k^t \mathbf{a}_k.$$

Because $\mathbf{a}_k^T(\mathbf{x}^{t,k} - \mathbf{x}^{t,k-1}) = \lambda_k^t \mathbf{a}_k^T \mathbf{a}_k$, the dual variable λ_k^t reflects the sign and size of the violation of constraint k by $\mathbf{x}^{t,k-1}$. Since the constraint becomes satisfied by the projection $\mathbf{x}^{t,k}$, this parameter can also be interpreted as the amount or size of the adjustment to $\mathbf{x}^{t,k-1}$ induced by constraint k which will also be referred to as the adjustment parameter.

The successive projection algorithm performs, within each iteration t, the step (10.7) for $k = 1, \ldots, K$. Note that the constraint matrix $\mathbf{A}$ need not be of full row-rank for this algorithm. Even if a row of $\mathbf{A}$ is duplicated, the algorithm still works; however, in that case, one of the constraints is used twice in each iteration. Since the algorithm must start with the original vector $\mathbf{x}^0$ to find a minimum distance solution, it is initialized by setting $\mathbf{x}^{1,0} = \mathbf{x}^0$. In order to carry over the last approximation $\mathbf{x}^{t-1,k}$ in iteration $t-1$ to iteration t, we define $\mathbf{x}^{t,0} = \mathbf{x}^{t-1,K}$.

Because the algorithm uses each of the constraints (or, equivalently, rows of $\mathbf{A}$) separately, iterative algorithms of this kind are also called *row-action* algorithms. Censor (1981) and Censor and Zenios (1997) review many such algorithms from which the algorithms presented in this section can be derived as special cases. Proofs of convergence, omitted here, also follow from the more general results in Censor and Zenios (1997).

Extension to Inequality Constraints. The problem of finding an $\mathbf{x}$ that is closest to $\mathbf{x}^0$, in the least squares sense, and satisfies inequality constraints amounts to the following minimization problem:

$$\min_{\mathbf{x}} \tfrac{1}{2}(\mathbf{x} - \mathbf{x}^0)^T(\mathbf{x} - \mathbf{x}^0) \tag{10.8}$$
$$\text{subject to } \mathbf{A}\mathbf{x} \leq \mathbf{b}.$$

Each of the K inequality constraints defines, as an admissible region, a half-space bordered by the hyperplane $\mathbf{a}_k^T\mathbf{x} = b_k$. The solution is the projection of $\mathbf{x}^0$ on the intersection of these half-spaces, which is a polyhedron. The successive projection algorithm defined in (10.7) can be extended to arrive at a remarkably simple algorithm to solve this problem. This method exploits the fact that although minimizing (10.8) subject to all of the constraints is not an easy problem, minimization of (10.8) subject to each of the constraints separately is a simple subproblem which can be solved explicitly. This extension of algorithm (10.7) leads to an algorithm first proposed by Hildreth (1957) that can be described as follows:

$$\begin{aligned} \delta &= \min\left(z_k^{t-1}, \lambda_k^t\right), \\ \mathbf{x}^{t,k} &= \mathbf{x}^{t,k-1} + \delta\mathbf{a}_k, \\ z_k^t &= z_k^{t-1} - \delta. \end{aligned} \tag{10.9}$$

The algorithm is initialized by setting z_k^0 to 0 for all k and λ_k^t is as defined in (10.7). This algorithm differs from the algorithm for the equality constraint case in that it keeps track of the parameters δ of the adjustments to $\mathbf{x}^0$ induced by constraint k in the successive iterations t and accumulates the negative of these parameters in z_k^t. After t iterations, the adjusted $\mathbf{x}^{t,K}$ can be written as

$$\mathbf{x}^{t,K} = \mathbf{x}^0 - \sum_{k=1}^{K} z_k^t\mathbf{a}_k, \tag{10.10}$$

which shows the additive structure of the adjustments. The minimum function in (10.9) has the effect that δ is smaller than or equal to z_k^{t-1} and hence the z_k^t will remain nonnegative during the iterations and $-z_k^t$ will be nonpositive, which is in line with the constraints being $\mathbf{a}_k^T\mathbf{x} \leq b_k$.

The working of the algorithm can be described by considering the following three cases that can occur at an iteration t, when adjusting for constraint k (Censor and Zenios, 1997):

I. The current value of $\mathbf{x}^{t,k-1}$ violates the constraint and lies outside the admissible half-space. The new value $\mathbf{x}^{t,k}$ is then the projection onto the hyperplane bordering this half-space and will satisfy the constraint with equality. In this case λ_k^t is negative, δ is negative, and the accumulated adjustment parameter z_k^t will increase.

The next two cases apply if $\mathbf{x}^{t,k-1}$ is in the interior of the admissible half-space and λ_k^t is positive. The constraint is then satisfied and there may be room to improve on the objective function by reducing the amount of adjustment induced by the constraint. The two values that δ can assume, depending on whether λ_k^t is smaller or larger than z_k^{t-1}, lead to the following two cases:

II. When λ_k^t is smaller than the accumulated adjustment parameter z_k^{t-1}, $\delta = \lambda_k^t$ and z_k^t will be smaller than z_{k-1}^t. The adjustment parameter δ projects in this case $\mathbf{x}^{t,k-1}$ from the interior of the half-space onto the bordering hyperplane. This is the largest adjustment that can be made without violating the constraint.

III. When λ_k^t is larger than z_k^{t-1}, $\delta = z_{k-1}^t$ and the cumulative adjustment parameter z_k^t will be reduced to zero.

The last two cases reduce the amount of adjustment due to a constraint to the extent that either the amount of adjustment is reduced to zero or the constraint becomes satisfied with equality. Note that for equality constraints the admissible region is the hyperplane itself and an adjustment will always consist of a projection onto this hyperplane, from either half-space, or leaving the current iterate unchanged. In this case it is therefore unnecessary to keep track of previous adjustments.

Weighted Least Squares Adjustments. For various reasons, it can be be advantageous to use weights in the least squares criterion. In its most general form the resulting criterion is then the matrix-weighted, or generalized, least squares criterion, given by

$$\min_{\mathbf{x}} \tfrac{1}{2}(\mathbf{x} - \mathbf{x}^0)^T \mathbf{W}(\mathbf{x} - \mathbf{x}^0)$$

$$\text{subject to } \mathbf{Ax} = \mathbf{b}. \tag{10.11}$$

For the problem of adjustment of imputed values, a diagonal weighting matrix can be used with confidence weights. These weights reflect the confidence one has in the imputed values; values with large weights will be adjusted less because their adjustment will have a larger impact on the weighted least squares criterion than values with smaller weights. Such confidence weights can be determined subjectively, taking the quality of the imputation models into account, but a more formal assessment of confidence weights is possible as well. One approach could be to first estimate the prediction error of the imputation model for each variable using the fully observed data and then use the inverse of this error as a confidence weight.

Another reason for using weights for the imputed variables is that it is considered appropriate to adjust large values more than small values. Instead of minimizing the squared adjustments, one could for instance choose to minimize the relative squared adjustments $(\check{x}_j - x_j^0)^2/x_j^0$, corresponding to the choice $\mathbf{W} = \text{diag}(\mathbf{x}^0)^{-1}$.

In the adjustment procedures for macro-level estimates, a natural weighting matrix would be the precision matrix—that is, the inverse of the covariance matrix of the estimates. However, in many cases this matrix may not be easy to estimate, especially for complex input–output tables based on various sources including knowledge of experts. Moreover, nonsampling errors play an important role in such adjustment procedures. Given these difficulties, weights in such adjustment procedures are often (partly) based on expert judgment.

Given the weights, obtained in one way or another, we can follow the same steps as applied in deriving (10.5) to obtain the solution to the equality constraint weighted least squares problem:

$$\check{\mathbf{x}} = \mathbf{x}^0 - \mathbf{W}^{-1}\mathbf{A}^T(\mathbf{A}\mathbf{W}^{-1}\mathbf{A}^T)^{-1}(\mathbf{A}\mathbf{x}^0 - \mathbf{b}). \tag{10.12}$$

When $\mathbf{x}^0$ is projected onto a single constraint, the hyperplane $\mathbf{a}_k^T\mathbf{x} = b_k$, the solution reduces to

$$\check{\mathbf{x}} = \mathbf{x}^0 + \frac{b_k - \mathbf{a}_k^T\mathbf{x}^0}{\mathbf{a}_k^T\mathbf{W}^{-1}\mathbf{a}_k}\mathbf{W}^{-1}\mathbf{a}_k, \tag{10.13}$$

from which it follows that the equality and inequality constrained successive projection algorithms can be modified to take weights into account by setting $\lambda_k^t = (b_k - \mathbf{a}_k^T\mathbf{x}^{t,k-1})/(\mathbf{a}_k^T\mathbf{W}^{-1}\mathbf{a}_k)$ and replacing $\lambda_k^t\mathbf{a}_k$ and $\delta\mathbf{a}_k$ by $\lambda_k^t\mathbf{W}^{-1}\mathbf{a}_k$ and $\delta\mathbf{W}^{-1}\mathbf{a}_k$, respectively.

10.2.3 MULTIPLICATIVE ADJUSTMENTS AND THE RAS/IPF ALGORITHM

In the previous subsection the original vector $\mathbf{x}^0$ has been adjusted to satisfy the constraints by additive adjustments. This is a consequence of the least squares criterion. An alternative is to multiply the components of $\mathbf{x}^0$ with factors such that an admissible $\check{\mathbf{x}}$ results. Multiplicative adjustment algorithms arise when the discrepancy between the unadjusted original vector and the adjusted one is measured by the Kullback–Leibler divergence [cf. Kullback (1959), Ireland and Kullback (1968)]. This objective function, also referred to as Kullback–Leibler discriminating information or relative entropy, can be written as

$$\sum_j x_j(\ln x_j - \ln x_j^0) - \sum_j x_j + \sum_j x_j^0. \tag{10.14}$$

In statistics, this measure is often used to compare probability distributions in which case the last two sums in (10.14) vanish. Since the last sum is a constant that can be ignored in minimizing (10.14), we take as our objective function

$$\sum_j x_j(\ln x_j - \ln x_j^0 - 1). \tag{10.15}$$

Clearly, (10.15) is defined for positive x_j and x_j^0 only. However, as will be discussed below, the optimization algorithms derived from (10.15) can handle most situations with zero x_j^0 values by setting the corresponding adjusted values to zero as well. Still, the multiplicative adjustment algorithms remain limited to nonnegative variables.

Equality Constraints. In order to minimize (10.15) subject to the equality constraints $\mathbf{Ax} = \mathbf{b}$, we set up the Lagrangian for this problem, which is

$$L(\mathbf{x}, \boldsymbol{\lambda}) = \sum_j x_j(\ln x_j - \ln x_j^0 - 1) + \boldsymbol{\lambda}^T(\mathbf{Ax} - \mathbf{b}). \tag{10.16}$$

Equating the partial derivatives of $L(\mathbf{x}, \boldsymbol{\lambda})$ to zero gives

$$L'_{x_j}(\mathbf{x}, \boldsymbol{\lambda}) = \ln x_j - \ln x_j^0 + \boldsymbol{\lambda}^T \mathbf{a}_j = 0, \tag{10.17}$$

with $\mathbf{a}_j$ the jth column of $\mathbf{A}$ and

$$L'_{\boldsymbol{\lambda}}(\mathbf{x}, \boldsymbol{\lambda}) = \mathbf{Ax} - \mathbf{b} = \mathbf{0}. \tag{10.18}$$

From (10.17) we obtain

$$\check{x}_j = x_j^0 \exp(-\boldsymbol{\lambda}^T \mathbf{a}_j) = x_j^0 \prod_{k=1}^{K} \exp(-\lambda_k a_{kj}). \tag{10.19}$$

This expression shows the multiplicative structure of the adjustments. To calculate the adjustment factors, it remains to find an expression for $\boldsymbol{\lambda}$. Substituting (10.19) in (10.18) gives

$$\sum_j a_{kj} x_j^0 \exp(-\boldsymbol{\lambda}^T \mathbf{a}_j) - b_k = 0 \qquad \text{for } k = 1, \ldots, K. \tag{10.20}$$

With a solution of (10.20) for $\boldsymbol{\lambda}$ the adjustments can be carried out according to (10.19). Since equation (10.20) cannot be solved explicitly, some iterative algorithm is needed to obtain the solution. In the following we will discuss the important special case in which all a_{kj} are either one or zero. This corresponds to problems where sums of certain elements of $\mathbf{x}$ are constrained to be equal to known constants. To simplify matters further, we consider an algorithm of the row-action type—that is, an algorithm that uses one of these sum-constraints at a time. For a single constraint k and all a_{kj} equal to 1 or 0, (10.20) reduces to

$$\begin{aligned} \exp(-\lambda_k a_{kj}) &= \frac{b_k}{\sum_j a_{kj} x_j^0} && \text{for } a_{kj} = 1, \\ &= 1 && \text{for } a_{kj} = 0, \end{aligned} \tag{10.21}$$

and minimizing (10.14) under a single sum-constraint results in

$$\check{x}_j = x_j^0 \exp(-\lambda_k a_{kj}). \tag{10.22}$$

The adjustment factor equals $\exp(-\lambda_k)$ for the elements of $\mathbf{x}$ for which $j \in M_k$ with M_k the index set defined by $M_k = \{j \mid a_{kj} = 1\}$ and 1 for x_j for which $j \notin M_k$. The sum of the variables with $j \in M_k$ is b_k. If $b_k = 0$, then $\lambda_k = \infty$ and the adjustment factor is zero. In this case all variables with $j \in M_k$ are adjusted to zero. If $\sum_j a_{kj} x_j^0 = 0$ all variables x_j^0 with $j \in M_k$ are zero and the adjustment factor is undefined. If in this case b_k is positive, there cannot be a solution and a multiplicative adjustment procedure cannot be applied (note that an additive adjustment procedure, such as those presented in Section 10.2.2 can lead to a solution in such cases). If $\sum_j a_{kj} x_j^0 = 0$ and $b_k = 0$, the constraint is satisfied by $\mathbf{x}^0$ and the adjustment factor is defined to be 1.

The $\check{\mathbf{x}}$ resulting from (10.22) minimizes the relative entropy subject to the constraint k and is called the entropy projection of $\mathbf{x}^0$ onto the hyperplane $\mathbf{a}_k^T \mathbf{x} = b_k$. A row-action algorithm for solving (10.15) now proceeds according to

$$\begin{aligned} \exp(-\lambda_k^t) &= \frac{b_k}{\sum_{j \in M_k} x_j^{t,k-1}}, \\ x_j^{t,k} &= x_j^{t,k-1} \exp(-\lambda_k^t) \qquad \text{for } j \in M_k, \\ x_j^{t,k} &= x_j^{t,k-1} \qquad \text{for } j \notin M_k. \end{aligned} \tag{10.23}$$

Similar as with the least squares algorithms in Section 10.2.2, the algorithm must be started with the original vector $\mathbf{x}^0$ to find a minimum distance solution and is therefore initialized by setting $\mathbf{x}^{1,0} = \mathbf{x}^0$. We also again define $\mathbf{x}^{t,0} = \mathbf{x}^{t-1,K}$. If the algorithm is run for S iterations, the adjusted x values can be expressed as

$$x_j^S = x_j^0 \prod_{k: j \in M_k} \tau_k \tag{10.24}$$

with

$$\tau_k = \prod_{t=1}^{S} \exp(-\lambda_k^t),$$

When the algorithm described above is applied to a rectangular matrix (written as a vector to conform with the notation above) and the sum constraints are constraints on the row sums and column sums, a well-known and frequently applied special case with a long history results. In the macroeconomic literature this algorithm is called the RAS algorithm, with the adjustment of input–output matrices as the most prominent application [cf. Stone, Champerowne, and Maede (1942)]. In various other disciplines this algorithm is called the IPF algorithm, which stands for Iterative Proportional Fitting. The IPF algorithm is often applied to adjust multi-dimensional tables of counts (contingency tables) to new univariate or multivariate margins [see, e.g., Bishop, Fienberg, and Holland (1975, Chapter 3)].

Extension to Inequality Constraints. The algorithm defined in (10.23) can be generalized to handle inequality constraints in much the same way as with the least squares algorithm. This generalization can be formulated as

$$\begin{aligned} \delta &= \min\left(z_k^{t-1}, -\lambda_k^t\right), \\ x_j^{t,k} &= x_j^{t,k-1}\exp(\delta) \qquad \text{for } j \in M_k, \\ x_j^{t,k} &= x_j^{t,k-1} \qquad \text{for } j \notin M_k, \\ z_k^t &= z_k^{t-1} - \delta \end{aligned} \tag{10.25}$$

with λ_k^t as defined in (10.23). The algorithm is initialized by setting $\mathbf{x}^{1,0} = \mathbf{x}^0$ and $z_k^t = 0$. After S iterations the adjusted data vector can in this case be written as

$$x_j^S = x_j^0 \prod_{k:j\in M_k} \exp(-z_k^S). \tag{10.26}$$

Since the minimum function in (10.25) ensures that the $z_k^S \geq 0$, the cumulated adjustment factors $\exp(-z_k^S)$ for each of the K constraints are ≤ 1, which is in line with the restrictions being that $\mathbf{a}_k^T\check{\mathbf{x}}$ must be less than or equal to some constants b_k. If, during the iterations, $\mathbf{a}_k^T\mathbf{x}^{t,k-1}$ is larger than b_k, the current value of the cumulated adjustment factor will be reduced, since λ_k^t will be positive and δ will be negative. This reduction is such that $\mathbf{a}_k^T\mathbf{x}^{t,k}$ is equal to b_k. On the other hand, if $\mathbf{a}_k^T\mathbf{x}^{t,k-1}$ is smaller than b_k, the adjustment factor will remain the same (if $z_k^{t-1} = 0$) or will increase but no further than the value that sets $\mathbf{a}_k^T\check{\mathbf{x}}$ equal to b_k.

More on algorithms derived from entropy minimization can be found in the book by Fang, Rajasekra, and Tsao (1997). Several applications of such algorithms to estimation problems with survey data have been described by Blien and Graef (1992).

10.2.4 EXAMPLES REVISITED

The algorithms developed in Sections 10.2.2 and 10.2.3 are now applied to the examples of adjustments problems introduced in Section 10.2.1.

EXAMPLE 10.1 Adjustment of Imputations in a Business Record (*continued*)

For the first example, the business record with imputed values, the adjustment problem is to ensure consistency with all edit rules by changing the imputed values as little as possible. The edit rules for this record are

e_1: $x_1 - x_5 + x_8 = 0$ (*Profit = Total turnover − Total costs*),

e_2: $x_5 - x_3 - x_4 = 0$ (*Total turnover = Turnover main + Turnover other*),

e_3: $x_8 - x_6 - x_7 = 0$ (*Total costs = Labor costs + Costs of purchases*).

Only one edit rule is violated for this record ($e_1 = -130$), but in order to prevent violations of the other edit rules as a result of the adjustment process, the other edit rules must also be taken into account.

TABLE 10.3 **Adjusted Imputed Business Record**

	Description	Original Value	Least Squares	LS Nonnegative	Weighted LS
x_1	Profit	200	200	200	200
x_2	Number of employees	20	20	20	20
x_3	Turnover main activity	1000	1000	1000	1000
x_4	Turnover other activities	***30***	***−10***	***0***	***23.7***
x_5	Total turnover	***1030***	***990***	***1000***	***1023.7***
x_6	Labor costs	***500***	***590***	***600***	***623.7***
x_7	Costs of purchases	200	200	200	200
x_8	Total costs	***700***	***790***	***800***	***823.7***

Three adjustment algorithms have been applied to this problem, and in Table 10.3 the original values and the adjusted values according to each algorithm are shown. The first adjustment procedure was the least squares procedure described in Section 10.2.2, taking the three equality constraints into account. The results are in the second column, labeled "Least Squares." By decreasing *Total turnover* by 40 and increasing *Total costs* by 90, edit e_1 is now satisfied. As a consequence, the values of *Labor costs* and *Turnover other activities* also must be changed, with the result that *Turnover other activities* has become negative, which is not valid for this variable. To remedy this problem, the least squares procedure was run again but now with the additional inequality constraint *Turnover other activities* ≥ 0. The results are in the column labeled "LS nonnegative" and show that the adjustment to *Turnover other activities* has become smaller, and as a consequence the adjustment to *Total turnover* has also become smaller and thus the adjustments to the other two variables has increased. A weighted least squares adjustment with equality constraints only and weight equal to the inverse of the original values was also applied and the results are in the last column. These results show that *Turnover other activities*, which has a much larger weight than the other variables, is adjusted much less than by the other procedures. The adjustment of *Total turnover* must then also be small, and consequently the adjustment of the other values is the largest among the three procedures applied. ■

EXAMPLE 10.2 Micro-level Adjustment to New Data (*continued*)

For the second example the same variables are used as for the first one. In this case the original values for *Number of employees*, *Total turnover*, and *Total costs* have been replaced by more recent ones from a reliable administrative source. The purpose of the adjustment procedure is to update the other values to conform to this new information. The profit x_1 can be deduced directly from the new values of x_5 and x_8. The variables x_3 and x_4 must be updated to sum up to the new value of x_5, and the variables x_6 and x_7 must be updated to sum up to the new value of x_8. These two adjustment problems are very simple in this case because they are disconnected, they have no variables in common, and they can be treated separately, without requiring an iterative solution. Updated values, obtained by the least squares and RAS algorithms, are in Table 10.4.

TABLE 10.4 Business Record Adjusted to New Data

	Description	Original Value	Least Squares	RAS
x_1	Profit	***330***	***400***	***400***
x_2	Number of employees	25	25	25
x_3	Turnover main activity	***1000***	***1085***	***1165.0***
x_4	Turnover other activities	***30***	***115***	***35.0***
x_5	Total turnover	1200	1200	1200
x_6	Labor costs	***500***	***550***	***572.4***
x_7	Costs of purchases	***200***	***250***	***228.6***
x_8	Total costs	800	800	800

Since, in this case, the problem has such a simple structure, the difference between the least squares approach and the RAS approach is directly apparent. The least squares adjustments are additive adjustments, adding for instance 85 to both variables x_3 and x_4 to make them sum up to x_5. The RAS method is a multiplicative method that solves the same problem by multiplying both x_3 and x_4 by 1.165. This multiplicative method can also be interpreted as a form of ratio-imputation. For instance, the new value for x_4 is calculated or "imputed" by the ratio model $x_{4,new} = x_{4,old} \times (x_{5,new}/x_{5,old})$. ■

EXAMPLE 10.3 Macro-level Adjustment to New Data (*continued*)

For the third example, adjusting an input–output table to new margins, the new margins are taken to be those in Table 10.5.

TABLE 10.5 Old and New Margins for Input–Output Table

Row Totals		Column Totals	
Old Values	New Values	Old Values	New Values
600	500	600	500
150	50	200	150
400	350	350	250

All three adjustment procedures that were applied to the first example were applied to this example as well and in addition the RAS procedure was also applied. It was assumed that the two zero values in the original data are "structural zeros" that are zero for logical reasons and should remain zero. They were therefore left out of the adjustment process for all three least squares adjustment procedures. For the RAS method, these originally zero values remain zero due to the multiplicative nature of the adjustments.

TABLE 10.6 Input–Output Table Adjusted to New Margins

Economic Activity	Input to	Original Value	Least Squares	LS Nonnegative	Weighted LS	RAS
Manufacturing	Manufacturing	200	170.0	175	186.4	188.7
	Transportation	100	73.3	75	75.7	76.0
	Consumption	300	256.7	250	238.0	235.3
Transportation	Manufacturing	100	56.7	50	38.0	35.3
	Transportation	0	0	0	0	0
	Consumption	50	−6.7	0	12.0	14.7
Labor	Manufacturing	300	273.3	275	275.7	276.0
	Transportation	100	76.7	75	74.3	74.0
	Consumption	0	0	0	0	0

The results are in Table 10.6. The column "Least Squares" again shows a problem with a negative value for the value that was originally 50. But this problem can easily be solved by adding a nonnegativity constraint as reflected in the column labeled "LS Nonnegative." The weighted least squares adjustments show a strong resemblance with the RAS adjusted values. This last result can be explained by the property, shown by Kadas and Klafsky (1976), that the weighted least squares criterion with weighting matrix $\mathbf{W} = \text{diag}(\mathbf{x}^0)^{-1}$ can be viewed as an approximation to the relative entropy criterion. ■

10.3 Adjustment of Mixed Continuous and Categorical Data

In this section we describe an algorithm to adjust imputed data in a mix of continuous and categorical data such that all edits become satisfied. We also extend this algorithm to include integer-valued data. Section 10.3.1 describes the mathematical problem we are trying to solve. Section 10.3.2 discusses a heuristic algorithm for categorical and continuous data to obtain consistently imputed data that are close to the data that have been imputed using a statistical imputation model. Section 10.3.3 illustrates the proposed algorithm by means of some examples. Sections 10.3.4 and 10.3.5 extend the algorithm to encompass integer-valued data. This algorithm is illustrated by means of a simple example in Section 10.3.6. Section 10.3.7 concludes the section by explaining how the proposed adjustment algorithm can also be used to find (potential) errors in highly contaminated data.

10.3.1 THE ADJUSTMENT PROBLEM FOR CATEGORICAL AND CONTINUOUS DATA

We denote the categorical variables by $v_j (j = 1, \ldots, m)$ and denote the continuous variables by $x_j (j = 1, \ldots, p)$. For categorical data we denote the domain—that is, the set of the possible values, of variable v_j by D_j. As in Section 4.2 we assume that every edit $k (k = 1, \ldots, K)$ is written in one of the following two forms:

$$\text{IF } v_j \in F_j^k \quad \text{for } j = 1, \ldots, m,$$

(10.27) $$\text{THEN } (x_1, \ldots, x_p) \in \{\mathbf{x} \mid a_{k1}x_1 + \cdots + a_{kp}x_p + b_k \geq 0\}$$

or

$$\text{IF } v_j \in F_j^k \quad \text{for } j = 1, \ldots, m,$$

(10.28) $$\text{THEN } (x_1, \ldots, x_p) \in \{\mathbf{x} \mid a_{k1}x_1 + \cdots + a_{kp}x_p + b_k = 0\}.$$

To measure how close an adjusted record is to the original record, we now need a distance function that includes both categorical and numerical variables. In this section we consider distance functions of the type

(10.29) $$\sum_{j=1}^{m} w_j^C d(v_j, \check{v}_j) + \sum_{j=1}^{p} w_j^N |x_j - \check{x}_j|,$$

where the record after the imputation phase is given by $(v_1, \ldots, v_m, x_1, \ldots, x_p)$, the final record is given by $(\check{v}_1, \ldots, \check{v}_m, \check{x}_1, \ldots, \check{x}_p)$, the w_j^C's are nonnegative user-specified weights for the categorical variables, the w_j^N's are nonnegative user-specified weights for the numerical variables, and $d(v_j, \check{v}_j)$ is a nonnegative matrix satisfying $d(v_j, \check{v}_j) = 0$ if $v_j = \check{v}_j$. Note that $\check{v}_j = v_j$ and $\check{x}_j = x_j$ for variables (categorical and continuous, respectively) that have not been imputed in the imputation step, because we only modify the imputed values.

The adjustment problem for categorical and continuous data can be formulated concisely as: Minimize (10.29) by modifying the *imputed* values so that all edits (10.27) and (10.28) are satisfied.

Note that for purely continuous data, (10.29) reduces to

$$\sum_{j=1}^{p} w_j^N |x_j - \check{x}_j|. \tag{10.30}$$

10.3.2 AN ADJUSTMENT ALGORITHM FOR CATEGORICAL AND CONTINUOUS DATA

The problem of minimizing (10.29) subject to the constraint that all edits (10.27) and (10.28) become satisfied can be formulated as a mixed integer programming problem [see Kartika (2001)]. This mixed integer programming problem may be solved by using standard software. Unfortunately, this mixed integer programming problem is usually rather large, so solving it by means of standard mixed integer programming software is likely to be rather time-consuming.

In this section we will not make an attempt to solve the adjustment problem to optimality and restrict ourselves to describing a heuristic that is likely to give acceptable results in practice. In any case the heuristic will lead to consistent data that satisfy all edits.

Denote the set of variables that have been imputed by S. Let s_0 be equal to $|S|$, the number of variables in S. We assume that the variables in S can be imputed consistently. We start by filling in the original values for all variables not in S into the set of explicit edits. This leads to a reduced set of edits involving only the imputed variables. We eliminate these variables from the reduced set of edits by applying the elimination technique described in Section 4.4.2. We keep track of the corresponding sets of (implicit) edits after q variables in S have been eliminated ($q = 0, \ldots, s_0$). We denote the set of (implicit) edits after q variables in S have been eliminated by Φ_q. The set of edits for $q = 0$, Φ_0, is the reduced set of explicit edits.

After all s_0 variables in S have been eliminated, the set Φ_{s0} of relations not involving any unknowns is consistent; that is, none of the relations is self-contradicting. This follows from our assumption that the variables in S can be imputed consistently and from Theorem 4.4 in Chapter 4. (Φ_{s0} may be the empty set, which is consistent by definition.) Hence, according to Theorem 4.3 in Chapter 4 there is a value $\check{v}_q$ for the qth variable that has been eliminated such that Φ_{q-1} is consistent if we fill in this value. If the qth variable is categorical, we choose $\check{v}_q$ such that $d(v_q, \check{v}_q)$ is minimal. If there are several possible values $\check{v}_q$ for which $d(v_q, \check{v}_q)$ is minimal, we select one randomly. For the $(q-1)$th variable we apply the same approach, and so on.

We continue this process until all values of imputed categorical variables have been modified in the above way. We are then left with a set of imputed continuous variables (if any) and a current set of (implicit) edits involving only these variables. The final values for these continuous variables are then found by minimizing (10.30) subject to the constraint that the current set of implicit edits is satisfied. This minimization problem can simply be formulated as a linear programming (LP) problem and can, for example, be solved by means of the simplex algorithm.

THEOREM 10.1

The heuristic described above leads to a record that satisfies all edits.

Proof. That the value of each imputed categorical variable can be modified in such a way that the imputed values that have not yet been modified can later be adapted in such a way that all explicit edits can be satisfied is a direct consequence of Theorem 4.3 in Chapter 4. After the imputed values of the categorical variables have been modified, that theorem also implies that the remaining imputed continuous variables can be modified such that all explicit edits become satisfied. The problem of finding modified continuous values that are as close as possible, in the sense of (10.30), to the imputed values can be found by solving an LP problem. ∎

When only continuous variables have been imputed, our method solves the adjustment problem to optimality. When categorical variables have been imputed, optimality of the method is not guaranteed, because the optimal modified value is sequentially determined for each *individual* categorical variable separately. Optimality of the method would only have been guaranteed if the optimal modified values had been determined for all variables *simultaneously*. However, as we have already mentioned, this is a very difficult problem. The method described is "only" a heuristic. It is, however, much simpler and faster than an optimal method.

10.3.3 EXAMPLES

To illustrate the algorithm of Section 10.3.2, we give two examples. The first one involves only categorical variables. This example is taken from Kartika (2001). The second example involves a mix of continuous and categorical variables.

EXAMPLE 10.4

Suppose we have four imputed, categorical variables with domains $D_1 = \{1,2,3,4\}$, $D_2 = \{1,2,3\}$, $D_3 = \{1,2,3\}$, and $D_4 = \{1,2\}$ and no imputed continuous variables. Suppose also that the reduced edit set is given by

(10.31) IF ($v_2 = 3$) AND ($v_3 \in \{1,2\}$) AND ($v_4 = 1$), THEN ∅.

(10.32) IF ($v_2 \in \{2,3\}$) AND ($v_4 = 2$), THEN ∅.

IF ($v_1 \in \{1,2,4\}$) AND ($v_2 \in \{1,3\}$) AND ($v_3 \in \{2,3\}$),

(10.33) THEN ∅.

(10.34) IF ($v_1 = 3$) AND ($v_3 \in \{2,3\}$) AND ($v_4 = 1$), THEN ∅.

Here we use the convention that if a categorical variable is not mentioned in an IF condition, this variable may take any value in its domain in order to trigger the edit. The matrix element $d(v_j, \check{v}_j)$ in objective function (10.29) equals 1 if $v_j \neq \check{v}_j$ and 0 otherwise for all $j = 1, \ldots, 4$. Suppose that the vector of imputed values is given by $\mathbf{v}_0 = (3, 3, 2, 2)$. This vector fails edit (10.32).

We apply our algorithm to obtain a consistent record. We start by selecting a variable, say v_1. We eliminate this variable and obtain a set of implicit edits without v_1. This set of implicit edits is given by (10.31), (10.32), and

(10.35) IF ($v_2 \in \{1,3\}$) AND ($v_3 \in \{2,3\}$) AND ($v_4 = 1$), THEN ∅.

We again select a variable, say v_2, and eliminate this variable from the current set of edits. As a result, we obtain an empty set of implicit edits. This means that we may assign arbitrary values to v_3 and v_4. Because our aim is to keep the final record close to the imputed record, we assign to both variables their original imputed values, that is, 2. Now, a value has to be assigned to v_2 such that (10.31), (10.32), and (10.35) become satisfied given that to both the third and the fourth variable the value 2 has been assigned. Filling in the value 2 for both the third and fourth variable into (10.31), (10.32), and (10.35) gives the edit

(10.36) IF ($v_2 \in \{2,3\}$), THEN ∅.

The only possibility to satisfy (10.36) is to assign the value 1 to v_2. Finally, we assign a value to v_1 such that (10.31) to (10.34) are satisfied given the values that have already been assigned earlier. Filling in the values assigned to v_2, v_3, and v_4 in (10.31) to (10.34) gives the edit

(10.37) IF ($v_1 \in \{1,2,4\}$), THEN ∅.

The only way to satisfy (10.37) is to assign the value 3 to v_1, which happens to be its original imputed value. So, we obtain a new record $\check{\mathbf{v}}_0 = (3, 1, 2, 2)$ with target value $\sum_{j=1}^{m} d(v_j, \check{v}_j) = 1$.

If the variables are eliminated in a different order, one may arrive at a different solution with a different target value. To illustrate, we now assume that we start by eliminating v_4 instead of v_1. The set of implicit edits is then given by (10.33)

(10.38) $$\text{IF } (v_2 = 3) \text{ AND } (v_3 \in \{1,2\}), \text{ THEN } \emptyset.$$

and

(10.39) $$\text{IF } (v_1 = 3) \text{ AND } (v_2 \in \{2,3\}) \text{ AND } (v_3 \in \{2,3\}), \text{ THEN } \emptyset.$$

We now eliminate variable v_1. The set of implicit edits is given by (10.38) and

(10.40) $$\text{IF } (v_2 = 3) \text{ AND } (v_3 \in \{2,3\}), \text{ THEN } \emptyset.$$

We eliminate v_3 and obtain

(10.41) $$\text{IF } (v_2 = 3), \text{ THEN } \emptyset$$

as the only implicit edit. We eliminate v_2 and obtain the empty set as the set of implicit edits, which is consistent by definition.

To satisfy (10.41), we have to change the value of v_2. Suppose we make v_2 equal to 2. We now have to satisfy (10.38) and (10.40) given the value assigned to v_2. For this we do not have to change the value of v_3. Next, we have to satisfy (10.33), (10.38), and (10.39) by changing the value of v_1 given the values already assigned. We make v_1 equal to one of the feasible values, say to 4. Finally, we have to satisfy (10.31) to (10.34) by changing the value of v_4 given the values already assigned. We make v_4 equal to the only feasible value—that is, to 1.

So, we obtain a new record $\check{\mathbf{v}}_0 = (4, 2, 2, 1)$ with target value $\sum_{j=1}^{m} d(v_j, \check{v}_j) = 3$. This solution is clearly not optimal. ■

EXAMPLE 10.5

Suppose we have a data set with four categorical variables C_j ($j = 1, \ldots, 4$) and three continuous variables N_j ($j = 1, 2, 3$). The domains of the categorical variables are given by $D_1 = \{1, 2\}$, $D_2 = \{1, 2, 3\}$, $D_3 = \{1, 2\}$, and $D_4 = \{1, 2, 3\}$. Suppose that in a certain record C_3, C_4, N_2, and N_3 have been imputed and that the entire record after imputation is given by $C_1 = 2$, $C_2 = 1$, $C_3 = 1$, $C_4 = 1$, $N_1 = 24$, $N_2 = 3000$, and $N_3 = 66{,}000$. Suppose furthermore that the edits are given by

$$\text{IF } (C_1 = 1, C_4 \in \{1,3\}), \text{ THEN } \emptyset.$$

$$\text{IF } (C_2 = 1, C_3 = 1), \text{ THEN } \emptyset.$$

$$\text{IF } (C_1 = 2, C_2 \in \{1,3\}, C_4 \in \{1,3\}), \text{ THEN } \emptyset.$$

$$1,250N_1 \geq 15{,}000.$$

$$\text{IF } (C_2 \in \{1,3\}), \text{ THEN } N_2 = 0.$$

$$\text{IF } (C_2 = 2), \text{ THEN } 12N_2 \geq 15{,}000.$$

$$\text{IF } (C_2 = 2), \text{ THEN } 12N_2 - 875N_1 \geq 0.$$

$$\text{IF } (C_2 = 2), \text{ THEN } 1250N_1 - 8.4N_2 \geq 0.$$

$$\text{IF } (C_2 \in \{1,3\}), \text{ THEN } 1250N_1 - N_3 = 0.$$

$$\text{IF } (C_2 = 2, C_3 = 2), \text{ THEN } 1250N_1 + 12N_2 - N_3 = 0.$$

$$\text{IF } (C_2 = 2, C_3 = 1), \text{ THEN } 1250N_1 + 12N_2 - N_3 = -1250.$$

If a categorical variable C_j is not mentioned in the IF condition, we mean in fact that C_j can assume any value in its domain D_j. We start by filling in the values of the nonimputed values into the edits. We obtain the following reduced set of edits for the four imputed variables:

(10.42) $$\text{IF } (C_3 = 1), \text{ THEN } \emptyset.$$

(10.43) $$\text{IF } (C_4 \in \{1,3\}), \text{ THEN } \emptyset.$$

(10.44) $$N_2 = 0.$$

(10.45) $$N_3 = 30{,}000.$$

We eliminate a continuous variable from (10.42) to (10.45), say N_3, and obtain edits (10.42) to (10.44) for the remaining variables. We eliminate the remaining continuous variable N_2 and obtain edits (10.42) and (10.43) for the categorical variables. We eliminate a categorical variable, say C_4, and obtain edit (10.42) for the remaining categorical variable. Finally, we eliminate the final variable C_3. We do not obtain a self-contradicting relation. This implies that the imputed variables can indeed be imputed in a consistent manner.

Now we are going to adjust the imputed values. The categorical variables are adjusted in reverse order of elimination. Hence we start with adjusting the value of variable C_3. As we have just determined, this variable had to satisfy only one edit, namely (10.42). The only way to satisfy this edit is to set the value of C_3 equal to 2. Next we adjust the value of C_4. Variables C_4 and C_3 have to satisfy edits (10.42) and (10.43). Taking the adjusted value of C_3 into account, only one edit for C_4 remains, namely (10.43). The only way to satisfy this edit is to set the value of C_4 equal to 2.

We now adjust the values of the imputed continuous variables. We do this by solving an LP problem. Variables C_3, C_4, N_2, and N_3 have to satisfy edits (10.42) to (10.45). Taking the adjusted values for C_3 and C_4 into account, N_2 and N_3 have to satisfy (10.44) and (10.45). In principle, we solve the LP problem of minimizing (10.30) subject to (10.44) and (10.45), for instance by applying the simplex algorithm. In this case, however, the answer to the LP problem is trivial: The value of N_2 should be set to 0, and the value of N_3 should be to 30,000. We have now found an adjusted, consistent record. This record is given by $C_1 = 2$, $C_2 = 1$, $C_3 = 2$, $C_4 = 2$, $N_1 = 24$, $N_2 = 0$, and $N_3 = 30{,}000$. ■

10.3.4 THE ADJUSTMENT PROBLEM FOR CATEGORICAL, CONTINUOUS AND INTEGER DATA

The adjustment problem for a mix of categorical, continuous, and integer data is, as one would expect, very similar to the adjustment problem for a mix of categorical and continuous data. The only difference is that some numerical variables have to attain an integer value.

Let I denote the index set of the integer-valued variables. The adjustment problem problem for a mix of categorical, continuous, and integer data can then be formulated concisely as: Minimize (10.29) by modifying the *imputed* values so that all edits (10.27) and (10.28) are satisfied and x_j is integer for $j \in I$.

10.3.5 AN ADJUSTMENT ALGORITHM FOR CATEGORICAL, CONTINUOUS AND INTEGER DATA

The adjustment algorithm described in Section 10.3.2 can be extended to integer-valued data. This extension is quite complicated if we aim to find all optimal solutions for all possible sets of edits of type (10.27) and (10.28) (see also Chapter 5 for an extension of a branch-and-bound algorithm for the related error localization problem for categorical, continuous, and integer-valued data). However, if we restrict ourselves to certain sets of edits and use a conservative approach, the extension of our algorithm for the adjustment problem to integer-valued data becomes quite easy. The price we have to pay for this is that we do not always find a solution to the adjustment problem, even if such a solution exists.

Handling Equalities. In edit sets arising in practice, balance edits—that is, edits of type (10.28)—usually form a so-called totally unimodular matrix [see, e.g., Nemhauser and Wolsey (1988) for more on totally unimodular matrices]. This is, for instance, the case when the balance edits form a subset of a (nonhierarchical) two-dimensional table, where the values of the variables corresponding to the internal cells of this table have to sum up to the values of the variables corresponding to the row and column totals.

When the balance edits indeed form a totally unimodular matrix, any equality generated by the algorithm described in the previous subsections can be scaled so that all nonzero coefficients are equal to either $+1$ or -1. This implies that if a variable x_r has been eliminated by using (3.16), that is, by

$$x_r = -\frac{1}{a_{sr}}\left(b_s + \sum_{j \neq r} a_{sj}x_j\right), \tag{10.46}$$

this variable is guaranteed to have an integer value if b_s and the other variables involved in (10.46) have integer values. That is, elimination by using (10.46) will not lead to additional problems when integer-valued instead of continuous data are processed.

In our heuristic, we scale the b_k's and the coefficients of the integer-valued variables in all edits in Φ_0 so they are integer and have a greatest common divisor equal to 1. Newly created implied edits are also scaled in this manner. Whenever we eliminate an integer-valued variable x_r using a balance edit, we check whether the coefficient of x_r equals $+1$, 0, or -1. Like we already mentioned, this is guaranteed if the balance edits form a totally unimodular matrix, but this may also be the case for other kinds of matrices. If the coefficient of x_r in a balance edit is not equal to $+1, 0$, or -1, we output that our heuristic cannot handle this record. Hence the record is not adjusted by the heuristic and has to be adjusted in another manner—for instance, manually.

Eliminating an Integer Variable from a Set of Inequalities. Each time we eliminate a variable, the current set of edits is transformed to a new set of edits. The new set of edits involves at least one variable less than the current set of edits. In the case that a continuous or categorical variable is eliminated, the current set of edits can be satisfied if and only if the new set of edits can be satisfied. When eliminating an *integer-valued* variable from a set of inequalities, we use a conservative approach where the current set of edits can be satisfied if the new set of edits can be satisfied. There are cases, however, where the current set of edits can also be satisfied even if our new set of edits cannot be satisfied. Our heuristic does not detect these cases. This is the price we have to pay in order to be able to handle integer data in a simple manner.

When an integer-valued variable is eliminated from a pair of inequalities

$$\text{IF } v_j \in F_j^s \text{ (for } j = 1, \ldots, m),$$

$$\text{THEN } (x_1, \ldots, x_p) \in \left\{\mathbf{x} \mid \sum_{j=1}^{p} a_{sj}x_j + b_s \geq 0\right\}$$

and

$$\text{IF } v_j \in F_j^t \text{ (for } j = 1, \ldots, m),$$

$$\text{THEN } (x_1, \ldots, x_p) \in \left\{\mathbf{x} \mid \sum_{j=1}^{p} a_{tj}x_j + b_t \geq 0\right\},$$

we do not use our standard elimination technique (see Section 4.4.2), where the THEN condition of the resulting edit is given by

$$\tilde{a}_1 x_1 + \cdots + \tilde{a}_{r-1} x_{r-1} + \tilde{a}_{r+1} x_{r+1} + \cdots + \tilde{a}_p x_p + \tilde{b} \geq 0$$

with

$$\tilde{a}_j = |a_{sr}| a_{tj} + |a_{tr}| a_{sj} \qquad \text{for } j = 1, \ldots, r-1, r+1, \ldots, p$$

and

$$\tilde{b} = |a_{sr}| b_t + |a_{tr}| b_s$$

and the IF condition of the resulting edit is given by

$$v_j \in F_j^s \bigcap F_j^t \qquad \text{for } j = 1, \ldots, m.$$

Instead, we compute the so-called "dark shadow" [see, e.g., Pugh (1992), De Waal (2003, 2005), and especially Chapter 5 of the present book]. For convenience, we reintroduce some notation from Chapter 5: $x_0 = 1$ and $a_{k0} = b_k$ for $k = 1, \ldots, K$. Now, if $a_{sr} > 0$ and $a_{tr} < 0$, the dark shadow is given by

(10.47) $$\text{IF } v_j \in F_j^s \bigcap F_j^t \text{ (for } j = 1, \ldots, m),$$

$$\text{THEN } x \in \{\mathbf{x} \mid \sum_{j=0}^{p} (a_{sr} a_{tj} - a_{tr} a_{sj}) x_j \geq (a_{sr} - 1)(-a_{tr} - 1)\}$$

The resulting inequality is sometimes stricter than when the standard elimination technique would have been used. When the resulting edit (10.47) is satisfied, a feasible integer value for the eliminated variable x_r is guaranteed to exist. Note that if $|a_{sr}| = 1$ or $|a_{tr}| = 1$, (10.47) reduces to the edit that would result from the standard elimination techniques. In other cases the THEN condition of (10.47) is stricter than the THEN condition that would result for the standard elimination technique. This guarantees that a feasible integer value for the eliminated variable exists if the resulting edits for the remaining variables can be satisfied.

The Adjustment Algorithm. The adjustment algorithm for categorical, continuous, and integer data is similar to the adjustment algorithm for categorical and continuous data described in Section 10.3.2. While eliminating variables, there are two differences. First of all, continuous variables are eliminated first, followed by integer-valued ones and finally the categorical ones. While selecting suitable values for the eliminated variables, we apply the following procedure. Suppose we are handling the qth eliminated variable. If this qth variable is categorical, we choose the value such that the set of edits for this variable becomes satisfied and $d(v_q, \check{v}_q)$ is minimal. If there are several feasible values v_q for which $d(v_q, \check{v}_q)$ is minimal, we select one randomly. If the qth variable is integer-valued, we choose $\check{x}_q$ such that it is integer, Φ_{q-1} becomes satisfied, and $|\check{x}_q - x_q|$ is minimal. For the $(q-1)$th variable we apply the same approach, and so on. We continue this process until all values of imputed categorical and integer-valued variables have been modified in the above way. We are then left with a set of imputed continuous variables (if any) and a current set of (implicit) edits involving only these variables. The final values for these continuous variables are then again found by minimizing (10.30) subject to the constraint that the current set of (implicit) edits is satisfied.

When only continuous variables have been imputed, our method solves the problem of modifying the imputed values as little as possible to optimality. When categorical or integer-valued variables have been imputed, optimality of the method is not guaranteed, because the optimal modified value is sequentially determined for each *individual* categorical or integer-valued variable separately. Optimality of the method would only have been guaranteed if the optimal modified values had been determined for all variables simultaneously. However, as we have already mentioned, this is a very difficult problem. The method described is "only" a heuristic. It is, however, much simpler and faster than an optimal method.

Theorem 10.1 and the theory developed in Chapter 5 of this book guarantee that if the set of relations after all imputed variables have been eliminated does not contain any self-contradictions, the imputed variables can be adjusted in a consistent manner. As we said, the price we have to pay for this is that sometimes a solution to the adjustment problem is not found, even if such a solution does exist. If this occurs often, one can instead of using (10.47) to determine the dark shadow simply use the standard elimination method by treating all integer-valued as if they were continuous ones. This may find more solutions to the adjustment problem, but it does not guarantee that a solution exists if the set of relations after all imputed variables have been eliminated does not contain any self-contradictions.

10.3.6 AN EXAMPLE FOR INTEGER-VALUED DATA

We illustrate the algorithm of the previous section by means of a simple example involving only integer-valued data. The edits are given by

$$T = P + C, \tag{10.48}$$

$$P \leq 0.5T, \tag{10.49}$$

$$-0.1T \leq P, \tag{10.50}$$

$$T \geq 0, \tag{10.51}$$

$$T \leq 550N. \tag{10.52}$$

We assume that the values of variables P and C have been imputed, say the imputed value of P equals 60 and the imputed value of C is 90. The original values of T and N are 100 and 5, respectively.

We start by filling in the values for the variables that have not been imputed, that is, T and N. We then obtain the following set of reduced edits:

$$100 = P + C, \tag{10.53}$$

$$P \leq 50, \tag{10.54}$$

$$-10 \leq P, \tag{10.55}$$

$$100 \geq 0, \tag{10.56}$$

$$100 \leq 2750. \tag{10.57}$$

Edits (10.56) and (10.57) are satisfied and can obviously be discarded.

We select a variable occurring in (10.53) to (10.55), say C. We eliminate this variable from these edits. Because edit (10.53) cannot be combined with the other two edits to eliminate C, we only have to copy (10.54) and (10.55) to the new set of edits. We obtain the system given by (10.54) and (10.55).

To check whether the set of edits (10.48) to (10.52) can be imputed consistently by modifying the values of P and C, we eliminate P from (10.54) and (10.55) by determining the dark shadow. The edit we obtain, $-10 \leq 50$, is satisfied, which shows that (10.48) to (10.52) can indeed be imputed consistently by modifying the values of the integer-valued variables P and C.

We now select a value for P such that (10.54) and (10.55) become satisfied; that is, we select a value for P between -10 and 50. We try to keep the final (integer) value of P as close as possible to the imputed value. We therefore select $P = 50$.

Given this value for P, the set of edits (10.53) to (10.55) reduces to

$$100 = 50 + C, \tag{10.58}$$

$$50 \leq 50, \tag{10.59}$$

$$-10 \leq 50. \tag{10.60}$$

For the final (integer) value of C we select a value that satisfies (10.58) to (10.60). In this case there is only one allowed value, namely $C = 50$. We therefore select this value. The resulting record passes all edits.

10.3.7 USING THE ADJUSTMENT ALGORITHM FOR LOCALIZATION OF RANDOM ERRORS

The algorithm(s) for localization of random errors developed in Chapters 3 to 5 are quite time-consuming. As a result, they are sometimes unable to find solutions for records containing many random errors. Alternatively, one can apply the adjustment algorithm as a heuristic method to localize random errors for such highly contaminated records. In this subsection we explain the main idea.

The heuristic method aims not to solve the error localization problem(s) for random errors described in Chapters 3 to 5. Instead the heuristic method aims to construct a new, synthetic record that satisfies all edits (10.27) and (10.28) and that is as close as possible to the original record. The distance between the new, synthetic record and the original record is measured by means of (10.29).

Given a record not satisfying all edits or with missing values, we start by filling in arbitrary values for the missing values. For a missing numerical field we propose to fill in the value 0 because in many cases this is the correct value for missing numerical values. For a missing categorical field we fill in an arbitrary value from the corresponding domain. Subsequently, we try to minimize

$$\sum_{j=1}^{m} w_j^C d(v_j^0, \check{v}_j) + \sum_{j=1}^{p} w_j^N |x_j^0 - \check{x}_j| \tag{10.61}$$

subject to the condition that the new, synthetic record $(\check{v}_1, \ldots, \check{v}_m, \check{x}_1, \ldots, \check{x}_p)$ satisfies all edits (10.27) and (10.28). Here $(v_1^0, \ldots, v_m^0, x_1^0, \ldots, x_p^0)$ denotes the original record after filling in arbitrary values for the missing fields. The other symbols have the same meaning as in Section 10.3.1.

To minimize (10.61) subject to the condition that the new, synthetic record $(\check{v}_1, \ldots, \check{v}_m, \check{x}_1, \ldots, \check{x}_p)$ satisfies all edits (10.27) and (10.28) we temporarily treat all fields in the record $(v_1^0, \ldots, v_m^0, x_1^0, \ldots, x_p^0)$ as imputed ones. Next, we apply the algorithm described in Section 10.3.2 or 10.3.5, depending on whether integer-valued variables are involved or not.

After successful completion of this algorithm all fields of which the value was missing in the original record are considered as erroneous as well as all fields for which the value in the new, synthetic record $(\check{v}_1, \ldots, \check{v}_m, \check{x}_1, \ldots, \check{x}_p)$ differs from the corresponding value in the record $(v_1^0, \ldots, v_m^0, x_1^0, \ldots, x_p^0)$. Since the objective function (10.61) is minimized subject to the condition that the new, synthetic record satisfies all edits, the fields identified by this heuristic as being erroneous can obviously be imputed such that all edits are satisfied.

The objective function (10.61) does not measure the (weighted) number of fields that needs to be changed as the objective function corresponding to the Fellegi–Holt paradigm [see Fellegi and Holt (1976) and Chapters 3 to 5 of the present book] does. Instead (10.61) measures the distance between the values of the original record (after filling in arbitrary values for the missing fields) and a synthetic record that satisfies all edits. The closest—in terms of (10.61)—synthetic record that satisfies all edits determines which fields are considered as being erroneous.

For purely continuous data the problem of minimizing (10.61) subject to the condition that the synthetic record satisfies all edits reduces to an LP problem. This LP problem can generally be solved in a fraction of the time required to solve the error localization problem based on the Fellegi–Holt paradigm exactly. At Statistics Netherlands, this approach for continuous data based on solving an LP problem has been evaluated and compared to the results of algorithms based on the Fellegi–Holt paradigm. As could be expected, the approach based on solving an LP problem instead of algorithms based on the Fellegi–Holt paradigm resulted in more fields that were considered erroneous. However, after imputation and adjustment of the imputed data to satisfy the edits, the results of both approaches were similar. For more details on the approach based on solving an LP problem and a comparison with the results of algorithms based on the Fellegi–Holt paradigm, we refer to Harte (2000).

For highly contaminated data containing many errors, the heuristic described above seems appropriate. The Fellegi–Holt paradigm aims to identify fields that are likely to be erroneous. For records that are not very contaminated with errors, the Fellegi–Holt paradigm is often successful; that is, the fields localized as being erroneous are indeed erroneous. However, for records containing many errors the Fellegi–Holt paradigm rarely achieves its aim. For such records, the fields localized as being erroneous are only rarely (all) truly erroneous fields. For such highly contaminated records, one may as well use the heuristic described above. In our view, it is not very important that a record that could be made consistent by changing the values of, say, 12 fields is in fact made consistent by changing, say, 15 fields. In general, the probability that the 12 or 15 fields localized as being erroneous are indeed erroneous is negligible. For records containing only a few erroneous values, it seems sensible not to apply the above heuristic, but an algorithm based on the Fellegi–Holt paradigm. For such records the Fellegi–Holt paradigm does succeed in a reasonable fraction of cases to localize the truly erroneous fields.

REFERENCES

Bishop, Y. M. M., S. E. Fienberg, and P. W. Holland (1975), *Discrete Multivariate Analysis: Theory and Practice*. The MIT Press, Cambridge, MA.

Blien, U., and F. Graef (1992), ENTROP: A General Purpose Entropy Optimizing Method for the Estimation of Tables, the Weighting of Samples, the Disaggregation of Data, and the Development of Forecasts. In: *Softstat '91. Advances in Statistical Software 3*, F. Faulbaum, ed. Gustav Fisher Verlag, Stuttgart.

Censor, Y. (1981), Row-Action Methods for Huge and Sparse Systems and Their Applications. *SIAM Review 24*, pp. 444–466.

Censor, Y., and S. A. Zenios (1997), *Parallel Optimization. Theory, Algorithms, and Applications*. Oxford University Press, New York.

De Waal, T. (2003), *Processing of Erroneous and Unsafe Data*. Ph.D. Thesis, Erasmus University, Rotterdam (see also www.cbs.nl).

De Waal, T. (2005), Automatic Error Localisation for Categorical, Continuous and Integer Data. *Statistics and Operations Research Transactions 29*, pp. 57–99.

Fang, S. C., J. R. Rajasekra, and J. Tsao (1997), *Entropy Optimization and Mathematical Programming*. Kluwer Academic Publishers, Boston.

Fellegi, I. P., and D. Holt (1976), A Systematic Approach to Automatic Edit and Imputation. *Journal of the American Statistical Association 71*, pp. 17–35.

Harte, P. (2000), *Testing Automatic Editing Based on the Simplex Method* (in Dutch). Report 3296-00-RSM, Statistics Netherlands, Voorburg.

Hildreth, C. (1957), A Quadratic Programming Procedure. *Naval Research Logistics Quarterly 4*, pp. 79–85.

Ireland, C. T., and S. Kullback (1968), Contingency Tables with Given Marginals. *Biometrika 55*, pp. 179–188.

Kaczmarz, S. (1937), Angenäherte Auflösung von Systemen linearer Gleichungen. *Bulletin International de l'Academie Polonaise des Sciences et des Lettres, series A, 35*, pp. 335–357.

Kadas, S. A., and E. Klafsky (1976), Estimation of the Parameters in the Gravity Model for Trip Distribution: A New Model and Solution Algorithm. *Regional Science and Urban Economics 6*, pp. 439–457.

Kartika, W. (2001), *Consistent Imputation of Categorical and Numerical Data.* Report, Statistics Netherlands, Voorburg.

Kullback, S. (1959), *Information Theory and Statistics.* John Wiley & Sons, New York.

Nemhauser, G. L., and L. A. Wolsey (1988), *Integer and Combinatorial Optimisation.* John Wiley & Sons, New York.

Pugh, W. (1992), The Omega Test: A Fast and Practical Integer Programming Algorithm for Data Dependence Analysis. *Communications of the ACM 35*, pp. 102–114.

Shlomo, N., and T. De Waal (2008), Protection of Micro-data Subject to Edit Constraints Against Statistical Disclosure. *Journal of Official Statistics 24*, pp. 229–253.

Stone, J. R. N., D. A. Champerowne, and J. E. Maede (1942), The Precision of National Income Accounting Estimates. *Reviews of Economic Studies 9*, pp. 111–125.

CHAPTER ELEVEN

Practical Applications

11.1 Introduction

In this chapter we give three examples of cases where editing and/or imputation techniques are applied in practical situations. The first application is completely focused on automatic editing for one particular survey, namely the Dutch survey on Environmental Costs. This application is described in Section 11.2, which is based on Houbiers, Quere, and De Waal (1999). The second application study, described in Section 11.3, has a much wider range. It considers the use of various editing and imputation techniques in order to edit and impute several data sets in the so-called EUREDIT project. That section is based on Pannekoek and De Waal (2005). Finally, Section 11.4 describes an application of selective editing in the Dutch Agricultural Census.

11.2 Automatic Editing of Environmental Costs

11.2.1 INTRODUCTION

In this section we describe the results obtained by automatically editing the 1997 survey on Environmental Costs held by Statistics Netherlands. The central issue of this survey is to investigate the investments made by enterprises to decrease environmental pollution. Using mainly deductive and mean imputation in CherryPi—a computer program for automatic edit and imputation based on the Fellegi–Holt paradigm (see Chapter 3 of this book) developed by Statistics Netherlands [see De Waal (1996)]—the automatically edited data are found to reflect the manually edited data for almost all variables in all publication classes surprisingly well.

Handbook of Statistical Data Editing and Imputation, by T. de Waal, J. Pannekoek, and S. Scholtus

The survey consists basically of two parts. The first one deals with the net environmental charges due to levies, fines, licenses, and payments to external businesses for disposal of industrial and hazardous waste. Furthermore, expenses in research to decrease environmental damage are also included. The second part of the survey is concerned with industrial investments for the protection of the environment, and the running costs of utilities that are put into use during the previous year. Since in this part firms are asked to describe their investments on a more qualitative basis, it is for our purpose (automatic editing) not very useful. We shall therefore focus on the first part of the questionnaire only.

Traditionally, the data of this survey have been edited in an interactive manner. The data concerning each firm (i.e., a record) should satisfy certain edits. These constraints can be hard or soft, meaning that they definitely should be (hard edits) or likely might be (soft edits) satisfied. With the help of a tailor-made computer program, possible errors in a record, such as inconsistent column totals (hard errors) or very large deviations from historical data (soft errors), were shown as a warning message to the editor, who then decided whether or not to update the corresponding variable with a more plausible value by, for example, recontacting the company in question. As we have described in Chapter 6 of this book, this is a rather inefficient approach.

In 1999, it was investigated whether the survey data could be automatically edited to a satisfying degree. Obviously, an automatic editing and imputation system can never be expected to give as good results as a real editor would obtain. Nonetheless, one can expect to obtain at least consistent and hopefully reliable values for the final publication figures. Indeed, although after automatic editing not all records may contain the actual true values for every single variable, they will all be consistent, and reasonably well-imputed records can further, on an aggregated level, prove to be a very decent base for calculating publication figures. Only selected results of the 1999 research are explained here. We refer to Houbiers, Quere, and De Waal (1999) for a comprehensive discussion of the research results.

This section is organized as follows. In Section 11.2.2, we show preliminary information on the data, and also a selection of the hard and soft edits used by the editors, some of which we discuss in greater depth. In Section 11.2.3 we give a short description of the automatic editing program CherryPi together with the input used—that is, the chosen edits, reliability weights, and imputation method. In Section 11.2.4 we show results of the automatic editing process, and we conclude the case study in Section 11.2.5.

11.2.2 RAW AND MANUALLY CLEANED SURVEY DATA

Description of the Available Data. The data for the Environmental Costs survey were obtained through a paper questionnaire sent to enterprises involved in industrial activities. Enterprises that take part in this survey are involved in mining and quarrying, manufacturing (except construction), or public utilities. Furthermore, only those enterprises with five or more employees (defined as size classes 3 to 9) are investigated. In some branches of industry, all companies are integrally observed, since for companies belonging to such branches, the level of environmental disposal charges is usually high. The remaining companies (that is, most of them) are approached by means of a sample survey. In 1997, from the approximately 16,000 enterprises in the population, the questionnaire was sent to about one-third, of which 4122 responded to the survey.

Estimates of publication figures for the whole population (the 16,000 companies) are calculated from the figures obtained for those companies that replied to the survey, using weights assigned to each "cell" defined by size class times branch of industry. The weights are given by the following expression: total number of employees in the whole population of companies belonging to that cell divided by the total number of employees in companies surveyed and belonging to that very same cell. The weights are thus always at least equal to 1. Publication figures are presented in 12 clusters of branches of industry. Furthermore, small companies with 5 to 19 employees (size classes 3 to 4) are separated from the larger companies (those with 20 or more employees, size classes 5 to 9).

In the survey, some 25 variables regarding waste disposal costs are under investigation. These variables are listed in Table 11.1. Each company in the survey is identified by an identification number. Furthermore, its branch of industry, corresponding weight (used to calculate publication figures), size class, and number of employees are all known. The raw data (4122 records), as well as the manually edited data (also 4122 records), are available, so that results obtained through automatic editing can be compared to the manually cleaned data. Out of the 4122 records, some 1984 firms were also present in the 1996 survey, of which we only have the manually edited (clean) data set.

Manual Editing. Generally, editors follow certain prescribed rules when checking a record. We report in Table 11.2 the hard edits and most frequently violated soft edits used in the manual editing process. Obviously, it should be kept in mind that those edits do not cover all possible occurrences justifying a change of value for some variable(s), because the human checking process cannot be modeled exactly.

The last soft edit compares historical data with present-day data; that is, the deviation of $C11_{97}$ from $C11_{96}$ is presumed unlikely to exceed 50% if $C11_{96}$ is nonzero. Should the situation occur anyhow, a warning message is shown to the editor. This soft edit has an equivalent for all variables with the exception of $C31$, $C33$, $C35$, $C36$, $C80$, $C81$, and $C82$, where an acceptable deviation amounts to at most 30%. If data lie outside of the 50% (30%, respectively) margin around the 1996 data, this can, for example, be a sign that costs were given in millions of euros instead of in thousands of euros.[1] Note that the variables $C80$ and $C81$ are the only variables that are assumed to be related to each other in the sense that knowledge of either of them can give a reasonable estimate for the other one (cf. the first soft edit).

To get some feeling for the manual editing process, we compared the cleaned data with the raw data. It turns out that out of 4122 raw records, 1174 records were indeed changed. In 601 cases there was only a single change made, in 311

[1] Actually, the Dutch currency at the time of data collection and at the time of the evaluation study was the guilder. In this section, we will, however, express all financial figures in millions of euros.

TABLE 11.1 Description of the Variables

Variable	Description
$C11$	Costs for disposal of hazardous waste
$C13$	Costs for disposal of slurry from waste water treatment
$C14$	Costs for disposal of other slurry and fertilizer
$C15$	Costs for disposal of radioactive waste
$C16$	Costs for disposal of waste from air purification plant
$C17$	Costs for disposal of remaining waste
$C18$	Total of $C11$ to $C17$
$C20$	Part of $C18$ paid to privately owned businesses
$C21$	Part of $C18$ paid to local authorities
$C22$	Part of $C18$ paid to the state
$C31$	Legal fees for drainage
$C33$	Special tax 1
$C35$	Special tax 2
$C36$	Costs for waste water disposal
$C40$	Legal fees for environmental licenses
$C60$	Environmental fines
$C62$	Costs for environmental damage restoration
$C63$	Costs for purging soil pollution
$C70$	Costs paid to outside organizations for environmental research
$C71$	Costs for internal environmental research
$C72$	Costs paid to outside organizations for research on ecological products
$C73$	Costs for internal research on ecological products
$C80$	Number of working weeks spent on environmental coordination
$C81$	Costs of $C80$
$C82$	Costs for environmental coordination paid to outside organizations

TABLE 11.2 Hard Edits and Some of the Soft Edits Used in Manual Editing

Hard Edits

$$C11 + C13 + C14 + C15 + C16 + C17 = C18$$
$$C20 + C21 + C22 = C18$$
$$C80 > 0 \Leftrightarrow C81 > 0$$

Soft Edits

$$1000 \times C80 \leq C81 \leq 4500 \times C80$$
$$C31 + C33 + C35 + C36 \neq 0$$
$$-50\% \leq \frac{C11_t - C11_{t-1}}{C11_{t-1}} \times 100\% \leq 50\% \text{ if } C11_{t-1} > 0$$

there were two changes made, in 122 records three variables had been changed, and in 140 records four or more fields had been adjusted. So, in this data set, over 70% of the records were clean right from the very beginning! This does not mean that they were not checked. Actually, data concerning those firms spending large amounts of money to waste disposal were taken in correctly right from the start (thus, in the raw data file). We also examined the number of times each single variable was changed as a result of the manual editing process, when compared to its value in the raw data set. The more times a variable was manually changed, the less confidence the editor apparently had in the actual value of this variable. In the automatic editing process, we can reflect this tendency by giving appropriate values to the reliability weights assigned to each variable.

Edits Based on Historical Data. Using the data concerning those companies present in both the 1996 and the 1997 survey, we considered how many changes were in fact made to the 1997 raw data if the deviation was more than 50% (30% for the C-thirties and C-eighties) compared to the 1996 survey data. For each variable we counted the number of records falling inside or outside the 50% (respectively 30%) margin if the particular variable was nonzero in 1996. For the records falling outside the margin, we counted the number of records that indeed were changed and the number of records that were not changed. The same was done for those records falling within the margin. Furthermore, the number of corrected and uncorrected records with a zero entry for a particular variable in the 1996 data set was also counted for each variable.

The main conclusion drawn from comparing the 1997 data to historical 1996 data was that relatively few fields falling outside the margins were indeed corrected. In other words, editors received unnecessarily many warning messages, out of which only a very small fraction did indeed require attention. Moreover, it turned out that changing the margin levels did not change the outcome. If the upper 5 percentile of largest deviations from historical data was considered, a similar large fraction of fields was left unchanged.

We therefore concluded that for this particular survey, historical values are unreliable predictors for present-day data, at least when using automatic editing. This is probably a consequence of the fact that companies get rid of certain kinds of waste every once in a few years instead of every single year, that law enforcement may change quite rapidly, or that firms may have done some one-time investments to decrease environmental pressure.

Nonetheless, historical data *can* be used to track down records in which costs have been given in wrong units (in this case in millions of euros instead of thousands of euros). Because, if amounts have been given in incorrect units, the total sum of all costs in a record will be relatively small when compared to both historical totals and to totals of similar companies in the survey.

11.2.3 AUTOMATIC EDITING

The objective of automatic editing is to replace, at the push of a button, the editors' time-consuming task of leafing through all respondent forms (usually on a computer screen) to check for implausible and inconsistent data and possibly correcting it. The most difficult part of automatic editing is to formalize the intelligence, knowledge, and experience of editors into mathematical rules. In interactive editing, the editor has in mind a set of constraints, which may or may not be satisfied, and decides, based upon previously acquired knowledge or possibly upon recontacting the company under investigation, whether or not a certain field should be corrected. In automatic editing, the rules are all hard. They are either satisfied or not, and in the latter case the data *will* definitely be changed. There is no real option to make any further distinction between cases. As a consequence, it must be decided whether soft edits used in manual editing may or may not be used as hard edits in automatic editing. Before going into further detail, we shall first give a brief description of CherryPi, an automatic editing and imputation system developed at Statistics Netherlands [see De Waal (1996)].

CherryPi was developed in the spirit of GEIS, the Generalized Edit and Imputation System of Statistics Canada [see Kovar and Whitridge (1990)]. Data from a survey are required to satisfy a set of linear edits. If a certain record satisfies all constraints, the record is clean. If not, the fields that should be changed to make the record consistent (error-free) were, in the version of CherryPi that was applied for this evaluation study, determined using an adapted version of Chernikova's algorithm [see, e.g., Schiopu-Kratina and Kovar (1989) and Chapter 3 of this book]. One of the assumptions made is the generalized principle of Fellegi and Holt [see Fellegi and Holt (1976) and Chapters 3, 4, and 5 of this book].

After determining which fields need to be changed, the program starts to impute more plausible values. If this cannot be done deductively (see Section 9.2 of this book), the program uses the imputation model specified by the user. In CherryPi, regression imputation (and related techniques, such as mean imputation or ratio imputation) is the only possibility for imputation.

Table 11.3 shows some of the edits that were used by CherryPi. We used the hard edits from Table 11.2, together with some extra edits. For all variables we set a maximum value, which we obtained from the manually cleaned data, so as to force reasonable imputations for really large values, and forced a positivity requirement on some variables.[2] The first soft edit from Table 11.2 was also used, except for the fact that the upper and lower bounds of $C81$ in terms of $C80$ were slightly adapted. Due to the results obtained earlier on the use of historical data for editing, we decided not to include any edits referring to the 1996 edited data.

We used the 1996 data to check for thousand-errors. In this survey, a thousand-error occurs when amounts are reported in millions of euros instead of thousands of euros, which means that the reported values are too low. In a first automatic editing step—designed to detect and correct systematic thousand-errors—we added all costs for each company in 1997, and, if available, also for 1996, to obtain the derived variables $TOTAL_{97}$ and $TOTAL_{96}$. Then we corrected a thousand-error in a record by multiplying all variables (except $C80$)

[2] These univariate edits are not shown in Table 11.3, to save space. The interested reader is referred to Houbiers, Quere, and De Waal (1999).

TABLE 11.3 Edits Used in CherryPi

$C11 + C13 + C14 + C15 + C16 + C17 = C18$
$C20 + C21 + C22 = C18$
$400 \times C80 \leq C81 \leq 5400 \times C80$
IF $(C80 + C81 > 0)$ AND $(C80 > 0)$, THEN $C81 > 0$
IF $(C80 + C81 > 0)$ AND $(C81 > 0)$, THEN $C80 > 0$
IF $(TOTAL_{97} > 0)$, THEN
$C18 + C31 + C33 + C35 + C36 + C40 + C60 + C62+$
$C63 + C70 + C71 + C72 + C73 + C81 + C82 > 0$

with a factor of 1000 if $TOTAL_{97}$ and $TOTAL_{96}$ satisfied one of the following cases:

- Size class 3 or 4 and $TOTAL_{96}/TOTAL_{97} > 100$
- Size class 5 to 9 and $TOTAL_{96}/TOTAL_{97} > 60$
- Size class 3 or 4 and $TOTAL_{96}$ missing and $0 < TOTAL_{97} < 100$
- Size class 5 to 9 and $TOTAL_{96}$ missing and $0 < TOTAL_{97} < 200$

Those cases seemed to agree well with the manual editing procedure. This pre-editing can easily be done in, for example, SPSS. The variable $TOTAL_{97}$ was eventually used in CherryPi again, in the shape of a new edit: If $TOTAL_{97} > 0$, then the sum of all variables must be larger than 0, in order to ensure that once one of the fields involved in a sum is nonzero, then it either remains nonzero or some other field(s) become(s) nonzero. For example, suppose that in some record the only nonzero variable is $C20$, then the record obviously does not satisfy the constraints. If we did not include the $TOTAL_{97}$ edit, CherryPi would set $C20$ to zero, whereas with the $TOTAL_{97}$ edit, CherryPi changes both $C18$ and $C17$ and gives them some values, as desired.

Reliability weights assigned to each variable were derived from the number of times that the variable was changed manually by the editors: The more times it was changed, the lower the weight it was given. We tried to keep those weights as simple as possible.

For all variables except for $C80$ and $C81$, we used mean imputation (see Chapter 7), because none of the variables could be predicted with any auxiliary variable such as number of employees or size class. For simplicity's sake, we decided to impute records regardless of their branch of industry or size class. Of course, imputation by publication class should improve the final result somewhat. Variables $C80$ and $C81$ are, in contrast, related to one another, so for those two fields we used linear regression imputation with $C80$ as predictor for $C81$, and vice versa (see Chapter 7 for a description of regression imputation).

11.2.4 RESULTS

Below we present the results of the automatic editing procedure with CherryPi, when using the reliability weights, imputation methods and edits described in Section 11.2.3.

Comparing Manual and Automatic Editing. Before looking at the publication figures on an aggregated level, we compared the most frequently changed combinations of variables when using automatic editing with the most frequently occurring combinations in the manual editing process. It turned out that the most frequently changed combinations of variables resembled each other more or less. Larger differences were found for combinations of variables that occur less frequently, due to the fact that many more soft edits are used in manual editing and the final decision as to whether or not change some field in a record may depend on a-typical considerations. In fact, CherryPi does not impute the records in many different ways, whereas the manual editors do. This confirms our earlier remark, that it is very difficult indeed to model the human editing process in great detail.

We also examined the number of times each variable was changed in automatic editing, both during the pre-editing stage as well as due to CherryPi. The most important observation was that only few fields have been changed. With the sole exception of one (incorrect) change in *C*36, the fields *C*31 to *C*73 and *C*82 were not changed at all during the editing with CherryPi. This is a consequence of the fact that in this survey only few edits were available to relate variables to each other, and it shows again that as long as only a poor set of edits is available, the work of manual editors can never be reproduced completely. In this respect we stress again the importance of converting as much of the knowledge of the editors into mathematical formulae as possible. But, as we will show below, even a simple set of constraints for automatic editing, such as the one we used, can give quite reasonable results on an aggregated level.

Publication Figures. We now examine the publication figures on an aggregated level. Although only preliminary weights to calculate aggregates were available to us, the numbers below give a fairly good impression of the quality of the automatic editing and imputation process used in this survey. In Table 11.4 we compare the clean (manually edited) aggregates, the automatically edited (pre-editing plus CherryPi) aggregates and the aggregates calculated from the raw data. For all fields we made a distinction between companies in size classes 3 to 4 (5 to 19 employees) and size classes 5 to 9 (20 employees or more). Comparisons were also made on a less aggregated level, for different branches of industry, but these are omitted here to save space. We refer to Houbiers, Quere and De Waal (1999) for all results.

As can be seen from Table 11.4, the results of the automatic editing procedure are quite good on this aggregated level. Note again that, apart from *C*80 and *C*81, only the simplest form of imputation, mean imputation, was used. An important reason why mean imputation gave good results in this case is that the records contributing most to the final figures were apparently taken in correctly from the start. This suggests that the use of a plausibility index (see Chapter 6 of this book) to select, say, the 5% or 10% most important records to be checked carefully (that is, manually) while automatically editing the remaining records, can lead to representative, consistent, and reliable publication figures for this survey. Of course, the final results can always be improved by

TABLE 11.4 **Comparison of Manually Edited, Automatically Edited, and Raw Data on an Aggregated Level: Total Expenditures of Smaller and Larger Companies across All Branches of Industry**[a]

	Size Classes 3–4			Size Classes 5–9		
Variable	Clean	CherryPi	Raw	Clean	CherryPi	Raw
$C11$	8.7	9.8	9.8	178.9	176.6	177.0
$C13$	2.5	2.7	2.7	58.8	60.7	59.9
$C14$	2.0	2.0	2.0	16.4	20.3	20.3
$C15$	0.0	0.0	0.0	9.5	9.2	7.1
$C16$	0.2	0.2	0.2	4.9	4.9	4.8
$C17$	39.5	37.3	37.7	267.9	272.8	262.4
$C18$	52.9	52.0	48.6	537.6	545.7	515.0
$C20$	45.0	44.8	45.0	491.1	496.7	542.3
$C21$	6.8	6.1	6.1	21.1	23.9	21.7
$C22$	1.1	1.1	1.1	25.4	25.1	25.4
$C31$	4.8	5.2	5.2	22.8	25.1	22.8
$C33 + C35$	12.7	13.7	13.2	214.4	214.7	203.0
$C36$	0.2	0.5	0.5	17.0	17.1	15.9
$C40$	2.8	2.7	2.7	13.3	13.4	13.2
$C60 + C62$	0.8	0.8	0.8	4.7	5.2	5.1
$C63$	1.1	1.0	1.0	105.3	103.9	103.5
$C70$	2.1	2.1	2.1	42.5	40.7	40.4
$C71$	1.8	1.8	1.8	45.8	45.0	44.3
$C80$	4719	5061	15527	62773	61758	58995
$C81$	7.8	8.5	8.3	137.1	136.2	204.8

[a]All values are in millions of euros, except for the values of $C80$, which are in number of working weeks.

searching for more and better edits, and by using more sophisticated methods for imputation.

11.2.5 CONCLUSION AND RECOMMENDATIONS

We explained in this section how publication figures for the 1997 Dutch survey on Environmental Costs were obtained using an automatic editing procedure. What a team of editors working on a data set wishes to obtain are statistics that are as close as possible to their true values. Ultimately, it does not matter whether or not every single bit of data in the data set is really carrying its true value, because on an aggregated level such errors do generally cancel out. A "cleaned" data set—that is, a data set ready for statistical analysis—should be consistent (totals do add up, etc.), and should not contain gross, obvious errors. As was shown throughout, provided that the edits used in the cleaning procedure are carefully chosen, raw data sets can indeed be processed automatically so as to obtain sufficiently clean data, ready for statistics that are sufficiently close to the ones obtained through manual editing. It is thus likely that automatic editing, preferably in combination with a plausibility index for selecting those records which should be dealt with manually due to their importance, can (at least partially) replace the time-consuming, manual editing for this kind of survey. We further believe that the use of more sophisticated imputation methods such as group mean imputation or hot deck imputation techniques would lead to even better results than the current ones.

To be completely honest, we have to add a word of warning: Although the results were good, it should be kept in mind that quite some work was put into making appropriate edits and into the preprocessing of the data. Furthermore, we had the chance to be able to compare our results with those from the editing team, which inevitably means that we had the tendency to try to match our results to theirs, by adapting our edits, say, or the reliability weights. Moreover, it should be reminded that the major companies' data were keyed in correctly right from the start, but their data are surely the ones having the biggest influence on the publication figures. In practice, we do not have all this knowledge in advance (we do not even have a complete raw data set in advance, as forms are received over a period of several months), and the task of automatic editing thus takes another dimension when working in real time.

11.3 The EUREDIT Project: An Evaluation Study

11.3.1 INTRODUCTION

The EUREDIT project (see http://www.cs.york.ac.uk/euredit) was a large international research and development project aimed at improving the efficiency and the quality of the statistical data editing and imputation process at national statistical institutes (NSIs). It involved twelve institutes from seven countries. Six of those institutes were NSIs, namely Office for National Statistics UK (overall project coordinator), Statistics Finland, Swiss Federal Statistical Office, Istituto Nazionale Di Statistica, Statistics Denmark, and Statistics Netherlands. Four universities participated in the project: Royal Holloway and Bedford New College, University of Southampton, University of York, and University of Jyväskylä. Finally, two commercial companies, the Numerical Algorithm Group Ltd. and Quantaris GmbH, were involved in the project. The project lasted from March 1, 2000 until February 28, 2003.

For Statistics Netherlands, the main aims of the project were:

1. To evaluate current "in-use" methods for data editing and imputation and to develop and evaluate a selected range of new or recent techniques for data editing and imputation;
2. To compare all methods tested and develop a strategy for users of edit and imputation leading to a "best practice guide."

The EUREDIT project concentrated on automatic methods for editing and imputation. Other editing methods, such as selective editing [see Lawrence and McDavitt (1994), Lawrence and McKenzie (2000), Hedlin (2003), and

Chapter 6 of this book] where part of the data are edited manually, were not or hardly examined.

It has been argued that the role of statistical data editing should be broader than only error localization and correction [see Granquist (1995, 1997), Granquist and Kovar (1997), Bethlehem and Van de Pol (1998).] We fully agree with this point of view and, for instance, consider the feedback provided by the edit process on the questionnaire design at least as important as error localization and correction. However, within the EUREDIT project the role of editing was strictly limited to error localization and correction, and in the present section on the EUREDIT project we will therefore follow this.

This section describes the approach applied by Statistics Netherlands on the two business surveys used in the EUREDIT project. This approach mimics part of the currently used approach at Statistics Netherlands for editing and imputing data of annual structural business surveys. We describe the development of our edit and imputation strategy and give results supporting the choices we have made.

Although the methods and tools we consider in this section are automatic ones, they require quite a bit of expert knowledge and statistical analysis to set up. In practice, the tools have to be set up only once, however. For future versions of the same survey, one only needs to update the parameters of the various methods. This updating process can to a substantial extent be automated. So, for future versions of the same survey, preparing and using the methods and tools we consider in this section are almost fully automated, much more than for a first time.

In practice, to edit and impute a data set, one often uses corresponding cleaned data from a previous year. In the EUREDIT project, however, data of only one year were available. In our edit and imputation methods we therefore had to restrict ourselves to using only data from the data set to be edited and imputed itself. Our general strategy can in a natural way be extended to the case where cleaned data from a previous year are available.

In the literature, there is quite a scarcity of articles on the combined application of editing and imputation techniques in practice. The main articles we are aware of are by Little and Smith (1987) and Ghosh-Dastidar and Schafer (2003). Little and Smith (1987) focus on outlier detection and outlier robust imputation techniques. Ghosh-Dastidar and Schafer (2003) focus on outlier detection and multiple imputation based on a regression model. The present section focuses on automatic editing and imputation techniques for two surveys that are considerably more complex than the ones considered by the aforementioned authors. Moreover, whereas the edit and imputation techniques applied by these authors do not ensure internal consistency of individual records, such as component variables summing up to a total, our procedures ensure such consistency.

The remainder of this section is organized as follows. Section 11.3.2 describes how the evaluation experiments were carried out within the EUREDIT project. The two data sets we consider in this section are discussed in Section 11.3.3. Section 11.3.4 sketches the edit and imputation methodology applied by Statistics

Netherlands to these data sets. The general outline of our approach is the same for both data sets. Section 11.3.5 describes the development of our edit and imputation strategy and how we have tried to optimize various aspects of this strategy. Section 11.3.6 ends this section by drawing some conclusions.

11.3.2 THE EVALUATION EXPERIMENTS

For each data set used in the EUREDIT project six different versions were, in principle, constructed: three evaluation data sets and three development data sets. These six data sets are given in Table 11.5.

The evaluation data sets were used to evaluate the edit and imputation procedures applied by the participants in EUREDIT. All three versions of the development data—that is, including the "true" data—were sent to all participants. The development data sets could be used to train neural networks or to optimize parameter settings of statistical methods, for instance. The development data represent the fact that in a real-life situation one can learn from past experience. The records and information in the development data sets differed from the records and information in the evaluation data sets.

A Y^* data set contains "true" values, the corresponding Y_2 data set the data with missing values but with no errors, and the corresponding Y_3 data set the data with both missing values and errors. The Y^*, Y_2, and Y_3 data sets can, respectively, be interpreted as cleaned data, edited but not yet imputed data, and raw data. A Y_2 data set allows one to evaluate imputation methods, a Y_3 data set allows one to evaluate a combination of editing and imputation methods. Constructing a data set with only errors but no missings, a Y_1 data set, was considered to be too unrealistic a scenario.

The $Y_{2,E}$ data and the $Y_{3,E}$ data were sent to all participants in the EUREDIT project. These participants then applied their methods to these data sets. The $Y_{2,E}$ data only had to be imputed, and the $Y_{3,E}$ data had to be both edited and imputed. The "true" evaluation data were not sent to the participants in the project. These data were retained by the coordinator of the project, the Office for National Statistics (UK), for evaluating the data sets "cleaned" by the various methods applied.

In the ideal situation, one would have a data set with true values, a corresponding data set with actual missings without errors, and a data set with actual missings and actually observed errors. This would allow one to evaluate edit and imputation methods by comparing edited and imputed data sets to the true data. Unfortunately, data sets with true values are very rare. In the EUREDIT

TABLE 11.5 **The Six Versions of Each Data Set**

Type	"True"	With Missing Data	With Missing Data and Errors
Evaluation	Y_E^*	$Y_{2,E}$	$Y_{3,E}$
Development	Y_D^*	$Y_{2,D}$	$Y_{3,D}$

project, data sets with true values were not available. Out of necessity, we defined the "true" data as the data that the provider of the data set considered to be satisfactorily cleaned according to their edit and imputation procedures. The errors in the Y_3 data are not actual errors; neither are the missing values in the Y_2 and Y_3 data the actual missing values. These missing values and errors were synthetically introduced in the corresponding Y^* data set by the coordinator of the EUREDIT project. In this way the mechanisms that generated the missing values and the errors were fully controlled by the coordinator, while remaining unknown to the participants in the EUREDIT project. By the full control of the coordinator over the error generation mechanism and the missing data mechanism, it was possible to ensure that the Y_2 and Y_3 data sets provide sufficient challenges to the participants, while at the same time remaining as realistic as possible. The fact that the error generation mechanism and the missing data generation mechanism were unknown to the participants mimics reality, where these mechanisms are also unknown to NSIs.

Along with the data sets sent to the participants—the $Y_{2,E}$ data, the $Y_{3,E}$, data and the three development data sets—metadata related to these data sets, such as edits and data dictionaries, were provided.

Each participant in the project was allowed to submit several cleaned versions of the same data set, where for each version other parameters or another method was used. The results of the evaluation experiments—that is, the quality of the cleaned data sets—were assessed by applying a large number of evaluation criteria. These evaluation criteria measured many different aspects of an edit and imputation approach, such as its ability to identify errors, to identify the large errors, to accurately impute individual values, to preserve the distributional aspects of the data, and to estimate publication totals and averages. In the section on the results of our approach, Section 11.3.5, we describe a number of such evaluation criteria. We refer to Chambers (2004) for more details regarding the evaluation criteria.

11.3.3 THE DATA SETS

UK Annual Business Inquiry. The UK Annual Business Inquiry (ABI) is an annual business survey containing commonly measured continuous variables such as turnover and wages. The development data sets contain 6099 records and the evaluation data sets contain 6233 records. A long and a short version of the questionnaire have been used in the data collection. As a consequence, in the evaluation data sets for 3970 businesses, scores on only 17 variables are available (the short version); and for 2263 records, scores on 32 variables are available (the long version). Three variables, *class* (anonymized industrial classification), *turnreg* (registered turnover) and *empreg* (registered employment size group), were not obtained from the questionnaires but from completely observed registers. These variables could be used to construct suitable imputation strata, for instance. In the long questionnaire, 26 variables contained errors or had missing values; in the short questionnaire, only 11 variables. The names and brief descriptions of the main variables are given in Table 11.6.

TABLE 11.6 The Main Variables in the ABI Data Set

Name	Description
turnover	Total turnover
emptotc	Total employment costs
purtot	Total purchases of goods and services
taxtot	Total taxes paid
assacq	Total cost of all capital assets acquired
assdisp	Total proceeds from capital asset disposal
employ	Total number of employees
stockbeg	Value of stocks held at beginning of year
stockend	Value of stocks held at end of year
capwork	Value of work of a capital nature

The variables in the ABI data set can be subdivided according to a three-level hierarchy. The first level consists of the key economic variables *turnover*, *emptotc*, *purtot*, *taxtot*, *assacq*, and *assdisp* and the main employment variable *employ*. The six key ABI economic variables are highly skewly distributed. The second level consists of the secondary variables *stockbeg*, *stockend*, and *capwork* measuring business activity. For the long questionnaire, the third level consists of variables corresponding to components of three key economic variables, namely the components of *purtot*, *taxtot*, and *emptotc*. For the short questionnaire, the third level consists of two component variables for *purtot*, but no components for the other key economic variables.

For the ABI data both hard and soft edits were provided. In total, 24 hard edits are specified for the ABI data: 20 nonnegativity edits and 4 balance edits. Some of these hard edits are only applicable for the long questionnaire, some others only for the short questionnaire, the rest for both types of questionnaire. In total, 25 soft edits are specified for the ABI data: 12 ratio edits and 13 upper/lower bound rules. Some soft edits are conditional on the type of questionnaire and/or on the values of certain variables. An example of a conditional edit is

$$\text{IF } employ > 0, \text{ THEN } emptotc/employ \geq 4.$$

The edit is satisfied if *employ* is not larger than zero, irrespective of the value of *emptotc*. Both the value of *employ* and the value of *emptotc* may be incorrect. It is possible that an observed positive value of *employ* should in fact be zero.

Swiss Environmental Protection Expenditures Data. The Swiss Environmental Protection Expenditures (EPE) data consist of information on expenditure related to environmental issues. The data are the responses to an environmental questionnaire plus additional general business questions, distributed to enterprises in Switzerland in 1993.

The data sets contain 71 variables in total. The development data sets contain 1039 records, and the evaluation data sets 200. There are four main groups of financial variables (in thousands of Swiss Francs): variables related to investments, expenditures, subsidies, and income. The nomenclature of the variables in these four groups follows a logical structure. The last two letters indicate which aspect of environmental protection the variable refers to. The letters *wp* indicate "water protection," *wm* "waste treatment," *ap* "air protection," *np* "noise protection," *ot* "other," and *tot* (or *to*) "(sub)total." Variables related to subsidies begin with *sub*, variables related to income with *rec* (abbreviation for "receipts"). Variables related to investments and expenditures start with two blocks of each three letters. If the last block of three letters is *inv*, the variable refers to investments. If the last block of three letters is *exp*, the variable refers to expenditures. The first block of three letters subdivides the variable further: *eop* indicates "end-of-pipe," *pin* "process-integrated," *oth* "other," and *tot* "(sub)total." For instance, *eopinvwp* indicates the end-of-pipe investments with respect to water protection, and *eopinvtot* indicates the total end-of-pipe investments. Tables 11.7 and 11.8 will further clarify the nomenclature of the variables.

As for the ABI data, the variables in the EPE data sets can be subdivided according to a three-level hierarchy. The first level consists of four key economic

TABLE 11.7 Edits that Apply to Investments for the EPE Data

Investments	Water Protection	Waste Treatment	Air Protection	Noise Protection	Other	(Sub)total
End-of-pipe	*eopinvwp*	*eopinvwm*	*eopinvap*	*eopinvnp*	*eopinvot*	*eopinvtot* (vi)
Process integrated	*pininvwp*	*pininvwm*	*pininvap*	*pininvnp*	*pininvot*	*pininvtot* (vii)
Other	*othinvwp*	*othinvwm*	*othinvap*	*othinvnp*	*othinvot*	*othinvtot* (viii)
(Sub)total	*totinvwp* (i)	*totinvwm* (ii)	*totinvap* (iii)	*totinvnp* (iv)	*totinvot* (v)	*totinvto* (ix) C (x) R (xi) T

TABLE 11.8 Edits that Apply to Expenditures for the EPE Data

Expenditures	Water Protection	Waste Treatment	Air Protection	Noise Protection	Other	(Sub)total
Current expenditure	*curexpwp*	*curexpwm*	*curexpap*	*curexpnp*	*curexpot*	*curexptot* (xvii)
Taxes	*taxexpwp*	*taxexpwm*	*taxexpap*	*taxexpnp*	*taxexpot*	*taxexptot* (xviii)
(Sub)total	*totexpwp* (xii)	*totexpwm* (xiii)	*totexpap* (xiv)	*totexpnp* (xv)	*totexpot* (xvi)	*totexpto* (xix) C (xx) R (xxi) T

variables: *totinvto*, *totexpto*, *subtot*, and *rectot*. These variables are highly skewly distributed. The second level consists of 20 component variables corresponding to these four total variables, namely the components of *totinvto* (*totinvwp*, *totinvwm*, *totinvap*, *totinvnp*, and *totinvot*), the components of *totexpto* (*totexpwp*, *totexpwm*, *totexpap*, *totexpnp*, and *totexpot*), the components of *subtot*, and the components of *rectot*. Finally, the third level consists of 30 variables that correspond to the components of *totinvwp*, *totinvwm*, *totinvap*, *totinvnp*, *totinvot*, *totexpwp*, *totexpwm*, *totexpap*, *totexpnp*, and *totexpot*.

All edits specified for the EPE data are hard ones. In total there are 54 nonnegativity edits and 23 balance edits. Two of the balance edits can be deleted because they are logically implied by the other balance edits. So there are 21 nonredundant balance edits. The balance edits follow a complex pattern, basically consisting of two two-dimensional tables and two one-dimensional tables of which the internal cell values have to add up to the marginal totals. The two two-dimensional tables are shown in Tables 11.7 and 11.8. For each table, a column with component variables has to add up to a subtotal variable. For instance, in Table 11.7, *eopinvwp*, *pininvwp*, and *othinvwp* have to add up to *totinvwp* [edit (i)]. A row with component variables has to add up to a subtotal variable. For instance, in Table 11.7, *eopinvwp*, *eopinvwm*, *eopinvap*, *eopinvnp*, and *eopinvot* have to add up to *eopinvtot* (edit (vi)). All component variables, all column subtotal variables, and all row subtotal variables have to add up to a total variable (e.g., *totinvto* in Table 11.7). This is indicated in the tables by C [sum of column totals; e.g., edit (ix) in Table 11.7], R [sum of row totals; e.g., edit (x)] and T [sum of component variables; e.g., edit (xi)]. The two one-dimensional tables state that the components of *subtot* have to add up to *subtot*, respectively, that the components of *rectot* have to add up to *rectot*.

11.3.4 APPLIED METHODOLOGY

Overview. In this section a number of "standard" edit and imputation methods that were applied by Statistics Netherlands to the ABI and EPE data are briefly described. For the EUREDIT project, we have subdivided the edit and imputation problem into three separate problems:

1. The error localization problem. Given a data set and a set of edits, determine which values are erroneous or suspicious, and either correct these values deductively or set these values to missing (see Chapters 2 to 5);
2. The imputation problem. Given a data set with missing data, impute these missing data in the best possible way (see Chapters 7 to 9);
3. The adjustment problem. Given an imputed data set and a set of edits, adjust the imputed values such that all edits become satisfied (see Chapter 10).

For the first and the last problem, algorithms and prototype software have been applied that were developed at Statistics Netherlands and that were extended as part of the EUREDIT project. For the imputation problem we have used a combination of regression and hot deck methods implemented in S-Plus scripts.

At Statistics Netherlands, we aim to let edited and imputed data sets satisfy all specified edits. The edits therefore play a prominent role in our methods.

Error Localization. We now describe our methodology for localizing the errors in a data set. We distinguish between the localization of systematic errors and random errors, because these kinds of errors require a different treatment.

Finding Systematic Errors. In this project, the only systematic errors we aimed to detect and correct were thousand-errors. As discussed in Chapter 2, thousand-errors can often be detected by comparing a respondent's present values with those from previous years, or by comparing the responses to questionnaire variables with values of register variables. For the experiments in the EUREDIT project, only the second option is a possibility. Using the ABI development data, it appeared that a considerable number of thousand-errors occurred in all financial variables. Most of these errors could be found by calculating the ratio of *turnover* (the reported turnover) to *turnreg* (the turnover value from the register) and deciding that a thousand-error was present if this ratio was larger than 300. All financial variables in such records were then divided by 1000. In the EPE development data, no thousand-errors occurred.

Using the Fellegi–Holt Paradigm. To detect random errors, we adopted the Fellegi–Holt paradigm (see Chapter 3). In the EUREDIT project we have applied a prototype version of the *Cherry Pie* module of SLICE [see De Waal (2001)] that is based on the branch-and-bound algorithm described in Chapter 3. The *Cherry Pie* module was a successor of CherryPi (without an "e" at the end), which was a stand-alone application based on the vertex generation approach that was also described in Chapter 3. The most important output of *Cherry Pie* consists of a file that contains for each record a list of all optimal solutions to the error localization problem—that is, all possible ways to satisfy the edits by changing a minimum (weighted) number of fields. One of these optimal solutions is selected for imputation (see below). The variables involved in the selected optimal solution are set to missing and are subsequently imputed by the methods described later. In general, *Cherry Pie* also generates a file with records for which it could not find a solution, because more fields in these records would have to be modified than a user-specified maximum allows. In our experiments, however, we used Cherry Pie to determine all errors in each record.

Selection of Cherry Pie *Solutions.* In practice it is quite common that application of the Fellegi–Holt paradigm yields several optimal solutions. *Cherry Pie* simply returns all these solutions. Each solution consists of a set of suspicious observed values. To select one of these solutions, we have implemented a relatively simple approach.

The general idea is to determine the most suspicious set of observed values. To this end we first calculate crude predictions or "anticipated values" for all the variables in the solutions generated by *Cherry Pie* (see Section 6.3 for similar uses of anticipated values). These anticipated values are based on register variables only, since these are (assumed to be) without errors. Subsequently, distances are calculated between the observed values in a record and the corresponding anticipated values in each of the solutions for that record. The optimal solution returned by *Cherry Pie* for which this distance is maximal is the one involving the variables that deviate most from their predicted values. The variables in this maximal distance solution are, in some sense, the variables with the most outlying values, and these values are hence considered to be the erroneous ones. Thus, we use error localization by outlier detection as a means to single out one of the several solutions to the error localization problem based on the Fellegi–Holt paradigm. The maximal distance solution will be processed further; that is, the variables in this solution will be set to missing, and these missing values will subsequently be imputed.

The distance function used is the sum of normalized absolute differences between the observed values and the predicted values in a record, that is,

$$D_k = \sum_{i \in I_k} \left| \frac{y_{ij} - \hat{y}_{ij}}{\sqrt{\hat{var}(e_{ij})}} \right|,$$

where y_{ij} denotes the observed value of variable j in record i, $\hat{y}_{ij}$ the corresponding predicted value, I_k the index set of the variables in the kth optimal solution returned by *Cherry Pie*, and $\hat{var}(e_{ij})$ an estimate for the variance of the prediction error. More involved distance measures could have been used instead—for instance, a Mahalanobis distance that takes the correlations between the residuals into account. The anticipated values that we used in applying this approach were ratio-type estimators of the form

$$\hat{y}_{ij} = x_{ij} \frac{\bar{y}_j}{\bar{x}_j},$$

where x_{ij} is the value of the (register) predictor variable for variable y_j in record i, $\bar{y}_j$ is the mean over all clean records (records that do not violate any of the edits) of variable y_j, and $\bar{x}_j$ is the mean over the same clean records of x_j. Actually, we used separate ratio estimators within strata, which is a richer model that replaces the single parameter estimate $\bar{y}_j/\bar{x}_j$ by similar estimates for each stratum separately, but for notational simplicity we only describe the unstratified case here. In the applications the predictor used was the only relevant continuous register variable (registered turnover) in combination with stratification by industry type.

Imputation. Next, we sketch the imputation methods that were applied. For more details we refer to Pannekoek (2004a) and Pannekoek and Van Veller (2004).

Deductive Imputation. For a number of missing values in the ABI and EPE data, the value can be determined unambiguously from the edits provided for these data sets. For these missing values the deductive imputation techniques described in Sections 9.2.2 and 9.2.3 have been applied. It should be noted that deductive imputations will be in error if the observed values from which these imputations are derived contain errors. Nevertheless, these deductive imputations are the only values that are consistent with the edit rules. So, given that the possibilities for finding and correcting errors are exhausted, deductive imputation is a logical first imputation step. In Section 11.3.5 we discuss the influence of errors on other (nondeductive) imputation methods.

Multivariate Regression Imputation. As described in Chapter 8, a standard technique for imputing several continuous variables simultaneously is to employ a multivariate linear regression model to derive predictions for the missing values. We refer to Chapter 8 for a description of multivariate regression imputation.

Other Imputation Methods. The regression method is based on a linear additive model for the data. When such a model is not a realistic approximation for the data, regression imputation may give poor results. In the ABI and EPE data there are a number of nonnegative variables with many zero values (often 50% or more). For such variables, the assumption of a linear model for a continuous dependent variable is problematic. The regression imputations will never be zero (unless all predictor variables are) and negative predictions will often occur. With only a few exceptions, these variables are component variables that should satisfy certain balance edits, a requirement that will not be satisfied by regression imputed values. For these variables, nearest-neighbor hot deck methods have been applied that (1) will not impute negative values, (2) will impute zero values, and (3) ensure that at least some of the balance edits are satisfied by the imputed values. In our application of nearest-neighbor imputation, we have used the minimax distance function. For details on nearest-neighbor imputation, we refer to Section 7.6.3.

For variables that are part of a balance edit such as subtotals or component variables, we have applied the ratio hot deck method described in Section 9.3.

For the ABI data the ratio hot deck method ensures that all hard edits are satisfied because they are either balance edits or nonnegativity edits, and each variable occurs only once in a balance edit. The situation is different for the EPE data where many variables are part of more than one balance edit. This is illustrated in Table 11.7 of Section 11.3.3. Suppose that the subtotals of Table 11.7—that is, *totinvwp*, *totinvwm*, *totinvap*, *totinvnp*, *totinvot*, *eopinvtot*, *pininvtot*, and *othinvtot*—are observed or already imputed. Then we can use the ratio hot deck method and the subtotals *totinvwp*, *totinvwm*, *totinvap*, *totinvnp*, and *totinvot* to impute all component variables, in which case these imputed values will not necessarily sum up to the subtotals *eopinvtot*, *pininvtot*, and *othinvtot*, or vice versa. In such cases where the imputation method does not ensure that edits are satisfied, we have adjusted the imputed values such that they do satisfy all edits. For this we have applied the adjustment algorithm of Section 10.3, which in this case—where only continuous data are involved—reduces to simply solving a small linear programming problem.

11.3.5 RESULTS

In this section we present some results of our approach. The performance of our approach as applied to the evaluation data was measured by a number of evaluation criteria, developed in the EUREDIT project. We begin by introducing the evaluation criteria that will be used in the remainder of this section. We shall then present some results using the development ($Y_{2,D}$ and $Y_{3,D}$) data that were used to decide on questions such as: how to detect systematic errors, which stratification to use for imputation within strata, and which imputation method to use (regression, hot deck, ratio hot deck) for which variables. The result of these choices was a final edit and imputation strategy to be applied to the ABI and EPE evaluation data sets. We conclude by presenting some evaluation results for our methods.

We present only a limited number of statistical results in this section. For many more results, we refer to Pannekoek and De Waal (2005) and the underlying reports of that paper: Pannekoek and Van Veller (2004) for the $Y_{2,D}$ data, Pannekoek (2004b) for the $Y_{2,E}$ data, Vonk, Pannekoek, and De Waal (2003) for the $Y_{3,D}$ data, and Vonk, Pannekoek, and De Waal (2004) for the $Y_{3,E}$ data. For a detailed comparison of the results of our strategies with those of the other partners in the EUREDIT project, we refer to Chambers and Zhao (2004a, 2004b).

Evaluation Criteria. To evaluate the editing and imputation methods, we use a limited subset of the many evaluation criteria defined by Chambers (2004) for use in the EUREDIT project.

To measure the error-finding performance of our approach, we use an alpha, a beta, and a delta measure. Consider the following 2 × 2 contingency table:

		Detected Error	No error
True	Error	a	b
	No Error	c	d

where, for a particular variable, a, b, c, and d denote the number of cases falling into each category. The alpha measure equals the proportion of cases where the value for the variable under consideration is incorrect but is still judged acceptable by the editing process:

$$\alpha = \frac{b}{a+b}.$$

It is an estimate for the probability that an incorrect value for a variable is not detected by the editing process. The beta measure is the proportion of cases where a correct value for the variable under consideration is judged as suspicious by the editing process,

$$\beta = \frac{c}{c+d},$$

and estimates the probability that a correct value is incorrectly identified as suspicious. The delta measure is an estimate for the probability of an incorrect outcome from the editing process for the variable under consideration,

$$\delta = \frac{b+c}{a+b+c+d},$$

and measures the inaccuracy of the editing procedure for this variable.

To measure the imputation performance, we use a d_{L1}, an m_1 and an *rdm* measure. The d_{L1} measure is the average distance between the imputed and true values defined as

$$d_{L1} = \sum_{i \in M} w_i \left| \hat{y}_i - y_i^* \right| \Big/ \sum_{i \in M} w_i \, ,$$

where $\hat{y}_i$ is the imputed value in record i of the variable under consideration, y_i^* denotes the corresponding true value, M denotes the set of records with imputed values for variable y, and w_i is the raising weight for record i.

The m_1 measure, which measures the preservation of the first moment of the empirical distribution of the true values, is defined as

$$m_1 = \left| \sum_{i \in M} w_i (\hat{y}_i - y_i^*) / \sum_{i \in M} w_i \right| .$$

Finally, the *rdm* (relative difference in means) measure is defined as

$$rdm = \frac{\sum_{i \in M} \hat{y}_i - \sum_{i \in M} y_i^*}{\sum_{i \in M} y_i^*}.$$

It is important to note here that these imputation performance measures are only used in a relative way—that is, to compare different imputation methods in an experimental setting. Smaller values of the measures indicate better imputation performance. These measures are not necessarily appropriate or sufficient to measure the impact of imputation on the quality of survey estimates in general. For an actual production process, it depends on the intended use of the data whether record level accuracy (d_{L1}) or more aggregated measures of imputation bias like m_1 or *rdm* are more important. Furthermore, to assess the importance of bias caused by imputation, it should be related to another quality aspect such as sampling variance.

Developing a Strategy. We now present some results and general considerations that motivated our choices on the following issues: (1) a threshold value to detect thousand-errors; (2) whether or not to use soft edits in the error localization step; (3) an effective stratification for regression imputation; and (4) hot deck versus regression imputation for component variables and variables with many zero values. We then use the development data to demonstrate the influence of errors on imputation performance.

Detecting Thousand-Errors. Using the development data, we developed a strategy for detecting thousand-errors. With the true values available, these errors were detected by dividing all perturbed values by their true values. When these ratios are close to 1000, they point to thousand-errors. In 191 records of the ABI $Y_{3,D}$ data, thousand-errors were made in *all* financial variables. As mentioned before in Section 11.3.4, we consider a record to contain a thousand-error if the ratio between *turnover* and *turnreg* is larger than 300. This threshold value of 300 has been determined by minimizing the number of misclassifications. For a threshold value of 300, 187 thousand-errors were correctly detected, 4 thousand-errors were not detected, and 5905 records were correctly considered not to contain a thousand-error, and for 3 records it was incorrectly concluded that they contain a thousand-error. The number of misclassifications is small, especially if we take into consideration that 2 thousand-errors could never be detected by this approach, given their values of zero on *turnreg*.

Edits. Our approach explicitly uses edits specified by subject-matter specialists. The performance of the approach is therefore directly dependent on the quality of the specified edits.

As discussed in Section 11.3.3, the edit rules for the ABI Y_3 data consist of hard (logical) edits and soft edits. The data should at least satisfy all hard edits, but it is likely that a considerable number of errors remain undetected when using these hard edits only. On the other hand, the soft edits are designed by subject-matter specialists for interactive editing and may be too strict for automatic editing, possibly resulting in a considerable number of correct records that are identified as incorrect. For the application to the $Y_{3,E}$ data, we have chosen not to make a selection of edit rules that we expect to perform best but to run two experiments: one that uses all edit rules for error localization (Strategy I) and one that uses only the hard edit rules (Strategy II). For the EPE data, only hard edits were specified by the subject-matter specialists.

Different Stratifications for the Multivariate Regression Imputation Procedure. As is common for business surveys, the ABI data include an indicator for the type of industry: the variable *class*. Imputation procedures (as well as other estimation procedures) for business surveys are often applied separately for different types of industry, thus allowing the parameters of the imputation model to vary between different types of industry. For the ABI data, we considered multivariate regression imputation within 14 strata based on the variable *class*. As an alternative, we also considered a stratification suggested by ISTAT as a result of their experiments with the ABI data (Di Zio, Guarnera, and Luzi, 2004). This stratification is based on the register variables *turnreg* and *empreg* and consists of the following three strata for each type of questionnaire: (1) $turnreg < 1000$; (2) $turnreg \geq 1000$ and $empreg \leq 3$; (3) $turnreg \geq 1000$ and $empreg > 3$. The resulting number of strata is six for variables that are on both questionnaires and three for variables that are only part of either the long questionnaire or the short questionnaire. This last stratification variable will be referred to as *strat*.

In order to decide which stratification to use, the multivariate regression imputation method was applied to the variables *turnover*, *emptotc*, *purtot*, *taxtot*, *stockbeg*, and *stockend* (see Table 11.6 for a description of these variables) and *pursale* (purchases of goods bought for resale) of the ABI $Y_{2,D}$ data set, using each of these stratifications. To compare the results, we computed for each variable the relative difference between the mean of the imputed values and the mean of the corresponding true values. The results showed that stratification by *strat* leads to a better preservation of the mean than stratification by *class* for six of the eight variables, even though the number of classes is much less. Based on these results, stratification by *strat* was used in the evaluation experiments.

The EPE data include a variable *act* (industrial activity) that is comparable in meaning to the variable *class* in the ABI data as well as a variable *emp* (number of employees) without missing values. However, since the number of records for the EPE data is much smaller than for the ABI data, the possibilities for stratification are much more limited. As an alternative to full stratification, we have included *emp* and eight dummy variables for the categories corresponding to the first digit of *act* in the multivariate regression imputation procedure. In this way the regression model used for imputation always includes additive effects of *emp* and *act* (along with other predictor variables, depending on their availability for a particular record), thus providing a differentiation in imputations between industry type and numbers of employees.

Hot Deck Imputation Versus Regression Imputation. One of the imputation methods considered for component variables was the ratio hot deck imputation method, but for some component variables we investigated the performance of regression imputation as well. Application of these two methods to the six component purchase variables—that is, the six component variables of *purtot*—of the ABI $Y_{2,D}$ data showed that, with respect to the *rdm* criterion, multivariate regression imputation is better for three variables; but for the other three variables, ratio hot deck is better.

These results do not point strongly to one of the imputation methods as the method of choice. The regression imputation method has some disadvantages not shared by the ratio hot deck imputation method. In particular, some imputed values are negative while the corresponding variables should only assume nonnegative values and, contrary to the ratio hot deck method, the regression imputed component variables will not satisfy the corresponding balance edit. Similar experiments were carried out on the EPE data with comparable results. For these reasons we decided to use ratio hot deck imputation for all component variables.

Some variables such as *assacq* and *assdisp* are not component variables and can therefore not be imputed by the ratio hot deck imputation method; but regression imputation is also not well-suited, because these variables contain a large number of zero values. For these variables a standard nearest-neighbor hot deck imputation method was used with a distance function based on the variables *turnreg* (registered turnover) and *empreg* (registered number of employees) and stratification by *class*.

Several alternatives to this hot deck imputation method were investigated and evaluated on two criteria:

1. The relative difference between the means of the imputed values and the true values for the missing data.
2. The difference between the number of imputed zero values and the true number of zero values among the missing data.

One alternative was a two-step approach, using a hot deck method to impute whether or not the missing value is zero and subsequently a regression imputation approach to impute only the nonzero values. Negative imputations by the regression step of this method were set to zero. The preservation of the mean value for this approach was a little bit better than for the hot deck imputation method. The number of zero values, however, appeared to be much too large because of the extra zeros introduced by the regression part of this method (besides the zeros that had already been imputed by the hot deck). To prevent these extra zeros, the regression imputation was also applied with a log transformation of the target variable. This resulted in the same, rather accurate, number of zero values as the hot deck method, but the performance with respect to the preservation of the mean was much worse. So, if it is important to have the number of firms that have nonzero values for assets disposed (*assdisp*) or assets acquired (*assacq*) about right and at the same time preserve the means reasonably well, the hot deck imputation method seems to be a good compromise.

Influence of Errors on Imputation Performance. So far, the development data have been used to decide on an edit and imputation strategy to be applied to the evaluation data. These development data, for which the true values for both missing values and erroneous values are available, also give us the opportunity to explore the effect of errors on the imputation performance.

In Tables 11.9 and 11.10 some imputation results are given for the four overall total variables (imputed by multivariate regression and deductive imputation) for the EPE $Y_{3,D}$ data set and the EPE $Y_{2,D}$ data, respectively. These results include the true mean of the imputed values (*mean true*), the mean of the imputed values themselves (*mean imp.*), the relative difference between these two means (*rdm*), and the number of imputations (*# imp.*).

TABLE 11.9 **Preservation of Mean Values for the Four Overall Total Variables of the EPE $Y_{3,D}$ Data**

Variable	*mean true*	*mean imp.*	*rdm*	*#imp.*
totinvto	1872.67	1073.41	−0.43	21
totexpto	206.38	46.38	−0.78	36
subtot	15.00	21.88	0.46	2
rectot	743.18	210.80	−0.72	11

TABLE 11.10 Preservation of Mean Values for the Four Overall Total Variables of the EPE $Y_{2,D}$ Data

Variable	*mean true*	*mean imp.*	*rdm*	*#imp.*
totinvto	1509.42	1413.06	−0.06	19
totexpto	1083.45	1083.45	0.00	33
subtot	15.00	19.48	0.30	2
rectot	743.18	362.90	−0.51	11

Two of the variables in Tables 11.9 and 11.10 (*totinvto* and *totexpto*) contain more imputations for the $Y_{3,D}$ data set than for the $Y_{2,D}$ data set because *Cherry Pie* found errors in these variables. Four errors in *totinvto* and *totexpto* were not detected. For the other two variables, no errors are present or detected. The values of *rdm* show that the means for the $Y_{3,D}$ data are less well preserved than for the $Y_{2,D}$ data, for all variables. In general, the quality of imputations of a regression procedure can be influenced adversely by errors for two reasons. First, the values of some of the predictor variables in the records with missing values can be erroneous. Second, errors in any of the variables in the records with missing values as well as in fully observed records can lead to biased estimates of the regression coefficients. In our case, the four undetected errors are all in records with no missing values. Thus the lesser quality of the imputations for the $Y_{3,D}$ data can be explained entirely by the influence of the errors on the estimated regression coefficients.

Application to the Evaluation Data. We shall now discuss the results of the application of our edit and imputation strategy to the evaluation data. First we show results related to the edit rules: results that show the effectiveness of deductive imputation and the amount of adjustment that is necessary to let the imputed values satisfy the edit rules. Next, we give some results for the error localization performance (alpha, beta, and delta measures) and imputation performance (d_{L1} and m_1 measures) for the ABI and EPE data.

Deductive Imputation. In total, about 42% of the values to be imputed in the EPE $Y_{2,E}$ data could be deductively imputed, and about 45% of the values to be imputed in the EPE $Y_{3,E}$ data could be deductively imputed. So a substantial amount of the values to be imputed can be deductively imputed by using the edits. These numbers are slightly lower for the ABI $Y_{2,E}$ and $Y_{3,E}$ data, but there too a substantial number of deductive imputations were carried out. Note that the total number of the fields to be imputed in each of the $Y_{3,E}$ data sets (ABI and EPE) depends on the number of implausible values that have been identified.

Adjustment of Imputed Values. As mentioned in Section 11.3.4, the imputation methods for the ABI data already take the hard edits into account, and adjustment of imputed values is therefore not necessary. For the EPE data sets, not all hard edits are taken into account and the imputed values have been adjusted such that the final records satisfy all hard edits. But since most of the hard edits for the EPE data sets are taken into account in the original imputations, the effect of adjusting imputed values is limited. For the EPE $Y_{2,E}$ data, only 111 of the 2230 imputed values are adjusted—that is, about 5.0%. The sum (over all variables in the EPE $Y_{2,E}$ data) of the absolute differences of the means of the imputed values and the means of the adjusted imputed values is 70.6, and the sum (over all variables) of the means of the imputed values is 2,855.9. So the "average" change to the imputed values owing to the adjustment procedure is about 2.5%. For the $Y_{3,E}$ data, 95 values of the 2362 imputed values were adjusted (i.e., about 4.0%) and the average change is 1.1%.

Edit and Imputation Results for the ABI Evaluation Data. In Table 11.11 the error localization results for Strategies I and II are presented. The variables *taxrates* (amounts paid for national nondomestic rates) and *taxothe* (other amounts paid for taxes and levies) in Table 11.11 are the two components of *taxtot*.

The alphas are quite high for both strategies, pointing to a large proportion of undetected errors. Because fewer edits apply to the variables, it is evident that the alphas are larger for Strategy II than for Strategy I. Conversely, the betas are smaller, because using less edits results in fewer correct values considered implausible by the editing process. Most deltas are similar or smaller for Strategy II than for Strategy I, showing that the amount of misclassification is smaller with fewer edits.

In Table 11.12, imputation results for the ABI evaluation data are presented. These results pertain to the $Y_{3,E}$ data set with errors localized by either Strategy I or Strategy II and the $Y_{2,E}$ data set (missings only).

The results show, with a few exceptions, that the results are much better for the $Y_{2,E}$ data than for both experiments with the $Y_{3,E}$ data. An exception

TABLE 11.11 **Error Localization Results for the ABI Evaluation ($Y_{3,E}$) Data Set Using Strategy I (all edits) and Strategy II (Hard Edits Only)**

	Strategy I			Strategy II		
Variable	alpha	beta	delta	alpha	beta	delta
turnover	0.529	0.055	0.096	0.628	0.000	0.054
emptotc	0.378	0.274	0.284	0.613	0.001	0.059
purtot	0.696	0.016	0.117	0.708	0.006	0.111
taxrates	0.585	0.004	0.027	0.654	0.002	0.027
taxothe	0.589	0.000	0.023	0.647	0.000	0.025
taxtot	0.569	0.045	0.107	0.679	0.001	0.082
stockbeg	0.599	0.002	0.059	0.636	0.001	0.062
stockend	0.589	0.002	0.059	0.636	0.001	0.062
assacq	0.630	0.001	0.049	0.662	0.000	0.050
assdisp	0.619	0.001	0.038	0.651	0.001	0.040
capwork	0.559	0.001	0.009	0.559	0.001	0.009
employ	0.678	0.133	0.159	1.000	0.000	0.048

TABLE 11.12 Imputation Results for ABI Evaluation $Y_{3,E}$ and $Y_{2,E}$ Data Sets

	$Y_{3,E}$ Data Strategy I		$Y_{3,E}$ Data Strategy II		$Y_{2,E}$ Data (No Errors)	
Variable	d_{L1}	m_1	d_{L1}	m_1	d_{L1}	m_1
turnover	428.43	169.40	74.81	55.51	126.39	60.47
emptotc	59.39	56.68	42.50	36.29	12.42	3.52
purtot	858.10	834.12	331.30	306.74	4.56	1.96
taxtot	7.92	5.94	40.25	36.17	3.41	0.58
taxrates	6.64	0.77	20.02	15.69	1.20	0.87
taxothe	6.70	5.72	52.49	46.35	0.82	0.71
assacq	36.19	29.57	33.91	27.67	115.37	105.20
assdisp	66.08	60.96	71.08	65.44	3.46	1.94
employ	3.33	0.97	2.66	2.00	4.21	1.02
stockbeg	30.36	14.01	190.97	177.21	45.82	6.07
stockend	25.90	3.13	27.56	15.89	47.16	6.96
capwork	19.40	18.06	19.40	18.06	2.69	2.59

where the imputations for the $Y_{3,E}$ data are much better than for the $Y_{2,E}$ data occurs for *assacq*. From the results in Table 11.11 it was concluded that the error localization Strategy II performed better than Strategy I. The imputation results, however, show that the difference in imputation performance between these two experiments is not so clear cut. For three of the main variables, *turnover*, *emptotc*, and *purtot*, the imputation results are better for Strategy II than for Strategy I, but for *taxtot* and the components thereof (*taxrates* and *taxothe*) as well as *stockbeg*, Strategy I is better.

Edit and Imputation Results for the EPE Evaluation Data. Results for the error localization performance for the EPE evaluation data set ($Y_{3,E}$) are summarized in Table 11.13. A striking result is that the alphas are often 1, indicating that none of the errors has been correctly localized. It should be noted, however, that there are only a few errors in each variable. But still, the overall

TABLE 11.13 Error Localization Results for the EPE $Y_{3,E}$ Data Set

Variable	#errors	alpha	beta	delta
totinvto	12	0.833	0.003	0.014
totexpto	14	0.500	0.009	0.017
subtot	1	1.000	0.000	0.001
rectot	1	1.000	0.000	0.001
totinvwp	5	1.000	0.001	0.006
totinvwm	8	0.625	0.001	0.006
totinvap	6	1.000	0.000	0.006
totinvnp	5	1.000	0.001	0.006
totinvot	2	0.500	0.000	0.001

error detection performance is not very good; only 13 (24%) of the 54 errors in these variables have been detected correctly.

The imputation results for both EPE evaluation data sets [not presented here; see Pannekoek (2004b), and Vonk, Pannekoek, and De Waal (2004)] show that the imputation performance is more often better for the $Y_{2,E}$ data than for the $Y_{3,E}$ data.

11.3.6 CONCLUSIONS

From the analyses carried out by Chambers and Zhao (2004a, 2004b)—not reported in this book—we conclude that the approach used by Statistics Netherlands performed well in comparison with the methods of the other participants in the EUREDIT project. Our approach could be applied to edit and impute both the ABI and EPE data, something that many edit and imputation approaches evaluated under the EUREDIT project were unable to do. Another strong point of our approach is that it leads to data that satisfy the specified edits. Other approaches that lead to acceptable results for either the ABI or the EPE data do not guarantee that edits are satisfied by the edited and imputed data sets. Finally, our approach is a very flexible one. Individual steps, such as the detection of systematic errors and the imputation of erroneous and missing values, can, if desired, be modified separately without having to change the other steps in the approach. Furthermore, more steps can easily be added. For instance, other experiments on the ABI data indicate that for these kinds of data, it is useful to identify outliers and impute them by means of an outlier-robust method. Such an outlier detection step can, for instance, be added to our approach immediately after the detection and correction of systematic errors. The imputation method we have applied can be replaced by outlier-robust versions of the regression and hot deck imputation methods.

Despite the above-mentioned strong points of our approach, we are aware that automatic editing and imputation is a potentially dangerous approach. Our methodology correctly identifies only a low fraction of the errors in the observed data. Moreover, although the imputation performance of our methodology is good for the $Y_{2,E}$ data sets, it is less good for the $Y_{3,E}$ data sets. This leads us to the conclusion that the edit and imputation process should not be fully automated in practice.

We advocate an edit and imputation approach that consists of the following steps:

1. Correction of obvious systematic mistakes, such as thousand-errors.
2. Application of selective editing to split the records into a critical stream and a noncritical stream [see Lawrence and McDavitt (1994), Lawrence and McKenzie (2000), Hedlin (2003), and Chapter 6 of this book].
3. Editing of the data: the records in the critical stream are edited interactively, the records in the noncritical stream are edited and imputed automatically.
4. Validation of the publication figures by means of macro-editing.

The above steps are used at Statistics Netherlands in the production process for structural annual business surveys [see De Jong (2002)]. At Statistics Netherlands, so-called plausibility indicators [see Hoogland (2002)] are applied to split the records into a critical stream and a noncritical stream. Very unreliable or highly influential records achieve a low score on the plausibility indicators. Such records constitute the critical stream and are edited interactively. The other records—that is, the records in the noncritical stream—are edited automatically. Each year we edit and impute the same business surveys. To apply our automated approach to a new version of a business survey, we therefore only have to update the parameters. This updating process is to a substantial extent automated too. Edit and imputation of the records in the noncritical stream hence requires hardly any human intervention. This is in stark contrast with our experiences in the EUREDIT project, where we had to develop edit and imputation strategies for the ABI and EPE data sets from scratch. For some evaluation results on the combined use of selective editing and automatic editing on business surveys at Statistics Netherlands, we refer to Hoogland and Van der Pijll (2003).

The final validation step is performed by statistical analysts, who compare the publication figures based on the edited and imputed data to publication figures from a previous year, for instance. In this final step the focus is more on the overall results than on the correctness of individual records. Influential errors that were not corrected during automatic (or interactive) editing can be detected during this final, important, step, which helps to ensure the quality of our data.

At Statistics Netherlands, outlier detection techniques are used during the selective editing step and the macro-editing step. Large errors that were undetected by our approach in the EUREDIT project would in the production process at Statistics Netherlands probably be detected in either the selective editing or the validation step. In contrast to our approach in EUREDIT, where we had to restrict ourselves to edit and imputation methods using only data from the data set to be edited and imputed itself, in our production process for structural annual business surveys we use cleaned auxiliary data from a previous year throughout the entire editing process.

One could argue that with selective editing the automatic editing step is superfluous. At Statistics Netherlands, we strongly advocate the use of automatic editing, even when selective editing is used. We mention three reasons. First, the sum of the errors in the noncritical records may have an influential effect on the publication figures, even though each error itself is noninfluential. Provided that the set of edits used is sufficiently powerful, application of the Fellegi–Holt paradigm generally results in data of higher statistical quality. This is confirmed by various evaluation studies such as Hoogland and Van der Pijll (2003) and the evaluation study in Section 11.2. Second, many noncritical records will be internally inconsistent if they are not edited, which may lead to problems when publication figures are calculated or when microdata are released for public use. Finally, automatic editing provides a mechanism to check the quality of the selective editing procedures. If selective editing is well-designed and well-implemented, the records that are not selected for manual editing need no or only slight adjustments. Records that are substantially changed during the automatic editing step therefore point to an incorrect design or implementation of the selective editing step. We feel that automatic editing, when used in combination with other editing techniques, can only improve the quality of the data, not deteriorate it.

We also feel that only a combined approach using selective editing, interactive editing, automatic editing, and macro-editing can improve the efficiency of the traditional interactive edit and imputation process while at the same time maintaining or even enhancing the statistical quality of the produced data. To some extent our intuition is confirmed by our experiences in the EUREDIT project where our approach to automatic edit and imputation, a mix of several different methods for automatic edit and imputation, led to good results in comparison with the methods of other participants in the EUREDIT project.

11.4 Selective Editing in the Dutch Agricultural Census

11.4.1 INTRODUCTION

The Dutch Agricultural Census is conducted annually to monitor the structure of the agricultural sector in The Netherlands. The target population consists of all Dutch agricultural businesses with a net production income of at least 4260 euros; there are about 80,000 of these businesses. During the census, data are collected on each element of the target population, and they are subsequently edited interactively. Since the manual editing process is very costly and resources are limited, some form of selective editing is needed.

In 2008, three simple score functions were developed and evaluated, to replace the existing ad hoc selection procedure. In this section, we describe the three suggested score functions and the results of an evaluation study.

11.4.2 THREE SCORE FUNCTIONS

As discussed in Chapter 6, a common way to define a score function is to take the product of two components,

$$S_i = F_i \times R_i, \tag{11.1}$$

where the influence component F_i measures the contribution of business i to target parameters, and the risk component R_i measures the size and likelihood of a potential error in the data on business i. The three score functions we consider for the Dutch Agricultural Census are all of the general form (11.1).

To measure the influence of a business in the Dutch Agricultural Census, two survey variables are obvious candidates: the total production income TI and the total number of employees TE. Since both variables are only observed in the survey itself, they may contain errors. For the purpose of constructing an influence measure, we especially do not want to underestimate the size of a business. For this reason, it was decided to take the maximum of TI and TE as an influence measure, where the variables are first transformed to the same scale by dividing each variable by its maximum value:

$$F_i = \max\left\{\frac{TI_i}{\max_i TI_i}, \frac{TE_i}{\max_i TE_i}\right\}.$$

In this way, if one of the values TI_i and TE_i is observed erroneously, the value of F_i cannot be too low.

As the target population is rather heterogeneous, the measure we just defined may give a false impression of the influence of some businesses. For instance, a small farm that specializes in certain uncommon crops is likely to have a negligible impact on most target parameters for the population as a whole, but it may have a high impact on target parameters related to the production of particular crops. Since estimates on such more detailed levels are also published, it is important to take this heterogeneity into account in the selective editing procedure. We have therefore adopted a stratification of the target population based on business classification codes. Let $U_1, \ldots, U_H$ denote the strata. We refer to variable scores of the ith business in stratum U_h by TI_{hi}, TE_{hi}, and so on. An improved version of the influence measure is now defined as follows:

$$F_{hi} = \max\left\{\frac{TI_{hi}}{\max_{i \in U_h} TI_{hi}}, \frac{TE_{hi}}{\max_{i \in U_h} TE_{hi}}\right\}.$$

For the risk component, the survey variables TI and TE are again obvious candidates, since they provide a summary of many other survey variables, and since the corresponding target parameters are the main outcome of the Agricultural Census. Thus, it is especially important to find influential errors in these two variables, because records with errors in TI and TE are likely to also contain errors in other variables and because it is important to obtain reliable population estimates of TI and TE. For the purpose of detecting anomalous values, a particularly interesting quantity is the ratio of these two variables:

$$x_{hi} = \frac{TI_{hi}}{TE_{hi}}.$$

It is expected that this quantity—that is, the production income per employee—should be more or less constant in time for each business, and also more or less the same for businesses from the same stratum. Here, by "more or less constant" and "more or less the same" we mean that large differences in x_{hi} for a business from one year to another, or across businesses with the same type of production, are highly suspicious.

Since the census is conducted annually, there are reference data from the previous year available for each business. Thus, as a possible measure of suspicion, we can compare the value of x_{hi} in the current year to its value in the previous year. If we denote these values by $x_{hi,t}$ and $x_{hi,t-1}$, respectively, the following risk measure seems appropriate:

$$R_{hi}^{(1)} = \max\left\{\frac{x_{hi,t}}{x_{hi,t-1}}, \frac{x_{hi,t-1}}{x_{hi,t}}\right\} - 1. \tag{11.2}$$

We take the maximum of the ratio $x_{hi,t}/x_{hi,t-1}$ and its inverse in (11.2), because a large decrease in x_{hi} from one year to another is considered just as suspicious as a large increase. The constant term -1 is added so that the range of $R_{hi}^{(1)}$ becomes the interval $[0, \infty)$.

A potential problem with risk measure (11.2) is that the reference value $x_{hi,t-1}$ might also be based on erroneous information. If a respondent makes the same error in both years, resulting in, for instance, a too high value of $x_{hi,t}$ *and* $x_{hi,t-1}$, then a score function based on (11.2) will probably not consider these data as suspicious. We shall now consider two risk measures that do not have this drawback, because they compare x_{hi} to (a robust version of) its average value in the stratum U_h. We use the stratum median $\text{med}_{i \in U_h} x_{hi}$ as a robust estimate of the average production income per employee in stratum U_h.

Two possible ways to measure deviations from the stratum median are

$$A_{hi} = \left| x_{hi} - \underset{i \in U_h}{\text{med}}\, x_{hi} \right|$$

and

$$B_{hi} = \max\left\{\frac{x_{hi}}{\underset{i \in U_h}{\text{med}}\, x_{hi}}, \frac{\underset{i \in U_h}{\text{med}}\, x_{hi}}{x_{hi}}\right\} - 1.$$

By adding the constant term -1 in the definition of B_{hi}, both measures have the same range, namely $[0, \infty)$. A possible advantage of B_{hi} over A_{hi} is that B_{hi} considers large deviations above and below the stratum median as equally anomalous, whereas A_{hi} attains higher values for deviations above the stratum median than below. For example, suppose that a stratum has a median x value of 20, and suppose that we encounter two businesses with $x_{hi} = 5$ and $x_{hi} = 80$, respectively. According to A_{hi}, the deviations are 15 and 60, while according to B_{hi}, the deviation equals 3 for both businesses.

In order to interpret deviations from the stratum median as a measure of suspicion, we need to determine what kind of deviations are common for that stratum; that is, we need to take the spread of the deviations in U_h into account. Since the ordinary standard deviation is sensitive to outlying values (which are precisely the values that we are trying to identify), we use the more robust inter-quartile range (iqr) to measure the spread. By dividing A_{hi} and B_{hi} by their iqr, we obtain sensible risk measures:

$$R_{hi}^{(2)} = \frac{A_{hi}}{\underset{i \in U_h}{\text{iqr}}\, A_{hi}}$$

and

$$R_{hi}^{(3)} = \frac{B_{hi}}{\underset{i \in U_h}{\text{iqr}}\, B_{hi}}.$$

A large value of $R_{hi}^{(2)}$ or $R_{hi}^{(3)}$ corresponds with a deviation of x_{hi} from its stratum median that is large compared to its iqr for that stratum and therefore suspicious.

Using the general form (11.1), we have now obtained three possible score functions:

$$S_{hi}^{(1)} = F_{hi} \times R_{hi}^{(1)},$$
$$S_{hi}^{(2)} = F_{hi} \times R_{hi}^{(2)},$$
$$S_{hi}^{(3)} = F_{hi} \times R_{hi}^{(3)}.$$

To find the most appropriate score function for the Dutch Agricultural Census, an evaluation study was conducted. The results of this study are given in the next subsection.

11.4.3 EVALUATION STUDY

To evaluate the success of a selective editing strategy based on a particular score function, two properties are important. First, of the records identified as suspicious by the score function, the fraction of records containing errors should be as high as possible; that is, the number of records that are reviewed unnecessarily should be low. Second, of the records not identified as suspicious by the score function, the fraction of records containing errors should be as low as possible; that is, the number of missed errors should be low. These two properties have an obvious analog in hypothesis testing, where one wishes to minimize the occurrence of type I and type II errors.

Consider a set of M records, of which m are considered suspicious by a score function (in combination with some cutoff criterion—for example, a threshold value). Moreover, let M_e denote the number of records containing errors, and let m_e denote the number of records containing errors that are also considered suspicious by the score function. We define the following evaluation measures:

$$P_{\text{wrong}}(m) = \frac{m - m_e}{m}, \tag{11.3}$$

$$P_{\text{mis}}(m) = \frac{M_e - m_e}{M_e}. \tag{11.4}$$

The interpretation of these measures is as follows: $P_{\mathrm{wrong}}(m)$ is the fraction of records that are considered suspicious by the score function, but are in fact correct, and $P_{\mathrm{mis}}(m)$ is the fraction of records containing errors that are not considered suspicious by the score function. Clearly, a good score function should lead to low values of both $P_{\mathrm{wrong}}(m)$ and $P_{\mathrm{mis}}(m)$, since high values of $P_{\mathrm{wrong}}(m)$ are detrimental to the efficiency of the editing process, while high values of $P_{\mathrm{mis}}(m)$ are detrimental to the quality of the editing process.

The purpose and calculation of these measures is similar to the misclassification errors α and β presented in Section 11.3.5. In particular, α is identical to $P_{\mathrm{mis}}(m)$, while β is, in terms of the quantities defined here, equal to $(m - m_e)/(M - M_e)$.

Both evaluation measures (11.3) and (11.4) depend on m, the number of suspicious records identified by the score function. For an effective score function, the number of erroneous records that remain to be identified will become smaller as m increases, so it is expected that $P_{\mathrm{wrong}}(m)$ increases with m. In fact, as $m \to M$, it follows that $m_e \to M_e$, so $P_{\mathrm{wrong}}(m)$ converges to $(M - M_e)/M$, the fraction of correct records in the data set. For any score function, $P_{\mathrm{mis}}(m)$ is monotonically decreasing in m, with the minimum value 0 attained for $m = M$. Thus, there are two conflicting interests when trying to choose m such that $P_{\mathrm{wrong}}(m)$ and $P_{\mathrm{mis}}(m)$ are both as small as possible. A good score function should show, for small values of m, a sharp decrease in $P_{\mathrm{mis}}(m)$ combined with a slow increase in $P_{\mathrm{wrong}}(m)$.

To test the three score functions from Section 11.4.2, we conducted an experiment with data from the Dutch Agricultural Census of 2008. The data set used in the experiment contained about 40,000 records—that is, about half of the total number of records processed annually. The three score functions were calculated for this data set, and for each score function the records were ranked in descending order of their scores. Next, a subset of 305 records was selected and submitted to four subject-matter specialists for review. This subset contained both records that were identified as suspicious *and* records that were not identified as suspicious by some or all of the score functions. Records from the latter category were added to the test data, to see whether many serious errors would be missed if selective editing were to be based on one of the score functions. The subject-matter specialists did not receive the outcome of the score functions for the test data; in fact, no information on the selection criteria was communicated to them at all until after the experiment.

For each record from the evaluation data set, the verdicts of the subject-matter specialists were combined into a final verdict. This yielded the following result: of the 305 records, 212 were considered correct and 93 were considered erroneous, or at least suspicious enough to warrant follow-up action. (Since the subject-matter review was only done for the purpose of the experiment, no follow-up actions were actually taken.)

Based on the outcome of the experiment, we calculated the evaluation measures (11.3) and (11.4) for each of the score functions, with $M = 305$ and $M_e = 93$, and $m = 1, \ldots, 305$. For each value of m, m_e represents the number of erroneous records identified as suspicious among the m records of the test data set with the highest score.

Figure 11.1 shows plots of $P_{\text{wrong}}(m)$ and $P_{\text{mis}}(m)$ for each of the score functions $S_{hi}^{(1)}$, $S_{hi}^{(2)}$, and $S_{hi}^{(3)}$. In order to interpret the results with 305 records in terms of the original data set of 40,000 records, the horizontal axes of the plots

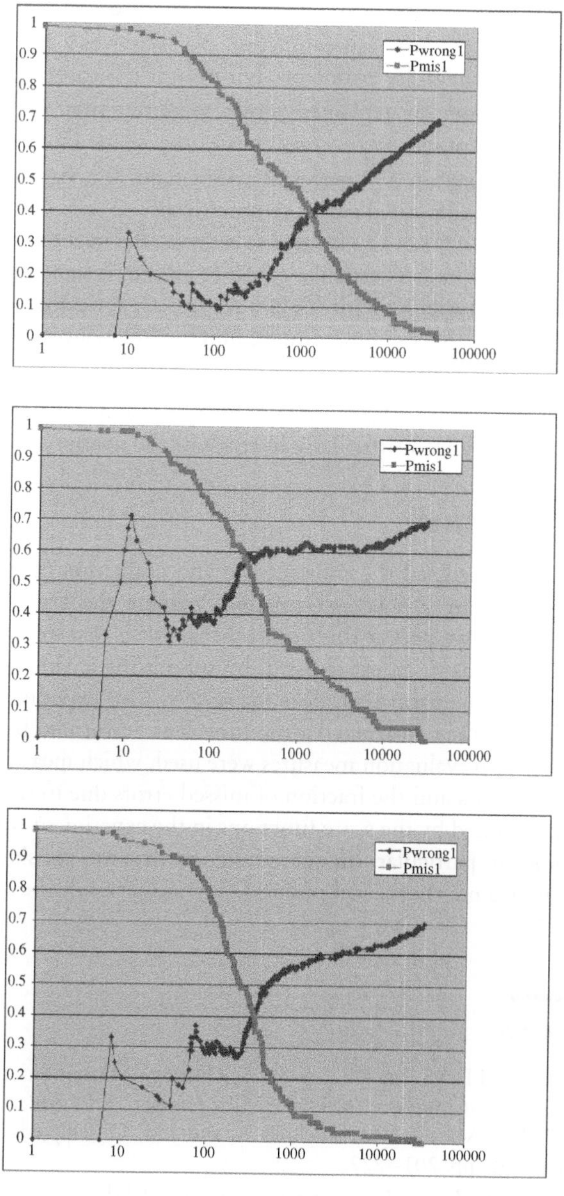

FIGURE 11.1 Plots of P_{wrong} and P_{mis} for score functions $S_{hi}^{(1)}$, $S_{hi}^{(2)}$, and $S_{hi}^{(3)}$ in the experiment.

in Figure 11.1 display not m, but the rank number of the mth record in the original data set of 40,000 records, sorted in descending order of S_{hi}. This gives a more accurate feel of the number of records that need to be reviewed to achieve a particular combination of P_{wrong} and P_{mis}. Note that for each score function, P_{wrong} converges to 212/305 $\approx$ 0.70, as explained above.

The plots indicate that score function $S_{hi}^{(3)}$ achieves the best result, since for this function we find the best combination of a sharp decrease in $P_{\text{mis}}(m)$ and a slow increase in $P_{\text{wrong}}(m)$. Score function $S_{hi}^{(1)}$ achieves a lower fraction of records considered wrongfully erroneous, but only at the cost of a much higher fraction of missed errors. As explained above, score function $S_{hi}^{(1)}$ tends to miss errors that are made consistently in two consecutive years. Finally, as expected, we find that score function $S_{hi}^{(2)}$ performs worse than $S_{hi}^{(3)}$, because it does not treat overstated and understated values symmetrically.

Plots like Figure 11.1 can also be used to estimate the fractions of unnecessary reviews and missed errors at a particular cutoff point. For instance, if only the first 500 records of the full data set with 40, 000 records are reviewed, when sorted in descending order of each score function, we find the following estimates: for $S_{hi}^{(1)}$, $P_{\text{wrong}} \approx 0.25$ and $P_{\text{mis}} \approx 0.55$; for $S_{hi}^{(2)}$, $P_{\text{wrong}} \approx 0.60$ and $P_{\text{mis}} \approx 0.35$; and for $S_{hi}^{(3)}$, $P_{\text{wrong}} \approx 0.50$ and $P_{\text{mis}} \approx 0.25$. This again shows that the third score function is the best choice, particularly in terms of the fraction of missed errors.

11.4.4 CONCLUSION

In this section we discussed the construction and evaluation of score functions for selective editing in a practical situation, namely the Dutch Agricultural Census. An interesting aspect of this practical example is that we were able to test various score functions in an experiment, by submitting a carefully selected set of test data to subject-matter specialists for review. This experiment allowed us to analyze which of the constructed score functions performed best in practice. For this analysis, two evaluation measures were used, which measure the fraction of unnecessary reviews and the fraction of missed errors due to selective editing. The behavior exhibited by the score functions in the experiment was in line with our expectations; in particular, the score function that was considered to be the most promising from a theoretical point of view achieved the best results in the experiment.

REFERENCES

Bethlehem, J. G., and F. Van de Pol (1998), The Future of Data Editing. In: *Computer Assisted Survey Information Collection*, M.P. Couper, R.P. Baker, J. Bethlehem, C.Z.F. Clark, J. Martin, W.L. Nicholls II, and J.M. O'Reilly eds. Wiley John & Sons, New York, pp. 201–222.

Chambers, R. (2004), Evaluation Criteria for Statistical Editing and Imputation. In: *Methods and Experimental Results from the EUREDIT Project*, J. R. H. Charlton, ed. (http://www.cs.york.ac.uk/euredit/).

Chambers, R., and X. Zhao (2004a), Evaluation of Edit and Imputation Methods Applied to the UK Annual Business Inquiry. In: *Towards Effective Statistical Editing and Imputation Strategies—Findings of the EUREDIT Project*, J. R. H. Charlton, ed. (http://www.cs.york.ac.uk/euredit/).

Chambers, R., and X. Zhao (2004b), Evaluation of Edit and Imputation Methods Applied to the Swiss Environmental Protection Expenditure Survey. In: *Towards Effective Statistical Editing and Imputation Strategies—Findings of the EUREDIT Project*, J. R. H. Charlton, ed. (http://www.cs.york.ac.uk/euredit/).

De Jong, A. (2002), *Uni-Edit: Standardized Processing of Structural Business Statistics in the Netherlands*. Working Paper No. 27, UN/ECE Work Session on Statistical Data Editing, Helsinki.

De Waal, T. (1996), *CherryPi: A Computer Program for Automatic Edit and Imputation*. UN/ECE Work Session on Statistical Data Editing, Voorburg.

De Waal, T. (2001), SLICE: Generalised Software for Statistical Data Editing. In: *Proceedings in Computational Statistics*, J. G. Bethlehem and P. G. M. Van der Heijden, eds. Physica-Verlag, New York, pp. 277–282.

Di Zio, M., U. Guarnera, and O. Luzi (2004), Application of GEIS to the UK ABI Data: Editing. In: *Methods and Experimental Results from the EUREDIT Project*, J. R. H. Charlton, ed. (http://www.cs.york.ac.uk/euredit/).

Fellegi, I. P. and D. Holt (1976), A Systematic Approach to Automatic Edit and Imputation. *Journal of the American Statistical Association 71*, pp. 17–35.

Ghosh-Dastidar, B., and J. L. Schafer (2003), Multiple Edit/Multiple Imputation for Multivariate Continuous Data. *Journal of the American Statistical Association 98*, pp. 807–817.

Granquist, L. (1995), Improving the Traditional Editing Process. In: *Business Survey Methods*, B.G. Cox, D.A. Binder, B.N. Chinnappa, A. Christianson, M.J. Colledge, and P.S. Kott, eds. John Wiley & Sons, New York, pp. 385–401.

Granquist, L. (1997), The New View on Editing. *International Statistical Review 65*, pp. 381–387.

Granquist, L., and J. Kovar (1997), Editing of Survey Data: How Much is Enough? In: *Survey Measurement and Process Quality*, L.E. Lyberg, P. Biemer, M. Collins, E.D. De Leeuw, C. Dippo, N. Schwartz, and D. Trewin, eds. John Wiley & Sons, New York, pp. 415–435.

Hedlin, D. (2003), Score Functions to Reduce Business Survey Editing at the U.K. Office for National Statistics. *Journal of Official Statistics 19*, pp. 177–199.

Hoogland, J. (2002), *Selective Editing by Means of Plausibility Indicators*. Working Paper No. 33, UN/ECE Work Session on Statistical Data Editing, Helsinki.

Hoogland, J., and E. Van der Pijll (2003), *Evaluation of Automatic versus Manual Editing of Production Statistics 2000 Trade and Transport*. Working Paper No. 4, UN/ECE Work Session on Statistical Data Editing, Madrid.

Houbiers, M., R. Quere, and T. de Waal (1999), *Automatically Editing the 1997 Survey on Environmental Costs*. Report 4917-99-RSM, Statistics Netherlands, Voorburg.

Kovar, J., and P. Whitridge (1990), Generalized Edit and Imputation System, Overview and Applications. *Revista Brasileira de Estadistica 51*, pp. 85–100.

Lawrence, D., and C. McDavitt (1994), Significance Editing in the Australian Survey of Average Weekly Earnings. *Journal of Official Statistics 10*, pp. 437–447.

Lawrence, D., and R. McKenzie (2000), The General Application of Significance Editing. *Journal of Official Statistics 16*, pp. 243–253.

Little, R. J. A., and P. J. Smith (1987), Editing and Imputation of Quantitative Survey Data. *Journal of the American Statistical Association 82*, pp. 58–68.

Pannekoek, J. (2004a), (Multivariate) Regression and Hot-Deck Imputation Methods. In: *Methods and Experimental Results from the EUREDIT Project*, J. R. H. Charlton, ed. (http://www.cs.york.ac.uk/euredit/).

Pannekoek, J. (2004b), Imputation Using Standard Methods: Evaluation of (Multivariate) Regression and Hot-Deck Methods. In: *Methods and Experimental Results from the EUREDIT Project* J. R. H. Charlton, ed. (http://www.cs.york.ac.uk/euredit).

Pannekoek, J., and T. de Waal (2005), Automatic Edit and Imputation for Business Surveys: the Dutch Contribution to the EUREDIT Project. *Journal of Official Statistics 21*, pp. 257–286.

Pannekoek, J., and M. G. P. Van Veller (2004), Regression and Hot-Deck Imputation Strategies for Continuous and Semi-Continuous Variables. In: *Methods and Experimental Results from the EUREDIT Project*, J. R. H. Charlton, ed. (http://www.cs.york.ac.uk/euredit/).

Schiopu-Kratina, I., and J. G. Kovar (1989), *Use of Chernikova's Algorithm in the Generalized Edit and Imputation System*. Methodology Branch Working Paper BSMD 89-001E, Statistics Canada.

Vonk, M., J. Pannekoek, and T. de Waal (2003), *Development of (Automatic) Error Localisation Strategy for the ABI and EPE Data*. Research paper 0302, Statistics Netherlands, Voorburg.

Vonk, M., J. Pannekoek, and T. de Waal (2004), Edit and Imputation Using Standard Methods: Evaluation of the (Automatic) Error Localisation Strategy for the ABI and EPE Data Sets. In: *Methods and Experimental Results from the EUREDIT Project*, J. R. H. Charlton, ed. (http://www.cs.york.ac.uk/euredit/).

Index

Handbook of Statistical Data Editing and Imputation, by T. de Waal, J. Pannekoek, and S. Scholtus